# Dinosaurs

## AN ADVENTIST VIEW

David C. Read

International Standard Book Number (ISBN): 9780982030509
Library of Congress Control Number (LCCN): 2009901986

Published by Clarion Call Books, LLC
300 South Old Betsy Road
Keene, Texas, 76059

Printed in the United States of America

Edited by Ken McFarland and David Read
Design and page layout by Steve Trapero
Cover design by Steve Trapero
Illustration clearance and production coordinated by Christine Cummings-Weis

FOR MY PARENTS

# Contents


| | | |
|---|---|---|
| | Foreword | i |
| | Introduction and Acknowledgments | v |
| 01 | Digging Dinosaurs | 1 |
| 02 | Bones of Contention | 19 |
| 03 | King of the Tyrant Lizards | 35 |
| 04 | When Did the Dinosaurs Live? The "Scientific" View | 45 |
| 05 | Can a Christian Follow Lyell? | 85 |
| 06 | Can a Christian Follow Darwin? | 115 |
| 07 | Ellen White on Lyellism & Darwinism | 135 |
| 08 | When Did the Dinosaurs Live? The Biblical View | 149 |
| 09 | Ice Age—The Flood's Aftermath | 183 |
| 10 | The Flood in History | 203 |
| 11 | Where Did the Dinosaurs Come From? The Darwinian View | 229 |
| 12 | The Copying Error Theory of Evolution | 239 |
| 13 | Does the Fossil Record Support Darwinism? | 267 |
| 14 | What Are Dinosaurs? | 285 |
| 15 | Other Strange Creatures in Flood-laid Strata | 311 |
| 16 | Did Birds Evolve from the Dinosaurs? | 327 |
| 17 | Ancient Man—the Darwinian View | 353 |
| 18 | Ancient Man—the Biblical View | 387 |
| 19 | How Advanced Were the Ancients? | 415 |
| 20 | Genetic Engineering | 447 |
| 21 | An Alternative Theory of the Dinosaurs | 483 |
| 22 | Defending the Amalgamation Theory | 509 |
| 23 | The Asteroid Theory | 541 |
| 24 | What Killed the Dinosaurs? | 555 |
| 25 | Were all Dinosaurs Destroyed in the Flood? | 589 |
| 26 | A Choice of Paradigms | 605 |
| | Appendix—The Dating Game | 611 |
| | Selected Bibliography | 647 |
| | Photography & Illustration Credits | 653 |
| | Index | 657 |

# Foreword

Over the course of many years, I have read widely on the issue of the intersection of science and religion. As a member and past Chairman of the Board of Directors of the Geoscience Research Institute at Loma Linda, California, I have more than a passing interest in this subject. A few months ago, I had the opportunity to read the manuscript of a new book—Dinsoaurs: An Adventist View. I was favorably impressed with the manuscript's compelling argument. I was told that the author, David Read, is an attorney; hence, I was not surprised to find that the document had the earmarks of a trial lawyer building a persuasive case. I was also impressed with the scholarly approach. The manuscript was replete with footnotes documenting the author's research. His sources included a long list of scholars to whom he gave due credit. After reading it, I felt strongly that the manuscript should be published and made available to Adventists and other interested readers.

The discovery of the remains of dinosaurs in the earth's rocks, culminating with the unearthing of Tyrannosaurus rex in 1902, awakened widespread interest in these very strange, extinct creatures. The curiosity about them, and fascination with them, has never been more intense than it is today. It is generally held that dinosaurs appeared about 220 million years ago and died out about 65 million years ago and have since then been extinct—why? Obviously, we do not have complete answers to every question, and we never will in this life. But by starting with (1) the most reliable scientific facts available, (2) the biblical record, and (3) the profound and very revealing statements of Ellen G. White on the subject, the author works his way toward conclusions that seem reasonable.

The author carefully and very convincingly argues his case that dinosaurs lived before the universal flood documented in the Bible and that the flood destroyed and buried dinosaurs and all humans then living. God brought the flood because of the exceedingly sinful and rebellious circumstances that prevailed on the earth at that time. In this cleansing process, all life, human and animal, was given a new beginning.

The author very judiciously counters the arguments of creationists who believe that such grotesque animals as dinosaurs were part of God's original creation. He argues that dinosaurs could have been the result of human intervention, in a process Ellen White called "amalgamation." This and other complex and controversial issues are openly and candidly discussed. Among the issues analyzed are: Uniformitarianism, which the author calls "Lyellism"; catastrophism; Flood legends; the Ice Age; radiometric dating; Darwinism and evolution; genetic mutations; DNA; the Darwinian versus the biblical view of early man; the "sons of God and the daughters of men"; genetic engineering; the "confused species"; and three possible meanings of Ellen White's use of the term amalgamation.

The origins of the dinosaurs, and indeed all theories about origins, seem to be inconclusive and somewhat speculative. Thus, it is fair to counter the materialistic speculation of scientists with reasoning that builds upon a biblical foundation. I am very pleased that this book is being made available to the public. It will strengthen any reader's faith in the truthfulness and accuracy of Scripture and the Spirit of Prophecy.

In the fascinating book, The Great Controversy Between Christ and Satan, we find the statement, "We can never by searching find out God. . . . While God has given ample evidence for faith, He will never remove all excuse for unbelief. All who look for hooks to hang their doubts upon will find them." After quoting this passage, David Read concludes with the following forthright statement:

"It is not possible to approach the facts of nature in a completely neutral and unbiased manner and allow the facts to inform us whether we were created or evolved. There are too many facts, far too much data, to ever sort out in a hundred life times. One chooses one's model, creation or evolution, before approaching the data. That choice is a religious choice. It is made with the heart, not the intellect. Doubts about creation can only be removed by getting to know, love, and trust the Creator God."

Neal C. Wilson
Past Chairman, Board of Trustees
Loma Linda University and Medical Center

# Introduction and Acknowledgments

In this age of intense competition for our time and attention, a short book is preferable to a long one, and I apologize for having written such a long book. Were this book written only for Seventh-day Adventists, it could have been much shorter, because Adventists already understand and take for granted the "young-earth creationist" worldview, which chapters 4 through 13 are devoted to explaining. But my goal was to make the argument accessible to non-Adventist and even non-Christian readers, and to achieve that goal it was necessary to include this background material. Although this material is well rehearsed among creationists, I hope I have presented it in a fresh and compelling manner.

This work's central thesis regarding the origin and nature of dinosaurs is not original; based upon informal "polling," it seems to be held by about 75% of Adventists. The possible connection between Ellen White's references to antediluvian "amalgamation" and modern genetic engineering has been obvious to many Adventists for more than 30 years. (I first heard of it from my elementary school science teacher in 1978.) But this thesis has never been explored at book length, and the time is ripe for such an exploration.

The first three chapters recount the history of the discovery of dinosaur fossils during the 19th Century, and describe some of the personalities involved. Chapter four discusses the layers in the earth's crust—the geologic column—and how geologists came to believe that the geologic column represents long eons of time. Chapters five through seven discuss the biblical and doctrinal reasons Adventist reject this interpretation of the column. Chapter eight argues that much of the geologic column was deposited during the Genesis Flood, and chapter nine explores the climatic/geologic after-effects of that deluge. Chapter ten surveys some of the flood legends that are so widespread.

Chapters eleven through thirteen discuss, and reject, Darwinian evolution as an explanation for the origin of dinosaurs, while chapter fourteen focuses on dinosaur physiology. Chapter fifteen discusses the physiology of some of the other creatures that were buried along with the dinosaurs, while chapter sixteen critiques the currently fashionable theory that dinosaurs evolved into birds. Chapters seventeen through nineteen step back from the dinosaurs and explore two opposing conceptions of the early human race, the biblical versus the Darwinian. Chapter twenty is a brief survey of genetic engineering and its probable future course, while chapters twenty-one and twenty-two explore the central thesis. Chapters twenty-three and twenty-four provide further argument that a deluge—not an asteroid strike—destroyed the dinosaurs. Chapter twenty-five defends the Adventist view that all dinosaurs were buried in the deluge (not taken aboard the ark), and chapter twenty-six is a brief conclusion.

With a few exceptions, each chapter advances the argument and lays a logical foundation for the next chapter. Those unfamiliar with creationist thought and Adventist doctrine will find it helpful, if not necessary, to read the book from front to back. Those already conversant with young-earth creationist thought may wish to skip around, reading chapters of interest.

Notes and references are provided at the end of each chapter, rather than at the end of the book. I have used general, rather than scientific, cite form. As frequently as possible, I have cited Darwinian sources for factual assertions, for the obvious reason that my case is stronger when built upon facts admitted by my ideological opponents. In many instances when I have cited creationist sources, I have gone the extra mile by including my creationist source's non-creationist source. In some cases, only creationist sources were available because the issue concerns some aspect of the creationist model.

I am indebted to many friends and acquaintances who reviewed the manuscript at various stages and provided helpful comments. Among them, I want to thank Dr. Kurt Wise, Dr. Arthur Chadwick, and Dr. Lee Spencer, who read all or portions of the manuscript, made many helpful comments and criticisms, and corresponded with me via e-mail. Their input helped me avoid serious errors and caused me to strengthen the argument in some places. Obviously, their help does not constitute an endorsement of the theories presented, some of which they vehemently disagreed with. Any remaining errors are, of course, entirely my own, as is my writing style, which in places irritated them and probably will irritate other working scientists, as well.

I also want to thank everyone who participated in the production of the book, including Ken McFarland, Steve Trapero, and Christine Weis. Most of all, I want to thank my father, Dr. John Read. Without his steadfast encouragement and support, I would never have finished this book, much less seen it published.

David C. Read
September, 2008
Glendale, California

CHAPTER 01

# Digging Dinosaurs

Georges Cuvier's parents were Huguenots—French Protestants who were often furiously persecuted. During one of these periods of persecution, which began in the early 1500s and continued into the late 1700s, four hundred thousand Huguenots fled France for other countries. To worship unmolested, Georges' parents moved to Montbeliard, an independent principality northwest of Switzerland that only later became part of France. There, in 1769, Georges was born. When he grew old enough, his father sent him to Normandy to be educated by Protestants.

Everyone around Georges Cuvier recognized that he was an extraordinary child with a promising intellect. He was given the opportunity to attend an elite academy in Stuttgart. Cuvier's great passion was nature, and particularly the study of fossils—a study that came to be known as paleontology. Cuvier began studying marine life and the fossils of marine animals. The quality of his work brought him to the attention of France's leading scientists, and he was given a research post at the Museum of Natural History in Paris.

Extinction—the idea that species die out leaving no living representatives—was a controversial concept. One school of thought—"natural theology"—held that God would not permit a species to become extinct, because such a loss would diminish an already perfect creation.[1] As Cuvier (**Plate 1**) studied the mastodon and mammoth remains unearthed around Paris, he realized that these fossil elephants were significantly different from living African and Indian elephant species. He assumed that the mastodon and mammoth were now extinct and would not be found living in some remote corner of the world. As he studied more types of fossils, Cuvier realized that in the past, many groups of animals had become extinct, seemingly wiped out all at once.

Some of these extinct animals were very strange compared to living types. In 1770, quarrymen in Holland had discovered a bizarre fossil—a jaw more than three feet long with a full set of sharp teeth. Like a sacred relic, the fossil was kept in a glass case at the home of a local clergyman. During Napoleon's reign, Imperial French forces seized the fossil and sent it to Paris. Cuvier examined the jaw and compared it to known animals, discerning that it had belonged to an extinct marine reptile. The reptile was later named *Mosasaurus,* or "Meuse Lizard" after the Meuse River region in which it was found. (See, **Plate 2**, for a similar skull)

The Mosasaur (**Plate 3**) was the first large, extinct marine reptile to become widely known in scientific circles. Cuvier also identified a puzzling fossil that other experts had described merely as an "unknown marine animal." Based only upon a drawing—the fossil's owners would not send it to Cuvier—Cuvier correctly surmised that the specimen was not a marine animal but a land-based flying reptile. He observed that the leading edge of each wing was composed of a greatly elongated finger. He christened it *Pterodactyl,* after the Greek words for "wing" and "finger." (**Plate 4**)

Cuvier's acquaintance with the anatomy of a wide variety of creatures allowed him to develop general principles of anatomy.

One of these was that every anatomical structure is functionally related to all the other structures in the body. For example, biologists can tell whether an animal is a carnivore or an herbivore by its teeth, which are specialized, and based upon the teeth they can predict features such as claws and the configuration of the digestive system. Cuvier rejected the theory of evolution, because any significant change in one part of an animal would necessitate complementary changes in many other parts of the animal's anatomy. The various organs of each species, he believed, are too closely coordinated to allow significant evolutionary change. Cuvier remained the world's leading comparative anatomist until his death in 1832.

Before the latter half of the nineteenth century, there were very few professional scientists. But clergyman, doctors, and lawyers pursued paleontology as an avocation. Such men were educated and had an income to support their hobby. Dr. James Parkinson, famous for having described Parkinson's disease, is a good example. Dr. Parkinson was an avid fossil collector and published a three-volume work entitled *Organic Remains of a Former World*, in which he described all the fossil reptiles known up to that time. Such amateurs made real and significant advances in paleontology.[2]

Perhaps the greatest amateur fossilist was not a doctor or a clergyman but a self-educated woman from the working class. Mary Anning (1799-1847, **Plate 5**) has been called "the greatest fossilist the world ever knew." Anning was born in Lyme Regis, on the southern shore of England. Richard Anning, Mary's father, was a poor cabinetmaker and, to make matters worse, a Dissenter in a time when only members of the Church of England could hold significant civil or military offices. The Dissenters were Christians—typically Presbyterians, Baptists, Congregationalists, and Quakers—who could not go along with some of the Anglican Church's doctrines.

The cliffs along the beach at Lyme Regis contain many fossils.

Mary's parents collected fossils to sell to affluent summer tourists, thereby supplementing the family's meager income. Richard Anning died when Mary was young, leaving the family in debt and without a provider. The remaining Annings scrounged a living from the fossils that they found and sold, but their situation was never comfortable.

When she was 11 years old, Mary's brother, Joseph, discovered a fossil *Ichthyosaurus* ("fish lizard"), which Mary collected the following year. It was the first specimen of that species to become known to English science.[3] Ichthyosaurs were marine reptiles shaped like porpoises. They had porpoise-shaped bodies, a dorsal fin, two flippers on each side, and a long, snout-like jaw full of sharp teeth.

**Figure 1.1: Ichthyosaurus**

By the time she was in her mid-20s, Anning had taken the lead in the family fossil business. Museums, scientists, and European nobles sought out the fossils that Anning found and prepared. Anning's most important find was her 1828 discovery of a large, extinct marine reptile unknown to science. Anning sold the specimen to a collector named Birch, who made it available to Rev. William Conybeare to study. Conybeare, who had also named *Mosasaurus,* named it *Plesiosaurus* ("near lizard"). Plesiosaurs were marine reptiles with large bodies, four large flippers, and long necks.

**Figure 1.2: Plesiosaurus**

Anning was not just a skilled fossil hunter; she understood what she was collecting. She read everything she could get her hands on about fossils. Being without means, she usually borrowed the publications, which she would then painstakingly copy by hand. She even drew copies of the illustrations. Some found it hard to believe that a young woman from the lower class could possess the knowledge and skills that Anning displayed. One highborn lady wrote of Anning:

> The extraordinary thing in this young woman is that she has made herself so thoroughly acquainted with the science that the moment she finds any bones she knows to what tribe they belong. She fixes the bones on a frame with cement and then makes drawings and has them engraved. . . . It is certainly a wonderful instance of divine favor—that this poor, ignorant girl should be so blessed, for by reading and application she has arrived to that degree of knowledge as to be in the habit of writing and talking with professors and other clever men on the subject, and they all acknowledge that she understands more of the science than anyone else in this kingdom.

Many gentlemen fossilists picked Anning's brain, sometimes publishing her insights as their own. Anning was aware of this and resented it. In her later years, however, Anning did receive recognition for her work, including monetary recognition. The British Association for the Advancement of Science paid her a small annuity, and the Geological Society of London collected a stipend for her.

When Mary died of breast cancer, her obituary was published in the *Quarterly Journal of the Geological Society*—an organization that did not admit women until 1904. The historians of paleontology long ignored Anning, probably because of her sex, her low social status, and the lack of documentation of her finds. The majority of Anning's fossil finds ended up in museums and personal collections without her being credited as the discoverer. Ironically, the tongue twister Anning inspired, "she sells sea shells by the sea shore" is better remembered today than she is.[4]

By the 1820s, the concept of extinction was well established, and science had begun to discover some of the strange reptiles of the past. The stage was set for the discovery of dinosaurs. At about the same time, two English naturalists discovered two different kinds of dinosaurs. An English country doctor named Gideon Algernon Mantell (1790-1852, **Plate 6**) discovered and named the first herbivorous (vegetarian) dinosaur.

Gideon Mantell's profession was medicine, but his passion was fossils. He devoted every spare moment to his hobby. He married Mary Ann Woodhouse, and the couple had four children. Mary Ann was a competent sketch artist, a skill more often cultivated in the days before photography than at present. She was soon drawn into her husband's hobby; she sketched the fossils he found or purchased. In 1822, the Mantells published a volume entitled *The Fossils of the South Downs*, written by Gideon and illustrated by Mary Ann with 42 plates and 364 separate drawings.

That same year, the Mantells made a fascinating discovery. The traditional story is that Dr. Mantell had gone to the country to make some house calls and had brought Mary Ann along for the ride. While Gideon was inside with a patient, Mary Ann strolled along the road where the carriage was parked. The road was under repair and, at one place where paving material was piled up, Mary Ann spotted a stone containing a large fossil tooth, which she showed her husband when he returned. (More recent research has cast doubt on this charming story, but it has become part of dinosaur lore.[5])

The paving stones had been quarried at the nearby town of Cuckfield. Mantell had often collected fossils from that quarry and was familiar with the typical fossils found there, but he had never seen anything like this tooth before. It looked as if it had been used for chewing coarse vegetation, and was the size one would expect in a large pachydermal mammal. But Mantell knew that there were no large mammal fossils in the rocks quarried at Cuckfield. He suspected the tooth was from a reptile, but one very different from modern reptiles, which do not chew their food.

Mantell took the tooth to a meeting of the Geological Society in London and showed it to several experts, but they believed it was either from a fish or from a mammal. Mantell had a friend named Charles Lyell—a lawyer whose real interest was geology—who was soon to visit Paris, so Mantell asked him to show the tooth to Cuvier. Upon returning from Paris, Lyell reported that Cuvier had taken one look at the tooth and stated that it was the upper incisor of a rhinoceros. Despite discouragement from all the leading authorities of his time, the 32-year-old Mantell continued to trust his own hunch that the tooth was from an unknown, extinct reptile.

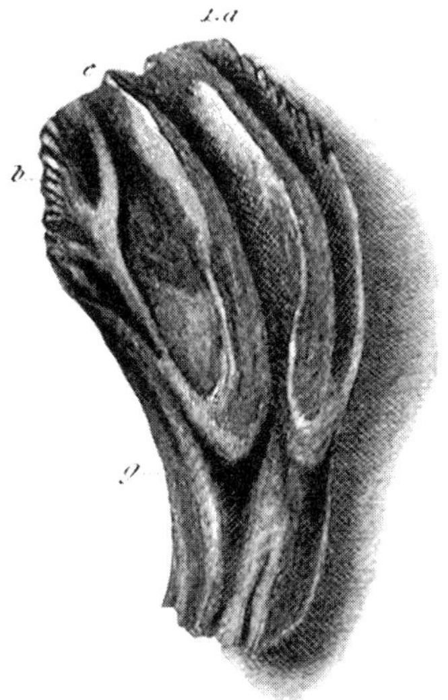

**Figure 1.3: Mantell's Fossil Tooth**

Further collecting in the Cuckfield quarry turned up several more teeth similar to the one Mary Ann Mantell had found. Gideon had drawings of these teeth sent to Cuvier, asking him to reconsider his hasty initial diagnosis. At length, Cuvier wrote back:

> These teeth are certainly unknown to me; they are not from a carnivorous animal, however I think that they belong . . . to the order of Reptiles. From their external appearance you could mistake them for fish, . . . but their internal structure is greatly different from these. . . . Is it possible that we have here a new animal, an herbivorous reptile? . . . Time will confirm or invalidate this idea.[6]

This was confirmation from the world's greatest anatomist that Mantell had discovered something new and different—a gigantic herbivorous reptile.

To search for a modern tooth analogous to his fossil, Mantell visited the Hunterian Museum in London. The museum's collection of thirteen thousand human and animal anatomical specimens had been accumulated by the English surgeon John Hunter (1728-1793) and donated to the Royal College of Surgeons on the condition that the collection be made available to the public for research purposes.[7] Mantell began poring over the museum's reptilian teeth and bones. He showed his fossil to an assistant curator named Samuel Stutchbury, who had just prepared an iguana skeleton for display. Stutchbury noted the similarities between iguana teeth and Mantell's fossil tooth. The fossil tooth turned out to be very similar to an iguana's tooth, but several times larger. Here was the modern analog for which Mantell was looking.

Mantell was preparing to name the animal *Iguanosaurus*, but the Rev. Conybeare objected, noting that *Iguanosaurus* would mean "iguana reptile," which would not distinguish the extinct animal from a modern iguana, which is also a reptile. He suggested the name *Iguanodon*, which means "Iguana tooth," and Mantell agreed with this suggestion. In 1825, Mantell published, in the *Philosophical*

*Transactions* of the Royal Society of London, a paper entitled, "Notice on the *Iguanodon*, a newly-discovered Fossil Reptile, from the Sandstone of Tilgate Forest, in Sussex." Like most scientists, Mantell craved recognition for his work, and *Iguanodon* ensured that he would receive it. "I shall ride on the back of my *Iguanodon* into the temple of immortality!" he wrote in his diary.

**Figure 1.4: Iguanodon**

Fifty-three years later, in 1878, coal miners in Bernissart, Belgium, stumbled upon a spectacular deposit of *Iguanodon* bones. The bones were impregnated with iron disulphide—pyrite, or "fool's gold"—and the miners at first thought they had found tree trunks of gold. Paleontologists were called in and began to extricate the fossils from the depths of the mine. Because there were so many bones and because the pyrite caused them to disintegrate when exposed to dry air, any attempt to dig out individual bones was abandoned. Instead, the rock was quarried in square blocks, covered with plaster, reinforced with steel bands, and brought to the surface. The blocks were meticulously numbered and catalogued so that the three-dimensional puzzle could be reassembled on the surface.

The preparation of the Bernissart Iguanodons took twenty-five years. To prevent the bones from disintegrating, each bone had to be chemically treated immediately upon being freed of its rock matrix.[8] Belgian paleontologist Louis Dollo supervised the task, devoting a long and very productive scientific career to the Bernissart find. At the time, the Belgian Royal Museum of Natural History was a young institution, lacking a permanent building. The only place large enough to house the dinosaurs was Brussels' Chapelle Saint-Georges, a fifteenth-century cathedral. So with fitting irony, the "dragons" of Bernissart were prepared and mounted in a church named after a dragon-slaying saint. Forty separate *Iguanodons* were recovered, including twenty-four complete skeletons. The Belgian Royal Museum now owns an impressive diorama of eleven standing *Iguanodons*, with a further twenty recumbent upon the floor—a museum display unique in the entire world.

The honor of having discovered the first carnivorous dinosaur goes to the Rev. William Buckland. Buckland (1784-1856) was a minister and a minister's son. While studying theology at Oxford, he became interested in geology. He stayed on at Oxford as a graduate student and later as a "reader," or assistant professor, eventually becoming the world's first full professor of geology, as opposed to mineralogy. Buckland was also pursuing a parallel career in the Church of England, advancing through the ecclesiastical ranks.

**Figure 1.5: Rev. William Buckland (1784-1856)**

Buckland was a remarkably unstuffy character. His lectures "would keep his audience in roars of laughter, as he imitated what he thought to be the movements of the *Iguanodon* or *Megatherium*, or, seizing the ends of his long, clerical coat tails, would leap about to show us how the pterodactyl flew." Buckland kept a bear, named Tiglath Pileser (after the ancient Assyrian monarchs), at Christ

Church College, Oxford. He would dress Tig in Oxonian cap and gown and bring him to college parties.

Buckland was a bachelor until age 42. One summer, he was riding in a coach, reading Cuvier's most recent volume on fossils, when he noticed that a young lady sharing the coach with him had the same book among her possessions. When he commented on their shared interest, the young lady, Mary Morland, told him that she had drawn the illustrations for Cuvier's book. Buckland married her the following year. It is remarkable that Mantell and Buckland, who were good friends, were both married to competent illustrators named Mary.

The Buckland home was a menagerie of exotic animals. One visitor reported having to keep his feet tucked under him on the sofa out of fear of being bitten by a jackal that was wandering around the room. Presently, the visitor heard a crunching sound under the sofa, which turned out to be the jackal eating a stray guinea pig. Guests at the Buckland home might be served such exotic dishes as crocodile or ostrich meat.

In 1815, Buckland was given some fossil remains from the Stonesfield quarry, which supplied Oxford with most of its roofing slate. The fossils consisted of several vertebrae, some hind limbs, fragments of a pelvis, and part of a jaw that sported one very formidable fang. Hoping that more parts of the animal would be recovered, Buckland delayed almost a decade before writing a description. He discussed the find with other fossilists, however, and it became common knowledge among them. For example, William Conybeare, Buckland's friend and fellow Reverend in the Anglican Church,[9] published a book on fossil marine reptiles in which he mentioned the "huge lizard" of Stonesfield. It was Conybeare, always ready to help with the naming of things, who suggested the name *Megalosaurus* ("big lizard"). (Thus, Conybeare ended up naming *Mosasaurus, Plesiosaurus, Iguanodon,* and *Megalosaurus*—

a remarkable accomplishment.) In 1819, when Cuvier came to England to view the Hunterian collection, he inspected the Megalosaurus material that Buckland had on hand.

In 1824, some nine years after it was originally discovered, Buckland finally published his "Notice on the Megalosaurus or Great Fossil Lizard of Stonesfield." This was the first scientific description of a dinosaur. Buckland placed *Megalosaurus* within the order of saurians or reptiles, but was amazed at the beast's gigantic size. Based on Cuvier's expertise in comparative anatomy, Buckland estimated that the animal was more than forty feet long and as bulky as an elephant. As of this writing, no complete *Megalosaurus* skeleton has ever been found, and its dimensions are not definitively known. Ironically, the Iguanodon, which Mantell was urged to name after only a tooth, has been attested by dozens of complete skeletons, whereas Buckland's Megalosaurus, named for a whole animal, is still based only upon fragmentary remains.

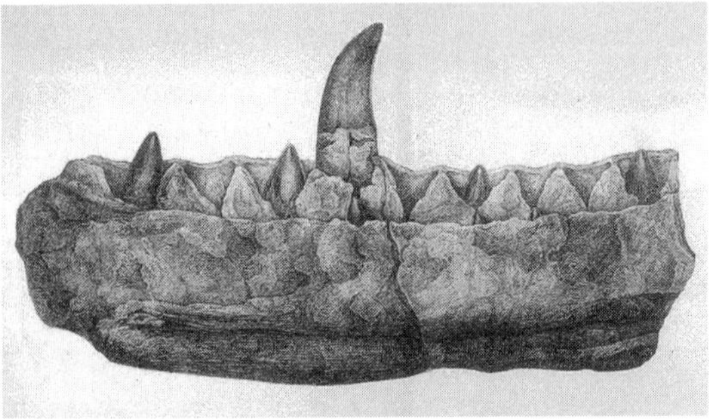

**Figure 1.6: Buckland's Megalosaurus Jaw from the Stonesfield quarry**

Once Buckland and Mantell had blazed the trail, other scientists began to name and describe dinosaurs. The succeeding years saw the description of Mantell's *Hylaeosaurus* (1833), Zbarzevski's *Macrodontophion* (1834), Riley and Stutchbury's *Thecodontosaurus*

**Figure 1.7: Megalosaurus**

and *Palaeosaurus* (1836), Meyer's *Plateosaurus* (1837), and Richard Owen's *Cladeidon* and *Cetiosaurus* (1841).

Richard Owen (1804-1892, **Plate 7**) was to become the leading naturalist of the 1840s. Like Mantell, Owen was a surgeon by trade but, unlike Mantell, Owen was well connected and politically astute. He managed to turn his passion for natural history into a lucrative career. As a young man, he became an assistant to William Clift, conservator of the Hunterian Museum, where Mantell had linked his fossil tooth to the modern iguana. Owen's job included cataloging the museum's collection of anatomical specimens, which had grown to twenty thousand in number. This task was greatly complicated by the fact that the previous steward of the collection had burned most of Hunter's documentation because he had been claiming Hunter's discoveries as his own. Owen identified and labeled every specimen, reorganized the entire collection, and published a catalogue. He eventually married his boss's daughter, Caroline Clift.

Owen was appointed Hunterian Professor at the Royal College of Surgeons and tasked with presenting lectures on anatomy that made use of the Hunterian Collection. One of Owen's many offices entailed dissecting all the animals that died at the zoo. The zoo provided no laboratory, so Owen used his own house. On one occasion, the autopsy of part of an elephant created such a stench that Caroline asked Richard to smoke cigars all over the house to mask the smell. If you know how bad cigar smoke smells, you can appreciate what a foul odor the putrefying elephant was giving off. This job did have the advantage, however, of allowing Owen to acquire an incomparable knowledge of comparative anatomy.

It was not by accident that Owen came to be called the "English Cuvier;" some influential British naturalists felt that their country needed a comparative anatomist of Cuvier's stature, and Owen was their candidate. Owen had much in common with Cuvier, including French Huguenot ancestry and an unstinting opposition to the theory of evolution. The reptiles that Mantell and Buckland had found and described confirmed Owen's belief that evolutionary theory was wrong. These ancient reptiles were much larger and far more complex than any modern reptile. Clearly, the reptiles had made no evolutionary progress, in the sense of becoming more complex. To the contrary, they had gone backward.

At an August 2, 1841 meeting of the British Association for the Advancement of Science, Owen lectured for over two hours on various fossil animals, including *Megalosaurus* and *Iguanodon.* The lecture helped clarify Owen's thinking that these remarkable animals were a group distinct from modern reptiles. In an 1842 article based on the lecture, Owen coined the name *Dinosaur,* from the Greek *deinos*, meaning "terrible" or "fearfully great," and *sauros*, meaning "lizard" or "reptile." In the article, Owen touched upon topics that are still vigorously debated today: Did dinosaurs have four-chambered hearts, like modern mammals? Was dinosaurian metabolism closer to modern warm-blooded mammals than to modern cold-blooded reptiles?

Owen's career continued on its upward trajectory, as he succeeded Clift as Conservator of the Hunterian Museum. He was a personal friend of Queen Victoria, and was often invited to the palace to lecture on natural history. The Queen gave him a large house, and he was appointed Superintendent of the British Museum (Natural History). He was instrumental in establishing a separate building in South Kensington for the natural history department of the British Museum. In addition to naming the dinosaurs, Owen authored over six hundred scientific papers. In his later years, however, Owen's reputation among younger scientists began to suffer, primarily because Owen never assented to Darwinism. After he died, the Darwinists all but wrote him out of the history books.

---

[1] See, e.g.,Wheeler, Gerald W., *The Two-Taled Dinosaur,* (Nashville, TN: Southern Publishing, 1975), p. 88; Christopher McGowan, *The Dragon Seekers,* (Cambridge, MA: Perseus Publishing, 2001), pp. 2-3. This view is not biblically informed; it ignores the fact that, after Adam's sin and the resulting fall of mankind, the creation is under a curse and no longer perfect. Gen. 3:17-19; Rom. 5:12; Rom. 8:18-22; Rev. 22:1-3

[2] The facts set out in chapters one through three are drawn from the following sources: David A. E. Spalding, *Dinosaur Hunters,* (Rocklin, CA: Prima Publishing, 1993); John Noble Wilford, *The Riddle of the Dinosaur,* (New York: Alfred A. Knopf, 1985); Edwin H. Colbert, *The Great Dinosaur Hunters and their Discoveries,* (New York: Dover Publications, 1968); Douglas J. Preston, *Dinosaurs in the Attic,* (New York: St. Martin's Press, 1986); David Rains Wallace, *The Bonehunters' Revenge,* (New York: Houghton Mifflin, 1999); and Christopher McGowan, *The Dragon Seekers,* (Cambridge, MA: Perseus Publishing, 2001).

[3] In 1719, English antiquarian William Stukeley had described a fossil subsequently determined to be an Ichthyosaur. Stukeley was one of the first men to carefully study Stonehenge, which we will discuss in Chapter 19.

[4] Wilford, supra, p. 27; Spalding, supra, p. 14. See, also, Torrens, Hugh, "Mary Anning (1799-1847) of Lyme: 'the greatest fossilist the world ever knew,'" *British Journal for the History of Science,* 25: 257-84, 1995; http://www.ucmp.berkeley.edu/history/anning.html; http://www.sdsc.edu/ScienceWomen/anning.html; http://www.lymeregismuseum.co.uk/fossils.htm.

[5] The story is from Sidney Spokes' 1927 biography of Mantell, which was long the standard work on Mantell's life. Mantell himself may have told the story on several occasions. More recent scholarship has shown, however, that Mantell was already in possession of several dinosaur teeth and bone fragments by 1821, a year before his wife's legendary stroll. It is possible that Mantell was confused about dates, and that his wife found the fossil in 1820, not 1822. In later years, after Mary Ann left him, Mantell ceased giving her credit for finding the original *Iguanodon* tooth. See, Dean, Dennis R., *Gideon Mantell and the Discovery of Dinosaurs* (Cambridge University Press, 1999); Wilford, supra, pp. 28-31.

[6] McGowan, supra, p. 84. See, also, John R. Horner and James Gorman, *Digging Dinosaurs,* (New York: Workman Pub. 1988), p. 170, citing, Philippe Taquet, "Cuvier-Buckland-Mantell and the Dinosaurs," in *Actes du Symposium Paleontologique,* E. Buffetaut, J.M. Mazin and E. Salmon eds. (Monbeliard: G. Cuvier, 1983), pp. 476-494).

[7] The Hunterian Museum of the Royal College of Surgeons in London, founded on the collection amassed by Dr. John Hunter (1728-1793) is not to be confused with the Hunterian Museum of the University of Glasgow, Scotland, based on a collection amassed by another eminent 18th century British surgeon and anatomist, Dr. William Hunter (1718-1783), although a mere American should be forgiven some confusion.

[8] Unfortunately, preservation techniques were never perfected, and the bones continued to slowly deteriorate. A paleontologist working at the Belgian Museum of Natural History in the late 1970s determined that the *Iguanodon* skeletons were too delicate to be of much help to researchers.

[9] Henry Adams, in *The Education of Henry Adams*, wrote, "Every curate [churchman] in England dabbled in geology." Conybeare did more than dabble, however. Along with Buckland and the Rev. Adam Sedgwick, Buckland's counterpart at Cambridge, he is remembered for significant geological contributions.

CHAPTER 02

# Bones of Contention

In 1858, William Parker Foulke unearthed some large fossil bones near Haddonfield, New Jersey. Foulke brought the fossils to Dr. Joseph Leidy, a Philadelphian who was America's first professional paleontologist. Leidy was one of the few men in the world capable of understanding that the bones were from a dinosaur similar to *Iguanodon*. Leidy described the find, and named it *Hadrosaurus foulkii*, after its discoverer, Foulke, and the place of its discovery, Haddonfield.[1] Hadrosaurs are called "duckbilled dinosaurs" because their snouts somewhat resemble the bill of a duck.

One of Leidy's students was a brilliant young man named Edward Drinker Cope. Cope was born in Philadelphia, a Quaker from the Quaker state, and the grandson of a wealthy shipping magnate. Cope's father wanted him to become a farmer—and even bought him a farm—but Edward was always intent on pursuing science. Philadelphia was a center of scientific study, with a long tradition of independent, non-university affiliated naturalists. In addition to studying with Leidy, Cope took a tour of European universities and museums. He published his first scientific paper at the age of 19 and

**Figure 2.1: Hadrosaurus**

named his first dinosaur in 1866, at age 26.[2] Cope's scientific accomplishments were legion, but he is best remembered for his bitter and public feud with fellow paleontologist Othniel Charles Marsh.

Marsh was born to a New England farm family of modest means, but his mother's brother, George Peabody, became very wealthy.[3] With inherited money he did not receive until he was 21, Marsh entered an elite preparatory school, graduating at age 25, seven or eight years older than most of his classmates. He next studied at Yale University for six years, and then made an academic tour of Europe, studying at Berlin and Breslau.

While in Europe, Marsh made a side trip to London, where he convinced his uncle to donate money for a new science building at Yale, which became the Peabody Museum. Marsh was shrewdly using his uncle's money to smooth his own career path; upon his return from Europe, he was unanimously voted to the professorship in paleontology at Yale. Although this position paid no salary, it carried no teaching duties either, and Marsh was free to do research and write.

Part of the reason for the rivalry between Cope and Marsh is priority: the first naturalist to publish a description of a new species gets to name it and gets the glory for having "discovered" it. Cope and Marsh were finding many of the same dinosaurs, and either man would often rush into print just a few days after examining a new fossil specimen. To facilitate speedy publication, Cope purchased a controlling interest in the journal *American Naturalist*.[4] But priority cannot fully explain the bitterness of the feud, which must be blamed upon the arrogant and domineering nature of the protagonists.

Largely as a result of the Cope-Marsh rivalry, dinosaur collecting, describing, and naming accelerated to a near-frantic pace. Before the rivalry began, only nine species of dinosaur had been found in North America; Joseph Leidy had described and named most of those. Cope and Marsh added 136 new species to the roster; Marsh named 80, Cope 56. Not all of these "species" have stood the test of time; Cope and Marsh occasionally gave two or more names to fossils that turned out to be from the same species. In paleontology jargon, they were "splitters." Dinosaurs were just one aspect of both men's wide-ranging interest in vertebrate fossils, but dinosaurs were more of a specialty for Marsh than for Cope.[5]

Pacifism is a Quaker tenet, but where it concerned Marsh, Cope was no pacifist. Cope and Marsh had first met in Berlin as students, and were apparently on good terms for several years. In 1867, Cope honored Marsh by naming a fossil amphibian *Ptyonius marshii* and Marsh reciprocated the following year by naming a species of mosasaur *Mosasaurus copeanus*.

Many years later, when the two traded insults in the pages of the *New York Herald*, Cope claimed that his feud with Marsh began in 1868 when Cope was collecting near Haddonfield, New Jersey. Marsh came down to visit, and Cope took him on a tour of the quarries.

Not long afterward, Cope claimed, Marsh paid the quarry owners to send their finds to him instead of Cope. "In endeavoring to obtain fossils from those localities," wrote Cope, "I found everything closed to me and pledged to Marsh for money considerations."

In Marsh's version, he endured Cope's trespasses until "the number was approaching the Biblical limit of seventy times seven, when a break occurred between us, and since then we have not been friends." Marsh claimed that the bad blood began in 1869, when Marsh visited Cope at the Philadelphia Academy of Sciences to view a reconstructed *Elasmosaurus* ("thin-plated lizard")—a marine reptile similar to *Plesiosaurus* but with a longer neck and tail. Marsh claimed to have pointed out that the specimen had been reconstructed with the skull and some of the neck vertebrae attached to the tail, instead of the neck, and Cope took great offense at being corrected. (Although Cope had indeed made this mistake, Leidy, not Marsh, pointed it out; Marsh's claim in the *New York Herald*, decades later, appears to have been *post hoc* opportunism.[6])

**Figure 2.2 Othniel Charles Marsh (1831 to 1899)**

Both men knew that the western interior of North America was largely unexplored by fossil hunters and likely a fossil bonanza. Marsh's professor at the University of Breslau, Ferdinand Roemer, had said so, and had even written a book on the Cretaceous fossils of Texas. When Marsh was ready to go west, however, it still was not safe to do so alone, so he arranged for a military escort. Marsh became friends with Gen. Philip Sheridan, and obtained a letter of introduction from Gen. William Tecumseh Sherman that "became an open sesame at all army posts." Marsh's 1870 fossil hunting expedition, which consisted of Marsh and fifteen Yale students, was escorted by a detachment of Cavalry and led by the renown scout "Buffalo Bill" Cody.

On another foray, in 1874, Marsh collected in the White River badlands of what is now Nebraska and South Dakota, an area controlled by the peaceful Oglala Sioux. During the course of the dig, Marsh befriended Chief Red Cloud of the Sioux. When Marsh was preparing to leave, Red Cloud gave him samples of bad treaty rations, including moldy flour, maggoty pork, and spoiled sugar, asking Marsh to show them to the Great Father in Washington as evidence of the local Indian agent's malfeasance. A few months later, when Marsh was in Washington for scientific meetings, he brought the matter to the attention of the commissioner of Indian Affairs, as well as the Secretary of the Interior, Columbus Delano. These men ignored Marsh, so he went over their heads to President Ulysses S. Grant.

**Figure 2.3: Edward Drinker Cope (1840-1897)**

Several influential newspapers took up the cause, prompting an investigation. The matter became known as "the Indian ring"—yet another scandal in the scandal-plagued Grant administration. Delano was forced to resign amid allegations of mismanagement. An ironic footnote to this episode is that Gen. George Armstrong Custer was very much on Marsh's (and Red Cloud's) side. Custer wrote to Marsh stating, "I simply desired to take you by the hand and thank you for the considerable and disinterested service you have rendered the government and the country . . . by your exposing the well known frauds and irregularities of the Indian Ring. . . . The Red Cloud Agency . . . is not an isolated instance. I know that the Indians at other agencies are the victims of similar mistreatment."

Meanwhile, Cope had undertaken his own western field trips. On one trip in 1872, Cope hired a "scout" named Sam Smith. Unfortunately for Cope, Smith was secretly working for Marsh, and he sabotaged Cope's expedition. Smith wrote Marsh, stating, "My motive in going with Cope was to keep him off some places as I think is good bone country close here." The resulting fiasco literally sickened Cope; his wife had to come west to nurse him back to health. Despite this setback, Cope collected in the west during the next two summer seasons, in association with government-sponsored geological surveys. In 1875, Cope's father died, leaving him an inheritance of nearly $250,000.00, which at the time was a considerable fortune.

In 1876, shortly after the Sioux under Sitting Bull and Crazy Horse wiped out Custer's command at Little Big Horn, Cope decided to go fossil hunting in the Judith River[7] badlands of Montana, the borderland between Sioux and Crow territory. He was warned that either tribe could kill him and blame it on the other. But Cope reasoned that most of the Crow were peaceful, and that every able-bodied Sioux would be fighting with Sitting Bull two hundred miles away and would not return before he could haul out his fossils.

He was right. The few Crow they encountered were peaceful, and Cope entertained them by popping his false teeth in and out of his mouth. One Indian rode two days on horseback just to see this trick. Later, Cope's scout and cook deserted after they learned that Sitting Bull's camp was too close for comfort. Cope loaded his fossils onto a riverboat at Cow Island, an army outpost on the Missouri River. A few days later, Sitting Bull crossed the Missouri at Cow Island, igniting a battle in which five soldiers were killed.

By 1877, both Cope and Marsh had the money to buy fossils. They could hire agents to find the fossils, and the agents could hire crews to dig them out of the ground. They spent freely. The resulting flow of bones to Philadelphia and New Haven was staggering. Too much was not enough. Even if both men had more fossils than they could do justice to, they would collect to keep the other man from getting the specimen. A century later, the Peabody Museum staff was still unpacking crates of fossils that Marsh had collected.

In that year of 1877, Arthur Lakes, an Oxford-educated clergyman and college professor, was fossil hunting in the mountains near Morrison, Colorado, when he found an extraordinarily large vertebra and humerus bone, apparently from "some gigantic saurian." Lakes was well informed and knew that Marsh or Cope would pay handsomely for the fossils. He first wrote to Marsh, then dug up the bones and sent ten boxes worth to Marsh and a similar shipment to Cope.

Marsh acted first, sending his chief field collector, Benjamin Mudge, to Colorado to secure Lakes' loyalty. (In the correspondence between Marsh and Mudge, they referred to Cope by the code name "Jones.") Mudge successfully recruited Lakes to the Marsh camp, and Lakes asked Cope to send his share of the fossils on to Marsh, which must have angered Cope. Based upon these bones, Marsh named *Titanosaurus,* which he later changed to *Atlantosaurus,* because, as Cope helpfully pointed out, the name *Titanosaurus* had already been used. This was the first dinosaur from the Morrison Formation, a layer of rock that has since yielded countless dinosaur fossils.

Although Cope lost out on the Morrison site, he had just learned of an equally spectacular find. O. W. Lucas, a naturalist and teacher, was collecting plants near Cañon City, Colorado, when he came upon some large fossil bones. Lucas shipped several of the specimens to Cope—enough for Cope to name the dinosaur *Camarasaurus* ("chambered lizard") (**Plate 8**). Both *Atlantasaurus* and *Camarasaurus* were sauropods. The sauropods ("lizard foot," a term coined by Marsh) were herbivorous (vegetarian) dinosaurs with gigantic bodies and long necks and tails. They were quadrupedal, meaning that they walked on all fours.

When he learned of the Cañon City discovery, Marsh sent one of his agents, Samuel Williston to snoop around, but the Marsh party failed to get any fossils from the area. In July 1877, however, Marsh received a letter from "Harlow and Edwards," stating that they knew where Marsh could find huge fossil bones in abundance. "We measured one shoulder blade and found it to measure four feet eight inches (4 ft. 8 in.) in length. One joint of the vertebrae measures two feet and one half in circumference and ten inches (10) in length." As proof of their good faith, they sent Marsh a few boxes of bones, which were just as they had described. Marsh instructed Williston to leave Cañon City and head for Harlow and Edwards' location—Como Bluff, near Medicine Bow, Wyoming Territory.

"Harlow and Edwards" turned out to be William Harlow Reed and William Edward Carlin, both of whom worked for the Union Pacific Railroad. Reed had discovered the dinosaur bones while hunting for game. They had not exaggerated the richness of the find; it was the mother lode. "Cañon City and Morrison are simply nowhere in comparison with this locality both as regards perfection, accessibility and quantity," reported Williston. The dinosaur remains "extend for seven miles & are by the ton. . . . The bones are very thick, well preserved, and easy to get out."

Marsh hired Carlin and Reed at the then handsome salary of $90.00 per month (and they stayed on the railroad's payroll, as well). Their contract with Marsh required them to take all reasonable precautions to keep out "all other collectors not authorized by Prof. Marsh," meaning Cope. During the first year, they gathered thirty tons of dinosaur fossils. Among these were the large sauropods, *Diplodocus* ("double beam") and *Brontosaurus* ("thunder lizard").

They also found a peculiar creature that had rows of triangular bony plates running along its spine. Marsh named this creature *Stegosaurus*, which means "roofed lizard" (**Plate 9**), because the plates reminded Marsh of roofing tiles. It is still not known whether the plates stood up straight or lay flat along the animal's body, or what purpose they served, if any. Stegosaurus grew up to thirty feet long and weighed several tons, yet it had a brain the size of a golf ball.

*Stegosaurus* also had a sacral plexus, sort of like a junction box in the pelvis where the nerves from the tail and hind legs enter the spinal cord. The sacral plexus was twenty times larger than the walnut-sized brain, and scientists speculated that it served as a second "brain" to control the activities of the back of the body.[8] The idea that *Stegosaurus* had two "brains" inspired Bert L. Taylor, a columnist for the Chicago *Tribune*, to write his immortal dinosaur poem:

Behold the mighty dinosaur,
Famous in prehistoric lore,
Not only for his power and strength
But for his intellectual length,
You will observe by these remains
The creature had two sets of brains—
One in his head (the usual place),
The other in his spinal base.
Thus he could reason *a priori*
As well as *a posteriori*.
No problem bothered him a bit;

He made both head and tail of it.
So wise was he, so wise and solemn,
Each thought filled just a spinal column.
If one brain found the pressure strong
It passed a few ideas along.
If something slipped his forward mind
'Twas rescued by the one behind.
And if in error he was caught
He had a saving afterthought.
As he thought twice before he spoke
He had no judgment to revoke.
Thus he could think without congestion
Upon both sides of every question.
Oh, gaze upon this model beast,
Defunct ten million years at least.

Como Bluff could not long be kept secret from Cope. Soon, a man named Haines visited the dig. He claimed to be selling groceries, but he asked more questions about fossils than food, and Williston knew he was an agent for Cope. Within a couple of years, Cope had managed to hire Carlin away from Marsh (mainly because Marsh did not remunerate regularly and promptly), and since Carlin outranked Reed in the railroad hierarchy, he forced Reed out of the Como Bluff station. Reed set up camp a few miles away and endured the winter in a tent, which he called "camp misery." Reed's loyalty was commendable, but it went too far: Once, after he had recovered a quarry's best specimens, Reed smashed the remaining fossils to prevent them falling into Cope's hands.

There was so much suspicion, distrust, greed, and treachery

among the various agents of Cope and Marsh at Como Bluff that we cannot chronicle all of it here. David Rains Wallace has described them as "characters in a Gilded Age *Treasure of the Sierra Madre.*" But these characters were digging up previously unknown animals. One of the animals recovered by Cope's agents was an almost complete skeleton of the bipedal carnivore *Allosaurus* ("different lizard") (**Plate 10**).[9] This remarkable specimen lay crated up in one of Cope's warehouses until it was bought by the American Museum of Natural History. Marsh kept quarries open at Como Bluff for twelve years; his agents discovered twenty-six new species of dinosaur there.

One of Reed's best finds was a nearly complete *Brontosaurus* skeleton, which Marsh reconstructed and mounted in the Peabody Museum at Yale. Because the skeleton of this massive animal impresses everyone who sees it, *Brontosaurus* became one of the most famous dinosaurs. The Sinclair Oil Company made it their logo, and it was the star of the Disney movie "Fantasia."

In 1877, Reed and Carlin had found part of a pelvis and some vertebrae, which Marsh named *Apatosaurus* ("deceptive lizard"). Two years later, Marsh published a description of *Brontosaurus* ("thunder lizard") (**Plate 11**). A generation later, Elmer Riggs, a paleontologist with Chicago's Field Museum, pointed out that *Apatosaurus* was merely a juvenile *Brontosaurus,* and that the name *Apatosaurus* had priority. No one doubted Riggs, but everyone ignored him. Even after 120 years of brontosaurian fame, the pedants and legalists are still urging us to call this animal *Apatosaurus.* It will not be easy to erase *Brontosaurus* from the popular lexicon, however, a couple of generations of us having been raised watching television's Fred Flintstone eat "Brontosaurus burgers." Even the U.S. Postal Service ignored the pedants when, in 1989, it issued a series of dinosaur stamps that included a *Brontosaurus* stamp.[10]

A substantive problem with *Brontosaurus* was that for over a century the specimens were displayed with the wrong skull. Reed's beautiful Como Bluff skeleton lacked a skull, so Marsh attached a *Camarasaurus* skull that had been found four miles away. Many years later, Earl Douglass, a paleontologist with the Carnegie Museum in Pittsburgh, found a *Brontosaurus* skeleton with a skull right beneath it that looked less like the *Camarasaurus* skull Marsh had used and more like a *Diplodocus* skull. This was the first indication that something was wrong, but it was not until 1975, when two scientists published an article on the subject, that the scientific establishment finally admitted Marsh's error and corrected the museum displays by mounting the proper skulls.

In 1887, a government geologist sent Marsh some bison-like horn cores. Marsh assumed these were from an extinct buffalo and, against the advice of some of his assistants, published a notice christening the species Bison alticornis. A few months later, one of Marsh's fossil hunters, John Bell Hatcher, sent Marsh more horn cores, this time from confirmed dinosaur rocks that did not contain large fossil mammals. When Hatcher's horn cores matched the previous set, Marsh realized he had made a mistake and that "*Bison alticornis*" was a dinosaur, not a buffalo. He told Hatcher to collect the rest of the skull, which weighed more than a ton. Marsh named the creature *Triceratops horridus* ("horrid three-horned face," **Plate 18**). Hatcher became an expert at finding horned dinosaurs; over the course of four seasons, he found more than fifty ceratopsian skulls, with their beaks, horns, and bony neck frills. *Triceratops* has become one of the most famous and easily recognizable dinosaurs.

The feud between Marsh and Cope worsened in later years. By 1889, Cope had lost his fortune because of imprudent mining investments, and he could not land a paying job. The politically well-connected Marsh, on the other hand, had gotten himself appointed vertebrate paleontologist to the U.S. Geological Survey. Marsh then used his influence to block Cope from government employment and prevent the Geological Survey from funding a second volume of Cope's massive treatise on fossil vertebrates.

Cope finally secured a professorship at the University of Pennsylvania, and the feud should have died down. But neither man could let it go. Marsh was probably behind the U.S. Secretary of the Interior's claim that Cope's fossils were government property, having been collected on various government geological surveys. The Secretary demanded that Cope surrender them to the Government.

This was the last straw. Cope had been keeping a file on Marsh for over twenty years; he now took his "Marshiana" to the newspapers. In 1890 the *New York Herald* published a series of articles chronicling Marsh's every scientific shortcoming, real and imagined. Marsh responded in kind, and the feud that had been the embarrassing secret of the scientific community was splashed across the headlines. Cope died of kidney failure in 1897, Marsh of pneumonia in 1899. But the money flowing from New Haven and Philadelphia had enticed capable men into the dinosaur hunting trade, and these men continued to hunt dinosaurs long after Cope and Marsh went extinct.

**Figure 2.4: Triceratops**

[1] Authorities differ about the name *Hadrosaurus*, with some saying it means "bulky lizard" or "sturdy lizard," and other saying it means "Haddon lizard."

[2] The dinosaur was found near Haddonfield and Cope named it *Laelaps aquilunguis.* It was a carnivorous theropod. Marsh named a similar dinosaur *Dryptosaurus*, and Koch may have already discovered the genus in 1839.

[3] George Peabody's is a remarkable rags-to-riches story. His father died when he was 16, leaving 8 children, debts and a mortgaged house. George worked in a dry goods shop, later as a traveling salesman, and served in the army during the war of 1812. After the war, Peabody went into the importing and merchandising business with one of his army friends, Elisha Riggs. Peabody, Riggs, and Co. was profitable, and Peabody grew wealthy. On overseas business trips, Peabody took a liking to London, and established a home and a brokerage business there. Another Peabody partner, Junius Spencer Morgan, father of J. P. Morgan, later assumed control of the London business. George Peabody devoted his retirement years to philanthropy on both sides of the Atlantic, establishing institutes for the promotion of learning and culture in seven American cities, as well as endowing university chairs, establishing a college for teachers, and founding museums. The fact that Peabody, a Connecticut Yankee, donated $1.5 million dollars to build housing for low-income tenants in London may have been instrumental in keeping Great Britain from recognizing the Confederacy during the American Civil War.

[4] Two books on the Cope/Marsh feud have recently been published: David Rains Wallace's *The Bonehunters' Revenge: Dinosaurs, Greed, and the Greatest Scientific Feud of the Gilded Age*, (New York: Houghton Mifflin, 1999 (paperback edition by Mariner Books)), and Mark Jaffe's *The Gilded Dinosaur: The Fossil War Between E.D. Cope and O.C. Marsh and the Rise of American Science*, (New York: Crown Books, 2000 (paperback edition by Three Rivers Press)). Older books on this subject include Elizabeth Noble Shor's *The Fossil Feud* (Hicksvillle, NY: Exposition Press, 1974) and Robert Plate's *The Dinosaur Hunters* (New York: David McKay Co., 1964).

[5] Cope is credited with having named 1,282 new species of fossil vertebrates, Marsh with 536.

[6] See, e.g. Mike Everhart, *The Tale of a Tail: or how easy it was to put the correct head on the wrong end of Elasmosaurus platyurus Cope 1868*; Everhart, Michael J., *Oceans of Kansas - A Natural History of the Western Interior Sea* (Bloomington: Indiana University Press, 2005) pp. 122-123.

[7] Named by William Clark, of the Lewis and Clark expedition, after his sweetheart and future wife Julia "Judith" Hancock.

[8] Similar spinal or ganglion expansions have been found in modern animals, and are common in migrating birds. In modern animals, these spinal expansions have nothing to do with controlling the body; hence it seems unlikely that the stegosaur's "second brain" was anything of the sort. See, e.g., Currie, Philip J., and Eva B. Koppelhus, *101 Questions About Dinosaurs*, (Mineola, NY: Dover, 1996), p. 49.

[9] As of this writing, about 37 Allosaurs have been found, and only 13 of these were more than 25% complete. Of those thirteen, only three complete skulls have been recovered. Recently, a group of creationists excavated a 70% complete *Allosaurus* skeleton measuring 22 feet in length, 10 feet in height, with a yard-long skull. "Up to now," notes Doug Phillips, "a well-funded and insular community of evolutionary theorists have dominated the field of paleontology, directing most of the large dinosaur finds to research and museums committed to interpreting the fossil evidence through the faith-driven assumptions of evolution. To have a dinosaur of this size and significance within the camp of scientists committed to the creation model is nothing short of a *coup d 'etat.*" Visionforum.com, "Home School Expedition Uncovers Rare Allosaur and Giant Sauropod," May 20, 2001, online at: http://www.visionforum.com. See, also, "Home-schoolers find intact dinosaur skeleton," *WorldNetDaily*, May 21, 2002. Online at: http://worldnetdaily.com/news/article.asp?ARTICLE_ID=27686.

[10] My word processing program's dictionary includes the word *Brontosaurus*, but considers *Apatosaurus* . . . well, *nomina dubia.*

CHAPTER 03

# King of the Tyrant Lizards

As public interest in dinosaurs grew, the world's great museums collected specimens and had them mounted for display. Morris K. Jesup, a self-made millionaire and president of New York City's American Museum of Natural History, found that, in the dinosaur department, his museum was falling behind other world-class institutions. To turn that situation around, and get ahead in the dinosaur game, Jesup needed to recruit someone who could organize a paleontology department for the museum. After trying unsuccessfully to lure Marsh away from Yale, Jesup hired Henry Fairfield Osborn (1857-1935), a wealthy and well-connected paleontologist who had studied under Cope.

To build the museum's fossil holdings, Osborn convinced Jesup to purchase Cope's collection. He also arranged for the museum to finance and organize its own fossil hunting expeditions. By fielding at least one such expedition every season since the 1890s (and usually more than one), the American Museum amassed the largest collection of dinosaur fossils in the world. Osborn stayed at the American Museum throughout his long career, following Jesup as president and eventually becoming known as "the father of American vertebrate paleontology."

Before Osborn could collect dinosaurs, he had to collect dinosaur hunters. His star was a man named Barnum Brown (1873-1963). Brown dug up more dinosaur bones than anyone else who ever lived. Osborn once remarked, "Brown is the most amazing collector I've ever known. He must be able to smell fossils. If he runs a test-trench through an exposure it will be right in the middle of the richest deposit. He never misses."

Brown was born in Carbondale, Kansas, a few days before the circus came to town. At his brother's suggestion, he was named after the great showman P. T. Barnum. As a boy, Barnum would follow his father's plow, picking up fossils as they fell off the moldboard. His collection outgrew the house and came to occupy the laundry shed. Brown attended the University of Kansas, where he studied with one of Marsh's former agents, Samuel Williston. He was admitted to graduate school at Columbia University, but never finished his Ph.D., going to work for Osborn instead.

Osborn sent Brown to quarry at Como Bluff. Brown found a few bones, but he thought Marsh and Cope had probably picked the location clean. When Osborn visited the site, however, he and Brown explored a nearby bluff and found dinosaur bones weathering out of the side of a hill. The bone-bearing layer was traced to an outcrop known as the Medicine Bow incline. Here, dinosaur bones were lying on the surface of the ground in great numbers. Years earlier, a shepherd dwelling in this tree-less land had built a small cabin out of dinosaur bones, using sod and adobe to fill in the interstices. The site became known as the Bone Cabin Quarry. Over the next three years, museum workers shipped over sixty tons of dinosaur bones from Bone Cabin Quarry to New York.

Five miles south of Bone Cabin Quarry, they found more bones. These turned out to be from a huge *Brontosaurus*. The skeleton was collected and shipped to New York, where it took two years to chip away the rock matrix surrounding the bones, and another three years to assemble and mount the specimen—the largest dinosaur ever displayed. Complete specimens are seldom found, and in specimens mounted for museum display the missing bones are usually replaced by plaster casts. The better practice is to make the casts a different color than the actual fossils, to avoid confusion as to which is which. In the nineteenth century, however, this was not always done. Hence Mark Twain, after seeing the Bone Cabin dinosaurs, could quip, "Professor Osborn has just reconstructed a 75-foot dinosaur. If the plaster had held out he would have made it 100 feet."

Barnum Brown's most spectacular find came at another Wyoming dinosaur graveyard. Cowboys had reported seeing strange bones weathering out of the badlands gullies of Hell Creek. In 1902, Brown took an expedition to Hell Creek, where he found the skeleton of a very large carnivorous dinosaur. He spent the next two summers digging and blasting it out of the hard sandstone. Osborn named the beast *Tyrannosaurus rex* ("king of the tyrant lizards") (**Plate 12**).

*Tyrannosaurus rex* has become the most famous of all dinosaurs—the star of "Jurassic Park" and other motion pictures. A few years later, Brown found another *Tyrannosaurus* at Hell Creek, in even better condition than the first. "I have seen nothing like it before," he wrote to Osborn. "This skull alone is worth the summer's work, for it is perfect." The American Museum mounted both specimens, but during the Second World War, when it was feared that the Germans might bomb New York City, the museum kept the second specimen and gave the first *T. rex* to the Carnegie Museum in Pittsburgh.

Brown had an exceptionally long career, with many notable finds. One came in 1931, when he learned of a dinosaur graveyard at Howe Ranch, Wyoming. To fund an expedition to collect these sauropod bones, Brown turned to the Sinclair Oil Company. During the 1930s and 40s, Sinclair gave free dinosaur booklets to the motorists who bought gasoline at its stations. Brown wrote those booklets, and Sinclair bankrolled many of Brown's dinosaur digs. A large expedition returned to Howe Ranch in 1934 and dug out thirty-five tons of fossils, including four thousand individual bones.

One of the American Museum's most ambitious collecting trips was the Central Asiatic Expedition, organized and led by Roy Chapman Andrews. From childhood, Andrews was a great enthusiast of natural history. "To enter the American Museum of Natural History was my life's ambition," he wrote. "Nothing else ever had a place in my mind." Immediately after college, Andrews headed for New York and the American Museum. The only job then available at the museum was scrubbing floors in the taxidermy department, but Andrews took it.

Andrews was a man of action and adventure, and lacked the patience for painstaking paleontological work. "I cannot curb my impatience sufficiently when a specimen has been found," he wrote.[1] When museum staffers would find a specimen that had been damaged through mishandling, they would say it had been "RCAed." Roy Chapman Andrews turned out to be a vivid writer and speaker, however, and his entertaining lectures raised much money for the museum. He wrote twenty-two books, most of them directed at younger readers, some of whom Andrews inspired to pursue paleontology as a career.

The Central Asiatic Expedition was designed to search for fossils that would show the supposed evolutionary transition from apes to humans. Osborn believed that man had evolved in Central Asia, a theory later replaced by the "out of Africa" theory popular today. Andrews gathered a multi-disciplinary group of specialists, including two geologists, a physician/anthropologist, a chief of motor transport, a photographer, a Mongolian government agent who could act as a guide and interpreter, and a retinue of mechanics, cooks, and helpers. Andrews' second-in-command was paleontologist Walter Granger, who had worked with Barnum Brown at Como Bluff.

In April 1922, a convoy of three Dodge automobiles and two Fulton trucks rolled through a gate of the Great Wall of China, headed for Mongolia. Because there were no gas stations along the route, a caravan of seventy-five camels had left two weeks earlier, leaving gasoline at pre-arranged caches. Four days into the expedition, Andrews started finding dinosaur fossils.

At a site called "Flaming Cliffs," the expedition found the remains of a *Protoceratops* ("before horned face"), a dinosaur similar to *Triceratops*, but without horns. They returned to Flaming Cliffs the next summer and found some fascinating items weathering out of the sandstone: three oblong fossil eggs, each about eight inches long and seven inches around. Because dinosaur eggs had never been identified before, the scientists could not be sure what kind of animal had laid these eggs. But Andrews' crew had a hunch that they had found dinosaur eggs. Further searching turned up twenty-five more eggs, two of which were broken, exposing the skeletons of embryonic dinosaurs.

The scientists cut out a section of sandstone and shipped it to the American Museum in one piece. Back in New York, preparators chipped away at the rock, revealing thirteen large, oblong eggs, in two layers, arranged in concentric circles with their ends pointing toward the center of the circle. The expedition also discovered the fossils of the predatory dinosaur, *Velociraptor* ("speedy thief"), a name Michael Crichton and Steven Spielberg would later make famous. Andrews took several more expeditions to Mongolia before deteriorating political conditions rendered the trips too dangerous.

Dinosaur digging continues apace. In fact, as with all areas of scientific inquiry, the pace of discovery has accelerated. One of the most celebrated finds of recent years came in 1990, when Sue Hendrickson discovered a 90 percent complete *Tyrannosaurus rex* skeleton—the largest and most complete T. rex found to that time. Hendrickson found the specimen (**Plate 13**), named "Sue" in her honor, in the Black Hills of South Dakota, within the boundaries of the Cheyenne River Sioux Indian Reservation. In 1997, "Sue"[2] was sold at auction to Chicago's Field Museum for the staggering sum of $8.36 million.

Dinosaur fossils are found on every continent, including Antarctica, as well as on the larger islands, such as Greenland, New Zealand, and Madagascar. *Encyclopedia Britannica* informs us that over one thousand dinosaur species have been named. This is true, but misleading. First, the definition of "species" is more subjective than most people realize. There are many definitions of "species," each tailored to the needs and purposes of a different scientific discipline. The popular conception is that a species consists of a group of animals that can breed and produce fertile offspring. But working biologists use the term to describe narrower groupings.

As defined by Ernst Mayr, one of the twentieth century's leading biologists, species are "groups of actually or potentially interbreeding natural populations that are reproductively isolated from other such groups." In other words, animals that could breed with each other are considered separate species if, because of geographical isolation or any other reason, they do not normally inter-breed. For example, coyotes are a species separate from the several species of wolves, yet in zoos coyotes will mate with wolves and produce fertile offspring.

Second, it is impossible to observe the breeding habits of the dinosaurs, so we cannot know which animals were mating with which. We have only bones upon which to base a judgment as to affinity. But differences among the skeletons might be due to sexual dimorphism—the difference between males and females— age differences, or simply the variety of form that is observed within every species. Imagine how easy it would be, if one had only bones to work with, to assign the Great Dane and the Toy Poodle to different species, when in fact they are both varieties of the domestic dog (*Canis familiaris*).

Third, despite the discovery of the great dinosaur graveyards of the American west and such spectacular finds as the *Iguanodons* of Bernissart, there are still relatively few complete or nearly complete dinosaur skeletons. They remain rare and valuable, as illustrated by the story of "Sue," and paleontologists often find themselves studying fragmentary remains.

Finally, there is a bias toward naming new species. The first person to describe a new species gets to name it. He can name it after his wife, the professor he needs to approve his doctoral dissertation, or anyone else. If there is a fig leaf of doubt about the correct classification of a fossil specimen, it will be given a new species name.[3] When it comes to fossil forms, the species concept has become almost meaningless. Species names are called "trivial" names. No one identifies a dinosaur by its species name alone; only the genus name has much currency. For example, no one calls *Triceratops horridus* by the species name "horridus," but everyone knows what a Triceratops is. If you talk about a "rex," people might understand that you are referring to *Tyrannosaurus rex*, but that is the exception that proves the rule.

Even at the level of genera, there is not a completely reliable count. Many genera are based upon fragments or single bones. Take the sauropods—those huge vegetarian dinosaurs with very long necks and tails (think *Brontosaurus*). As of 1990, paleontologists had named ninety different genera and 150 different species of sauropods, but only five genera are based upon virtually complete skeletons. The rest are based upon partial skeletons or, more commonly, single bones or bone fragments. Fourteen species are based only upon teeth, six species are based upon a single limb bone, and twenty-five are based on one or more vertebrae. Paleontologists consider about a dozen genera to be firmly established and another dozen "fairly well established."[4] The rest (sixty-six genera) are *nomina dubia* –doubtful names. There is enough information to believe that these sixty-six genera were separate from the established genera, but not enough to confirm they were separate from each other.

During the past couple of decades, there has been a resurgence of interest in all things related to dinosaurs. Readers can peruse a large stack of new dinosaur books at bookstores and libraries. Moviegoers flocked to see the computer-generated dinosaurs featured in box office hits such as the "*Jurassic Park*" trilogy.

Adults find dinosaurs mildly interesting; children are deeply fascinated by them. Many children go through a dinosaur phase in which they memorize dinosaur facts and lore. The Disney movie "*Dinosaur*" was directed at children, and a popular children's television show features "Barney," a purple dinosaur that sings, "I love you, you love me." Whenever I admitted I was writing about dinosaurs, almost without fail I was asked if the book was intended for children. Most children grow out of the dinosaur phase; those who do not outgrow it become paleontologists.

The existence of the dinosaurs raises some interesting questions. When did the dinosaurs live? Where did they come from? Why did they become extinct? The scientific establishment has a set of answers to these questions. But science is guided by a philosophy that excludes the possibility of God acting in the physical world. Many Christians, not wishing to appear "anti-science," have accommodated their beliefs to science's conclusions, even though those conclusions follow from atheistic premises. Other Christians defend the inspiration and accuracy of scripture.

Most Seventh-day Adventists are within the latter group. Although we recognize that the Bible is not a science textbook, we reject the view that the events described in the early chapters of Genesis did not really happen and are best understood as myth, allegory, or parable. Moreover, because Seventh-day Adventists embrace the prophetic gift of Ellen G. White, we have a perspective on such questions as the mystery of the dinosaurs that is unique even among creationists.

---

[1] Great patience must be exercised when extracting a fossil from the rock. Not everyone has this patience. After a week of observing Barnum Brown delicately chipping away on a specimen at Howe Ranch, for example, the 82-year old ranch owner had to be physically prevented from using a pickaxe to hack out the bones

[2] Not only was "Sue" on Sioux land, "Sue" turned out to be aptly named for another reason: she occasioned years of litigation involving the landowner, Maurice Williams, the federal government, a Sioux tribe, and the commercial fossil hunters who originally bought and excavated the specimen, who were doing business as Black Hills Institute for Geological Research. Although Black Hills had paid Williams $5,000.00 for the specimen, Williams was able, with the help of the government and his tribe, to void the sale on the ground that the land was held in trust for him by the federal government, and by federal statute he needed the government's permission to sell any part of it. The auction proceeds thus benefited Maurice Williams. In a federal criminal prosecution arising out of Sue's discovery, the U.S. Attorney charged the principals behind the Black Hills Institute with 39 felonies. They were later convicted on 8 felony charges, all unrelated to the discovery of "Sue" (and five of which were reversed on appeal), and forced to incur $1,000,000.00 in legal fees defending themselves. See, Steve Fiffer, *Tyrannosaurus Sue: The Extraordinary saga of the largest, most fought over T. rex ever found,* (New York: Freeman and Co., 2000)

[3] This is not a new phenomenon. In 1859, Charles Darwin wrote, "It is notorious on what excessively slight differences many palaeontologists have founded their species; . . ." *Origin of the Species,* (New York: Grammercy Books), p. 305.

[4] J.S. McIntosh, "Sauropoda," in *The Dinosauria* (Berkeley: University of California Press, 1990) David B. Weishampel, Peter Dodson, and Halszka Osmolska, editors, p. 345. At the time *The Dinosauria* was compiled in 1990, there were 285 genera of dinosaurs. In a recent magazine article, however, Peter Dodson states that there are now 527 named dinosaur genera. Stephen Ornes, "Move Over, T. rex," *Discover,* vol. 27, no. 12, December, 2006.

CHAPTER 04

# When Did the Dinosaurs Live? The "Scientific" View

"*An adventure 65 million years in the making,*" teased the posters and previews for the motion picture "Jurassic Park," reflecting the current belief that the dinosaurs became extinct sixty-five million years ago. This belief is frequently stated as fact. It is *not* fact. It is a conclusion that depends on the *interpretation* of facts. The dinosaur fossils do not tell us when the dinosaurs lived. The belief that the dinosaurs lived sixty-five million years ago is based upon the inferred age of the rocks in which dinosaur bones are found. The inferred age of the rocks is, in turn, based on an assumption about the rate at which geological processes occurred in the past. But the rate assumed is neither necessary nor probable. That is the executive summary. The long version follows.

## A. THE ROCKS

Layers of sedimentary rock cover about three-quarters of the earth's land surface. Sedimentary rock is formed when water or wind deposits sediment—silt, dirt, sand, gravel, etc. The sedimentary rocks are usually layered or stratified. Think of a layer cake, with layers of varied color and thickness, and you have an idea what the crust of the earth looks like. There are many good places in the arid American west to view exposed, stratified, sedimentary deposits. One of the best is at the rim of the Grand Canyon. (**Plate 14**) There, water has cut a channel down through numerous sedimentary layers of differing color and texture.

Water deposited the sedimentary layers, often burying plants and animals in the process. These buried plants and animals are called fossils, from the Latin word *fossilis*, meaning "dug up." The term *fossilization* is sometimes used to describe a chemical process in which organic material is replaced or fortified by hard minerals. This process is more correctly called *permineralization* or *petrifaction*. (Petrifaction is the complete replacement of the original bone; permineralization is the filling of open spaces in the bone with hard minerals.)

Scientists also use the term *fossil* to describe any ancient plant, animal, or human trace, regardless whether or to what extent mineralization has occurred. For example, although dinosaurs are believed to have died out sixty-five million years ago, unmineralized dinosaur bones—bones that look and feel just like old cow bones—have been found on Alaska's North Slope.[1] These bones are called fossils despite their youthful appearance and lack of mineralization. Fossils of animals with backbones, called vertebrates, are relatively rare; land-dwelling vertebrates are especially rare. The overwhelming majority of fossil animals are marine invertebrates—sea-dwelling animals that lacked a backbone but typically had a shell.

Under normal circumstances, an animal is not preserved as a fossil. Dead bodies decompose quickly under attack by scavengers and bacteria. Even the skeleton is soon scattered. The fossilization of a largely intact skeleton, with the bones arranged roughly as they would have been in the living animal (articulated), requires rapid burial. When we find a fossil animal whose skeleton has been preserved in articulation, we can be sure the animal had not been dead long when it was buried.

Occasionally, even soft body parts are found fossilized or preserved in contour by the surrounding rock. Fossilization of soft body parts requires extremely rapid burial in an environment with little oxygen—typically burial in thick silt under water. If the animal is buried in an oxygen-rich environment, microorganisms and other creatures will quickly break down the soft tissue, and it will not be preserved in the rock.

In the nineteenth century, William Smith, a British civil engineer and surveyor, learned the geology of England through his work on canal digging projects. His hobby was collecting and cataloging fossils from the sites where he worked. He discovered that each layer contained fossils unlike those in other layers. Using these "guide fossils" or "index fossils" as markers, Smith could identify a particular layer of rock wherever it was exposed. He produced a famous map of the rock layers of England and became known as "Strata" Smith. *Stratigraphy* is the sub-discipline of geology concerned with studying layers of rock, identifying layers, and tracing a given layer over a wide area. The index fossil method Smith pioneered continues to be the primary tool geologists use to identify rock strata.

By correlating the strata at various outcrops and exposures around the world, geologists developed a concept known as the *geologic column.* The geologic column is subdivided into several major divisions of strata, illustrating their hierarchical relationship to each other. I hate to spring a lot of technical jargon on the reader, but it is difficult to discuss geology, paleontology, and many of the other topics this book will wrestle with without being familiar with the basic nomenclature of the geologic column.

For our purposes, the important parts of the geologic column are the major divisions of the Phanerozoic, or fossiliferous strata, so called because they contain many fossils. The major divisions of fossiliferous strata are, from bottom to top, the Cambrian,

| | | |
|---|---|---|
| CENOZOIC ERA (age of recent life) | Quaternary Period | The several geologic eras were originally named Primary, Secondary, Tertiary, and Quaternary. The first two names are no longer used. Tertiary and Quaternary are retained but used as period designations. |
| | Tertiary Period | |
| MESOZOIC ERA (age of medieval life) | Cretaceous Period | Derived from Latin word for chalk (creta); first applied to extensive deposits that form white cliffs along the English Channel. |
| | Jurassic Period | Named for the Jura Mountains, located between France and Switzerland, where rocks of this age were first studied. |
| | Triassic Period | Taken from the word "trias," indicating the threefold character of these rocks in Europe. |
| PALEOZOIC ERA (age of ancient life) | Permian Period | Named for Perm, a former province in eastern Russia, where these rocks were first studied. |
| | Pennsylvanian Period | Named for the state of Pennsylvania where these rocks have produced much coal. |
| | Mississippian Period | Named for the Mississippi River valley where these rocks are well exposed. |
| | Devonian Period | Named after Devonshire, England, where these rocks were first studied. |
| | Silurian Period | Named after Celtic tribes, the Silures and the Ordovices, that lived in Wales during the Roman conquest. |
| | Ordovician Period | |
| | Cambrian Period | Taken from the Roman name for Wales (Cambria) where rocks containing the earliest evidence of complex forms of life were first studied. |
| PRECAMBRIAN | Ediacaran or Vendian Period (nomenclature not settled) | Ediacaran, after the Ediacara Hills of South Australia, where strange Precambrian fossil forms were discovered. |

**Figure 4.1: Major Divisions of Geological Time**

Ordovician, Silurian, Devonian, Carboniferous (which is subdivided into Mississippian and Pennsylvanian), Permian, Triassic, Jurassic, Cretaceous, Tertiary, and the Quaternary. The Tertiary is subdivided into (from bottom to top) the Paleocene, Eocene, Oligocene, Miocene, and Pliocene. The Quaternary is subdivided into the Pleistocene and Recent (or Holocene).

The eleven major divisions are often combined into three large groups: the Paleozoic, consisting of the Cambrian up through the Permian; the Mesozoic (the dinosaur strata), consisting of Triassic, Jurassic, and Cretaceous; and the Cenozoic, consisting of Tertiary and Quaternary. Figure 4.1, from the U.S. Geological Survey, is a graphic depiction of the geologic column. You will need to refer frequently to this chart unless and until you memorize the names and their hierarchical relationships.

Not all of these major divisions are present at any given location. Typically, they are *not* all present. But the divisions that are found are always found in the same vertical relationship to each other, unless there has been subsequent geological activity. For example, if the Ordovician and Silurian are missing in a given locality, the Devonian will lie on top of the Cambrian, and not vice versa. If the Permian and Carboniferous are missing, the Triassic will overlie the Devonian.[2]

Although there are fossils common to many of the sedimentary strata, usually at least one or a few types are unique to each division of strata. These fossils are used as index fossils or guide fossils for purposes of identification. For example, dinosaur fossils are unique to the Triassic, Jurassic, and Cretaceous.

The order of the sedimentary rocks is real; it has been established by countless observations. Knowledge of the order of strata is not merely of academic interest; it is of enormous value in many practical endeavors such as mining and fossil fuel production. Oil and gas exploration companies have geologists on staff, and have invested many millions of dollars in proprietary sub-surface stratigraphic mapping. They would not have invested this time and money if the concept of the geologic column were invalid.

You may have noticed, however, that Figure 4.1 is labeled "Major Divisions of Geological Time," and the subdivisions are described as "ages" and "periods." This is where the difficulty arises for conservative Christians. The geologic column has come to be regarded not simply as a description of the order in which the rocks are found, but as a chronicle of long ages in the history of the planet. In fact, the names of the divisions of strata have come to refer *primarily* to geologists' notions about the time when the rocks were deposited. When a geologist speaks of the Jurassic, for example, he may be referring to a division of sedimentary rock first studied in the Jura Mountains of Switzerland. But he is more likely to be referring to a notional period of the Earth's history that he believes began about 210 million years ago and lasted until about 145 million years ago, give or take a few million years.

The order in the geologic column is real, but the notion that the column represents hundreds of millions of years of time is based upon assumption and inference. How did geologists conclude that it took hundreds of millions of years to deposit these layers of rock? The rest of this chapter will be devoted to answering that question.

## B. WHEN DID THE ROCKS BECOME CLOCKS?

### 1. STENO AND THE NATURE OF FOSSILS

In the sixteenth and seventeenth centuries, there was no agreement as to the nature and origin of fossils. Some believed that fossils grew in the earth through a "plastic virtue" at work in nature, a view associated with followers of Plato.[3] Another opinion was that seeds or germs fell into the earth and there grew into organic shapes, a view associated with followers of Aristotle. It was also commonly believed that fossils grew inside the earth as *lusi naturae* ("games of nature").

From our vantage point, these opinions seem silly, but they made sense in those days. First, the term *fossil* meant everything that was "dug up," including gemstones and minerals (which do "grow" inside the earth). Second, many fossils are remnants of extinct animals, and the concept of extinction was not fully accepted. Third, the living representatives of some of the more important non-extinct forms had not yet been discovered. Finally, many fossils resembling marine organisms were found high in the mountains many miles from the ocean; it was difficult to conceive how marine animals came to be buried so far from the sea and at such a high elevation.

There were some, Leonardo da Vinci among them, who argued that fossils really were the remains of plants and animals. Even those who argued for organic origin, however, usually did so only with regard to the fossils of non-extinct marine organisms found close to sea level. In other words, the easy cases were attributed to organic origin, but some of the harder cases were still attributed to "plastic virtue" or "games of nature."[4]

Among those who were of two minds regarding the origin of fossils was Robert Plot (1640-1696), a professor of chemistry at Oxford. For most fossils, he leaned toward the "plastic virtue" explanation.[5] But in his 1677 book, *The Natural History of Oxfordshire*, Plot included an illustration of a stone that "has exactly the Figure of the lowermost part of the thigh bone of a man, or at least some other animal . . ." Plot conceded that the specimen was "a real bone now petrified," but since it was larger than a human thigh bone, he thought it might have come from an elephant. Based upon Plot's illustration, paleontologists now believe it was a *Megalosaurus* femur.[6]

The work of Nils Stensen (1638-1686) marks the beginning of modern geology.[7] Stensen, who Latinized his name to Nicolaus Stenonis (since anglicized to Steno), was born in Copenhagen, Denmark, and attended medical school in Holland. Steno excelled at dissection, and became one of the world's leading anatomists. In 1665, Steno moved to Florence, Italy, to join a group of scientists supported by the wealthy nobleman Ferdinando de Medici.

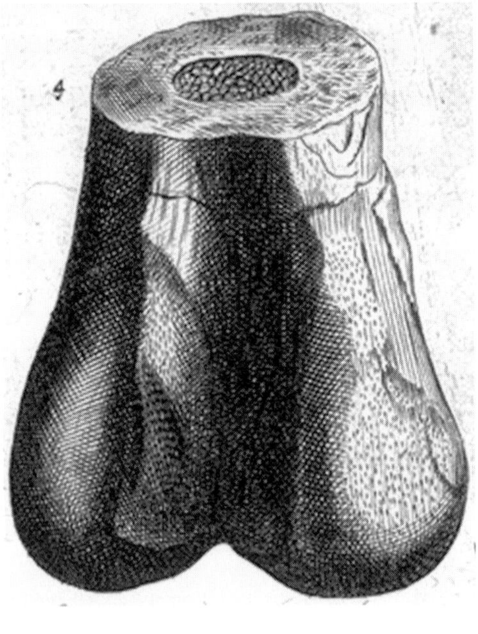

Figure 4.2: Plot's fossil – possibly part of a Megalosaurus femur

In 1666, when fishermen caught a large Great White shark, Ferdinando had the head sent to Steno for dissection. Steno was struck by how much the shark's teeth resembled fossils called glossopetrae or "tongue stones." (Some glossopetrae are more than twice the size of modern shark teeth.) Steno was not the first to notice the resemblance.[8] But Steno's study of fossil shark teeth led him to realize that, rather than argue the origin of individual fossils, he had to attack the fossil problem in a comprehensive way. He had to explain how any solid, such as a fossil, could come to be found *inside* another solid, such as a layer of rock.

Steno started with a simple axiom: A solid will cause anything that forms around it to conform to its shape. If the "tongue stones" had grown within solid rock, they would have been distorted by that rock in the same way a tree root molds itself around rock as it grows. But fossils often left a concave impression in the surrounding rock, proving that the fossils must have been solid before the surrounding rock became solid. The shark teeth must have been buried in soft sediment that later hardened into rock. Thus, Steno used rigorous logic to argue something that seems obvious today: Fossils were life forms that were buried in mud—the mud subsequently hardening into stone.

The manner in which fossils were preserved suggested—indeed, it almost proved—the manner in which the sedimentary rocks were formed. The rock strata were formed when muddy water deposited its particles on the bottom. This process left horizontal layers of mud that usually hardened into stone. Steno noticed that each layer conforms to the shape of the layer below it, leading to the inference known as Steno's law of superposition: Layers of rock are arranged in a relative time sequence, with each layer older than the one above it and younger than the one below. Steno's law is a statement of relative time only; the layers might have been deposited one after another in rapid succession, or with millions of years separating each layer.

The manner in which the sedimentary strata were deposited, full of fossils, led inexorably to a conclusion about *when* they were deposited. Steno believed that most of the fossiliferous strata were deposited during the Genesis Flood.[9] Studying the rocks of the Apennine Mountains of Tuscany, Italy, Steno noticed that the lower layers had no fossils, while the upper ones were rich in fossils. But both types were stratified sedimentary deposits apparently deposited by water. The strata were tilted, not horizontal, yet Steno believed they had been deposited horizontally.

To explain what he was observing, Steno theorized two separate episodes of sedimentation. Steno believed the non-fossiliferous strata had been deposited in the universal ocean of the second day of creation. These layers did not contain any fossils because God had not yet "let the water teem with living creatures," an event that occurred on the fifth day of the creation week. Gen. 1:9, 10, 13, 20, 21, 23. Steno attributed the second episode of sedimentation to the Genesis Flood. These strata were full of fossils because numerous living creatures were buried during the Genesis Flood.[10]

In 1669, Steno wrote out his ideas on fossil formation and stratigraphy in a Latin work entitled, "Preliminary discourse to a dissertation on a solid body naturally contained within a solid," better known by the short title, *Prodromus* ("forerunner"). Not long after writing *Prodromus*, Steno converted to Catholicism, abandoned science, became a priest, and spent the rest of his life ministering to the Catholic minority in northern Germany (for which the Catholic Church beatified him in 1988). He never wrote the larger work for which *Prodromus* was intended only as an introduction. Yet *Prodromus* was important enough that some have called Steno the "Father of Stratigraphy."[11]

By the end of the seventeenth century, John Harris (1666-1719), theologian, mathematician, and vice president of the Royal Society of London, could claim that, "all sober and judicious men are now convinced that the exuviae of sea animals, so plentifully found at this date in the strata of the Earth, and in the most hard and solid stone and marble, are the lasting proof of the Deluge itself and of its universality."[12] But Harris was not correct on either count. The fossil controversy continued for almost another century,[13] with both Christians and non-believers on each side of the issue.[14] Moreover, not all who conceded that fossils were the remains of buried organisms believed that those organisms had been buried during the Genesis Flood.

## 2. THE FLOOD GEOLOGISTS

A major stumbling block to acceptance of the truth that the Genesis Flood deposited the fossiliferous strata was the incredible depth and breadth of those strata. Throughout modern history, few Bible believers have entertained an adequate conception of the magnitude of the Flood—how destructive it was, how deeply it buried plants and animals, and how comprehensively it reshaped the surface of the earth. Many naturalists who were open to the historicity of the Genesis Flood searched for a single Flood stratum, not realizing that almost *everything* that we now observe in the crust of the earth is in some way a legacy of the Genesis Flood.

One of the few who realized the true scope and magnitude of the Genesis Flood was Martin Luther. In his *Commentary on Genesis*, Luther wrote that the location of Eden could not be determined by the location of modern rivers, because those rivers did not exist before the Flood:

> The awful Deluge destroyed all things. . . . Who can doubt, therefore, that the fountains of those rivers [of Eden] were also broken up and confounded? As therefore, mountains exist where fields and fruitful plains before flourished, so there can be no doubt that fountains and sources of rivers are now found where none existed before, and where the state of nature was changed by that mighty convulsion. Nor do I entertain the least doubt that all these wonders of nature, which are from time to time discovered, are the effects of that same awful visitation, the Deluge. In the metal mines which are now explored are found large logs of wood, hardened into stone; and in masses of stone themselves are perceived

> various forms of fish and other animals. With the same confidence I also believe that the Mediterranean Sea before the Deluge was not within the land. My persuasion is that the position that it now occupies was formed by the effects of the terrible Flood.[15]

Had everyone shared Luther's conception of the Flood, there would never have been a controversy over the origin and nature of fossils, and the early geologists would not have made the mistake of trying to identify a single Flood layer.

A major seventeenth century work on the Genesis Flood was Thomas Burnet's *A Sacred Theory of the Earth*, published in 1681. Burnet (1635?-1715) was an Anglican minister and chaplain to the king. (Like Buckland, Conybeare, and many other Anglican churchmen of the sixteenth through the early nineteenth centuries—the time before science became a profession—Burnet was more interested in science than in religion.) Burnet argued that during the Flood the earth's axis assumed its present 23.5-degree tilt relative to the plane of its orbit around the sun.[16] Since the axis tilt is what causes the seasons, the antediluvian world would have had a constant year-round climate.

Burnet also responded to one of the criticisms leveled at Flood geologists; namely, that the earth does not contain enough water for a worldwide deluge. (We now know that this was never a valid argument; if its topography were completely smooth, a boundless, mile-and-a-half deep ocean would cover the surface of the globe.[17]) Burnet realized that the pre-Flood earth might have had a smoother topography, but he grossly underestimated the depth of the current oceans and thus the amount of water he had to work with. In order to make his erroneously small volume of water go farther, Burnet theorized that the pre-Flood Earth was perfectly smooth, like an eggshell, having no mountains. During the Flood, the "eggshell" violently cracked up, creating the high mountains we see today.

Given what we now know about plate tectonics, there is a sense in which the earth's crust did violently crack up during the Flood. But Genesis speaks of the pre-Flood earth as having mountains (Gen. 7:20), so it could not have been as smooth as Burnet asserted.

Burnet tried to explain the Flood exclusively in terms of physics and other natural laws and phenomena. "As the centerpiece of his logic, Burnet insists again and again that the earth's scripturally specified history will be adequately explained only when we identify natural causes for the entire panoply of biblical events."[18] But the Bible clearly teaches that the Flood was a supernatural act of God, so Burnet ended up ranging freely in his interpretation of Genesis. Burnet's Flood was so much a matter of physics and natural law that it would have happened regardless of man's sin and God's judgment upon that sin.[19]

Strangely, Burnet did not mention fossils, even though they are good evidence of the Deluge. In later works, Burnet admitted that he considered science above scripture, stating that the Bible should be interpreted literally only on matters for which we have no evidence from nature to the contrary, but that when such natural evidence does exist, the sacred text must be reinterpreted to fit the evidence.[20] In this respect, Burnet was in step with his countryman and contemporary, John Locke, the early enlightenment philosopher.

About fifteen years later, another major work on the Flood appeared. John Woodward (1665? – 1728) has been called "the first major figure in English geology." Woodward came from a poor family and was apprenticed to a linen draper, but the king's physician discovered Woodward's high intellect and apprenticed him to practice medicine. After a visit to the Jurassic rocks of the Cotswolds, Woodward became interested in fossils and set out to formulate a theory of their origin. He became familiar with the geology of

England, as even an unsympathetic writer was forced to admit:

> Woodward . . . had acquired the most extensive information respecting the geological structure of the crust of the earth. He had examined many parts of the British strata with minute attention; and his systematic collection of specimens, bequeathed to the University of Cambridge, and still preserved there as arranged by him, shows how far he had advanced in ascertaining the order of superposition.[21]

In 1695, Woodward published an *Essay Towards a Natural History of the Earth*, in which he argued that fossils were the remains of animals buried during the Genesis Flood. Woodward believed that the Flood dissolved the antediluvian rocks, which then re-formed after the Flood. He argued that the sorting of fossils into the strata in which they are now found was a result of the specific gravity of the animals; heavier animals sank lower, lighter ones were buried higher. Woodward repudiated Burnet's approach of trying to explain the Flood solely by natural causes: "A deluge neither could then, nor can now, happen naturally."[22]

Followers of Woodward included Charles Leigh, John Morton, and William Derham in England, and Jean-Jacques Scheuchzer and D. S. Buttners on the continent. Several were uncomfortable with Woodward's theory that the fossils had been sorted by weight, but there was no hostility toward the Genesis Flood as an explanatory concept. Scheuchzer translated Woodward's book into Latin, and in 1708 published his own book, entitled *Complaint and Vindication of the Fishes*, in which he argued that the abundant fossil fish found in the strata of the Alps were entombed during the Flood.[23]

German geologist D. S. Buttners published *Signs and Witnesses of the Flood* in 1710 and *Fossil Corals* in 1714. Buttners argued that the pre-Flood world was in many respects different from, and better than, the current world. It was exceedingly beautiful, with low, rolling hills covered with a deep, fruitful soil. A daily mist watered the earth, and there was no rain, lightning, hail, snow, or drought. After the Flood, plants and animals degenerated in size and in quantity. Buttners understood the magnitude of the Deluge, calling it a stupendous affair and ridiculing those who would measure it in "thimblefuls."[24]

Buttners conception of the pre-Flood world was similar to that of William Whiston, a professor of geology at Cambridge who had published *A New Theory of the Earth* in 1696. Whiston also believed that the pre-Flood earth was watered by a gentle mist rather than by rainfall. He believed the antediluvian world was "much more fruitful than at present: and the multitude of the vegetable productions much greater."[25] In 1729, Etienne-Augustin Souciet, a Jesuit scholar, defended Flood geology in a three-part article in the French journal *Trevoux*, stating that to interpret the fossils as relics of the Flood confounded unbelief by providing concrete proof of "the most terrible and incredible event ever to take place." The Flood was not natural in its cause, argued Souciet, but it had natural effects.[26]

Between 1724 and 1730, John Hutchinson (1674-1737) published a book in three parts entitled *Moses' Principia* ("Moses' Principles"). Hutchinson argued that the order of the strata resulted from deposition by tidal waves flowing in different directions during the Genesis Flood, deluging and then exposing the earth, as the waters advanced and then receded.[27] He ridiculed Woodward's theory regarding the role of specific gravity in sorting the fossils.[28] In 1750, Patrick Cockburn (1678-1749) published a book entitled, *An Enquiry Into the Truth and Certainty of the Mosaic Deluge*, in which he rejected the argument that there was not enough water to create a Deluge.[29]

In 1761, Alexander Catcott (1725-1779), a professor at Cambridge, published *A Treatise on the Deluge,* which included a theory of how the mountains were formed. He argued that although the Genesis Flood laid down continuous sheets of strata, erosion subsequently washed away most of the material, leaving only mountains to mark the heights the strata had once attained. He argued that much of this erosion happened during the closing days of the Deluge, as great quantities of water drained off of previously formed strata.[30] Catcott's theory is a better explanation of mesas, or tabletop formations, than of the great mountain chains like the Rockies and the Alps.

Another noteworthy Flood geologist was Geneva, Switzerland's Jean-Andre De Luc (1727-1817). De Luc published a series of papers arguing that many currently observable geological processes cannot be traced very far into the past. He observed that the Rhône River, as it flowed into Lake Geneva, was building a delta out into the lake. But if this had been going on a very long time, the lake would have disappeared, turning into an alluvial plain. De Luc believed that, during the Flood, the antediluvian continents had collapsed, and the present continents had emerged from what had been the pre-Flood ocean basins.[31]

As geologists began to develop the idea of the geologic column, they pointed to the Genesis Flood to explain the fossiliferous portions. Giovanni Arduino (1714-1795), an inspector of mines in Tuscany and later director of agriculture and industry for the Republic of Venice, divided the rocks into Primary, Secondary, and Tertiary. Arduino's Primary included non-fossiliferous hard rocks, such as schist, gneiss, and quartz veins. The Secondary included most of the fossiliferous rocks, roughly corresponding to today's Paleozoic and Mesozoic. Arduino's Tertiary consisted of the higher fossiliferous rocks.

The terms *Primary* and *Secondary* are no longer used, but *Tertiary* and the resulting *Quaternary* survive to the present day, describing those rocks above the dinosaur strata. Johann Gottlob Lehmann (1719-1767), a mining engineer and teacher in Berlin, also divided the rocks in a scheme similar to Arduino's, and attributed some of the fossiliferous strata to the Genesis Flood.[32]

In 1789, John Williams (1732-1795) published a work entitled, *The Natural History of the Mineral Kingdom.* Drawing upon many years of practical experience in "mineral engineering" and coal mining, Williams noted that coal is found in stratified deposits that are just like other such deposits, and must have been deposited by water, just as the others were. Having observed the grain of the wood in many coals, he concluded that, "the antediluvian timber was the original of coal."[33]

The abundance of coal convinced Williams that the pre-Flood world was richly forested:

> I have already made it pretty evident that the greatest part of the surface of the earth, before the Deluge, was covered with a luxuriant growth of tall timber, that this antediluvian timber is the origin of our pit-coal; and that it was a sufficient and an adequate source of all coals in the world.[34]

Williams rejected Woodward's specific gravity argument, noting that it did not apply with regard to the coal measures. He concluded that only water, surging and retreating in successive tides, could have created the coal measures. Water came in and dumped sediment, then retreated, exposing the ground before returning to deposit more sediment:

> I am convinced, by the phenomena of the strata, that every stratum which we see regular was formed upon the then surface of the globe; that is, the very part of

> the globe where a particular part of the stratum was made, *began to be dry land* before the next tide brought with it the matter which produced that stratum, and so on, stratum super stratum, tide after tide, until all the strata were completed.[35]

Rhoda Rappaport, an historian of geologic thought, notes that no comprehensive history of the Flood geologists has ever been written. Byron C. Nelson's *The Deluge Story in Stone* is probably the closest we have to such a history. But the foregoing discussion demonstrates that during the seventeenth and eighteenth centuries, many naturalists incorporated the Genesis Flood into their geology. In their day, these men were in the mainstream of geological thinking. Although there never was a time when all serious students of the earth believed in a universal Deluge, as long as the culture of the Protestant Reformation and counter-reformation held sway, naturalists and scientifically minded people were willing to take the Genesis narrative seriously.[36]

## 3. THE ENLIGHTENMENT REJECTS BIBLICAL HISTORY

In the second half of the eighteenth century, the intellectual climate changed. The philosophy of the "enlightenment," which elevated human reason above divine revelation, became dominant. John Locke (1632-1704) wrote:

> In all things, therefore, where we have clear evidence from our ideas . . . reason is the proper judge; and revelation . . . cannot in such cases invalidate its decrees: nor can we be obliged, where we have the clear and evident sentence of reason, to quit it for

> the contrary opinion, under a pretence that it is a matter of faith . . .[37]

Locke was a Christian, but his philosophy tended to exalt reason above revelation, and he defended Christianity on the basis of its reasonableness.[38]

Later enlightenment philosophers, especially the French *philosophes*, formed a negative opinion of Christianity.[39] Voltaire (1694-1778), for example, regarded Christianity as a grotesque fairy tale. In the late eighteenth century, France was the world center of intellectual life and home to many geologists. Given the zeitgeist, these geologists wanted to base their science on reason, and reason alone. Over the course of a fifty-year period between 1750 and 1800, geology moved away from a scriptural interpretation of Earth's history.

By the end of this time, only one of the fossiliferous layers—the rocks then referred to as the "drift," but now called the Pleistocene—was still sometimes attributed to the Genesis Flood. This movement away from Flood geology was the result of a change in philosophy among geologists. It was a disregard for God and the Bible, *and not new geological evidence*, which led naturalists to discard the Flood theory of geology and revert to pagan Greek theories about origins.[40]

The enlightenment had a powerful influence in Scotland, such that the period between 1740 and 1790 is referred to as the Scottish Enlightenment. The most important Scottish philosopher was David Hume (1711-1776), who had great contempt for Christianity.[41] In his treatment of miracles in *An Enquiry Concerning Human Understanding* (1748), Hume singled out the first five books of the Bible for ridicule:

> . . . consider a book, presented to us by a barbarous and ignorant people, written in an age when they

> were still more barbarous, and in all probability long after the facts which it relates, corroborated by no concurring testimony, and resembling those fabulous accounts which every nation gives of its origin. Upon reading this book, we find it full of prodigies and miracles. It gives an account of a state of the world and of human nature entirely different from the present: Of our fall from that state: Of the age of man, extended to near a thousand years: Of the destruction of the world by a deluge: Of the arbitrary choice of one people, as the favourites of heaven; and that people the countrymen of the author: Of their deliverance from bondage by prodigies the most astonishing imaginable: I desire anyone to lay his hand upon his heart, and after a serious consideration declare, whether he thinks that the falsehood of such a book, supported by such a testimony, would be more extraordinary and miraculous than all the miracles it relates; which is, however, necessary to make it be received . . .[42]

Among Hume's friends were James Watt, the inventor of the steam engine, and Adam Smith, the founder of economics. Another member of this milieu was James Hutton (1726-1797), a gentlemen scholar interested in geology.[43]

## 4. UNIFORMITARIANISM & CATASTROPHISM

Working in an intellectual environment that excluded any appeal to scriptural history, James Hutton adopted the views held in ancient times by Xenophanes and during the renaissance by

Leonardo da Vinci and others. They believed that the sedimentary strata had been formed on the bottom of a sea as sediment washed off the continents.

The age-old weakness of this theory had been the inability to explain how strata that formed on the bottom of the ocean were now found high in the mountains. But Hutton proposed an explanation: Volcanism. Hutton believed that, as layer upon layer of sediment accumulated on the ocean floors, the increasing weight of the sediment eventually caused heat and pressure to melt the lower strata, allowing magma to intrude, expand, and create uplift. The magma lifted new continents at the sites of the old oceans, while the old continents slumped and became the new ocean basins. This cycle, argued Hutton, repeated itself endlessly.

One of Hutton's tenets was that no catastrophic floods were necessary to explain the strata. The system had always cycled at about the same speed. If so, the earth must be immensely old. It must have required unimaginably vast amounts of time to form the thick layers of sedimentary rock in the ocean basins and then lift those rocks to the height above sea level at which we now find them.

The geological feature to which Hutton pointed as a proof of "deep time" was an angular unconformity at Siccar Point on the coast of Scotland. There, horizontal strata rested upon up-tilted strata, at a nearly perpendicular angle. Hutton reasoned that this resulted from a multi-stage process lasting millions of years. First, the lower sediments were deposited as horizontal layers, then raised above water level and tilted upward during a violent tectonic event. Next, erosion planed a nearly horizontal surface across the perpendicular beds. Then, the land subsided below water level again, where a new series of strata were deposited horizontally over the up-tilted ones. Finally, the entire sequence was again uplifted, producing the outcrop seen today.

In Hutton's system, the earth was an endlessly cycling machine, with erosion wearing the continents down and re-depositing the sediment, and magma lifting the redeposited sediment to begin the cycle anew. The most famous line of Hutton's 1795 book, *Theory of the Earth,* was the final sentence: "The result, therefore, of our present enquiry is, that we find no vestige of a beginning, no prospect of an end." It was the only line many people ever read. The book was one thousand pages in length and poorly written. Most of Hutton's sources were French, and almost half of the book consisted of long, un-translated quotations from French authors.[44] But in 1802, John Playfair published a more reader-friendly book promoting Hutton's theories.

Even in the post-enlightenment intellectual environment, many geologists concluded that what they were observing in the crust of the earth could only be the result of catastrophe. To Georges Cuvier, for example, the strata did not indicate slow accumulation, but a *series* of catastrophes. In his 1812 *Discourse on the revolutions of the surface of the globe,* Cuvier argued that these catastrophes had been sudden and that no process observable today was adequate to account for them. Cuvier noted that, "none of the agents that [nature] employs today would have sufficed to produce its ancient effects."[45] In other words, the slow and constant processes of erosion and sedimentation taking place today could not account for what Cuvier observed in the stratified crust.

Some, like William Buckland and Yale University geologist Benjamin Silliman, correlated the last of Cuvier's "revolutions" to the Genesis Flood, but Cuvier himself was *not* motivated by a desire to support the biblical narrative. Though a lifelong Protestant, Cuvier was also a product of the French enlightenment and was dedicated to finding naturalistic explanations for the observed geological facts. He insisted upon keeping science and religion separate and regarded as futile and misguided all attempts to support religion from the findings of science.[46] He used the term *revolution* and avoided the term *catastrophe* precisely because he felt catastrophe would connote supernatural intervention.[47]

Cuvier's opinions carried much weight. For a long time after geology had been purged of biblical influence, there was still a willingness to consider catastrophic explanations for geological phenomena. Those who followed Cuvier came to be called *catastrophists*, and the followers of Hutton came to be called *uniformitarians*. The debate between catastrophists and uniformitarians continued for decades.

## 5. THE TRIUMPH OF UNIFORMITARIANISM

Eventually, another Scotsman, Charles Lyell (1797-1875), took up Hutton's cause. Lyell was born to a wealthy Scottish family. His father was an avid naturalist with an excellent library that included volumes on geology. Lyell studied at Oxford, where he attended William Buckland's colorful geology lectures. After Oxford, Lyell studied law and was admitted to the bar in 1825. With his father's financial support, however, Lyell was able to practice more geology than law and soon began planning a textbook. For nine months during 1828 and 1829, Lyell toured interesting geological sites in Europe.

Lyell admitted that his book was not to be a balanced treatment of the issues but, to the contrary, a sustained argument in favor of uniformitarianism:

> It will not pretend to give even an abstract of all that is known in geology, but it will endeavour to establish the *principle of reasoning* in the science; and all my geology will come in as illustration of my views of those principles, and as evidence strengthening the system necessarily arising out of the admission of

> such principles, which, as you know, are neither more nor less than that *no causes whatever* have from the earliest time to which we can look back, to the present, ever acted but those *now acting*; and that they never acted with different degrees of energy from that which they now exert.[48]

The first volume of *Principles of Geology* was published in 1830, with subsequent volumes appearing in 1831 and 1833. Lyell argued that all geological phenomena could be explained as the result of processes now in operation. The present is the key to the past. No old causes are extinct, Lyell argued, and no new ones have been introduced. Present causes have always operated at about the same rate and intensity as today. *Principles of Geology* was a "passionate brief" for uniformitarianism.[49]

It was also a winning brief. *Principles* went through eleven editions in Lyell's lifetime; he was at work on the twelfth when he died. Lyell's book had a much greater influence on geological thinking than any other book ever written.[50] *Principles* made uniformitarianism the prevailing view—and eventually the *only* view—held by geologists. Uniformitarianism became the basic assumption and organizing principle of historical geology. Indeed, it became a dogma that could not be challenged in geological circles for over one hundred years.

But why was Lyell such a successful champion of uniformity, and is uniformity necessary to geological science? To answer these questions, we turn to the writings of the late Stephen Jay Gould (1941-2002).[51] Gould was a Harvard paleontologist who believed all the usual things about natural history: that the world is about 4.5 billion years old, that complex life forms entered the fossil record about six hundred million years ago, and that Darwin's theory of evolution was largely correct. Gould was no creationist; to the contrary, he was contemptuous of creationists.[52] Keep that in mind while pondering his criticisms of Charles Lyell.

The reason for Lyell's success in promoting uniformitarianism was not the logical force of his arguments, but the superiority of his verbal skills. Most scientific writing is intended to convey technical information to other specialists; the prose tends to be serviceable, but drab and uninspiring. The scientific profession, wrote Gould, "self-selects for poor writing."[53] Charles Lyell, by contrast, was a great stylist with a gift for persuasive writing. "Truth is supposed to prevail by force of logical argument and wealth of documentation, not by strength of rhetoric," wrote Gould. "Yet we will never comprehend the reasons for Lyell's triumph unless we acknowledge the role of his verbal skills."[54]

**Figure 4.3: Charles Lyell (1797-1875), champion of uniformitarianism**

Lyell argued that when the rocks are observed firsthand and inductive reasoning employed, the facts lead to uniformitarian conclusions. Not true. Uniformitarianism is not a conclusion forced by study of the rocks, but an *a priori* assumption brought to the study of the rocks. The raw geological facts, encumbered with the least theoretical baggage, support catastrophism. "Read literally, then and now, the geological record is primarily a tale of abrupt transitions, at least in local areas," wrote Gould.[55]

Far from being opposed to fieldwork and induction, catastrophists studied the rocks, and they took at face value what they observed:

> The characteristic method of catastrophism, promulgated particularly by Cuvier, was empirical literalism—an approach diametrically opposed to Lyell's unfair characterization of these scientists as speculators opposed to field evidence. The catastrophists tended to accept what they saw as reality . . . The defense of catastrophism was rooted in the most direct (or minimally "interpretive") reading of geological evidence.[56]

The debate between Lyell and the catastrophists was never about the importance of fieldwork and observation—it was about how to *interpret* the observed facts. Yet Lyell mischaracterized catastrophists as armchair speculators opposed to fieldwork, and praised uniformitarians as the champions of fieldwork, observation, and inductive reasoning. Archibald Geikie promoted Lyell's myth in his influential history, *The Founders of Geology* (1897), and from there it found its way into most geology textbooks.[57]

Lyell's mischaracterization became what Gould called "cardboard history," or "textbook cardboard." Textbook authors typically include a brief history of their discipline, written in the "heroic tradition," meaning that they portray the pioneers in their discipline as heroically overcoming ignorance and superstition. In geology textbooks, the uniformitarians are the heroes of empiricism and reason, while the catastrophists represent ignorance and superstition.

The truth was the opposite. Cuvier was a very careful observer of the rocks, whereas Lyell's portrayal of Hutton as an advocate of fieldwork was "one of the most flagrant mischaracterizations ever perpetrated by the heroic tradition in the history of science."[58] Textbook history has gotten it exactly backward:

> I just find it deliciously ironic that cardboard history touts Lyell's victory as the triumph of fieldwork, while catastrophists were the true champions of a geological record read as directly seen. Lyell, by contrast, urged that theory . . . be imposed upon the literal record to interpolate within it what [Lyell's] theory expected but imperfect data did not provide.[59]

> If we buy the simplistic idea that uniformity triumphed by fieldwork, then we will never understand how fact and theory interact with social context, and we will never grasp the biases in our own thinking (for we will simply designate our cherished beliefs as true by nature's dictates).[60]

Gould was not alone in the opinion that Lyell was strong on theory and weak on fieldwork. British geologist Derek Ager came to the same conclusion, stating it even more bluntly:

> Uniformitarianism triumphed because it provided a general theory that was at once logical and seemingly "scientific." . . . But I would like to suggest that, in the first half of the [nineteenth] Century, the "catastrophists" were better geologists than the "uniformitarians."

* * *

> In fact I must emphasize again that, in my view, the early uniformitarians were the theoreticians and the catastrophists were the careful field observers. Lyell himself was an excellent man on principles and processes, but he did not take very much interest in rocks.
>
> * * *
>
> My excuse for this lengthy and amateur digression into history is that I have been trying to show how I think geology got into the hands of the theoreticians who were conditioned by the social and political history of their day *more than by observations in the field.*[61]

The 19th Century geologists, wrote Ager, allowed themselves to be "brain-washed" into avoiding catastrophic interpretations of the geological facts.[62]

Many have the naïve idea that science proceeds in a strictly "Baconian" or empirical manner, meaning that scientists first make careful observations, and only after piling up many such observations do they formulate a theory to explain the observed facts. This is not what happened in geology. Catastrophism is a more straightforward induction, or interpretation of raw geological data, than uniformitarianism. The 19th Century geologists adopted uniformitarianism not because the evidence forced them to adopt it—as we have seen, the evidence supported the catastrophists—but because uniformitarian theory appealed to them on philosophical grounds. But even the philosophical appeal of uniformitarianism was largely a product of Lyell's rhetorical skill, as we shall now see.

## 6. UNIFORMITARIANISM IS NOT NECESSARY TO GEOLOGY

Scientific research follows certain methods. For example, the scientist assumes that natural law is constant and that it operated in the same way in the past as it is observed to operate today. Another *methodological* assumption is that natural processes such as erosion and sedimentation occurred in the past just as they occur today. On the other hand, the idea that these processes have always worked at the same *rate* as they are observed to work at today is a *substantive* claim of uniformitarianism. One of the rhetorical tricks that Lyell used in *Principles of Geology* was to lump methodological principles together with substantive claims and call them all uniformity.[63]

The methodological claims were (1) natural law is constant and uniform, and (2) the same processes that have acted in the past are now acting. The substantive uniformitarian claims were (1) these processes have always acted at the same rate, and (2) these processes have no historical direction; rather, they endlessly cycle.

The methodological principles were not controversial; all geologists embraced them, including Cuvier and the other catastrophists. But the substantive claims were very controversial; prior to Lyell, most scientists did not accept them. With skillful rhetoric, Lyell argued that methodological uniformity implied substantive uniformity. Lyell "pulled a fast one—perhaps the neatest trick of rhetoric, measured by subsequent success, in the entire history of science. He labeled all these different meanings as 'uniformity,' and argued that since all working scientists must embrace the methodological principles, the substantive claims must be true as well."[64]

Very early in his career, Gould noted that geologists tended to confuse methodological principles with substantive assumptions:

When I was studying freshman geology as an undergraduate at Antioch College, our professor took us to a hill of travertine (limestone deposited by a spring), and told us that it was 15,000 years old, according to a principle he called uniformitarianism. A colleague, he told us, had measured the current rate of accumulation. The principle of uniformity permitted us to assume that this rate had been constant—and a millimeter per year (or whatever) extrapolated to the bottom of the pile yielded an age of 15,000 years. If we do not accept the constancy of nature's laws in space and time, my professor added, we will be unable to apply any science beyond the immediate present.

At that stage of juvenility, I rarely challenged professorial pronouncements, but this argument seemed wrong. I could grasp the part about nature's laws, but constancy in accumulation of a travertine mound in southwest Ohio was a thing, not a principle. Why shouldn't the travertine have built twice as fast ten thousand years ago, or not at all for long stretches between periods of deposition? I turned to Lyell and to the classical sources on uniformitarianism. . . . I soon discovered that [Lyell] had gathered a motley set of claims under the common umbrella of uniformity—in particular, he had made . . . the very same conflation of methodological principles with substantive claims that had fueled my professor's error about the travertine mound.[65]

Clearly, the assertion that geological processes have always proceeded at the same rate is *not* required by any legitimate scientific principle. To the contrary, a sound scientific method demands that we study the sedimentary rocks for indications of how fast sedimentation actually occurred. Many of the clues gathered by such study—for example, the exquisite preservation of many fossil fish—suggest that sedimentation once happened much more rapidly than we normally observe today. Nevertheless, owing to Lyell's advocacy, the assumption of uniformity of rate became the basic principle of interpretation in all subsequent geological fieldwork, study, and writing.

This gratuitous assumption, smuggled into geology by Lyell's rhetoric, is just where creationists disagree with mainstream geology. While believing that God supernaturally intervened to cause the Genesis Flood, most creationists would agree that science must assume uniformity of natural law. Creationists assume that the same physical laws in effect today–principles of hydraulics and hydrodynamics, for example—were also in effect during the Genesis Flood. Creation scientists carefully study processes such as local flooding, volcanism, earthquakes, plate tectonics, sedimentation, erosion, and fossil formation (taphonomy).[66] The important difference is that they *do not assume uniformity of rate.* They believe that these processes once worked much more rapidly and on a vastly grander scale than they work today.

Uniformity of rate is not an unreasonable starting assumption. But creationists need not join in this assumption. In rejecting it, we are *not* rejecting science or the scientific method. It is not scientifically necessary to assume, pending the discovery of evidence indicating catastrophic deposition, that most rocks were formed very slowly. Why not assume, pending convincing evidence of slow deposition, that most rocks were formed catastrophically?

The sedimentary rocks can be interpreted as the residue of the Genesis Flood. No one disputes that water deposited the great bulk of the geologic column, and almost all of the major divisions of sedimentary strata contain marine fossils. The disagreement between creationists and mainstream geologists relates primarily to when and how rapidly the strata were deposited.

## 7. HOW THE ROCKS BECAME CLOCKS

It is the assumption of uniformity of rate that leads to the inference that the world is millions of years old. Given the rate of sedimentation observable today, it would have taken millions of years to form the numerous, thick layers of sedimentary rock found all over the world.[67] Hence, geologists developed the idea that each of the major sedimentary divisions represented a vast "period" of "geologic time" of millions of years' duration. By the middle of the nineteenth century, the idea that the earth was unimaginably old was firmly entrenched.

Dinosaur fossils are found in the Mesozoic rocks, which consist of the Triassic, the Jurassic, and the Cretaceous rocks. Dinosaur fossils are not found in rocks overlying the Cretaceous rocks. Believing that each major division of sedimentary rock represents a vast period of time, geologists reason that dinosaurs must have become extinct at the end of the "Cretaceous Period." Lyell estimated that it took eighty million years to deposit the sediment that formed the rocks of the Tertiary and Quaternary, but the currently accepted figure is about sixty-five million years. Geologists therefore believe that the end of the Cretaceous, and the extinction of the dinosaurs, happened about sixty-five million years ago. Hence the slogan, "*An adventure 65 million years in the making.*"

Current notions of the age of the earth are not wholly dependent upon the assumption of uniformity of rate. Two later developments powerfully buttressed the Lyellian assumption. In 1859, about thirty years after uniformitarianism became the prevailing geological dogma, Charles Darwin published *On the Origin of Species by Means of Natural Selection.* Darwin's theory that plant and animal life evolved slowly, over the course of millions of years, reinforced the uniformitarian assumption that the rocks themselves formed slowly, over millions of years.

More recently, in the twentieth century, radiometric dating, a technique purporting to establish the age of rocks, was developed. A more detailed discussion of radiometric dating is found in Appendix One, "The Dating Game." But the basic geological world-view (or *weltanschauung*) that the fossiliferous strata represent vast ages of "deep time" or "geological time," was firmly established long before either Darwinism or radiometric dating was invented.

---

[1] Currie, Philip J., and Eva B. Koppelhus, *101 Questions about Dinosaurs*, (Minneola, NY: Dover Publications, 1996), pp. 11, 12 ("A more spectacular example was found on the North Slope of Alaska, where many thousands of bones lack any significant degree of permineralization. The bones look and feel like old cow bones, and the discoverers of the site did not report it for twenty years, because they assumed they were bison, not dinosaur bones."); K. Davies, "Duck-bill dinosaurs (*Hadrosauridae, Ornithischia*) from the North Slope of Alaska." *Journal of Paleontology* 61(1):198-200 (1987) ("The quality of preservation is remarkable. The bones are stained a dark red brown but otherwise display little permineralization, crushing, or distortion."); Norman, David B., *Dinosaur!*, (New York: Prentice Hall, 1991), p. 21.

[2] There are many places where the strata are out of order, but they all lie in mountainous regions. The stratigraphical order of non-mountainous regions has been verified by many thousands of well cores. The vertical order of the fossils reflects the order of burial of pre-Flood ecological zones or biomes during the Genesis Flood, and is not inconsistent with a young earth. Creationists now generally concede that out-of-order strata are legacies of orogeny (mountain building) during the Flood or within a few centuries after it. As we will discuss later, out-of-order strata support the creationist view that mountain building was a rapid, catastrophic process, not the gradual affair envisioned by mainstream geology.

[3] Adams, Frank Dawson, *The Birth and Development of the Geological Sciences*, (Baltimore, MD: Williams & Wilkins, 1938), p. 13; Rudwick, Martin J.S., *The Meaning of Fossils*, (New York: Neale Watson Academic Publications, 1972, 1976), pp. 1-45.

[4] Rudwick, Martin J.S., *The Meaning of Fossils*, (New York: Neale Watson Academic Publications, 1972, 1976), pp. 44, 45.

[5] Plot wrote: "This brings me to consider the great question, now so much controverted in the World, whether the Stones we find in the Forms of Shell-fish, be *Lapides sui generis*, naturally produced by some extraordinary plastic virtue, latent in the Earth or Quarries where they are found? Or, whether they rather owe their Form and Figuration to the Shells of the Fishes they represent, brought to the places where they are now found by a Deluge, Earth-quake, or some other means, and there being filled with Mud, Clay, and petrifying Juices, have in tract of time been turned into Stones, as we now find them, still retaining the same Shape in the whole, with the same Lineations, Sutures, Eminences, Cavities, Orifices, Points, that they had whilst they were Shells? I must confess I am inclined rather to the opinion of Mr. Lister, that they are *Lapides sui generis*; than that they are formed in an Animal Mould. The latter Opinion appearing at present to be

pressed with far more, and more insuperable difficulties than the former." Adams, at p. 258, quoting, *The Natural History of Oxfordshire*, Oxford, 1705, p. 112.

6 Spalding, David A. E., *Dinosaur Hunters* (Rocklin CA: Prima Publishing, 1993) pp. 10, 11.

7 Gould, Stephen Jay, *Time's Arrow, Time's Cycle*, (Cambridge, MA: Harvard University Press, 1987) p. 51.

8 More than a century earlier, Swiss naturalist Conrad Gesner (1516-1565) had noted that *glossopetrae* looked like shark teeth, and French naturalist Guillaume Rondelet (1507-1566) had argued that they looked like shark teeth because they *were* shark teeth. Cutler, Alan, *The Seashell on the Mountaintop* (New York: Dutton (Penguin Group), 2003) p. 56. Italian botanist Fabio Colonna (1567-1650?) had even stated that the shark teeth must have been buried during the Genesis Flood. Rudwick, at 44.

9 See, e.g., Gould, at 52; Rudwick, at 72, 73. Lyell wrote that, "Steno conceded, as Fabio Colonna had done before him, that all marine fossils might have been transported into their present situation at the time of the Noachian deluge."

10 Albritton, Claude C., *The Abyss of Time*, (San Francisco: Freeman, Cooper & Co., 1980), pp. 36, 37; Gould, *Time's Arrow, Time's Cycle*, at 52-56; Rudwick, at 66-68; Nelson, Byron C., *The Deluge Story in Stone*, (Minneapolis, MN: Bethany Fellowship, 1931, 1968), pp. 12-18.

11 Later geologists, including Darwinists like Stephen Jay Gould, have been in awe of what Steno accomplished, given his youth (he was only 31 when he wrote *Prodromus*) and the short time he devoted to science. Alan Cutler's *The Seashell on the Mountaintop* (New York: Dutton (Penguin Group), 2003) is a recent, readable biography of Steno.

12 Nelson, Byron C., *The Deluge Story in Stone*, (Minneapolis, MN: Bethany Fellowship, 1931, 1968), p. 51.

13 One humorous example of late 18th century confusion about fossils involves Robert Plot's illustration of a possible *Megalosaurus* femur (Fig. 4.2). In 1763, Richard Brooks published a book entitled *The Natural History of Waters, Earth, Stones, Fossils, and Minerals*, in which he copied Plot's illustration of the thighbone but identified the specimen as *Scrotum humanum*. French philosopher J.B. Robinet took Brooks seriously, describing the specimen in a 1768 book as a "stony scrotum." Robinet thought such objects were nature's attempts to form human organs in an effort to create a perfect human type. More recently, Beverly Halstead, writing in the *Journal for Insignificant Research*, has noted that Brooks' label was created after

the agreed starting date for scientific names, is in the correct Latin form for species names, and should have priority over subsequent names. By the same logic that *Brontosaurus* is now called *Apatosaurus, Megalosaurus* should now be called *Scrotum humanum.* See, Spalding, at pp. 11, 12; Christopher McGowan, *The Dragon Seekers,* (Cambridge, MA: Perseus Publishing, 2001), p. 1.

[14] The German Jesuit Athanasius Kircher (1602-1680), for example, believed the Genesis narrative was intended as literal history, as evidenced by his 1676 book, *The Ark of Noah.* Yet Kircher had also published in 1664 a work entitled, *The Subterranean World,* in which he argued that the fossils were the result of a "lapidifying virtue diffused through the whole body of the geocosm" and that their formation was controlled by a "*spiritus plasticus*" of the sort that controlled the development of plants and animals. (See, Nelson, at 19; Rudwick at 56.) Another example is the Swiss author, Pastor Elie Bertrand, an orthodox Christian who believed the earth to be about six thousand years old. In his 1752 work, *Memoires sur la structure interieure de la terre* ("Memoirs on the Structure of the Interior of the Earth"), Bertrand argued that God had created the fossils *in situ* in the rocks. Voltaire, the famous French *Philosophe* who despised Christianity, borrowed Bertrand's arguments to argue that the fossils could not have been of organic origin. (See, J. H. Brumfitt, *The French Enlightenment,* [Cambridge, MA: Schenkman, 1972], pp. 93, 113.)

[15] Nelson, at 10.

[16] The theory that the earth's axis tilted either at the Fall or at the Flood pre-dated Burnet. For example, John Milton included it in *Paradise Lost.*

[17] Roth, Ariel, *Origins,* (Hagerstown, MD: Review and Herald, 1998), p. 208.

[18] Gould, at 28.

[19] Rappaport, Rhoda, *When Geologists were Historians,* 1665-1750, (Ithaca, NY: Cornell University Press, 1997), p. 140-143.

[20] Rappaport, at 147-149. See, also, Gould, at 26-30.

[21] Charles Lyell, *Principles of Geology,* Ch. III.

[22] Rappaport, at 150-152.

[23] Nelson, at 51.

[24] Nelson, at 52.

[25] Nelson, at 55. Whiston believed that a large comet passing too close to the earth caused the Flood. The comet's vapor trail caused the rain, and the tides generated by a large body coming so near the earth stretched and eventually cracked the earth's solid surfacing, loosing the underlying "fountains of the great deep." (See, Stephen Jay Gould, "The Godfather of Disaster," in *Bully for Brontosaurus,* (New York: W.W. Norton & Co., 1991), pp. 367-381.) Gould notes that Isaac Newton personally chose Whiston as his successor as Lucasian professor of mathematics at Cambridge. Gould states that, "we must still grant him [Whiston] a major role in the history of science."

[26] Rappaport, at 171.

[27] Nelson, at 55.

[28] Apparently, there was some revenge in Hutchinson's criticisms of Woodward. Hutchinson had amassed a collection of fossils, which he entrusted, along with a manuscript, to Woodward with the understanding that Woodward would help publish it. Woodward did not go forward with publication. Hence, in his own book, Hutchinson vented his spleen at Woodward. Hutchinson was something of an eccentric. His title *Moses' Principia* was an intentional reference to the more famous work by Sir Isaac Newton. Not only did Hutchinson argue that Woodward was wrong about the role of specific gravity in the Flood, but he also tried to refute the theory of gravity as set out by Newton. It was said of Hutchinson that, "his works evince a strange combination of talent and eccentricity."

[29] Nelson, at 55, 56.

[30] Nelson, at 56-66.

[31] Rudwick, at 110, 111.

[32] Ritland, Richard, "Historical Development of the Current Understanding of the Geologic Column: Part I," *Origins* 8(2):59-76 (1981).

[33] Nelson, at 74-76. Nelson devotes an entire chapter to John Williams' views.

[34] Nelson, at 77, 78.

[35] Nelson, at 78, 79

[36] Clark, Harold W., *Fossils Flood, and Fire,* (Escondido, CA: Outdoor Pictures, 1968), pp. 17, 18.

[37] John Locke, "Faith and Reason: Their Distinct Provinces," in *Knowledge and Value* Sprague and Taylor, Eds. 2d Ed. (New York: Harcourt, Brace & World, Inc., 1959), p. 343.

[38] John Locke, *On the Reasonableness of Christianity* (1695).

[39] The anti-Christian and atheistic sentiments of the French intellectuals were a reaction against the unreformed, corrupt version of Christianity that was dominant in late eighteenth-century France. See, Ellen G. White, *The Great Controversy*, pp. 265-288.

[40] Nelson, at 137.

[41] Hume once claimed that the English were "relapsing fast into the deepest stupidity, Christianity and ignorance," and noted that a noblemen friend of his "has a hankering after the Bible, and is indeed little better than a Christian in a way of his own." David Hume, as quoted by Richard Wollheim, in *David Hume on Religion*, (Cleveland and New York: Meridian Books, 1964), p. 28.

[42] David Hume, "Can Miracles be Known to Happen?" from *An Enquiry Concerning Human Understanding*, Section X. (1748), in *Knowledge and Value*, supra, p. 405.

[43] See, e.g., Gould, *Time's Arrow, Time's Cycle*, at 17 ("to discuss James Hutton without the Edinburgh of David Hume, Adam Smith, and James Watt is (to cite the tale of another Scotsman) to rip a child untimely from its mother's womb.")

[44] Gould, *Time's Arrow, Time's Cycle*, at 66.

[45] Rudwick, at 132.

[46] Gould, *Time's Arrow, Time's Cycle*, at 112-115, 133; Rudwick at 134-136.

[47] Rudwick, at 132.

[48] Clark, Robert T. and James D. Bales, *Why Scientists Accept Evolution*, (Grand Rapids, MI: Baker Book House, 1966), p. 16, citing, *The Life, Letters, and Journals of Sir Charles Lyell*, Vol. I, (London: John Murray, 1881), p. 260 (Italics in Clark and Bales).

[49] Gould, *Time's Arrow, Time's Cycle*, at 104, 105 ("Lyell's great treatise is not, as so often stated, a textbook summarizing all prevailing knowledge in a systematic way, but a passionate brief for a single, well-formed argument, hammered home relentlessly. All sections of the text, including the introductory history, push the same theme, while the order of sections also records the smoothly unfolding brief. The famous sentence penned by Darwin to introduce the last chapter of the *Origin of Species* would serve as well for Lyell's three installments: 'this whole volume is one long argument.'"). See also, "An Asteroid to Die For," *Discover* (10)10:60-65 (1989)—("Charles Lyell was trained as a lawyer, and his book is more a brief for gradualism than an impartial account of the evidence."); *Ever Since Darwin*, (New York: W. W. Norton & Co., 1977), pp. 149, 150—("Charles Lyell was a lawyer by profession, and his book is one of the most brilliant briefs ever published by an advocate.")

[50] See, e.g., Clark and Bales, at 11-17.

[51] Stephen Jay Gould passed away in 2002. He was a Harvard professor whose academic work dealt mostly with land snails, but he was famous for his monthly essays in *Natural History* magazine and the best-selling books compiled of those essays. He was a man of immense learning and broad interests, and his essays are very readable.

[52] Creationists quote Gould more than any other scientist not because Gould was sympathetic to creationism, but because he wrote so beautifully, he made the scientific facts so accessible, and his academic credentials were impeccable. Gould was a Darwinist through and through. For example, "This View of Life," the byline of Gould's monthly column in *Natural History* magazine, was inspired by the last sentence in Darwin's *Origin of Species*: "There is grandeur in this view of life, with its several powers, having been originally breathed into a few forms or into one; and that, whilst this planet has gone cycling on according to the fixed law of gravity, from so simple a beginning endless forms most beautiful and most wonderful have been, and are being, evolved." When Gould wrote of creationists, it was frequently with bemusement and sometimes with contempt.

[53] Gould, *Time's Arrow, Time's Cycle,* at 107.

[54] Gould, *Time's Arrow, Time's Cycle,* at 107.

[55] Gould, *Time's Arrow, Time's Cycle,* at 133.

[56] Gould, *Time's Arrow, Time's Cycle,* at 133.

[57] Gould, *Time's Arrow, Time's Cycle,* at 66, 67.

[58] Gould, *Time's Arrow, Time's Cycle,* at 67.

[59] Gould, *Time's Arrow, Time's Cycle,* at 134. See also, *Ever Since Darwin,* (New York: W. W. Norton & Co., 1977), pp. 149, 150—("The geologic record does seem to record catastrophes: rocks are fractured and contorted; whole faunas are wiped out. To circumvent this literal appearance, Lyell imposed his imagination upon the evidence.").

[60] Gould, *Time's Arrow, Time's Cycle,* at 114, 115.

[61] Ager, Derek V., *The Nature of the Stratigraphical Record,* (New York: John Wiley & Sons, 1973, 1981), pp. 44-46 (emphasis added).

[62] Ager, at p. 46. Ager was also a Darwinist and denounced "fundamentalists" and creationists, just as Gould did.

[63] Gould, *Time's Arrow, Time's Cycle,* at 118-124.

[64] Gould, *Time's Arrow, Time's Cycle,* at 119.

[65] Gould, *Time's Arrow, Time's Cycle,* at 117, 118. In 1965, Gould published a paper on Lyell's confusion of methodological principles with substantive assumptions, a theme that he developed much more completely in the book *Time's Arrow, Time's Cycle.* "It shall always be one of my greatest satisfactions that, as a teeny neophyte scholar alone in a little Ohio College, I noted this central confusion at the same time as a major revisionist movement to reassess Lyell's cardboard history was brewing among professional historians." *Time's Arrow, Time's Cycle,* at 118. Gould's faculty advisor at Harvard, Norman Newell, recognized that the appearance of catastrophe in the earth's rocks was probably more than just an appearance. "[W]e have began to realize," Newell wrote, "that there were many catastrophic events in the past, some of which happened just once." Wheeler, Gerald W., *The Two-Taled Dinosaur,* (Nashville, TN: Southern Publishing, 1975), p. 93. (Recognizing the significance of the trend away from uniformity of rate, Wheeler cited Gould's 1965 article.)

[66] See, e.g., Leonard Brand, "Reactions," *Origins* 7(1):6-11 (1980) —("The suggestion . . . that flood geologists are methodologically uniformitarian (i.e., they assume that natural laws have always been the same) is definitely correct. Furthermore, that is not an inconsistency in flood geology methodology; it is simply being realistic and recognizing that science can study only those aspects of the flood process that have followed known natural laws. That is why we can study the erosional and depositional processes occurring during the flood, but not any 'miraculous' processes used to initiate the flood.")

[67] In current geological thinking, the rate of sedimentation does not dictate the age of the Earth. By some estimates, the fossiliferous portion of the geologic column (Cambrian to Quaternary) could have been deposited, at the present rate of sedimentation, in between 5.7 and 57 million years, rather than 570 million years. See Brand, Leonard, *Faith, Reason, and Earth History,* (Berrien Springs, MI: Andrews University Press, 1997), p. 217, citing T. H. van Andel, "Consider the incompleteness of the geological record," *Nature,* 294-397, 298 (1981).

CHAPTER 05

# Can a Christian Follow Lyell?

"Don't you and your husband fight all the time?" a fellow graduate student asked Patricia Kelley. Kelley was studying paleontology at Harvard under Stephen Jay Gould, while her husband Jonathan was in seminary preparing for the ministry. "The question surprised me," writes Kelley, "for the thought that my research on evolution might be incompatible with my religious faith had never entered my mind."

When she was growing up, evolution was not an issue at her United Methodist Church in Cleveland, Ohio. At college, she attended the campus Presbyterian Church, along with other geology majors and faculty. Jonathan Kelley is now a Presbyterian minister, and Patricia, who teaches geology and paleontology at the University of North Carolina, also teaches Sunday School and sings in the choir at her husband's church. Kelley's minister father-in-law once quipped, "Tricia studies the ages of rocks, and Jonathan studies the Rock of Ages."[1]

One hesitates to introduce discord into this picture of marital harmony, but is there really no conflict between the naturalistic view of earth's history, as expressed in Lyellism and Darwinism, and Christianity? It seems someone has not done his homework.

## A. THE IMPLICATIONS OF LYELLISM[i] FOR CHRISTIANITY

Laying aside Darwinism for the moment, Lyellism by itself drastically undermines the Bible and biblical Christianity. The Bible states that the earth, with its plants and animals, was created in six days. Gen. 1:1 - 2:3; Ex. 20:11; 31:17. If Lyell was correct, however, the plants and animals were created over the course of almost 600 million years. That is a big difference. Lyellian geological theory could not possibly be farther from the biblical teaching.

We tend to forget that God created the world in six days. God foresaw this tendency and told us to remember it. He hallowed the Sabbath as a memorial to His creation. "By the seventh day God had finished the work he had been doing; so on the seventh day he rested from all his work. And God blessed the seventh day and made it holy, because on it he rested from all the work of creating he had done." Gen. 2:2, 3 (NIV).[2] After the Exodus from Egypt, when God was teaching the Israelites His laws, He used the manna to teach Sabbath observance. One day's worth of manna was collected each morning, but it would not keep overnight. Ex. 16:15-20. On the sixth day, however, the Israelites were to collect two day's worth, and it would keep overnight and throughout the Sabbath day. Thus, they did not need to collect manna on Sabbath morning, and those who went out to collect it found none. Ex. 16:5, 6, 22-30.

---

[i] The term Lyellism is stylistically preferable to the ponderous *Uniformitarianism.* Neither Darwin nor Lyell invented the theories that made each famous. But what Darwin did for evolution, Lyell had previously done for uniformitarianism: he argued the doctrine in such a compelling way that all opposition in the scientific community was swept aside. Darwin would have been the first to acknowledge that if there is a "Darwinism" there should also be a "Lyellism."

In the fourth of the Ten Commandments, God commands us to observe the Sabbath: "Remember the Sabbath day by keeping it holy. Six days you shall labor and do all your work, but the seventh day is a Sabbath to the Lord your God. On it you shall not do any work, neither you, nor your son or daughter, nor your manservant or maidservant, nor your animals, nor the alien within your gates. *For in six days the Lord made the heavens and the earth, the sea, and all that is in them, but he rested on the seventh day.* Therefore the Lord blessed the Sabbath day and made it holy." Ex. 20: 8-11 (NIV, emphasis added).

God states within the commandment itself that the Sabbath is a memorial to the creation week. This would have been a remarkably dishonest thing for God to do if the world had been made over the course of hundreds of millions of years, rather than in six days.

The uniqueness and importance of the Ten Commandments cannot be overemphasized. Accompanied by thunder and lightning, God descended to Mount Sinai and audibly spoke the Ten Commandments within the hearing of the entire congregation of Israel. Ex. 20:1-21; Deut. 4:10-13. God Himself inscribed the Ten Commandments in stone tablets. Deut. 4:13; Ex. 31:18. God did not take these extraordinary measures with any other of the laws and regulations He gave to the Israelites. Although Moses, in indignation at Israel's idolatry, smashed the original stone tablets, God commanded him to chisel out two more tablets, upon which God re-inscribed the Ten Commandments, again with His own finger. Ex. 34:1, 28; Deut. 10:1-4.

The new tablets, called the tablets of the Testimony, were placed inside the Ark of the Testimony. Ex. 34:29; 40:20; Deut. 10:5. The ark of the Testimony, also called the Ark of the Covenant, was a wooden box overlain with gold, with a solid gold lid, called the mercy seat or atonement cover. Affixed to the lid on each side were golden sculptures of angels, or cherubim, with their wings overshadowing the middle. Ex. 25:10-22. The ark was the most sacred article of furniture in the sanctuary. Only the Levites were allowed to carry it, one man was struck dead for touching it without authority, and seventy men of Beth-Shemesh were slain for looking inside it. Deut. 10:8; 2 Sam. 6:6, 7; 1 Sam. 6:19.

The ark was placed in the most sacred compartment of the sanctuary, the Most Holy Place. Ex. 26:33, 34. The visible manifestation of God's presence, the glory of God that some rabbis called the "Shekinah Glory," was just above the ark, between the cherubim. Ex. 25:22; Lev. 16:2; 2 Sam. 6:2; Psalms 99:1; Ezek. 9:3. Only the high priest was allowed to look upon the ark, and then only once a year on the Day of Atonement. Even on the Day of Atonement, the high priest could not enter without incense and the blood of a sacrifice. Lev. 16.

To summarize, the most sacred compartment of the sanctuary, the Most Holy Place, contained the most sacred article of furniture, the Ark of the Testimony. Above the Ark dwelt the visible glory of God, and within the Ark were the tablets of the Testimony. The tablets contained the Ten Commandments, in the middle of which is the Fourth Commandment. The Fourth Commandment contains God's statement, written by His own finger, that He created the world and all its creatures in six days. But if Lyell was correct, God's statement, at the very center of the Hebrew religion, is at best a practical joke.

The symbolism of the sanctuary services is clear. Sin, the broken law, has resulted in man's alienation from God, with death as the inevitable result. 1 John 3:4; Rom. 4:15, 6:23. The blood of Christ is the only sacrifice that can secure mercy and atonement with God. 1 John 2:1, 2; Heb. 9:11-28. This is the very heart of Christianity, its core doctrine. The broken law creates the need for the atonement of Christ. Whatever denigrates the law also denigrates the need for the atonement. A Lyellian interpretation of the strata leads to the conclusion that the world was not created in six days. Therefore, the commandment to observe the Sabbath as a memorial to the six-day creation is void. This is how Lyellism nullifies part of the law and, to the same extent, nullifies the atonement.

Lyellism directly conflicts with the Bible regarding Noah's Flood. Chapters six through nine of the book of Genesis contain a description of the events leading up to the Flood, the Flood itself, and its aftermath. All humanity, all the land animals, and all the birds, with the exception of those on Noah's ark, were destroyed. Gen. 7:21-23. Jesus Christ attests to the historical reality of the Flood. Mat. 24:37-39; Luke 17: 26, 27 ("As it was in the days of Noah, so it will be at the coming of the Son of Man. For in the days before the Flood, people were eating and drinking, marrying and giving in marriage, up to the day Noah entered the ark; and they knew nothing about what would happen until the Flood came and took them all away.") Peter attests to the reality of the Flood, as does the author of Hebrews. 2 Peter 2:5; 3:5, 7; Hebrews 11:7. After the Flood, God made a covenant with Noah and the animals that He would never again destroy the world with water. The rainbow is the symbol of this covenant. Gen. 9:8-17.

But Lyellism denies that a worldwide flood ever occurred. If Lyell was correct, the Flood story, Noah, his ark, and God's covenant sealed and symbolized by the rainbow, were all invented out of whole cloth, and should not be understood as factual or historical.

Lyellism also seriously erodes the doctrine of the Fall. The Bible teaches that death entered the world because of Adam's sin, with the resulting fall of the human race. Rom. 5:12. ("[S]in entered the world through one man, and death through sin"). The Bible teaches that death entered not only the human race, but also the entire creation, as a result of Adam's sin. Rom. 8:18-22.[3] The sedimentary strata show man at the top, and the entombed remains of a large variety of animals far down into the crust of the earth. If interpreted according to Lyellian assumptions, the strata show that death reigned for hundreds of millions of years before man ever appeared on the scene. Thus, in the Lyellian system, death, as a general phenomenon, cannot have been caused by the fall of mankind.

This was noticed early in the Lyellian revolution. American geologist and clergyman Edward Hitchcock, writing in 1840, stated:

> The general interpretation of the Bible has been, that until the fall of man, death did not exist in the world even among the inferior animals. For the Bible asserts that by man came death (1 Cor. 15:21) and by one man sin entered into the world and death by sin (Rom. 5:12). But geology [Lyellism] teaches us that myriads of animals lived and died before the creation of man.[4]

If death is not a result of the fall, then it must be part of God's plan—part of the creation that God declared "very good." (Gen. 1:31.)

The Bible also teaches that the animals were created to eat grass and other vegetation, not each other. "And to all the beasts of the earth and all the birds of the air and all the creatures that move on the ground—everything that has the breath of life in it—I give *every green plant for food.*" Gen. 1:30 (NIV, emphasis added). In the earth made new, the animals will not prey on each other or on man. Isaiah 11:6-9. ("The wolf shall dwell with the lamb . . . the lion shall eat straw like an ox. . . . They shall not hurt or destroy in all my holy mountain.") If all animals were created herbivores, and in Heaven will again be herbivores, the predatory nature of animals must be a result of the Fall.[5]

If Lyell was correct, however, nature was "red in tooth and claw"[6] long before Adam sinned. The dinosaurs provide many examples of meat-eating predators. *Tyrannosaurus rex* had powerful jaws packing sixty teeth, the exposed portions of which were six inches long, sharp, and serrated. In a single bite, *T. rex* could snap off a chunk of meat weighing five hundred pounds.[7] *T. rex* was one of a group of carnivorous dinosaurs called theropods ("beast foot").

Another of the theropods was *Velociraptor*, discovered during one of Roy Chapman Andrews' expeditions to Mongolia and made famous in the motion picture "Jurassic Park." In 1971, a *Velociraptor* was found with its jaws clamped on the neck frill of a *Protoceratops*, and the deadly talon of its second toe ripping into the *Protoceratops*' belly. The *Protoceratops*, meanwhile, seems to have pierced the *Velociraptor*'s chest with its beak. The animals were buried and fossilized in this position, preserving the evidence of their deadly encounter.[8] If the dinosaurs really lived some five dozen million years before man was created, however, Adam's sin could not have caused these animals to become predatory carnivores. The Lyellian is forced to conclude—contrary to Scripture—that the violent, predatory nature of the animal kingdom was part of God's original plan.

A disturbing picture of God emerges if one accepts Lyellism. God claims to have created the world in six days and destroyed it by a universal Flood, neither of which is true. He placed the Sabbath, a memorial of the six-day creation, at the center of the Ten Commandments, despite knowing the world was not created in six days. If one accepts Lyell, the Bible is also wrong in teaching that sin caused violence and death. These things existed for many millions of years before man was created. If the lion is indeed to lie down with the lamb in the Earth made new, it will be a marked departure from the order of God's original creation. For the Lyellian, not as much was lost by the Fall as Christians had always believed, and therefore not as much has been redeemed by the atoning sacrifice of Christ. Lyellism severely damages the integrity of biblical Christianity.

## B. CHRISTIAN REACTION TO LYELLISM

There is no way to reconcile Lyellism with biblical truth. Unfortunately, many Christians have tried, offering several theories in an attempt to accommodate the Bible to Lyellian geology. These accommodationist theories have common characteristics. First, instead of interpreting the earth's crust in the light of biblical history, they reinterpret the Bible to make it fit geological theories. Second, the accommodationist theories typically interpret most of the fossiliferous strata as the residue of long ages that passed before man was created, which has the effect of decoupling disease, predation, suffering, and death from Adam's sin and the Fall of mankind. The terrible things that we find in the fossil record must therefore have been part of God's original creation, which He called "good."

### 1. THE DILUVIUM THEORY

One of the earliest accommodationist theories was William Buckland's "diluvium" theory. He and several of his contemporaries taught that much of the Pleistocene layer, then often called the "diluvium" or the "drift," was deposited during the Genesis Flood. But these men also taught that the lower strata were formed hundreds of thousands of years before Adam was created, which deviates from the biblical doctrine that God created the plants and animals in one week just a few thousand years ago.[9] After he was shown that the "diluvium" was primarily the work of glaciers, not floodwaters, Buckland abandoned his diluvium theory in favor of the "tranquil theory," which held that Noah's Flood left no residue.[10]

### 2. THE GAP THEORY

A more prominent accommodationist theory is the "gap theory," also known as the "active gap," "ruin and restoration," "pre-Edenic ruin," "ruin and reconstruction" and "interval" theory. It postulates that a very long interval of time elapsed between Genesis 1:1 ("In the beginning, God created the heavens and the earth.") and Genesis 1:2 ("And the earth was without form and void, and darkness was upon the face of the deep, and the Spirit of God was hovering over the waters."). The gap theory is attributed to Thomas Chalmers (1780-1847), a Scottish Presbyterian pastor. Although it has many variations and has changed over the centuries, its core is that, during the supposed "gap" between the first two verses of Genesis, there existed a previous creation on this Earth.

The earliest versions of the gap theory teach that Lucifer ruled this previous creation, which was peopled by a race of beings without souls. Desiring to become like God, Lucifer eventually rebelled (Ezekiel 28:11-19, Isaiah 14:12-20),[11] incurring God's judgment in the form of a flood, called the "Lucifer flood." The Lucifer flood is indicated in Scripture by the "waters" and the "deep" of Genesis 1:2. The rest of the first chapter of Genesis describes a *re-creation* or *reforming* of the earth from a chaotic state, not the initial creative effort. All the plant, animal, and human fossils in the rock strata date from the "Lucifer flood " and are not genetically related to the plants and animals living today, which are a result of the re-creation described later in the chapter.

The gap theory was given widespread circulation in 1909 when Cyrus Ingersoll Scofield (1843-1921) included it in the annotations to his Scofield Reference Bible.[12] While the gap theory does explain the presence of violence and death in the lower fossiliferous strata, the explanation is extra-biblical. The Bible divulges nothing of this hidden history, although gap theorists interpret several verses—usually only the King James translation of those verses—in such a way as to give veiled clues about the time of the "gap."

As an accommodation of geology, the gap theory was more

successful when geological opinion was open to Cuvier's multiple catastrophes. Some of the strata could then be attributed to the "Lucifer flood" and some to the Genesis Flood. In its modern incarnations, the gap theory is strictly a device to expand the biblical time frame in order to allow time for the strata to form over the hundreds of millions of years that Lyellian geology insists upon. Modern gap theorists do not include a "Lucifer flood," and believe that the Genesis Flood left no geological traces.

But some modern gap theorists still make Lucifer the lord of the purported gap creation.[13] One cannot help but wonder, however, why for 130 million years Lucifer's only earthly subjects were marine animals, or why it took more than 160 million additional years for Lucifer to get some dinosaurs to rule, or why poor Lucifer was allowed to rule over the men without souls for only the final 1/10,000th of his "gap" reign.[14]

### 3. THE DAY / AGE THEORY

Another popular attempt to harmonize Lyellism and Scripture is the "day/age" theory, which holds that the days of the first chapter of Genesis are not literal twenty-four-hour periods but long epochs of millions of years. The Hebrew word for "day," *yôm*, can refer to something other than a literal twenty-four-hour period. In Genesis one, however, each *yôm* of the creation week is modified by an ordinal number, as in "the first day," "the second day" etc. When so modified, *yôm* always means a literal twenty-four hour day.[15]

Moreover, the phrase "there was evening and there was morning" is used to describe each of the days of the creation week. This obviously refers to the dark and light portions of a literal, twenty-four-hour day. Generally, the Hebrew text weighs heavily in favor of interpreting the days of Genesis 1 as literal days.[16] Attempts to interpret the days of creation as ages or eras rather than literal days have been driven *not* by textual considerations but rather by a perceived need to accommodate Scripture to Lyellism.[17]

One has to wonder why, if the days of creation were intended to indicate multi-million-year epochs, the writer of Genesis didn't just say that? It isn't as though ancient people could not conceive of an earth millions of years old. On the contrary, most ancients believed the earth to be much older than the Bible indicates. For example, Plato believed that the Flood had occurred two hundred million years ago. The Babylonian historian Berosus placed the creation two million years ago. Hindu traditions that were committed to writing 1,500 years ago teach that the earth's history can be divided up into endlessly repeating cycles of 4.32 billion years, each of which can be further subdivided into 1,000 subcycles of 4.32 million years duration. The ancient Chinese adopted similar teachings of long cycles.[18] Obviously, the ancients were familiar with "old earth" histories and legends; it is the Bible's "young earth" narrative that is unique.

Furthermore, the animals created during a given day of the creation week do not always correspond to the fossils found in the analogous "age." For example, the Bible teaches that birds were created on the fifth day, and then the land animals on the sixth. Gen. 1:20-25. Obviously, birds come before land animals in the biblical narrative. By contrast, fossils of land animals are found in lower sedimentary strata than the fossils of birds, and therefore land animals are thought to have lived many millions of years before the birds ever appeared. Thus, the order in which creatures appear in the Genesis days does not correspond to the order in which they appear in the Lyellian ages.[19]

In the biblical creation account, God *speaks* the creation into being *ex nihilo* (out of nothing). There is no reason this should have taken hundreds of millions of years. The long ages are necessary to allow time for Lyellian geological "history" and Darwinian biological "history"—not because God needed more than an instant, much less more than a day, to speak the creation into being. Moreover, unless the day/age theory is coupled with a pre-creation creation and a pre-Fall Fall such as those of the gap theory, it presents violence, predation, and death as part of the creation that God declared "very good," and not as results of Adam's sin. It also leaves no geological work for the Genesis Flood to perform.

### 4. THE TRANQUIL THEORY

Swedish botanist Carl Linne (Latinized as Carolus Linnaeus) (1707-1778), who is famous for having developed the binomial naming system for species, and a classification scheme for the plant and animal kingdoms, suggested that Noah's Flood was a calm, quiet event in which the waters slowly rose and slowly subsided, without leaving any mark on the earth's crust. This theory became known as the "tranquil theory." Both Lyell and Buckland seized upon the tranquil theory as a way to harmonize Scripture and uniformitarian geology. But unless coupled with the dubious pre-creation creation and pre-Fall Fall of the gap theory, the tranquil theory cannot explain death prior to the Fall of man. Moreover, the Bible describes an event of considerable violence: "all the springs of the great deep burst forth, and the floodgates of the heavens were opened." Gen. 7:11, 12 (NIV).

The tranquil theory makes no geological sense. Think of the destruction that local floods, tidal waves, and tsunamis cause today. Bridges, houses, automobiles, boulders and trees can be swept away and destroyed as if they were pebbles and matchsticks. Powerful ocean currents move massive amounts of water. The south equatorial current in the Atlantic Ocean moves six million tons of water a second northward across the equator.[20] These currents would have been even more powerful in the unbounded ocean of the Flood, and the amount of erosion and sedimentation taking place would have been immense. Even if the Flood had risen slowly and calmly, tidal action alone would have stirred up staggeringly vast amounts of sediment that would have left an indelible mark on the Earth. It quickly became obvious that the "tranquil theory" was untenable, and it has largely been abandoned.[21]

### 5. THE LOCAL FLOOD THEORY

The most popular accommodationist position is that the Genesis Flood was a local, Mesopotamian valley flood. Since the mid-seventeenth century, various writers have advanced this theory. It gained wide circulation after John Pye Smith promoted it in an 1839 book entitled, *On the Relation Between the Holy Scriptures and Some Parts of Geological Science.* Smith argued that the Flood could have been "universal" in having wiped out all human beings then living without having affected the entire globe.

Over the years, various local floods have been put forward as having inspired the biblical flood story. British archeologist C. Leonard Woolley, while excavating the site of biblical Ur, discovered a thick layer of silt separating two layers of artifacts. But it was soon discovered that Woolley's flood was too local to be *the* local flood.[22] A recent local flood theory, promoted by Bob Ballard, who located the wreck of *Titanic*, is that the story of Noah's Flood refers to a catastrophic filling of the Black Sea some seven thousand years ago.[23]

There are a number of serious problems with the local flood theory. As with the other accommodationist theories, it leaves intact the Lyellian explanation of the *worldwide* sedimentary layers, meaning that predation and death reigned for millions of years before the creation of mankind and his Fall. Moreover, the local flood theory cannot be reconciled with a straightforward reading of Scripture. Genesis tells of a flood that covered "all the high mountains under the entire heavens." Moses was at pains to describe, in the most categorical language he could muster, a universal flood with universal consequences:

> The waters rose and increased greatly on the earth, and the Ark floated on the surface of the water. They rose greatly on the earth, and all the high mountains *under the entire heavens* were covered. The waters rose and covered the mountains to a depth of more than 20 feet.[24] *Every living thing that moved on the earth perished* —birds, livestock, wild animals, all the creatures that swarm over the earth, and all mankind. *Everything on dry* land that had the breath of life in its nostrils died. Every living thing on the face of the earth was wiped out; men and animals and the creatures that move along the ground and the birds of the air were wiped from the earth. Only Noah was left, and those with him in the Ark. Gen. 7:18-23 (NIV, emphasis added).

After examining the Hebrew syntax of this passage, Adventist theologian Gerhard Hasel concluded: "There is hardly any stronger way in Hebrew to emphasize total destruction of 'all existence' of human and animal life on earth than the way it has been expressed. The writer of the Genesis flood story employed terminology, formulae, and syntactical structures of the type that could not be more emphatic and explicit in expressing his concept of universal, worldwide flood."[25]

If the Flood were merely local, it would have been absurd for Noah to spend 120 years[26] constructing a giant, 150-yard long barge to save him, his family, and the animals, and it would have been absurd for God to order him to do so. Noah and his family could simply have moved out of the way. In fact, anyone could have moved out of the way, defeating the purpose of the Flood.

Those who believe that a local flood could have wiped out all humanity assume that the human race dwelt only in Mesopotamia. If that long-lived race had been reproducing normally during the (at least) 1,650 years between the creation and the Flood, their numbers could easily have been in the hundreds of millions, even the billions. There is no reason to believe that their civilization was contained within the area that later came to be known as Mesopotamia. Moreover, the Bible explicitly states that all land animals and birds were destroyed in the Flood. Are we to believe that the wild animals and birds were also limited to Mesopotamia? The only interpretation that does not shred the fabric of the biblical narrative is that the Flood was worldwide, not local.

All of the accommodationist theories—including the "diluvium" theory, the gap theory, the day/age theory, the tranquil theory, and the local flood theory—are failures. They do not merely fail to protect the integrity of the Bible, they fold, spindle, and mutilate plain biblical teachings in a procrustean attempt to make them fit Lyellian conclusions. The whole enterprise reflects a belief that the speculations of geologists are more reliable than God's word.

Young-earth creationists, who reject Lyellism, do not dispute a law of nature, or even a fact of geology. Creationists merely reject uniformity of rate, believing instead that rapid erosion and massive sedimentation occurred around the time of the Genesis Flood. The creationist assumption and the Lyellian assumption are on the same epistemological[27] footing: neither can be proven correct because they concern unobservable and unreproducible events of the very distant past. The Lyellian assumption is more scientific only if the term "science" is understood to include the *philosophy* of naturalism, which forbids any explanation, hypothesis or theory that appeals to the supernatural, or that is suggested by the Bible.[28]

The issue is stark: geologists have adopted an interpretive principle that is directly contrary to biblical revelation. To the uniformitarian, "everything goes on as it has since the beginning of creation." But, as the apostle Peter warned us, "they deliberately forget that long ago by God's word the heavens existed and the earth was formed out of water and by water. By these waters also the world of that time was deluged and destroyed." 2 Peter 3:4-6 (NIV). Lyellian geologists who insist that Earth's crust must be interpreted in light of presently observable processes and rates "deliberately forget" the Flood, just as Peter said they would.

## C. THE RELATIONSHIP BETWEEN LYELLISM & DARWINISM

Most of the geologists who agreed with Lyell were creationists. Many, such as Buckland, Sedgwick and Conybeare, were clergymen in the Church of England, trained in theology. They were creationists, and they were not in sympathy with Jean Baptiste Lamarck, Robert Chambers, and the other Pre-Darwinian proponents of the theory of evolution or, as it was then called, "transmutation."[29]

In the final analysis, however, it is irrelevant that many of the early Lyellians were churchmen and creationists. Lyellism sounds the death knell for any straightforward reading of Scripture. "If only the geologists would let me alone, I could do very well," lamented John Ruskin in 1851, "but those dreadful hammers! I hear the clink of them at the end of every cadence of the Bible verses."[30]

Many conservative scholars and churchmen understood the destructiveness of Lyellism and devoted much time and effort to defending the literal Genesis narrative.[31] Even before Darwin published, "there were many, very many, who felt that irreparable damage to faith" would result from Lyellian geology. They felt that Lyell and the uniformitarians were undermining the authority of Scripture, starting down a road that eventually would lead to infidelity and atheism.[32]

And it turned out exactly as they feared. Lyellism paved the way for Darwinism by rendering the Scriptures a compendium of fables. If the Bible is wrong about a six-day creation week, wrong about the Flood, and wrong about death being a result of the Fall, it is probably wrong about all aspects of origins.

Second, Lyellism supplies the incredibly vast amount of time necessary to Darwinian evolution. By the time Darwin published *Origin of Species* in 1859, the scientific community had for decades believed that the earth's fossiliferous rocks were formed gradually, over the course of millions of years. The scientific community was thus primed to accept Darwin's theory of how life had evolved by the slow and steady accumulation of small changes over the course of millions of years.

Third, not only did Lyell prepare the scientific world to accept Darwinism, Lyell prepared Charles Darwin to formulate his theory. Darwin took a copy of *Principles of Geology* with him on his voyage aboard the *Beagle*, studied it attentively, and reported that, "the book was of the highest service to me in many ways."[33] "The very first place which I examined," wrote Darwin, "showed me clearly the wonderful superiority of Lyell's manner of treating geology, compared with that of any other author whose work I had with me or ever afterwards read."

In *Origin of Species*, Darwin acknowledged his debt to Lyell:

> He who can read Sir Charles Lyell's grand work on the Principles of Geology, which the future historian will

> recognise as having produced a revolution in natural science, yet does not admit how incomprehensibly vast have been the past periods of time, may at once close this volume.[34]

Upon Lyell's death in 1875, Darwin wrote, "I never forget that almost everything which I have done in science, I owe to the study of his great works."[35] Gould has called Darwin "a disciple of Charles Lyell," and Lyell "Darwin's guru."[36]

Fourth, Lyellism and Darwinism mutually support each other. Assuming the strata took hundreds of millions of years to form, the pattern of fossil distribution seems to confirm evolution. The Precambrian levels contain the fossils of single-celled microbes. The Paleozoic strata contain mostly the fossils of sea-dwelling creatures at lower levels, with amphibians and then reptiles appearing at higher levels. The Mesozoic strata contain the fossils of land-dwelling reptiles, such as the dinosaurs, and smaller mammals. The Cenozoic strata contain a fossil fauna dominated by larger mammals. At the very top of the geologic column, we find man.

This fossil pattern is predicted by Darwin's theory: life originated in the seas; eventually a fish became an amphibian and crawled onto land, some amphibians evolved into reptiles, some reptiles evolved into mammals and birds, and ultimately some ape-like mammals evolved into humans.

Given a Lyellian interpretation of the rocks, something like Darwinian evolution seems very likely, even if we do not understand the mechanics of it. In the face of this broad pattern, to argue against evolution on the basis of the absence of transitional fossil forms and the lack of a plausible evolutionary mechanism seems like quibbling, like nitpicking. Thus, Lyellism strongly supports and reinforces Darwinism.

By the same token, if Darwin's theory is correct, it indicates that the rocks formed very slowly and gradually, just as Lyell said they did. The evolution of the animals whose fossils are contained in the rocks must have been a very long, slow process. Evolution must have taken many millions of years to change the lowermost mammals in the fossil record, the tiny shrew-like mammals of the Triassic/Jurassic boundary, for example, into the great variety of mammals we see today. Accordingly, the rocks separating the early mammal fossils from today's mammal specimens must have been formed over the course of many millions of years. Thus, Darwinism strongly supports and reinforces Lyellism.

It is sometimes asserted that Lyell only reluctantly accepted Darwin's theory of evolution. But Darwin published in 1859 and by 1865, only six years later, Lyell had become an enthusiastic supporter of Darwin's theory. In the 1865 edition of *Principles*, Lyell added his own original arguments in support of Darwin's theory. In later editions of *Origin of Species*, Darwin proudly noted Lyell's conversion to evolutionary views:

> The several difficulties here discussed, namely [the lack of transitional forms in the fossil record, the simultaneous and sudden appearance of several groups of species, the absence of fossils below the Cambrian], are all undoubtedly of the most serious nature. We see this in the fact that the most eminent palæontologists, namely Cuvier, Agassiz, Barrande, Pictet, Falconer, E. Forbes, &c., and all our greatest geologists, as Lyell, Murchison, Sedgwick, &c., have unanimously, often vehemently, maintained the immutability of species. But Sir Charles Lyell now gives

> the support of his high authority to the opposite side; and most geologists and palæontologists are much shaken in their former belief.[37]

To summarize, the view that evolution took place over the course of millions of years proves that the rocks are millions of years old, and the view that the rocks are millions of years old proves that life evolved slowly over the course of millions of years. It is thus pointless to reject either Darwinism or Lyellism without rejecting both. If one's model of origins rejects both Lyellism and Darwinism simultaneously, however, astonishingly little evidence remains to support either view. As we have noted, Lyellism is based on an assumption about the past rate of geological processes. Darwinism is based on the belief that beneficial genetic mutations can accumulate to form complex new organs and bio-chemical systems, a belief that is *not* well supported by observation.[38]

Yet many Christian apologists do not want to pick a fight with Lyell. In recent years, many writers have criticized Darwinism but, with few exceptions, have not challenged uniformitarian geology.[39] Prominent among the recent critics of Darwinism are the leaders of the "Intelligent Design" movement, including Phillip Johnson, Michael Behe, and William Dembski.[40] This movement does not criticize mainstream geology. Phillip Johnson said that his group wishes to criticize Darwinism on scientific grounds without becoming involved in a defense of "biblical literalism." This seems a good strategy: attack the adversary's model without exposing your own for him to shoot back at.

But no one on the other side is fooled. The concept of "Intelligent Design" implies that there was a Designer and, for most, that

Designer is the God of the Bible. Phillip Johnson is a Christian, and his faith motivated him to become involved in the origins controversy. Most other prominent members of the Intelligent Design movement are also Christians.

Moreover, Darwinists label any criticism of evolution as creationism, regardless how scrupulously the critic keeps religion out of the argument. For example, Jonathan Wells, who holds Ph.D. degrees from Yale and Berkeley, wrote a book called "Icons of Evolution," in which he argued that the Darwinian fairy tales told in science textbooks do not reflect current scientific knowledge as embodied in the peer-reviewed scientific literature. Wells never mentioned his religious beliefs, yet his Darwinist detractors immediately attacked him because of his membership in the Unification Church of Rev. Sun Myung Moon.[41] If you criticize Darwinism, you will be called a creationist, so if you are motivated by a high regard for Scripture, why not be honest about it?

Those who would reject Darwin without also rejecting Lyell make a strange distinction that scientists do not make. Geological gradualism (Lyellism) came, logically and historically, before biological gradualism (Darwinism), but today they have merged into one unified system of assumptions that scientists use to theorize about the history of the earth and its flora and fauna. The philosophy behind this system is naturalism. Naturalism, or scientific materialism, is the basis of all of the historical sciences. Darwinism is the philosophy of naturalism applied to historical biology, and Lyellism is naturalism applied to historical geology. It is absurd for Christians to adopt supernaturalism with regard to historical biology, but be reconciled to naturalism in the realm of historical geology.

There are intermediate theories between "young earth" or biblical creationism on the one hand, and Darwinian evolution on the other. These include "theistic evolution, the idea that God "created" through the use of evolution, largely allowing evolution to take its course, but giving it a helping nudge every now and again. Theistic evolution is a bizarre attempt to marry two opposing philosophies, theism and atheism, supernaturalism and naturalism. For the term "theistic evolution," substitute "theistic naturalism," as suggested by Phillip Johnson, and it is obvious that theistic evolution is an oxymoron. Another of the intermediate theories is "progressive creationism," the doctrine that God created by stages, each stage lasting—or separated from the next by—many millions of years.

The intermediate positions are illogical compromises that are faithful neither to the Bible nor to the philosophy of naturalism that guides modern science.[42] Scientists condemn both of these approaches as mere creationism. Bible-believers ought to condemn them as well; they contradict basic Bible doctrines such as the six-day creation and the worldwide Flood.

There are only two internally consistent models: 1) the biblical model—God created the world in six-days and later destroyed it in a global Flood that comprehensively reshaped the surface of the earth, forming most of the geologic column, and 2) the naturalistic model—life evolved by natural selection acting upon genetic mutations, unguided by God or Providence, over the course of hundreds of millions of years. Each of these models has integrity—meaning that it is internally consistent and faithful to its guiding principles—and is not the product of an illogical compromise or accommodation.

Those who subscribe to Lyellism already deny the reality of a literal creation week and a universal flood, two foundational biblical claims. Lyellians, if they are intellectually consistent, will be drawn toward the purely naturalistic model of origins, i.e., Darwinism. But is that a problem for the Christian? Might not a person be both a

Christian and a Darwinist? After all, Patricia Kelley, our Christian paleontologist, never saw a conflict between Darwinism and her Christian faith. Let us examine the implications of Darwinism and see whether they are consistent with Christianity.

---

[1] Patricia H. Kelley, "Studying Evolution and Keeping the Faith," *Geotimes,* December 2000, at *http://www.geotimes.org/dec00/faith_feature.html.*

[2] A year is the time it takes the earth to circle the sun, a day the time it takes the earth to revolve on its own axis, and a month is the time it takes the moon to pass from a particular phase (e.g., full) back to the same phase, averaging 29.53 days. By contrast, the week has no natural reason for existence; it exists because God established it at the creation. Atheistic regimes have thus sought to displace the weekly cycle. For example, French revolutionaries invented a calendar in which the months were divided into three ten-day cycles, and all the days were renamed. See, Schama, Simon, *Citizens: A Chronicle of the French Revolution,* (New York: Alfred A. Knopf, 1989), pp. 771-774.

[3] "For the creation was subjected to frustration, not by its own choice, but the will of the one who subjected it, in hope that the creation itself will be liberated from its bondage to decay and brought into the glorious freedom of the children of God. We know that the whole creation has been groaning as in the pains of childbirth right up to the present time." Although the phrase *subjected to frustration* seems vague, the phrase *bondage/slavery to decay/corruption* implies that the creation is subject to disease and death. The creation is not subjected to disease, death, and natural catastrophes as a result of its own choice but as a result of man's sin. See John T. Baldwin, *Creation, Catastrophe, & Calvary,* (Hagerstown, MD: Review and Herald, 2000), pp. 112, 113.

[4] Hitchcock, Edward, *Elementary Geology,* (Amherst, Mass: J.S. and C. Adams, 1840), p. 218, as quoted by John T. Baldwin, *Creation, Catastrophe, & Calvary,* (Hagerstown, MD: Review and Herald, 2000), p. 111.

[5] "While they remained true to God, Adam and his companion were to bear rule over the earth. . . .The lion and the lamb sported peacefully around them or lay down together at their feet." Ellen White, *Patriarchs and Prophets,* p. 50; "God gave our first parents the food He designed that the race should eat. It was contrary to His plan to have the life of any creature taken. There was to be no death in Eden." White, *Spiritual Gifts,* pp. 120, 121.

[6] Alfred Lord Tennyson *In Memoriam* (1850)

[7] Dowswell, Paul, John Malam, Paul Mason, and Steve Parker, *The Ultimate Book of Dinosaurs,* (Bath, UK: Dempsey Parr, an imprint of Parragon, 2000), p. 31.

[8] Wilford, John Noble, *The Riddle of the Dinosaur,* (New York: Alfred A. Knopf, 1985), p. 241.

[9] Whitcomb, John C., and Henry Morris, *The Genesis Flood,* (Phillipsburg, NJ: P & R Publishing, 1961), pp. 93, 94.

[10] Whitcomb, et al, at 97, citing William Buckland, *Geology and Mineralogy Considered with Reference to Natural Theology,* (Bridgewater Treatises, 1836), p. 94.

[11] It is unfortunate that many critics of the gap theory have denied that these passages refer prophetically to Lucifer, not just to the kings of Tyre and Babylon. Gap theorists are correct in holding that these verses describe the origin of sin within Lucifer, the covering cherub. This event took place in heaven, however, not on earth during the hypothetical time gap between Gen. 1:1 and Gen. 1:2. See, Rev. 12:7-9; Luke 10:18.

[12] In his book *Genesis Vindicated,* George McCready Price noted:

> In the early part of the nineteenth century this theory was quite popular. With the rise of Lyell's geology and the modern biological sciences, it went into eclipse for several decades; but it was revived a generation or so ago by C. I. Scofield and others . . . As it is the theory taught in the notes of the Scofield Bible, it is now very aggressively held by many who are emphatic in calling themselves fundamentalists, and who seem to regard the notes of Mr. Scofield as on a par with the Bible itself. . . . The early Reformers had some sad experiences with editions of the Bible with "explanatory" notes. They found that such notes tended to have an unfair and perverting influence on the readers; so they all finally abandoned them. Price, *Genesis Vindicated,* (Review and Herald, 1941), p. 291.

The Scofield notes also did much to popularize futurist views of prophecy, in which the Antichrist is described as a future bad individual who will desecrate a rebuilt Jewish temple in Jerusalem. See, e.g., Steve Wohlberg, *The Antichrist Chronicles,* (Ft. Worth: Texas Media Center, 2001), pp. 92-94.

[13] See Jack W. Provonsha, "The Creation/Evolution Debate in the Light of the Great Controversy Between Christ and Satan," in *Creation Reconsidered,* James L. Hayward, ed. (Roseville, CA: Association of Adventist Forums, 2000), pp. 303-311.

[14] See, Weston W. Fields, *Unformed and Unfilled: A Critique of the Gap Theory,* (Burgener Enterprises, 1994, republished by Master Books, Green Forest, AK); Jack C. Scofield, *The Gap Theory of Genesis Chapter One*; Russell Grigg, "From the Beginning of Creation: Does Genesis have a Gap?" *Creation Ex Nihilo* 19(2):35-38 March–May 1997, condensed version at the Answers in Genesis Ministry website, *http://www.answersingenesis.org/docs/1132.asp*; Younker, Randall W., *God's Creation,* (Nampa, ID: Pacific Press, 1999), pp. 22, 23 ("the Active Gap Theory has not gained

many supporters among Bible interpreters who have a good knowledge of Hebrew.").

[15] Younker, Randall W., *God's Creation,* (Nampa, ID: Pacific Press, 1999), p. 29.

[16] See, Hasel, Gerhard F., "The Days of Creation in Genesis Chapter 1: Literal Days or Figurative Periods/Epochs of Time?" *Origins* 21(1):5-38 (1994) (GRI website *http://www.grisda.org/origins/21005.htm* ) which is Chapter 2 of Baldwin, supra. See also Younker, at 28-36.

[17] See, e.g., Fritz Guy, "Negotiating the Creation-Evolution Wars," *Spectrum,* vol. 20, No. 1 (October 1989) ("The problem with all of these interpretations is that they are not indicated, much less demanded, by the biblical text; they are simply *ad hoc* attempts to make Genesis agree with geology. . . . The gap theory ignores the structure of Genesis 1, and the function of the first sentence as the thesis of the whole chapter. The day-age theory ignores the impact of the refrain, 'There was evening and there was morning' [Genesis 1:5, 8, 13, 19, 23, 31, NIV]").

[18] Coffin, Harold, *Origin by Design,* (Washington, D.C.: Review and Herald, 1983), p. 287.

[19] Some creation week events have no analog in naturalist origins theories. For example, on the second day God separated the "waters below" from the "waters above," creating a "firmament." Gen. 1:6-8. The meaning of this passage is not clear, but many creationists interpret it to mean that God created an atmosphere bounded from below by the ocean and from above by a water vapor canopy which separated the atmosphere from outer space. At the flood, the vapor canopy precipitated into part of the water that flooded the earth, and the canopy was never replaced. Evolutionists, of course, deny that the vapor canopy ever existed. The vapor canopy theory remains controversial even among creationists. See, e.g., Dillow, Joseph C., *The Waters Above: Earth's Pre-flood Vapor Canopy,* (Chicago: Moody Press, 1981); Coffin, Harold, *Origin by Design,* (Washington, D.C.: Review and Herald, 1980), pp. 10, 11; Oard, Michael J., *An Ice Age Caused by the Genesis Flood,* (El Cajon, CA: Institute for Creation Research, 1990), pp. 26-28; Marsh, Frank Lewis, *Life, Man, and Time,* (Escondido, CA: Outdoor Pictures, 1967) pp. 59, 60; Whitcomb, John C., and Henry Morris, *The Genesis Flood,* (Phillipsburg, NJ: P & R Publishing, 1961), pp. 215, 240-242, 253-258, 399-405.

[20] Whitcomb, et al, at 100, 101.

[21] Whitcomb, et al, at 97-106.

[22] Whitcomb, et al, at 110, 111, citing George A. Barton, *Archaeology and the Bible,* (1937), citing G.E. Wright, *Biblical Archaeology,* (Philadelphia: Westminster Press, 1957). Woolley dug five trenches at Ur, only two of which showed the flood strata. A logical inference is that the flood did not overwhelm even the entire city. There was

no break in occupation of the site. Today most archaeologists and language specialists doubt that Woolley's flood strata could be the source of the biblical Flood narrative.

[23] Ryan, William, and Walter Pitman, *Noah's Flood,* (New York: Simon & Schuster, 1998).

[24] 15 cubits. A cubit could be 18 inches or 21 inches. Even taking an 18 inch cubit, 15 cubits is "more than 20 feet."

[25]Gerhard F. Hasel, "The Biblical View of the Extent of the Flood," *Origins* 2(2): 77-95 (1975), p. 78, (See Geoscience Research Institute website, *http://www.grisda.org/origins/02077.htm* ) quoted by Richard M. Davidson in "Biblical Evidence for the Universality of the Genesis Flood," which is chapter 4 of Baldwin, supra. See, also Younker at 76-86.

[26] The 120 years comes from Genesis 6:3: "Then the LORD said, 'My Spirit will not contend with man forever, for he is mortal; his days will be a hundred and twenty years.'" This text, like the two it is sandwiched between, is difficult to interpret. It could be interpreted as a limitation on the human lifespan. Lifespans decreased rapidly after the flood, from 600 years (Noah's son Shem, Gen. 11:10-11) to 148 years (Nahor, Gen. 11:24-25) in just eight generations. (Abraham, the tenth generation from Noah, lived to be 175 (Gen. 25:7)). There are some problems, however, with interpreting the text as a limitation on the human lifespan. First, 120 years is not an absolute limitation on the human lifespan. The *Guinness Book of World Records* notes that Jeanne Louise Calment of Arles, France, died in 1997 at the age of 122. A woman in Brazil was believed to have lived even longer, but that case could not be documented. Second, while 120 years is not an absolute limitation, it is not a very good prediction, either. Very few people live past the age of 100. A much better prediction is found in Psalm 80:10: "The length of our days is seventy years—or eighty, if we have the strength . . ." Even with today's advanced medical knowledge, seventy to eighty years is the average lifespan.

Another interpretation of the text is that God was setting a probationary period for the antediluvian race. He would allow 120 years for Noah to preach and to build and provision the Ark, before the antediluvian world would be destroyed. This is the interpretation endorsed by Ellen White: "Connection with God made [Noah] strong in the strength of infinite power, while for one hundred and twenty years his solemn voice fell upon the ears of that generation in regard to events, which, so far as human wisdom could judge, were impossible." *Patriarchs and Prophets,* p. 96. The early interpreters Jerome and Augustine saw Genesis 6:3 as the length of time that God had allotted for the antediluvians to repent before He sent his judgment upon them. See, Jerome, *Hebrew* 6.3; Augustine, *The City of God* 15.24. One of the Gnostic texts discovered in Egypt in 1947 also teaches that this was a period of God's grace toward the antediluvians: "And he (Noah) preached piety for one hundred and twenty years. And no one listened to him. And he made a wooden ark, and whom he had found entered it. And the flood took place." From the Nag Hammadi Library, "The Concept of our Great Power," translated by Frederik Wisse. See, *http://www.webcom.com/gnosis/naghamm/cgp.html.*

[27] *Epistemology* is a branch of philosophy that investigates the origin, nature, methods, and limits of human knowledge, i.e., the study of how we know what we think we know.

[28] Honest Darwinists admit that their concept of science includes an *a priori* philosophical commitment to naturalism. See Phillip Johnson, "The Unraveling of Scientific Materialism," *First Things*, 77:22-25 (November, 1997).

[29] Ritland, Richard, "Historical Development of the Current Understanding of the Geologic Column: Part I" *Origins*, 8(2):59-76 (1981); Historical Development of the Current Understanding of the Geologic Column: Part II," *Origins*, 9(1):28-50 (1982).

[30] John Ruskin, quoted in, "Those Dreadful Hammers: Lyell and the New Geology," in *The Faber Book of Science*, John Carey, ed. (London and Boston: Faber & Faber, 1995), p. 71.

[31] Noteworthy authors who remained thoroughgoing Flood geologists after the triumph of Lyellism include Thomas Rodd, Granville Penn, William Kirby, George Fairholme, and George Young. See Byron C. Nelson, *The Deluge Story in Stone*, (Minneapolis, MN: Bethany Fellowship, 1931, 1968), pp. 83-110. Dr. Terry Mortenson has written an excellent book on the scriptural geologists, covering Granville Penn (1761-1844), George Bugg (1769-1851), Andrew Ure (1778-1857), George Fairholme (1789-1846), John Murray (1786?-1851), George Young (1777-1848), and William Rhind (1797-1874). See, Mortenson, Terry, *The Great Turning Point* (Green Forest, AK: Master Books, 2004).

[32] Ritland, supra, (1982).

[33] Clark, Robert T. and James D. Bales, *Why Scientists Accept Evolution*, (Grand Rapids, MI: Baker Book House, 1966), p. 33, citing Life and Letters of Charles Darwin, vol. I, p. 60.

[34] Darwin, Charles, *Origin of the Species*, (Reprint: New York: Gramercy Books, 1979), p. 293 (Chapter IX, "On the Imperfection of the Geological Record" par. 6). Later editions omit the word *incomprehensibly*.

[35] Clark and Bales, supra, p. 34, citing *Life and Letters*, Vol. II, p. 374.

[36] Gould, Stephen Jay, *The Panda's Thumb*, (New York: W.W. Norton and Co., 1980), p. 179.

[37] Darwin, Charles, *On the Origin of Species by Means of Natural Selection,* The Harvard Classics, Vol. 11, (New York: P.F. Collier & Son, 1909-1914; web edition by Bartleby.com, 2001) Chapter 10, par. 6. *http://www.bartleby.com/11/1006.html.* Earlier editions have a slightly different list, which includes Sir Richard Owen, and shows Lyell only beginning to come around to Darwin's way of thinking:

> The several difficulties here discussed . . . are all undoubtedly of the most serious nature. We see this in the plainest manner by the fact that all the most eminent palæontologists, namely Cuvier, Owen, Agassiz, Barrande, Falconer, E. Forbes, &c., and all our greatest geologists, as Lyell, Murchison, Sedgwick, &c., have unanimously, often vehemently, maintained the immutability of species. But I have reason to believe that one great authority, Sir Charles Lyell, from further reflexion entertains grave doubts on this subject. Darwin, Charles, *Origin of the Species,* (Reprint: New York: Gramercy Books, 1979), pp. 315, 316.

"When the first edition of this work was published," wrote Darwin, sarcastically, "I included Professor Owen with other paleontologists as being firmly convinced of the immutability of species; but it appears . . . that this was on my part a preposterous error." Darwin, Charles, *Origin of the Species,* (Reprint: New York: Gramercy Books, 1979), p. 59. Further research would probably show that Sir Richard Owen, like modern creationists, did not hold to the immutability of species, but at the same time did not accept the larger molecules-to-man evolution story.

[38] See chapter 512, infra. See also Spetner, Lee, *Not by Chance,* (Brooklyn, NY: The Judaica Press, Inc., 1997), pp. 159, 160.

[39] See, e.g., Michael Behe, *Darwin's Black Box*; Lee Spetner, *Not by Chance*; Phillip Johnson, Darwin on Trial; Michael Denton, *Evolution: A Theory in Crisis*; Jonathan Wells, Icons of Evolution; William Dembski, *Intelligent Design.*

[40] Phillip Johnson is a graduate of Harvard and the University of Chicago School of Law. He clerked for Earl Warren, then chief justice of the U.S. Supreme Court, then taught law at Cal-Berkeley, where he is now professor emeritus. His book *Darwin on Trial* (Washington, D.C.: Regnery Gateway, 1991) and his subsequent work have helped to establish the "Intelligent Design" movement as an intellectual force to be reckoned with. Johnson's very keen intellect and impeccable academic credentials have allowed him to take the critique of evolution into the academic mainstream, where others could not take it. He has been diligent about participating in debates and fostering friendships with scientists. His primary intellectual service to the critique of Darwinism has been to show that the real

force of Darwinism is not in the scientific evidence, which is not convincing, but in the philosophy of naturalism, which has an unshakable hold on modern academia. "Naturalism is not something about which Darwinists can afford to be tentative," he writes, "because their science is based upon it. As we have seen, the positive evidence that Darwinian evolution either can produce or has produced important biological innovations is nonexistent." *Darwin on Trial,* p. 115.

But the philosophical basis of Darwinism is not news to creationists, certainly not to Adventists. In his 1929 book *Back to Creationism,* Harold Clark spent a chapter (entitled "A Revival of Paganism") tracing evolutionary philosophy from Aristotle to Darwin. He wrote that "the pedigree of the evolution hypothesis ought to be sufficient to cause any cautious thinker to consider very carefully before he accepts it. Instead of being the science that it so vociferously calls itself, it is a pagan philosophy parading in the name of science; and it is for this reason that the Fundamentalists are waging such an active campaign against it. The acceptance of a bald naturalism in the place of supernatural religion is in its very nature an unthinkable act on the part of believers in Bible Christianity." Harold W. Clark, *Back to Creationism,* (Angwin, CA: Pacific Union College Press, 1929), in *Creationism in Twentieth-Century America, Vol. 8: The Early Writings of Harold W. Clark and Frank Lewis Marsh,* ed. Ronald L. Numbers, (New York and London: Garland Pub. Inc., 1995), p. 122.

[41] See, e.g., Padian, Kevin, and Alan D. Gishlick, "The Talented Mr. Wells," *Quarterly Review of Biology,* 77:1 (March, 2002).

[42] For a detailed critique of "progressive creationism", specifically the views of astronomer Hugh Ross, see, Jonathan Sarfati, *Refuting Comprimise* (Green Forest, AK: Master Books, 2004).

CHAPTER 06

# Can a Christian Follow Darwin?

In this discussion, *Darwinism* does not mean "change over time," diversification and speciation within the basic types of animals, or small evolutionary changes sometimes called "micro-evolution." As used in this chapter, the term Darwinism means the larger claims of Darwinism—the microbes-to-man theory of evolution, in which a cell assembled itself, breathed life into itself, and then evolved into ever higher and more complex life forms. Most importantly, Darwinism means that man evolved from the apes and was not specially created by God.

From beginning to end, the Bible claims that God created the heavens and the earth,[1] that God is the creator of the entire universe,[2] that God created the animals,[3] and that God created man in God's own image.[4] If the Bible is wrong in these claims, it is not a trustworthy witness, and none of its other claims are credible.

Kurt Wise convinced himself of this by conducting a remarkable experiment. Wise had become interested in science at a young age and, by the time he was in the eighth grade, aspired to earn a Ph.D. from Harvard. But he soon began to wonder whether the Darwinism and Lyellism of modern "science" were consistent with his deeply held faith in Christ and the Scriptures through which he had come to know Christ. To test the prominence of the doctrine of creation, Wise took a new Bible and a pair of scissors, and began reading and cutting:

> Beginning at Genesis 1:1, I determined to cut out every verse in the Bible that would have to be taken out to believe in evolution. Wanting this to be as fair as possible, and giving the benefit of the doubt to evolution, I determined to read all the verses on both sides of a page and cut out every *other* verse, being careful not to cut the margin of the page, but to poke the page in the midst of the verse and cut the verse out around that. In this fashion, night after night, for weeks and months, I set about the task of systematically going through the entire Bible from cover to cover. Although the end of the matter seemed obvious pretty early on, I persevered.[5]

At the end of the experiment, there was not much left of Wise's Bible. It fell apart in his hands as he tried to lift it. "I had to make a decision between evolution and Scripture. Either the Scripture was true and evolution was wrong or evolution was true and I must toss out the Bible."

Wise chose to believe in the veracity of Scripture. He did not give up his dream, however, and eventually earned that Harvard Ph.D. All through the many years of formal education at institutions steeped in Darwinism, Wise has remained a biblical creationist. He taught at Dayton, Tennessee, the site of the famous Scopes trial, at a college named after William Jennings Bryan.

If its foundational teaching on creation is wrong, the Bible is impeached beyond any hope of rehabilitation. There is no reason to believe it the inspired Word of God. There is no reason to treat it differently than any other ancient literature. We would sadly have to agree with Charles Darwin that the Scriptures are "no more to be trusted than the sacred books of the Hindus."[6] If Darwinism is true, the Bible has no better claim to our time than Plato, Aristotle, Homer, Chaucer, and Shakespeare. It is merely literature to which the educated should be exposed.

Those Christians who would so casually discard the doctrine of creation do not seem to appreciate its centrality to the core doctrines of the faith. It is central to the doctrine of Christ, because, first of all, Christ is the Creator. The opening chapters of Hebrews and the Gospel of John confirm that God created the world through Jesus Christ. "Through him all things were made; without him nothing was made that has been made." John 1:2. "In these last days he has spoken to us by his Son, whom he appointed heir of all things, and *through whom he made the universe.*" Heb. 1:2. The bible clearly teaches that Christ is God and was with the Father from the beginning, and the Father created all things through Christ. Yet if Darwinism is correct, the universe created itself, and Christ created nothing. Darwinism demotes Christ from His office of Creator.

Second, the doctrine of creation is central to the doctrine of Christ, because it is central to the purpose of Christ's death on the cross. The central Christian belief—a doctrine not peculiar to any sect or denomination—is that Christ, through His atoning sacrifice as the Lamb of God, is the Redeemer of mankind.[7] This is the *sine qua non* of Christianity. But why does mankind need redemption? Mankind is sinful and is in need of redemption because of Adam's sin and the resulting fall of the human race.[8] The Bible specifically teaches that Christ was the second Adam, who overcame where the first Adam failed.[9]

But according to the basic tenets of Darwinism, there never was an Adam. Darwinism teaches that man evolved from lower primates. Since there was no Adam, Adam never sinned, and there never was a fall of mankind. Far from suffering the effects of a "Fall," mankind has experienced a spectacular rise from bacteria to Beethoven, from microbe to Mozart, from single cell to Shakespeare, from ameba to Einstein. Although not created in the image of any god, mankind has somehow managed to separate himself from all other animals by developing self-awareness and civilization, with its government, religion, commerce, law, science, technology, art, literature, theater, philosophy, etc.

Since there never was a Fall, mankind does not need redemption or atonement. Christ's role as redeemer of a fallen humanity is obliterated. One commentator on the relationship between science and theology put it this way: "The traditional [Biblical] view of redemption as reconciliation and ransom from the consequences of Adam's fall is nonsense for anyone who knows about the evolutionary background to human existence."[10] Since Darwinism destroys Christ's role as Redeemer, it effectively destroys Christianity:

> Evolution . . . contradicts the basic foundations of Christianity. As Leonard Verduin asserted, "in the place of the story of a 'Fall' has come the story of an *ascent*." Christianity and evolution are diametrically opposed. Either our first parents were created in the image of God and experienced a fall into sin or they did not. If they did not, then why be Christian? Calvary most radically questions evolution. If there has been no fall, why would we need God to die in our behalf?[11]

Those who would delete the first ten chapters of Genesis would fatally undermine the entirety of Scripture. The creation, the Fall, the plan of salvation, the sacrificial system—these are all laid out in the first ten chapters of Genesis. Without this foundation, none of the rest makes sense. Jesus pointed out that Moses' writings were about Himself (Luke 24:25-27; John 5:39) and that faith in Moses' writings was a prerequisite to faith in Himself: "If you believed Moses, you would believe me, for he wrote of me. But if you do not believe his writings, how will you believe my words?" John 5:46, 47 (RSV).

Christ can still be a moral teacher and a moral example, of course, but Christianity consists of far more than morals and ethics. The field of ethics is open to anyone of any religious persuasion, including atheists and agnostics. Indeed, the eighteenth-century gentlemen who founded the United States believed that moral and ethical principles could be discerned without recourse to revealed religion; the truths of "natural law" were "self-evident" to men such as Thomas Jefferson. Atheists, agnostics, animists, ancestor worshipers, Buddhists, Hindus, Taoists, Muslims, and Jews can be moral and ethical. To be a Christian, however, one must believe that Christ died to save fallen humanity. If there was no Fall, there can be no Christianity.

All attempts to reconcile Darwinism and Christianity founder on the doctrine of the Fall, a doctrine that, as Herman Melville noted, is "now popularly ignored."[12] Michael Ruse made an attempt to reconcile Darwinism and Christianity in his book *Can a Darwinian Be a Christian?*[13] Ruse notes that an essential component of Christian theology "is that humans are descended from a unique pair (monogenism). That part of the Adam and Eve story cannot be interpreted symbolically." Having conceded that the Adam and Eve story must be taken literally for Christianity to work, Ruse admits that it "goes completely against our thinking about the nature of the evolutionary process. Successful species like humans do not pass through single-pair bottlenecks: there is certainly no evidence that this was true of Homo sapiens, a species which seems to have been well spread around the earth."[14]

After returning to the subject later in the book, Ruse admits that there had to be a fall in order for Christianity to make sense. "In the course of evolution, there must have been a first moment of conscious moral choice. That is the point at which the 'fall of humanity' began and humans were estranged from that natural fellowship with God which should have been theirs, and from their natural ability to relate unselfishly to one another."[15]

But Ruse is once again forced to admit that although this is necessary to Christianity, it is not compatible with Darwinism:

> If one is an evolutionist of any kind, then the Adam and Eve story must be modified in some respect. . . . the Garden of Eden scenario, with the lion lying down with the lamb, sticks in the craw of the evolutionist. And the whole business of an original, unique Adam and Eve goes flatly against modern evolutionary biology. . . . Is one supposed to believe that the parents of Adam and Eve—for they will have had such in the evolutionary story, if not in Genesis—were soulless or sinless or what? And what about their brothers and sisters, and the next generation of *Homo sapiens,* most of whom were not descended from Adam and Eve?[16]

"Obviously," writes Ruse in his epilogue, "if you are a fundamentalist Christian, then the Darwinian reading of Genesis is going to give you major problems—insoluble problems, I suspect." But all versions of Christianity have Christ in the office of Redeemer and mankind in a fallen condition from which he needs redemption. That is not "fundamentalism." That is mere Christianity.

Those Christians who, like Patricia Kelley, believe that Christianity is compatible with Darwinism are not thinking the problem through to its logical conclusion. Kelley's mentor, Stephen Jay

Gould—who was not a Christian, but Jewish in ethnicity and Marxist in politics—claimed that there is no conflict between science and religion because they have separate, non-overlapping "magisteria," or teaching domains. He wrote that, "science and religion should be equal, mutually respecting partners, each the master of its own domain, and with each domain vital to human life in a different way."[17]

This sounds reasonable, but it is wrong. Christianity and "science" do have overlapping teaching domains, because they both teach on origins, and without its doctrine of origins, Christianity is a deracinated, senseless mess. Those who would avoid the conflict between evolution and creation by consigning science and religion to separate domains are condemning Christians to a kind of schizophrenia in which they are asked to believe scientific "facts" that are in gross conflict with their Christian beliefs.

Not all Darwinists pretend there is no conflict between evolution and Christianity. In fact, many are refreshingly honest about the conflict.[18] Will Provine, a Darwinist professor at Cornell University, has written, "As the creationists claim, belief in modern evolution makes atheists of people. One can have a religious view that is compatible with evolution only if the religious view is indistinguishable from atheism."[19]

In a debate with Provine, Phillip Johnson stated it this way: "The implication of evolutionary biology . . . is perhaps not exactly that God does not exist. If God does exist, however, existing is about the only thing He has ever done."[20] Provine stated that Darwin developed the theory of natural selection with an expressly anti-Christian purpose in mind. "He believed that inventing the idea of natural selection was like committing murder. He knew exactly what he was doing. He was murdering the cultural tradition in which he had been raised, and in which Phil continues to live (rather belatedly)."[21]

Christians should not fool themselves about the atheistic implications of Darwinism. Before the triumph of Darwinism, atheism was a minority position among intellectuals, because atheists were forced to violate common sense by positing a self-created universe. Nineteenth-century intellectuals were anxious to substitute a naturalistic understanding of origins. By supplying the crucial biological thesis, Darwin paved the way for atheistic domination of intellectual life.[22]

Darwin was not a believer. Although he studied theology at Cambridge, he had lost faith long before he formulated his theory of evolution by natural selection. Interestingly, the false doctrine of an eternally burning hell was instrumental in turning Darwin away from Christianity. In his autobiography, Darwin wrote:

> I can hardly see how anyone ought to wish Christianity to be true: for if so, the plain language of the text seems to show that men who do not believe, and this would include my father, brother, and almost all my best friends, will be everlastingly punished. And this is a damnable doctrine.[23]

It certainly is, but Darwin was mistaken in believing that it is Scriptural.[24] It is regrettable that Christians have preached this doctrine, maligning the character of the loving God of the Bible:

> It is beyond the power of the human mind to estimate the evil that has been wrought by the heresy of eternal torment. The religion of the Bible, full of love and goodness, and abounding in compassion, is darkened by superstition and clothed with terror. When we consider in what false colors Satan has painted the character of God, can we wonder that our merciful Creator is feared, dreaded, and even hated? The

> appalling views of God which have spread over the world from the teachings of the pulpit have made thousands, yes, millions, of skeptics and infidels.[25]

The heresy of eternal torment made an atheist of Darwin, and Darwin has made atheists of millions of people. Christians should not say that doctrines do not matter, that only one's relationship with Christ matters. Doctrines matter enormously because of the picture of God they portray, which bears on whether God is Someone people would *want* to have a close relationship with.

But if Darwinism is incompatible with Christianity, why is Christianity still so healthy in the United States? The answer is simple: most American Christians ignore Darwinism. A Gallup poll taken in November 2004 showed that 45 percent of Americans, the largest group, believe that God created human beings pretty much in their present form within the past 10,000 years. Another 38 percent believe that God guided the process of evolution.[26] Only 13 percent said they believed in evolution with no guidance or participation by God.[27] This last position, held by only a small minority, is the position of "science" and the scientific community. These polling data, which have not changed significantly since Gallup began asking this question in 1982, are a source of frustration for Darwinists, who are constantly trying to increase public acceptance of evolution. Darwinists jealously guard their monopoly in the public schools, although 68 percent of Americans believe that creationism should be taught along with evolution in public schools.[28]

The effect that Darwinism has had upon Christianity is best seen in Western and Northern Europe, where Christianity exists largely as a cultural heritage, rather than as a living religion, and opinion leaders are bitterly hostile to creationism.[29] Church is very sparsely attended, with fewer than 10 percent of the people attending church regularly, as compared to about 40 percent in the United States.

We are also seeing the effects of Darwinism in the liberal protestant denominations of the United States. These have yielded to the Darwinists, undercutting the foundations of their faith. Since the 1960s, they have suffered a precipitous decline in membership:

> The most significant fact about the mainline Protestant churches in America in the past two decades has been the drastic decline in membership. This began in the 1960s and has continued to the present. . . . It is difficult to conceptualize the extent of the membership declines suffered by the mainline churches during the 1970s and early 1980s. Every week these denominations averaged a decline of over five thousand; this is the equivalent of mainline Protestantism's closing one local church of almost seven hundred members every day for a decade and a half.[30]

Between 1965 and 2001, the Episcopal Church lost more than a million members, shrinking from 3.6 million to 2.5 million. Between 1970 and 1984, the Methodists closed 2,665 local churches, an average of nearly four per week.[31] In a fifteen-year period between 1968 and 1983, the Presbyterians lost 25 percent of their members, and the Christian Church (Disciples of Christ) lost 29 percent. Attendance was also down.[32]

These churches' repudiation of the doctrine of creation, which occurred during the late nineteenth and early twentieth centuries, did not lead directly to the decline in membership. Rather, it contributed to a long process (arguably begun in the 18th Century by

German "higher criticism") whereby the authority of Scripture was eviscerated and pastors lost interest in biblical preaching.

The creation week and the Flood are unacceptable to modern, naturalistic "science" because they depend upon the supernatural, upon God's miraculous power. But every other miracle in the Bible is just as unacceptable to science, for exactly the same reason. If we toss out the doctrine of creation because "science" doesn't like it, we must soon jettison every miracle in Scripture. And that is just what happened in the mainline protestant seminaries, where naturalistic explanations were sought for every miracle in the Bible.

Throwing out its miracles robs the Bible of its power and spiritual sustenance. It compromises or destroys the most precious promises in Scripture. To choose one among many possible examples, the Bible teaches that those who have died in Christ will be resurrected with a new body, even if their ashes have been scattered to the winds.[33] Obviously, the bodily resurrection taught in Scripture is no less miraculous than the original creation. If God lacked the ability to create two of us in one day, how can He *recreate* millions of us at the resurrection? What is left of Christianity after the great truths concerning the creation, the Fall, the plan of salvation, Christ's miracles, Christ's redeeming sacrifice as the second Adam, and the resurrection of the dead, are all tossed out?

Seminary professors taught their students to be embarrassed about every miracle in the Bible. Since there is a miracle on almost every page, seminary-trained pastors soon lost interest in biblical preaching. By the 1960s, three generations later, the mainline protestant denominations were totally devoted to some form of social or political teaching, at the expense of religion.[34] As the Methodist writers quoted above describe it:

> Our mission has been reduced to politics, our Social Principles mirror, to a great extent, the political opinions from one part (generally the left) of the secular political spectrum. When our church speaks, it speaks mostly in political terms rather than religious ones. The programs of many of our local churches appear to be a mix of activities from the Y.M.C.A., the Garden Club, and the League of Women Voters.[35]

It is not difficult to imagine how this would adversely affect attendance. If politics is all a pastor has to offer, why not stay home on Sunday morning and watch *Meet the Press*?

By contrast, the more conservative protestant denominations—which have not abandoned miracles, the doctrine of creation, or biblical preaching—are relatively healthy and growing.[36] Obviously, many things other than doctrine play a role in church growth and decline; in fact, doctrine may not matter as much as other factors, such as whether the congregation is kind and loving. But for people who care about intellectual integrity, any compromise with Darwinism is extremely corrosive to faith, and renders plenary biblical faith impossible.

To summarize what we have learned in the last two chapters, Lyellism can be compatible with Christianity only if one accepts a special creation of man at the top of the geologic column. Even then, the Bible is drastically undermined because, (1) not all living things were created in six days, and thus the Sabbath commandment is something of a hoax, (2) death and predation were not caused by sin, and (3) the world-wide Flood described in Genesis never happened, and all references to it, including those from the mouths of Jesus and Peter, are misleading. Nevertheless, with those limitations, it is possible to be both a Lyellian and a Christian.

Modern Lyellism is for all practical purposes, however, inseparable from Darwinism, and it is not possible to be a Darwinist and an intellectually consistent Christian. If Darwinism is true, Christianity is utter nonsense. By natural selection acting upon random genetic mutations, microbes evolved by stages into humanity. The story is one of amazing progress from a lower to a higher plane of existence. There were no perfect "Adam and Eve," no "sin," and no "Fall." Since there was no Fall, we have no need of a Redeemer. Death is a natural, necessary, and indispensable part of life, not the result of a fallen condition from which we need redemption. If Darwinism is true, any thought of Christianity being true is out of the question.

---

[1] Gen. 1:1; Ex. 20:11; 31:17; 2 Kings 19:15; 2 Chron. 2:12; Neh. 9:6; Job 38:4; Psa. 90:2; 102:25; 104:5; 115:15; 124:8; 146:5, 6; Prov. 8:27-29; Isa. 37:16; 45:18; 66:1, 2; Jer. 10:12; 32:17; 51:15; Acts 14:15; Heb.11:3; 2 Pet. 3:5; Rev. 10:6; 14:7

[2] Job 9:6-9; 26:7; 38:31-33; Psa. 8:3; 68:33, 34; 136:5-9; Isa. 40:22, 26-28; 42:5; Amos 5:8; Acts 17:24, 25

[3] Gen. 1:24, 25; 2:19; Jer. 27:5; Psa. 104:24-26.

[4] Gen. 1:26, 27; 2:7; 5:1, 2; Deut. 4:32; Psa. 8:5, 6; 100:3; 119:73; 139:14, 15; Eccl. 7:29; Isa. 17:7; 45:12; 64:8; Jer. 27:5; Matt. 19:4; Mark 10:6; Acts 17:24-26; Heb. 2:7; I Tim. 2:13.

[5] Kurt Wise, in *In Six Days: Why 50 Scientists Choose to Believe in Creation,* John F. Ashton, Ed. (Sydney: New Holland Pub., 1999 (published in the U.S. by Master Books, Green Forest, AK), p. 330.

[6] Darwin, Charles, *The Life and Letters of Charles Darwin, Vol. I,* p. 278, cited in Clark, Robert T. and James D. Bales, *Why Scientists Accept Evolution,* (Grand Rapids, MI: Baker Book House, 1966), p. 30.

[7] Gen. 3:15; 4:3-7; 8:20; 22:8; Ex. 20:24; 24:4-7; Lev. 1-7; 16:1-34; 17:11; Num. 28, 29; 1 Sam. 7:9; 2 Chron. 29:20-24; 30:15-20; 35:1-11; Ezek. 6:9-20; 7:17; 8:35; Psa. 22; Isa. 53; Ezek. 46; Mark 10:45; 14:12; Luke 22:20; John 1:29, 36; Acts 8:32; Rom. 3:24-26; 5:10-11; 1 Cor. 5:7; 15:3; II Cor. 5:18-19; Gal. 1:3-4; 4:4-5; Eph. 1:7; 5:2; Col. 1:13, 14, 19-22; 1 Tim. 2:5, 6; Titus 2:13, 14; Heb. 2:17; 9:11-28; 10:10; 1 Pet. 1:18-20; 1 John 2:2; 4:10; Rev. 5:8-13; 7:13-17; 12:11; 19:6-9; 21:22-27

[8] Gen. 3; 1 Kings 8:46; Psa. 51:5; 130:3; 143:2; Prov. 20:9; Eccl. 7:20; Isa. 53:6; Hos. 6:7; John 8:23, 24; 1 Cor. 15:22; Rom. 5:12; Eph. 2:1-3; 1 John 1:8;

[9] Rom. 5:12-21; 1 Cor. 15:21, 22, 45-49

[10] Baldwin, John T., *Creation, Catastrophe, & Calvary,* (Hagerstown, MD: Review and Herald, 2000), p. 108, quoting Karl Schmitz-Moormann, president of the European Society for the Relation of Science and Theology.

[11] *Seventh-day Adventists Believe . . ., A Biblical Exposition of 27 Fundamental Doctrines* (Hagerstown, MD: Review and Herald, 1988), p. 92.

[12] Herman Melville, *Billy Budd,* (New York: Signet Classics), p. 17

[13] Ruse, Michael, *Can a Darwinian Be a Christian?* (Cambridge and New York: Cambridge University Press, 2001).

[14] Ruse, at 75, 76.

[15] Ruse, at 205.

[16] Ruse, at 209.

[17] Gould, Stephen Jay, "Dorothy, It's Really Oz" *Time*, 154:8, August 23, 1999. Gould expanded on this theme in the book, *Rocks of Ages: Science and Religion in the Fullness of Life* (New York: Ballantine Books, 1999). Gould viewed religion as part, along with philosophy, art, and literature, of "the magisterium of ethics and meaning." (p. 62). But beliefs about ethics are only part of religious beliefs. Most religious beliefs have to do with the ultimate nature of reality—the existence of God, His character, His nature, His provision for mankind, the extent to which He has revealed Himself to mankind through His word and through His creation, the sinful, fallen nature of mankind, God's grace shown to mankind, God's desire for mankind to live in a sin-free, disease-free, death-free world, whether there is an afterlife, what happens to a person when they die, whether there is a heaven or a hell, etc. These are the main questions with which religion is concerned, not questions of ethics. In fact, ethical issues tend to be far more susceptible to reason and logic—and hence to a broad social consensus—than purely religious issues. As some of his critics noted, Gould defined *religion* in a way that is unrecognizable to the religious.

And what about "meaning?" Gould was willing to cede questions of "meaning" to the magisterium of religion, but the meaning of life is different if we evolved from monkeys than it is if we were created in the image of an eternal and omnipotent God. If we were created, our purpose is to glorify our Creator; if we evolved, our purpose is . . . well, who can say? If science decides whether we evolved or were created, then meaning is within the magisterium of science, not the religious magisterium.

[18] See, e.g., Jacob Weisberg, "Darwinism and Religion: Quit Pretending they're Compatible" Slate, August 10, 2005 "That evolution erodes religious belief seems almost too obvious to require argument. It destroyed the faith of Darwin himself, who moved from Christianity to agnosticism as a result of his discoveries and was immediately recognized as a huge threat by his reverent contemporaries. . . . evolutionists should quit pretending their views are no threat to believers. This insults our intelligence . . ."

[19] Provine, Will, "No Free Will," in *Catching Up with the Vision*, Ed. By Margaret W. Rossiter, (Chicago: University of Chicago Press, 1999) p. 123.

[20] Debate between Phillip Johnson and Will Provine at Stanford University on April 30, 1994. See *http://www.arn.org/docs/orpages/or161/161main.htm.* During the debate, Johnson states, "This is the fourth time that Will Provine and I have met in debate, . . . so I feel qualified to say where we will tend to agree and disagree during this debate. First, where we agree. The modern neo-Darwinian theory of evolution is fundamentally inconsistent with any meaningful theism—with any meaningful God who acts as creator of the world."

[21] Ibid. Ariel Roth cited an even more pointed statement made by Will Provine:

"Scientist work closely with religious leaders to fight against the introduction of creationism into the classrooms of public schools. Liberal religious leaders and theologians, who also proclaim the compatibility of religion and evolution, achieve this unlikely position by two routes. First, they retreat from traditional interpretations of God's presence in the world, some to the extent of becoming effective atheists. Second, they simply refuse to understand modern evolutionary biology and continue to believe that evolution is a purposive process.

We are now presented with the specter of atheistic evolutionists and liberal theologians, whose understanding of the evolutionary process is demonstrable nonsense, joining together with the ACLU and the highest courts in the land to lambaste creationists, who are caught in an increasing bind. Evolutionary biology, as taught in public schools, shows no evidence of a purposive force of any kind. This is deeply disturbing to creationists. Yet in court, scientists proclaim that nothing in evolutionary biology is incompatible with any reasonable religion, a view also supported by liberal theologians and religious leaders of many persuasions. Not only are creationists unable to have their 'creation science' taught in the schools, they cannot even convince the court system that evolution is in any significant way antithetical to religion; thus, the courts are effectively branding their religious views as terribly misguided. No wonder creationists (somewhere near half of the population!) are frustrated with the system and want equal time for their views, or at least to be spared bludgeoning with evolution." William B. Provine, Academe 73(1):50-52 (a review of Edward J. Larson's book *Trial and Error: the American Controversy of Creation and Evolution* (Oxford University Press)) as cited in Roth *Origins: Linking Science and Scripture* (Hagerstown, MD: Review and Herald, 1998) p. 26.

[22] Bethell, Tom, "Darwin in the Dock" *The American Spectator*, June 1992, quoting from a Phillip Johnson speech to the Trinity Evangelical Divinity School, in Illinois.

[23] Ibid. The passage was at first omitted from the published work, but in 1958, it was brought to light by Darwin's granddaughter. Although Darwin went to Cambridge with the intention of becoming a clergyman, he soon abandoned this ambition and

lost faith in the Bible. "But I had gradually come by this time, i.e. 1836 to 1839, to see that the Old Testament was no more to be trusted than the sacred books of the Hindoos [sic]." In his letters, Darwin describes a long intellectual process in which he gradually lost faith in Christianity. "Thus disbelief crept over me at a very slow rate, but was at last complete. The rate was so slow that I felt no distress." Darwin, Charles, *The Life and Letters of Charles Darwin, Vol. I*, pp. 277-278, cited in Clark, Robert T. and James D. Bales, *Why Scientists Accept Evolution*, (Grand Rapids, MI: Baker Book House, 1966), p. 30-31.

24 "Hell" is not a place, but an event that takes place at the end of time. Matt. 13:39, 40; Matt. 16:27; Rev. 22:12. "Hell" is the destruction by fire of the unsaved after the millennium of Revelation. Rev. 21:7, 8; Rev. 20:7-9, 14, 15. The entire earth is purified by fire. 2 Pet. 3:10-12. The destruction of the unsaved is complete; they do not live on and on. Matt. 10:28; Mal. 4:1-3; Psa. 145:20; 37:20; 21:9; Isa. 47:14; Ezek. 28:18, 19. Terms like *unquenchable fire* and *eternal fire* do not mean that the fire burns on and on. Compare Mark 9:45-48 with Jer. 17:27; Matt. 25:41 with Jude 7. "Hell" is prepared for the devil and his angels; God does not want any of us to be lost. Matt. 25:41; 2 Pet. 3:9; Ezek. 18:32.

25 Ellen White, *The Great Controversy*, p. 536.

26 This group makes the mistake of believing that strong evidence supports Darwinism. In fact, the case for Darwinism is based on philosophy. It is what you're left with to explain the creation if you deny, *a priori*, the existence of God. If you believe, as this group apparently does, that there is a God capable of guiding evolution, there is no reason to believe in the Darwinian origin myth in the first place.

27 The Gallup Organization, which reports that, "The public has not notably changed its opinion on this question since Gallup started asking it in 1982." One change is that, since 1997, the totally atheistic position has inched upward from 10% to 13% of the population.

28 The Gallup Organization, as reported by ABC News. See, also, Laurie Goodstein, "Teaching of Creationism is Endorsed in New Survey," *The New York Times*, August 31, 2005, citing a poll conducted by July 7-17, 2005 by the Pew Forum on Religion and Public Life and the Pew Research Center for the People and the Press, stating "nearly two-thirds of Americans say that creationism should be taught alongside evolution in public schools."

29 In 2002, when it became public that Emmanuel City Technology College in London was teaching creationism as part of its curriculum, astonishing hatred poured forth from the British press. "My academic colleagues across the Atlantic

have to waste large parts of their first-year biology courses trying to put right the damage done by the educational Taliban who now rule all over the United States," wrote Steve Jones, a professor of genetics at University College London. "British youngsters are now being fed subsidised untruths about the nature of science," he wrote. "It is easy to dismiss such people as clowns or cranks, but they are dangerous." Commenting on the two-million pounds donated by an English businessman to found the school, Jones ranted, "Now we know the price of ignorance . . . That seems a pretty good bargain at just a couple of grand a kid. You get a new crop of innocent minds to poison each year and, with plans to set up yet more schools with the same funding, stupidity will soon be available wholesale." "Creationism is anti-science and it is an abuse of children's burgeoning intellect to teach them that it is as credible as evolution," said Keith Wood, president of the England's "National Secular Society." Prime Minister Tony Blair defended Emmanuel College: "I think it would be very unfortunate if concerns over that were seen to remove the very, very strong incentive to make sure we get as diverse a school system as we properly can." *London Telegraph,* March 22, 2002. But critics feared that "Mr. Blair's enthusiasm for 'faith schools' will result in more of this kind of nonsense being taught in our education system." Sarah Cassiday, *The Independent,* March 19, 2002. In fairness to the mother country, similar things are said in the U.S. whenever it is suggested that American public schools give equal time to some form of creationism or intelligent design, although the tone is usually not so hateful and vitriolic.

[30] Willimon, William H., and Robert L. Wilson, *Rekindling the Flame, Strategies for a Vital United Methodism.*

[31] Ibid.

[32] Ibid. See, also, Dave Shiflett, *Exodus: Why Americans are Fleeing Liberal Churches for Conservative Christianity* (Sentinel, 2005); Thomas C. Reeves, *The Empty Church: The Suicide of Liberal Christianity* (Free Press, 1996); Dean M. Kelley, *Why Conservative Churches Are Growing: A Study in Sociology of Religion* (Mercer University Press, 1986, 1995); Colleen Carroll, *The New Faithful: Why Young Adults are Embracing Christian Orthodoxy* (Loyola, 2002);

[33] Job 19:25-27; Matt. 22:30; Acts 4:2; 23:6-8; 24:15, 21; John 5:28, 29; 11:23, 24; 1 Cor. 15:12-23, 35-58; 1 Thess. 4:16, 17; Phil. 3:20, 21; Rev. 20:4-15

[34] See, e.g., Earle E. Cairns, *Christianity Through the Centuries,* (Grand Rapids, MI: Zondervan, 1954, 1981), pp. 425-428, 443-448.

[35] Ibid.

[36] See, e.g., the sources listed in note 32, above. This discussion leaves out the Catholic Church, whose numerous pronouncements on origins are varied, often ambiguous, and tailored to the different concerns of different audiences. But it is important to note that Protestants and Catholics are in a different position with regard to Darwinism. Since Luther's day, Protestants have claimed the Bible as the sole authority for their doctrines and teachings. Even if *Sola Scriptura* has often been more aspiration than actuality, Darwinism, which destroys the authority of Scripture, is a terrible blow to the vitality of Protestant Christianity. The authority of the Roman Catholic Church, on the other hand, is derived primarily from its own traditions and claims. In the Roman system, it is the church that confers authority on Scripture, not vice versa. Thus, Darwinism is far less threatening to the Catholic Church. Nevertheless, there have been very prominent Catholic creationists, such as Alfred Watterson McCann and George Barry O'Toole. Michael Behe, an important figure in the "Intelligent Design" movement, is a Roman Catholic.

CHAPTER 07

# Ellen White on Lyellism & Darwinism

The biblical arguments outlined in Chapter Five are the primary reason why Seventh-day Adventists reject Lyellism. Because Seventh-day Adventists keep the Sabbath, and God established the Sabbath as a memorial to the creation week, the creation week has special importance to Adventists. A Lyellian interpretation of the fossiliferous strata leads to the conclusion that there was no creation week, because the world was "created" over the course of hundreds of millions of years. Lyellism is thus grossly inconsistent with the most basic Seventh-day Adventist beliefs.

But Adventists have additional authority for rejecting Lyellian geology. Ellen Gould White (1827-1915) was one of the founding members of the Seventh-day Adventist Church and a prolific writer. She wrote more than forty books and five thousand periodical articles covering a broad range of subjects, including religion, evangelism, prophecy, publishing, education, the family, and nutrition. In addition, her personal correspondence was on a vast scale. Her writings have been translated more often than have those of any other female writer, and she is the most translated American author of either gender. Her book *Steps to Christ* has been published in more than 140 languages.

Seventh-day Adventists believe that Ellen White was more than a gifted writer. We believe that she was a special messenger of God. Over the course of her life, beginning when she was 17, God gave Ellen White approximately two thousand visions, varying in length from less than a minute to nearly four hours. Ellen White wrote down the counsel and wisdom revealed to her through the visions.

The Bible promises that believers will see visions and dream dreams, and teaches that the gift of prophecy is one of the spiritual gifts given to the church. Joel 2:28, 29; Acts 2:17, 18; 1 Cor. 12:27-30. Ellen White's prophetic ministry was a manifestation of the continued presence and guiding of the Holy Spirit, as promised in Scripture.

The writings of Ellen White are not a substitute for the Bible, nor do Adventists place them on the same level as Scripture. She referred to her writings as "a lesser light" pointing people to the "greater light" of the Bible. "The fact that God has revealed His will to men through His Word," White noted, "has not rendered needless the continued presence and guiding of the Holy Spirit. On the contrary, the Spirit was promised by our Savior, to open the Word to His servants, to illuminate and apply its teachings."[1] Adventists do not view Ellen White's writings as superseding or altering Scripture, but as shedding light on Scripture.

One of Ellen White's visions confirmed that the creation week was a literal week, just like every week since. She specifically repudiated the "day/age" theory, i.e., that the days of the creation week represent seven vast, indefinite periods:

> I was then carried back to the creation and was shown that the first week, in which God performed the work of creation in six days and rested on the seventh day, was just like every other week. The great

God in his days of creation and day of rest, measured off the first cycle as a sample for successive weeks till the close of time. . . . On the seventh day of the first week God rested from his work, and then blessed the day of his rest, and set it apart for the use of man. The weekly cycle of seven literal days, six for labor, and the seventh for rest, which has been preserved and brought down through Bible history, originated in the great facts of the first seven days.

When God spake his law with an audible voice from Sinai, he introduced the Sabbath by saying, "Remember the Sabbath day to keep it holy." . . . "For in six days the Lord made heaven and earth, the sea and all that in them is, and rested the seventh day, wherefore the Lord blessed the Sabbath day and hallowed it." This reason appears beautiful and forcible when we understand the record of creation to mean literal days. The first six days of each week are given to man in which to labor, because God employed the same period of the first week in the work of creation. The seventh day God has reserved as a day of rest, in commemoration of his rest during the same period of time after he had performed the work of creation in six days.

But the infidel supposition, that the events of the first week required seven vast, indefinite periods for their accomplishment, strikes directly at the foundation of the Sabbath of the fourth commandment. It makes indefinite and obscure that which God has made very plain. It is the worst kind of infidelity; for with many who profess to believe the record of creation, it is infidelity in disguise. It charges God with commanding men to observe the week of seven literal days in commemoration of seven indefinite periods, which is unlike his dealings with mortals, and is an impeachment of his wisdom.

Infidel geologists claim that the world is very much older than the Bible record makes it. . . . And many who profess to believe the Bible record are at a loss to account for wonderful things which are found in the earth, with the view that creation week was only seven literal days, and that the world is now only about six thousand years old. These, to free themselves of difficulties thrown in their way by infidel geologists, adopt the view that the six days of creation were six vast, indefinite periods, and the day of God's rest was another indefinite period; making senseless the fourth commandment of God's holy law. Some eagerly receive this position, for it destroys the force of the fourth commandment, and they feel a freedom from its claims upon them.[2]

It is one of Satan's devices to lead the people to accept the fables of infidelity; for he can thus obscure the law of God, in itself very plain, and embolden men to rebel against the divine government. His efforts are especially directed against the fourth commandment, because it so clearly points to the living God, the Maker of the heavens and the earth.

There is a constant effort made to explain the work of creation as the result of natural causes; and human reasoning is accepted even by professed Christians, in opposition to plain Scripture facts.[3]

Inferences erroneously drawn from facts observed in nature have . . . led to supposed conflict between science and revelation; and in the effort to restore harmony, interpretations of Scripture have been adopted that undermine and destroy the force of the word of God. Geology has been thought to contradict the literal interpretation of the Mosaic record of the creation. Millions of years, it is claimed, were required for the evolution of the earth from chaos; and in order to accommodate the Bible to this supposed revelation of science, the days of creation are assumed to have been vast, indefinite periods, covering thousands or even millions of years.

Such a conclusion is wholly uncalled for. The Bible record is in harmony with itself and with the teaching of nature. Of the first day employed in the work of creation is given the record, "The evening and the morning were the first day." Genesis 1:5. And the same in substance is said of each of the first six days of creation week. Each of these periods Inspiration declares to have been a day consisting of evening and morning, like every other day since that time. In regard to the work of creation itself the divine testimony is, "He spake, and it was done; He commanded, and it stood fast." Psalm 33:9. With

> Him who could thus call into existence unnumbered worlds, how long a time would be required for the evolution of the earth from Chaos? In order to account for His works, must we do violence to His word?[4]

Ellen White was very clear that, in cases of conflict between Scripture and "science," Bible history must be preferred over naturalistic speculations about origins. Even the most brilliant minds will lose their way on this topic, unless they hold fast to the biblical record:

> God has permitted a flood of light to be poured upon the world in both science and art; but when professedly scientific men treat upon these subjects from a merely human point of view, they will assuredly come to wrong conclusions. It may be innocent to speculate beyond what God's word has revealed, if our theories do not contradict facts found in the Scriptures; but those who leave the word of God, and seek to account for His created works upon scientific principles, are drifting without chart or compass upon an unknown ocean. The greatest minds, if not guided by the word of God in their research, become bewildered in their attempts to trace the relations of science and revelation. Because the Creator and His works are so far beyond their comprehension that they are unable to explain them by natural laws, they regard Bible history as unreliable. Those who doubt the reliability of the records of the Old and New Testaments, will be led to go a step further, and doubt the existence of God; and then, having lost their anchor, they are left to beat about upon the rocks of infidelity.

These persons have lost the simplicity of faith. There should be a settled belief in the divine authority of God's Holy Word. The Bible is not to be tested by men's ideas of science. Human knowledge is an unreliable guide. . . . Moses wrote under the guidance of the Spirit of God, and a correct theory of geology will never claim discoveries that cannot be reconciled with his statements.[5]

To many, scientific research has become a curse. God has permitted a flood of light to be poured upon the world in discoveries in science and art; but even the greatest minds, if not guided by the word of God in their research, become bewildered in their attempts to investigate the relations of science and revelation.

Human knowledge of both material and spiritual things is partial and imperfect; therefore many are unable to harmonize their views of science with Scripture statements. Many accept mere theories and speculations as scientific facts, and they think that God's word is to be tested by the teachings of "science falsely so called." I Timothy 6:20. The Creator and his works are beyond their comprehension; and because they cannot explain these by natural laws, Bible history is regarded as unreliable. Those who doubt the reliability of the records of the Old and New Testaments too often go a step further and doubt the existence of God and attribute infinite power to nature. Having let go their anchor, they are left to beat about upon the rocks of infidelity.[6]

> In this generation there are many whose eyes become dazzled by the glare of human speculations, "science falsely so called;" they discern not the net, and walk into it as readily as if blindfolded. God designed that man's intellectual powers should be held as a gift from his Maker and should be employed in the service of truth and righteousness; but when pride and ambition are cherished, and men exalt their own theories above the word of God, then intelligence can accomplish greater harm than ignorance.[7]

Ellen White was remarkably blunt in warning about the dangers of reasoning from naturalistic premises in derogation of Scripture—exactly the type of reasoning embraced by Lyellian geology. She called such reasoning "sophistry":

> We need to guard continually against the sophistry in regard to geology and other branches of science falsely so-called, which have not one semblance of truth. The theories of great men need to be carefully sifted of the slightest trace of infidel suggestion.[8]

> In the last days the earth will be almost destitute of true faith. Upon the merest pretense, the word of God will be considered unreliable, while human reasoning will be received, though it be in opposition to plain Scripture facts.[9]

Unfortunately, the tendency to elevate human reasoning over divine revelation is a serious problem even among professed

Christians. Ellen White warned about this, and also condemned the type of creative interpretation—i.e., spiritualizing or explaining away plain scriptural teachings—that must be resorted to in order to make Scripture accommodate Lyellian and Darwinian theories:

> Infidelity prevails to an alarming extent, not in the world merely, but in the church. Many have come to deny doctrines that are the very pillars of the Christian faith. The great facts of creation as presented by the inspired writers, the fall of man, the atonement, and the perpetuity of the law of God, are practically rejected, either wholly or in part, by a large share of the professedly Christian world. Thousands who pride themselves upon their wisdom and independence regard it as an evidence of weakness to place implicit confidence in the Bible; they think it a proof of superior talent and learning to cavil at the Scriptures and to spiritualize and explain away their most important truths.[10]

> The truths most plainly revealed in the Bible have been involved in doubt and darkness by learned men, who, with a pretense of great wisdom, teach that the Scriptures have a mystical, a secret, spiritual meaning not apparent in the language employed. These men are false teachers. . . . The language of the Bible should be explained according to its obvious meaning, unless a symbol or figure is employed.[11]

Ellen White plainly stated that the fossils found in the earth are relics of the Genesis Flood:

> It is true that remains found in the earth testify to the existence of men, animals and plants much larger than any now known. These are regarded as proving the existence of vegetable and animal life prior to the time of the Mosaic record. But concerning these things Bible history furnishes ample explanation. Before the Flood the development of vegetable and animal life was immeasurably superior to that which has since been known. At the Flood the surface of the earth was broken up, marked changes took place, and in the re-formation of the earth's crust were preserved many evidences of the life previously existing. The vast forests buried in the earth at the time of the Flood, and since changed to coal, form the extensive coal fields, and yield the supplies of oil that minister to our comfort and convenience today. These things, as they are brought to light, are so many witnesses mutely testifying to the truth of the word of God.[12]

Interestingly, White probably mentioned dinosaurs without calling them by that name. "I was shown," she wrote, "that very large, powerful animals existed before the flood which do not now exist."[13] The dinosaurs are most prominent among large, powerful animals that existed before the Flood that no longer exist.

White also clearly denounced the theory of evolution, especially the theory that mankind evolved from lower orders of animals:

> God created man in His own image. Here is no mystery. There is no ground for the supposition that man was

evolved by slow degrees of development from the lower forms of animal or vegetable life. Such teaching lowers the great work of the Creator to the level of man's narrow, earthly conceptions. Men are so intent upon excluding God from the sovereignty of the universe that they degrade man and defraud him of the dignity of his origin.[14]

Akin to the theory concerning the evolution of the earth is that which attributes to an ascending line of germs, mollusks, and quadrupeds the evolution of man, the crowning glory of the creation.

When consideration is given to man's opportunities for research; how brief his life; how limited his sphere of action; how restricted his vision; how frequent and how great the errors in his conclusions, especially as concerns the events thought to antedate Bible history; how often the supposed deductions of science are revised or cast aside; with what readiness the assumed period of the earth's development is from time to time increased or diminished by millions of years; and how the theories advanced by different scientists conflict with one another—considering all this, shall we, for the privilege of tracing our descent from germs and mollusks and apes, consent to cast away that statement of Holy Writ, so grand in its simplicity, "God created man in His own image, in the image of God created He him"? Genesis 1:27. Shall we reject that genealogical record—prouder than any treasured in the courts of kings—"which was the son of Adam, which was the son of God"? Luke 3:38[15]

She saw the need for faith and humility in the study of nature, as in the study of nature's God:

> He who studies most deeply into the mysteries of nature will realize most fully his own ignorance and weakness. He will realize that there are depths and heights which he cannot reach, secrets which he cannot penetrate, vast fields of truth lying before him unentered. He will be ready to say, with Newton, "I seem to myself to have been like a child on the seashore finding pebbles and shells, while the great ocean of truth lay undiscovered before me."[16]

Although this is not an exhaustive compilation of Ellen White's statements on origins, it is a representative sample. Clearly, the Seventh-day Adventist Church cannot turn its back on young-earth creationism without also repudiating the prophetic authority of Ellen White. This church will always believe that God created the world and its basic life forms in six days, and will ever remain a young-earth creationist denomination.[17]

Adventists have been at the forefront of creationism. Seventh-day Adventists who have contributed to a biblical understanding of origins and earth history include George McCready Price, Harold W. Clark, Frank Lewis Marsh, Ernest S. Booth, Harold G. Coffin, Robert H. Brown, Arthur V. Chadwick, Ariel A. Roth, and Leonard Brand, among many others. The church established the Geoscience Research Institute, currently housed on the campus of Loma Linda University, to facilitate origins research.

---

[1] Ellen G. White, *The Great Controversy,* p. vii.

[2] Ellen G. White, *Spiritual Gifts* (Battle Creek, MI: Steam Press of the Seventh-day Adventist Publishing Association, 1864), vol. 3, pp. 90-92. Reprinted in *Signs of the Times,* 5:90, March 20, 1879, and in *Spirit of Prophecy* (Battle Creek, MI: Review and Herald, 1884), vol. 4, pp. 85-89.

[3] Ellen G. White, *Patriarchs and Prophets,* p. 113.

[4] Ellen G. White, *Education,* pp. 128, 129.

[5] Ellen G. White, *Patriarchs and Prophets,* pp. 113, 114.

[6] Ellen G. White, *The Great Controversy,* p. 522.

[7] Ellen G. White, *The Great Controversy,* p. 573.

[8] Ellen G. White, *Review and Herald,* March 1, 1898.

[9] Ellen G. White, *Spiritual Gifts,* vol. 3, p. 94.

[10] Ellen G. White, *The Great Controversy,* p. 583.

[11] Ellen G. White, *The Great Controversy,* p. 598-599.

[12] Ellen G. White, *Education,* p. 129.

[13] Ellen G. White, *Spiritual Gifts,* (Battle Creek, MI: Steam Press of the Seventh-day Adventist Publishing Association, 1864), Vol. 3, pp. 92. See also White, *Patriarchs and Prophets,* p. 112.

[14] Ellen G. White, *Patriarchs and Prophets,* (Washington D.C.: Review and Herald Pub., 1890), p. 45.

[15] Ellen G. White, *Education,* (Nampa, ID: Pacific Press), p. 130.

[16] Ellen G. White, *Education,* p. 133.

[17] The Adventist attitude toward Darwinism is probably best summarized by Clifford Goldstein in his article, "Seventh-day Darwinians," *Adventist Review* (July 24, 2003) ("What amazes me isn't so much that people can believe in evolution (after all, I used to), but that those who do still want to be Seventh-day Adventists. I can respect someone who, believing in evolutionary theory, rejects the Adventist Church entirely. I have no respect for those who think they can meld the two. . . . I speak, I believe, for millions of Seventh-day Adventists when I declare that . . . we will *never* make room for anything other than a literal six-day creation for life here—never.")

CHAPTER 08

# When Did the Dinosaurs Live? The Biblical View

When researchers at Montana State University were studying the bones of a *Tyrannosaurus rex,* they noticed that one of the leg bones had not completely fossilized. Mary Schweitzer was peering through a microscope at a thinly sliced cross-section of the bone, scrutinizing the tiny blood vessels. She discovered something that none of the other researchers had seen.

"The lab filled with murmurs of amazement," stated Schweitzer, "for I had focused on something inside the vessels that none of us had ever noticed before: tiny round objects, translucent red with a dark center." A colleague taking a look at these objects shouted, "You've got red blood cells! You've got red blood cells!" Schweitzer could hardly believe she had found a blood cell in an animal that had supposedly been dead for sixty-five million years. Her boss, John R. "Dinosaur Jack" Horner, suggested she try to prove the objects were not really red blood cells. "So far," Schweitzer says, "we haven't been able to."[1]

The scientists began testing the *T. rex* tissue for hemoglobin, a blood protein that carries oxygen from the lungs to the body tissues. Hemoglobin consists of *heme,* which contains iron, and *globin,* which contains a long string of proteins. The evidence that hemoglobin had indeed survived in the *T. rex* bone is as follows:

- The tissue extracted from the *T. rex* was reddish brown—the color of hemoglobin.
- Chemical signatures unique to heme were found when certain wavelengths of laser light were applied to the *T. rex* tissue.
- Because it contains iron, heme reacts to magnetic fields differently from other proteins. Extracts from the *T. rex* reacted in the same way modern heme compounds react.
- Extracts from the *T. rex* were injected into rats. If there was hemoglobin in the extracts, the rats' immune systems should build up antibodies against it. This is exactly what happened in carefully controlled experiments.[2]
- Schweitzer and her colleagues isolated a heme molecule, a ring-like structure that is found in hemoglobin and similar compounds.[3]

The team concluded that the "most parsimonious explanation of this evidence is the presence of blood-derived hemoglobin compounds preserved in the dinosaurian tissues."

Schweitzer recently made an even more spectacular discovery, announced in March 2005, after Jack Horner found another *Tyrannosaurus rex* at Hell Creek, Montana (where Barnum Brown found the first *T. rex*). After packing a *T. rex* femur in a plaster jacket, they found it was too heavy for a helicopter to lift. They were too far from any road to haul it out by truck, so Horner reluctantly cracked the femur in half. The bone was hollow and unmineralized.

Schweitzer soaked samples from the femur in a solution that dissolved the calcium. What remained was "flexible vascular tissue that demonstrates great elasticity and resilience." The tissue was soft and elastic; after being stretched, it resumed its original shape.

Schweitzer also found red blood cells (again) and what appear to be osteocytes, specialized cells that build and repair bone. Remarkably, nucleated red blood cells could be seen within the blood vessels. (Both reptile and bird red blood cells possess nuclei in circulation, whereas mammalian red blood cells lose their nuclei prior to entering circulation.) The researchers were able to squeeze red blood cells out of the flexible blood vessels.[4]

Red blood cells in unfossilized dinosaur bone, and the survival of soft tissue and hemoglobin, constitute powerful evidence against the notion that the dinosaurs lived sixty-five million years ago. Unless our eyes deceive us, the dinosaurs must have been buried more recently.

## A. OUTLINE OF A FLOOD MODEL

This Flood model is only a skeletal outline, intended mainly to correlate conventional geological period names with stages of the Genesis Flood. Creationists agree with other geologists regarding the order in which the strata were deposited. The main difference—and it is a huge difference—is over how fast this geological work was done. Lyellian geologists reckon the fossiliferous strata were deposited over the course of about 530 millions of years, whereas creationists believe the bulk of theses strata were deposited during one year—the year of the Genesis Flood.[5]

Many do not understand how the Flood could have left the stratified rocks we observe today. Intuitively, it would seem that the Flood would have mixed all the sediments together, such that there should be only one thick stratum of sediment left. Experiments have shown, however, that when different sizes of particles are carried in flowing water, stratification *always* occurs.

In 1965, when Bijou Creek in Colorado flooded its banks due to forty-eight hours of torrential rain, the resulting sedimentary deposits were stratified and exhibited particle sorting and bedding planes. Geologists had thought that bedding planes were formed when sedimentation stopped long enough to allow a layer to harden before sedimentation resumed. In the Bijou Creek flood, there was no time for the surface of any stratum to harden before sediment was deposited on top of it, yet this local flood produced bedding planes.[6]

The Genesis Flood was not a tranquil event, with waters rising slowly and steadily, as when you turn on the faucet in your bathtub. It was a terrifically violent event. Its primary purpose was the complete destruction of the antediluvian world. Ellen White describes the violence as follows:

> As the violence of the storm increased, trees, buildings, rocks and earth were hurled in every direction. The terror of man and beast was beyond description. Above the roar of the tempest was heard the wailing of a people that had despised the authority of God. Satan himself, who was compelled to remain in the midst of the warring elements, feared for his own existence.[7]

Genesis indicates that the Floodwaters went back and forth. This is not widely understood, because the first part of Genesis 8:3 has been mistranslated. The King James Version renders it "And the waters returned from off the earth continually: . . ." The Hebrew word translated "from off" is usually translated "upon" or "above." Two Hebrew words are translated together as "continually." The first is *yalak*, which means, "to go." The second is *shuwb*, which means, "to return." The text might better be translated, "And the waters returned *upon* the earth, *going and returning*."

The floodwaters surged in and retreated in a repeating cycle. One commentator explains it as follows:

> In Genesis 8:3, "continually" is a translation of two Hebrew verbs: shuwb, which means "to turn about, to return," and yalak, which means "to go." Together they present us with a graphic picture of the powerful churning action of the flood waters! A going is followed by a returning. Both verbs are set in the infinitive absolute form, indicating emphasis and duration. The flood waters did not tranquilly seep into the soil. This was a moving Flood, carrying back and forth vast amounts of water, soil, vegetation, and sediments. Gradually, layer after layer of sediments, vegetation, and other materials were laid down and covered over. The infinitive form means that it kept happening over a period of time (instead of only once if the imperfect form had been used). Terrific hydraulic forces were at work. Massive erosional and depositional actions were taking place. Gradually, layer after layer of sedimentary deposits were laid down.[8]

Genesis 8:3 describes the middle of the Flood, after the rain stopped but before the waters receded to where mountaintops became visible. But there is every reason to believe that the back and forth motion was typical of the entire catastrophe.[9]

Water overwhelmed an area, deposited sediment, then receded, only to return later and deposit more sediment atop that already deposited. In many areas, this cycle must have repeated many times over the course of several weeks before the Flood reached its height. There are many evidences of inundation and deposition, followed by retreating water and subaerial exposure of the deposited sediments. Dinosaur tracks, and perhaps even dinosaur nests, are among the evidence that Flood-laid sediments were exposed during the Deluge.

During the Flood, the continental plates that form the earth's crust were moving rapidly. New mountain ranges were being pushed up with unimaginable force and violence. Places on the bottom of the sea at the start of the Catastrophe came to rest high in the post-Flood mountains. The Burgess Shale of Canada is an example of this, containing a seafloor fauna that was suddenly buried in fine silt, near the beginning of the Flood. Mountain building activity later raised this antediluvian seafloor high into the mountains.

There also appears to have been a "crossing over" phenomenon, in which sediment deposited on a seafloor early in the catastrophe became part of a stratigraphic pile, the upper layers of which were above water later in the Flood. This can be inferred from the fact that there are dinosaur tracks in places where, farther down the stratigraphic pile, the sediment was deposited in a marine environment. Thus, it is clear that the Flood was an enormously complex event, and all simple models and "bathtub" analogies must be discarded.

## 1. PRECAMBRIAN DEPOSITS

Because the Precambrian rocks are largely devoid of fossils, they probably existed before the Flood.[10] Precambrian rocks do contain some fossils, mostly microorganisms such as bacteria and algae.[11] It is not surprising that the pre-Flood rocks should contain some fossil microorganisms, however, because these life forms can survive deep in the ground. Living bacteria have been found at a depth of 1,000 meters, and algae have been found at 210 meters.[12]

There are exceptions to the rule that only microorganisms are found in the Precambrian rocks. The mysterious fossils known as the Ediacaran fauna, after the Ediacara Hills of southern Australia where they were first found, range up to one meter in size. Ediacaran fossils are found only in the upper Precambrian, just below the Cambrian. At first, scientists interpreted the Ediacaran fauna as primitive soft corals, jellyfish, and segmented worms. But as these fossils were studied more closely, it became apparent that most were not related to modern life forms. The current Darwinian thinking is that no other life forms descended from the Ediacaran fauna; they simply became extinct at the beginning of the Cambrian.[13] Some believe that the Ediacaran fauna may have been more analogous to lichens than to any animal.[14]

The Ediacaran fauna were buried under conditions that appear to herald the beginning of the Genesis Flood. They were rapidly buried in deep water beneath sand and volcanic ash, suggesting the beginning of the Flood in the oceans by the eruption of the "great fountains of the deep." Ediacaran forms found in Namibia were buried rapidly beneath moving masses of sand.[15] Thus, it is possible that even some of the late Precambrian sediments might have been Flood-laid.

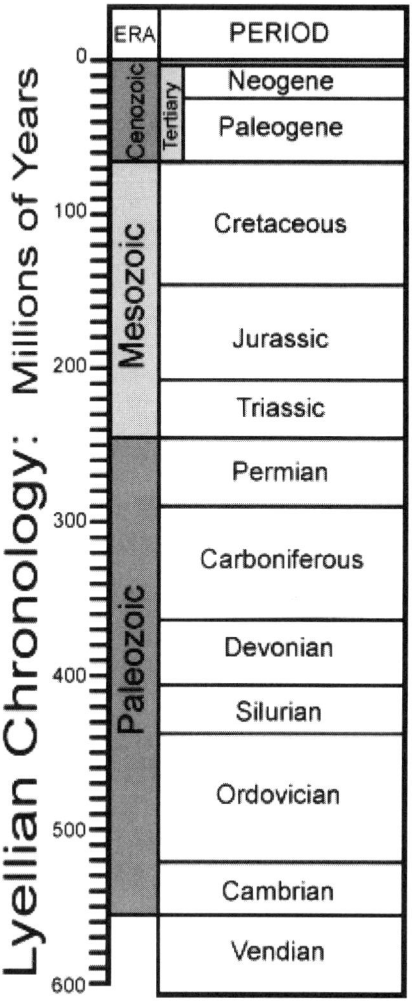

**Figure 8.1: Major Divisions of Strata, with notional Lyellian chronology**

## 2. PALEOZOIC DEPOSITS

Because the Flood began in the antediluvian seas, only marine fossils are found in Cambrian rocks. The bursting forth of the "great fountains of the deep" implies tectonic activity, which would have launched undersea turbidity currents. These turbidity currents carried sediment that buried animals living on the seafloor. All but three of the modern animal phyla have representatives in the lower Cambrian rocks.[16] This sudden appearance of numerous complex life forms in the lower Cambrian, which paleontologists call the "Cambrian explosion," is a riddle to Darwinists, but fits the Flood model perfectly.

As the Floodwaters continued to erode and deposit sediments, they buried mostly sea creatures. This is the case in the Ordovician, Silurian, and Devonian rocks, although some amphibians were buried in the Devonian strata. As the Flood continued, it buried incredibly vast amounts of vegetation, apparently including whole antediluvian forests. This buried vegetation would eventually turn into coal, oil, and natural gas. These fossil fuels are found throughout the fossiliferous strata, including vast amounts in Mesozoic and Cenozoic strata. But most of the coal in Great Britain is concentrated in one Paleozoic system, a fact that led early geologists to name that system the Carboniferous.[17]

It is worth slowing down, for a moment, to discuss the formation of coal. Coal is typically found in stratified deposits containing several types of sedimentary rock occurring in a regular, repeating sequence. The sequence will often include shale, freshwater limestone, clay, coal, marine limestone, and more shale. These repeating sequences are called *cyclothems* in America and *coal measures* in England.

Lyellians explain coal measures as the result of swamps that became peat bogs that were later inundated by the sea and buried under marine sediments. The scenario is as follows: (1) a forest grows up in a basin or plain beside the sea; (2) the forest becomes swampy; (3) a large peat bog forms from the swamp; (4) the peat bog is inundated by the sea, as a result of either subsiding land or a rising sea level, or both; (5) marine sediments are laid down on top of the peat bog; and (6) the area is exposed again, because of either rising land or a falling sea level.

The difficulty with this scenario is that many cyclothems are vertically piled on top of each other. In other words, the six-step process described above is supposed to have happened not once, twice, or three times, but as many as *one hundred times* in the same place![18] That is one hundred forests, one hundred swamps, one hundred peat bogs, one hundred inversions of land and sea level, one hundred periods of marine deposition, and one hundred more inversions of land and sea level—all in the same place. As one Lyellian geologist admitted, this explanation has "an element of unreality."[19]

To explain the many inundations, Lyellians have invoked a supposed Carboniferous-Permian ice age with glaciers fluctuating in mass, rhythmically trapping water on the continents, thus lowering the sea level, then melting and releasing water into the oceans, thus raising the sea level.[20] This facile explanation breaks down when it is considered that there are cyclical coal deposits ranging stratigraphically from the Devonian to the Miocene (upper Tertiary).[21] For example, cyclical coal deposits have been found in the upper Triassic of Sweden and the upper Cretaceous of Utah and Colorado.[22] No purported glaciation corresponds to these deposits.

The Lyellian explanation is even more unreal, given how laterally widespread cyclothems are. Some coal measures have been traced for four hundred miles along outcrops, and three hundred miles down basin from the outcrop, for a total area of over 100,000 square miles.[23] No modern swamp or peat bog is that large. The Lyellian scenario would require cyclical inversions of land and sea not just in restricted seaside areas but over many thousands of square miles, requiring unrealistically flat, low-lying continents.

One Lyellian textbook writer admits that "the whole problem [of how to interpret cyclothems] is in an interesting state of speculation, and no one theory is accepted by all students of coal geology."[24] But cyclical deposits are exactly what we would expect as a legacy of the "going and returning" waters of the Flood. Assuming varied times of uproot and rafting, variable rates of sinking and burial, and a tidal pulse of inundation, successive layers of sediment and organic debris would be deposited and accumulate in an alternating sequence.

Another problem with the Lyellian scenario is that it takes several feet of peat to form one foot of coal. Estimates vary from a few feet to twenty feet. Some coal seams in the U.S. are one hundred feet thick, and in Australia there are coal seams that are eight hundred feet thick. A conservative estimate of the ratio of peat to coal is ten to one and, although tectonic activity has thickened some coal seams, many seams would have required an unrealistically deep peat bog. Most peat bogs are less than fifty feet deep, and rarely exceed two hundred feet in depth; there is nothing like an eight-thousand-foot-deep peat bog in existence today.[25]

Forking coal seams pose another problem for the Lyellian explanation of coal. Sometimes, a diagonal coal seam connects parallel horizontal seams; this is called a "z-junction," for the obvious reason that, when viewed from the side, as in an outcrop or an open pit coal mine, the seams form the letter z. The Lyellian model can explain the parallel horizontal seams as the result of two peat bogs separated by time. No connecting diagonal seam, however, could form in the peat bog scenario. Catastrophic flooding, on the other hand, can explain z-junctions: vegetation is uprooted and accumulates as rafted debris, becomes waterlogged, sinks, and is buried under or *within* accumulating sediments. The organic debris can be buried at any angle or relationship to the enclosing sediments, including a z-junction.[26]

While on the subject of coal, we note that the casts of dinosaur footprints are frequently found protruding from the roofs of coal mines in Mesozoic strata. Coal miners in Price, Utah, have collected many such casts, and a local museum displays them. It appears that water deposited a mat of vegetation and then receded. Dinosaurs walked on top of this waterlogged mat, leaving footprints. Soon thereafter, the water returned and deposited sediment on top of the vegetation, filling the tracks with sand, thus creating natural casts. The vegetation turned into coal, and the sediment deposited over it hardened into the sandstone roof of the coal seam, complete with hanging footprint casts.

Obviously, these tracks could not have been preserved by the slow inundation of a transgressing sea over the course of thousands of years; rather, a sudden sediment dump preserved the tracks soon after they were made, as Lyellians admit:

> The footprints are depressions in the swamp peats that have been filled with flood-deposited sediment. . . . Peat is a water-saturated material that after compression tends to rebound to its undisturbed state. The abundance of well-defined tracks suggests that they were made just prior to the deposition of the sediment . . ."[27]

The geologist quoted above believed that a local flood in a river delta delivered the "flood-deposited sediment," but that is speculation. What is not speculation is that very soon after dinosaurs walked over the matted vegetation, a flood deposited sand in their tracks, thus preserving those tracks for the future wonderment of coal miners.

Some creationists have suggested that the Carboniferous deposits represent the remains of antediluvian floating forests. This would help explain many of the features of the Carboniferous deposits, including (1) the fact that many of the plants of the Paleozoic strata do not seem to have true roots, but instead have rhizomes, many of which are ill-adapted to penetrate the ground; (2) the fact that many of the trees had hollow branches and trunks, as if they were designed to weigh less in order to float better; and (3) the fact that fossil marine invertebrates and amphibians are frequently found with Carboniferous plant deposits.[28]

Paleocurrents are another set of data worth discussing in connection with the Paleozoic. Sedimentary rock is formed when water deposits sediment. Paleocurrents are indications of the direction in which the water was moving when it deposited the sediment. These indications include ripple marks, grooves, imbricated stones (stones piled like fallen dominoes in the direction in which they were transported), and fossils that were oriented in the current.

Dr. Arthur Chadwick of Southwestern Adventist University has studied paleocurrents for many years, obtaining data from all continents and all major sedimentary strata, Precambrian through Holocene/Recent. He has compiled data from over a million observed paleocurrent directions from over 25,000 sites. Chadwick made a remarkable discovery: for any given continent, the currents appear to be dominantly flowing in the same direction throughout the continent. This is particularly true for Paleozoic strata. In North America, Paleozoic currents flowed generally from the east and northeast to the west and southwest across the continent. Gradually, the current shifted from the westerly trend of the lower and mid Paleozoic to a southerly trend in the upper Paleozoic.

These results are difficult for a Lyellian to explain. Lyellism would predict that the direction of deposition would follow the contours of the land as it existed during a given geological period. The rivers, lakes, and oceans would not all be flowing and depositing sediment in the same direction. The observed pattern is more consistent with the Flood model. The Paleozoic strata would represent the rising floodwaters, characterized by intense, directional deposition.

Another topic worth mentioning in connection with the Paleozoic is the gaps in the strata. Throughout much of the Grand

Canyon, the Carboniferous Redwall Formation rests directly on a Cambrian formation. In the geologic column, the Ordovician, Silurian, and Devonian lie above the Cambrian but below the Carboniferous. These strata, which are missing at the Grand Canyon, represent one hundred thirty million years of "geologic time." *Yet we find very little evidence that the Cambrian layer was eroded during this time.* We find the Carboniferous lying on top of a flat Cambrian surface.

This is not a local phenomenon. The contact areas between strata purportedly separated in time by millions of years are nearly flat and sometimes "continent sized."[29] For example, the geological time periods missing at the Grand Canyon—the Ordovician, Silurian, and Devonian—are also missing at Dinosaur National Monument, over three hundred miles north of the Grand Canyon.[30] Many more examples could be adduced.[31]

How could so much time have passed with so little evidence of erosion or deposition? The only way for weathering to cease is for the weather to cease. Is it likely that the weather ceased for 130 million years? In fact, the postulated millions of years of "geological time" never happened. The layers were deposited one after another during the Genesis Flood, with little time intervening between deposits.

### 3. MESOZOIC DEPOSITS

In the sixteenth century, Dutch cartographer Abraham Ortelius noticed the jigsaw-puzzle fit of Africa and South America and suggested that the Americas had been torn away from Europe and Africa. "The vestiges of the rupture reveal themselves," he noted, "if someone brings forward a map of the world and considers carefully the coasts of the three continents."

In 1858, Antonio Snider-Pellegrini, a devout Christian living in Paris, published a book entitled *La Création et ses Mystères Dévoilés*

("Creation and its Mysteries Revealed"), in which he argued that the separation of the New World from the Old was the result of a rapid divergence of crustal plates during the Genesis Flood.[32] Snider-Pellegrini proposed a way that his theory could be tested: he suggested that the distribution of bands of distinctive, coal-rich, sedimentary rocks on one side of the Atlantic would perfectly match those on the other side. He also argued that the fossils on each side of the divide would match those on the other side. Later research vindicated his prediction.

In 1912, German meteorologist Alfred Wegener couched Snider-Pellegrini's theory in Lyellian terms, as opposed to creationist terms, and thus Wegener is usually credited with having invented it. It became known as the theory of "continental drift." Wegener had no credible explanation as to how the continents could have moved apart, and most geologists dismissed him as a crank.

As oceanographers began to map the oceans, they discovered underwater mountain ranges, including the Mid-Atlantic Ridge, which seemed to mark a line along which the African and South American continents separated from each other. American geologist Harry Hess, who had been a navy captain during the Second World War and had measured ocean floor depths, developed a theory that the ocean floors are spreading outward from the underwater ridges.

Molten magma comes up from the earth's mantle along these ridgelines, spreads, cools, and forms the basalt floors of the oceans. The ocean floors are constantly being formed and constantly spreading out from the underwater ridgelines. This raises the question, "If the Earth is not expanding, where is the old ocean floor going?" Hess had an answer, arguing that the older ocean floor plunges down into the earth's mantle at the locations of deep ocean trenches. These deep trenches are now understood as "subduction zones."

By the late 1960s, both continental drift and seafloor spreading theory had been placed within the larger context of plate tectonics theory. The gist of plate tectonics is that the solid outer covering of the earth, called the *lithosphere*, is divided into about a dozen major plates that move relative to each other, as they ride upon the earth's semi-molten mantle.

Plate tectonics theory has "explanatory elegance," meaning that it can explain a large array of geological facts in a straightforward manner.[33] One example is its explanation of earthquakes zones. Since the 1920s, geologists have recognized that earthquakes tend to be clustered in certain earthquake-prone zones. These zones are at the margins of plates—where the plates move against each other or where seafloor plate plunges beneath continental plate in subduction—and along the submarine ridges, where new ocean floor is extruded. Plate tectonics theory can predict the type and depth of an earthquake that will occur in a given location. It can also predict the type of magma

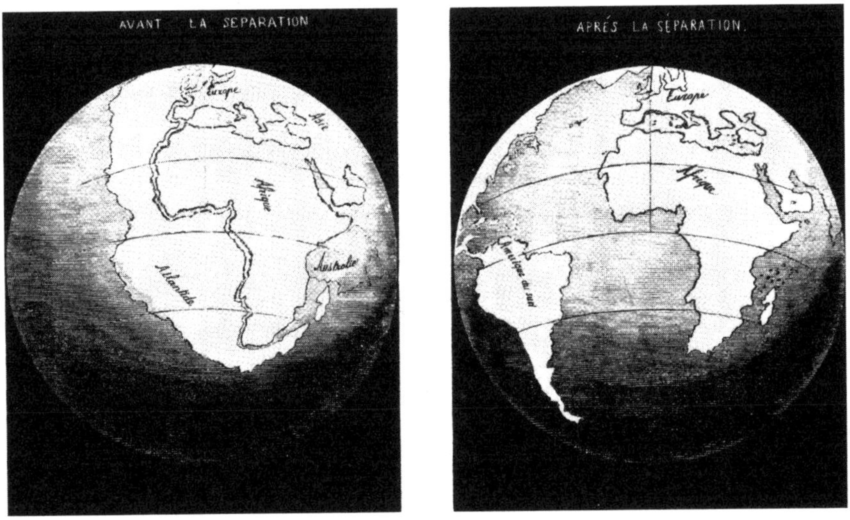


**Figure 8.2: Avant and Apres (Before and After). Crustal Divergence, as illustrated in 1858 by Antonio Snider-Pellegrini**

a volcano will eject based upon the volcano's proximity to a subduction zone.[34] Because of its remarkable explanatory power, plate tectonics has emerged as geology's grand unifying theory.

Creationist scientists, primarily Dr. John Baumgardner and Dr. D. Russell Humphreys, have been working on a variant of plate tectonics theory known as catastrophic plate tectonics. Catastrophic plate tectonics explains everything the conventional theory explains—and more.[35] The continental plates are now moving apart slowly, at a rate of only centimeters per year. The Lyellian assumption is that the plates have always moved at their present slow rates. Catastrophic plate tectonics, by contrast, postulates that during the Genesis Flood, continental plates moved apart extremely rapidly, at a rate of meters per second. In this instance, as in many others, creationists differ with mainstream geologist not over processes but over the rates at which those processes worked in the past.

There is evidence favoring the catastrophic model. If slabs of ocean floor had been sinking into the mantle along subduction zones over the course of many millions of years at the current slow rate, these slabs should have warmed to the temperature of the semi-molten mantle long before they reached the mantle's floor (which is over 1,700 miles below the surface). But if Baumgardner is correct in believing that large-scale, rapid subduction occurred just a few thousand years ago, there should be cooler, unincorporated ocean floor slab at the bottom of the mantle. Several studies have indicated that there are zones of cooler material reaching from the deep ocean trenches down to the bottom of the mantle.[36] This is a remarkable confirmation of catastrophic expectations, and difficult for the conventional theory to explain.

The catastrophic theory better explains the many instances of out-of-order strata in mountainous areas. For example, there is an area in the Rocky Mountains called the Lewis Overthrust, in which

Paleozoic strata have been pushed on top of Cretaceous strata for a distance of fifteen to thirty miles, along a five hundred mile front. It is difficult to imagine the energy that would have been necessary to lift billions of tons of rock and slide them many miles on top of younger rock strata. Adventist creationist George McCready Price refused to believe it happened; he argued that the strata must have been laid down in the order in which they are now found. But when plates collide at a speed of meters per second, they carry billions of times the energy that could be brought to bear in uniformitarian models, where the plates are moving at centimeters per year. This provides the unimaginably stupendous energy necessary to do the work of mountain building, with all its concomitant faulting, folding, and overthrusting.

There are unsolved problems with the catastrophic model, chief among them how the heat generated by the rapid creation of new ocean floor could have been dissipated. It would seem that magma coming up from the mantle and spreading out into seafloor at such a rapid rate would raise the ocean temperatures past the boiling point, killing all life on earth.[37] One author has suggested that the heat generated by rapid seafloor spreading was counteracted by cooling caused by volcanic ash blocking out the sun and reflecting away its heat.[38] Moreover, unsolved problems do not invalidate a model. Conventional plate tectonics theory, along with most other scientific theories, also has unsolved problems.

It appears that tectonic activity was occurring throughout the Flood. We mention plate tectonics theory in connection with the Mesozoic part of the Flood because that is when the continents began to separate. We know this because no part of the igneous slabs that comprise the current seafloors was formed before the Mesozoic, and all seafloor *sediment*, the detritus lying on top of the basalt slabs, is either Mesozoic or Cenozoic, and most of it is

Cenozoic.

As we enter the Triassic and move through the Jurassic and Cretaceous, the Flood is nearing its high-water mark. The areas of land remaining above water gradually shrink. Land animals, including the dinosaurs, are being herded together by rising water, drowned or otherwise destroyed, and buried by Flood-laid sediments. We still find plenty of marine fossils in the Mesozoic strata, and there are mixed fossil beds that contain both land and sea fauna. Chadwick also found discernable patterns in the paleocurrent data of the Mesozoic. In North America, the current gradually shifted from predominantly westerly in the Triassic (but with substantial scatter in the data) to easterly southeasterly in the Jurassic and lower Cretaceous, to strongly easterly in the upper Cretaceous.

## 4. CENOZOIC DEPOSITS

The Tertiary sediments are not as widespread as the Mesozoic strata, which, in turn, are not as widespread as the Paleozoic, which are often almost continent-wide in much of the world. The Tertiary deposits are found in inter-mountain basins and in great fans along the flanks of mountain ranges, and consist of material that has been eroded from older strata and redeposited in Tertiary strata. This suggests that after tectonic activity had elevated new mountains, water rapidly eroded these mountains and redeposited the sediments.

Other Tertiary deposits are found along coastlines. For example, Tertiary strata are found along the gulf coast of North America, from Texas to Florida. Some of these deposits are extremely thick. Exploratory wells in Louisiana have drilled down 25,000 feet without finding the bottom of the Tertiary. The fossil fauna of these gulf coast sediments alternate between land and marine forms, prompting geologists to describe the deposition as "a battle between torrential floods from the north and gigantic waves from the south."[39] This indicates vast erosion off of the newly formed North American continent, as the draining Floodwaters, blocked on one side by the Rockies and on the other by the Appalachians, drained down into the Gulf of Mexico.

With regard to the paleocurrent data, rock formations that cross the Cretaceous-Tertiary boundary, as well as rocks of the Paleocene (the lowest Tertiary "epoch" or subdivision), show evidence of an easterly flow. But, with the exception of the Paleocene epoch, the Tertiary paleocurrent data reflect local topography and no dominant directionality.[40] These data suggest that perhaps the Paleocene represents the Flood's high-water mark, and the higher Tertiary deposits correlate to the gradual draining of the waters.

Many opinions exist within creationism as to what was the last layer to be deposited during the year of the Genesis Flood. One scholar argues that it was as low as the Carboniferous, but that appears to be a minority position.[41] Several believe that the Cretaceous was the last Flood-laid period.[42] In this view, the Flood ended when the last of the dinosaurs was buried. Others would place the end of the Flood within the Tertiary.[43] Still others would place it within the Pleistocene.[44] "In our present state of knowledge," wrote Harold W. Clark, "it is not easy to draw a line between the last Flood deposits and the post-Flood sediments"—a statement that is as valid today as when Clark wrote it in 1968.[45]

One reason it is difficult to settle on a dividing line between Flood and post-Flood deposits is that the geological work done by the draining Floodwaters and that done during the early post-Flood period was the same—massive erosion and re-deposition. As they gradually receded, "going and returning," the Floodwaters eroded and redeposited previously laid, still-soft sediments. And in the centuries following the Flood, for reasons we will discuss in the next chapter, there was much greater precipitation. This greater rainfall rapidly eroded and redeposited sediment. Because of stupendous tectonic activity, both the Flood year and the early post-Flood era witnessed tremendous volcanic activity, with massive ejection of lava and volcanic ash. We would thus expect to find many layers of volcanic rock and tuff (hard volcanic rock composed of compacted volcanic ash) in both Flood and early post-Flood deposits.

One's opinion on the issue is also affected by whether one's focus is geology or paleontology. Although the Paleozoic and Mesozoic deposits are unlike anything being formed today, the Tertiary strata remind geologists of modern river and lake deposits. Thus, even creationist geologists tend to attribute the Tertiary to post-Flood activity. But many of the familiar modern mammal families do not have representatives in the fossil record until the mid to upper Tertiary. Although creationists accept a period of rapid diversification and speciation immediately following the Flood, it does not seem possible that all these animals could have evolved after the Flood within the time available in the biblical model. This consideration tends to push the Flood/post-Flood boundary to the upper Tertiary, even into the Pleistocene.[46]

## B. BIBLICAL CHRONOLOGY

The exact age of the dinosaur fossils depends upon the exact date of the Genesis Flood. The Bible contains a genealogy tracing the ancestry of Christ back to Adam. The portions of the genealogy found in Genesis Five and Genesis Eleven give the ages of the patriarchs when their sons were born. These age data and other chronological information in the Bible have allowed scholars to construct chronologies that date the Genesis Flood and the creation week.

Some argue that absolute dates cannot be derived from the genealogies, because it is not clear that all ancestors have been included. The term translated as *begat* does not necessarily specify a father-son relationship but sometimes only an ancestor-descendant relationship. Matthew's genealogy, for example, records that Jehoram begat Uzziah, but in the Old Testament, we learn that there were three generations between Jehoram and Uzziah—Ahaziah, Joash, and Amaziah. (Compare Matt. 1:8 with 2 Chron. 22-26).

However, a chain of statements that a patriarch lived a certain number of years before fathering a son, lived a certain number of years after that event, and lived a total number of years clearly implies that no generations have been skipped. Statements of this type are exactly what we find in Genesis 5 and 11. The genealogies in Genesis 5 and 11 are therefore unique. Adventist theologian Gerhard Hasel called them "chronogenealogies," because they preserve not only a genealogical record of descent, but chronological data as well.[47]

Over the millennia, thousands of biblical chronologies have been constructed. By far the most famous and influential chronology in the English-speaking world was that of Ireland's Archbishop James Ussher. Ussher's chronology, first published in 1650, became the standard, and it has appeared in the margins of the King James Bibles since 1679.[48] Ussher arrived at the date of 4004 B.C. for the creation week and 2349 B.C. for the year of the Flood.[49]

Seventh-day Adventist scientist Harold Coffin published a chronology that dates the Flood to 2522 B.C., about 170 years farther back in time than Ussher.[50] But Rene Noorbergen, another Adventist author, wrote a book entitled *Secrets of the Lost Races* in which, although he included no formal chronology, he seemed to place the Flood at approximately 3300 B.C., about one thousand years before Ussher's date.[51] How could authors with similar, conservative views of Scripture come to such widely differing results?

The main reason for such a wide divergence is that the

patriarchal age data in Genesis 5 and 11 vary among the different versions of Scripture. The Old Testament was assembled around the time of Ezra and Nehemiah in a text known as the "Vorlage," written in paleo-Hebrew. The Vorlage gave rise to three "recensions," the first of which was the *Samaritan Pentateuch,* committed to writing by 408 B.C., in paleo-Hebrew.

The second recension was the *Septuagint,* a Greek version translated in the third century B.C. by seventy Jewish scholars in Alexandria, Egypt. (It is called the "Septuagint," or the *LXX,* in scholarly shorthand, because it had seventy translators.) The Septuagint became the Bible of the Greek-speaking Jews living under the Ptolemaic Dynasty—rulers who inherited their domain from Alexander the Great.

The final recension was the *Masoretic text,* written around A.D. 100, possibly under the influence of Jewish Rabbis at Jamnia, where, according to some scholars, the Sanhedrin reconvened after the destruction of Jerusalem in A.D. 70. "A unified text suddenly became the standard at the end of the first century," wrote Seventh-day Adventist scholar Siegfried Horn, and the fact that "not one copy of a divergent text survived (except the Dead Sea scrolls that had already been hidden when Jamnia convened), indicates clearly that the council of Jamnia must have taken actions in this matter."[52] All of our English Old Testaments have been translated from the Masoretic text.

Frequently, the Septuagint gives an age for a patriarch at the birth of his son that is one hundred years older than the age given in the Masoretic. For example, the Masoretic states that Enoch was 65 years old at the birth of his son, Methuselah, whereas the Septuagint records that he was 165 years old. In some cases, the Samaritan Pentateuch agrees with the Septuagint's data, in some cases with the Masoretic, and in some cases with neither. Likewise,

Flavius Josephus (A.D. 37–c.100), in one section of his great work, Antiquities of the Jews, agrees much more often with the Septuagint than the Masoretic, but Josephus gives conflicting data in other parts of the same work.[53]

In cases where the Samaritan Pentateuch, the Septuagint, and Josephus all agree with the older patriarchal age, it would appear that the copiers of the Masoretic text dropped a digit that signified the extra one hundred years.[54] If Septuagint numbers are employed, the Genesis Flood can be pushed back beyond 3,500 B.C., whereas the shorter Masoretic numbers move the Flood forward to around 2,500 to 2,600 B.C. The Samaritan Pentateuch will yield a Flood date between these extremes, but closer to the Masoretic than the Septuagint.

The gospel writers and the apostles, especially the apostle Paul, frequently quoted the Septuagint in writings that are now part of the New Testament. Siegfried Horn argued that the Jews at Jamnia rejected the Septuagint precisely because it had become "the Bible of the Christians."[55] Although the differences are generally small, there are instances in which the Masoretic is significantly different from the Septuagint. Compare Paul's quotation of the Septuagint of Hosea 13:14 in 1 Corinthians 15:55 with the Masoretic in our Old Testament, or his quotation of Isaiah 64:4 in 1 Corinthians 2:9.

One could argue that the rabbis at Jamnia altered the Hebrew text just to spite Christians. Some passages that point forward to Christ seem to have been altered to remove their messianic import. Hebrews 1:6, for example, quotes the Septuagint of Deuteronomy 32:43 as saying, "Let all God's angels worship him." That sentence was completely omitted from the Masoretic (and hence from our English Old Testaments), but does appear in earlier Hebrew language texts—found among the Dead Sea scrolls—that pre-date the Masoretic. Another example is Hebrews 10:5-7, which quotes the Septuagint of Psalm 40:6-8 as saying, "Sacrifice and offering you did not desire, but a body you prepared for me," but our English Psalm 40, translated from the Masoretic, reads, "Sacrifice and offering you did not desire, but my ears you have dug [opened, pierced?] for me." Thus, there is reason for Christians to distrust the Masoretic.

The date of the Flood derived from Septuagint numbers accords better with historical data such as secular chronologies and king lists. Egypt's dynasties are believed to date back to approximately 3,000 B.C., Sumer's to 2,740 B.C., and China's to 2,852 B.C. (or 2,637 B.C., depending upon whether one accepts the "long count" or the "short count.") The Septuagint also better accommodates data from dendrochronology. For example, 4,600 rings, assumed to be annual growth rings, have been counted on one bristle cone pine tree. Thus, there are good reasons for preferring Septuagint numbers to the Masoretic numbers.

But there are serious problems with using the Septuagint numbers. There are many variations in the numerical data among the various Septuagint manuscripts.[56] By contrast, there are *no variations* among the extant Masoretic manuscripts in the age data of Genesis 5 and 11.[57] Because of the many variations, one cannot simply refer to "the Septuagint." It is necessary to specify which Septuagint manuscript is being used.

The most commonly used is the *Codex Alexandrinus*, one of the more complete manuscripts but not the oldest. (Another famous Septuagint manuscript, the *Codex Vaticanus*, is older than the *Alexandrinus* but contains the obvious mistake of having Methuselah live fourteen years beyond the Flood.) Some pro-Septuagint chronologists speak of "the Septuagint numbers," but no single Septuagint manuscript yields those numbers. In order to achieve a Flood date as far back as 3,500-3,600 B.C., one must cherry pick the largest numbers from among the various manuscripts.

Some Septuagint versions also include an extra "Cainan" (or "Kenan") in the genealogy of Genesis 11, between Arphaxad and Shelah. This extra Cainan does not appear in the Masoretic, the Samaritan, or in Josephus, which indicates that this is probably a case of scribal error. Yet Luke includes the extra "Cainan" in his genealogy (Compare Luke 3:35, 36 with Gen. 11:13), which is additional evidence that the Septuagint was the Bible of the New Testament writers.

Some scholars, Gerhard Hasel among them, argued that there was much more "scheme" or "system" in the Septuagint and the Samaritan Pentateuch than in the Masoretic. By "schematization," Hasel meant the altering of the numerical data in order to achieve patterns that make the numbers easier to memorize, or give the numbers some other aesthetic virtue. Hasel argued that, "the majority of scholarly opinion holds that the less schematized and more irregular chronology has claim to originality. It will have to be admitted that the LXX [Septuagint] has by far the more symmetrical, schematic chronology." And since the Masoretic was less schematized and more irregular, it more accurately reflected the unaltered original.[58]

Thus, there are excellent arguments for preferring the longer Septuagint numbers and excellent arguments for rejecting them in favor of the shorter Masoretic numbers. So we cannot say exactly when the Flood occurred. Scripture indicates that the Flood occurred sometime between 3,600 B.C. and 2,300 B.C. This range is sufficient for present purposes. There is such a wide chasm separating *any* biblically derived date for the death of the dinosaurs from the sixty-five million years ago postulated by Lyellian geology that it is pointless to labor over the details.[59]

---

[1] Wieland, Carl, "Sensational Dinosaur Blood Report" condensed version of *Creation Ex Nihilo* 19(4):42–43, September-November 1997. Found on AIG website at: *http://www.answersingenesis.org/docs/4232cen_s1997.asp* citing, M. Schweitzer and I. Staedter, "The Real Jurassic Park", Earth, June 1997, pp. 55-57. See also Wieland, Carl, "Evolutionist Questions AIG Report—Have Red Blood Cells Really been Found in *T. rex* Fossils?" On the web at: *http://www.answersingenesis.org/docs2002/0325rbcs.asp*; Schweitzer, M.H., Johnson, C., Zocco, T.G., Horner, J.H., Starkey, J.R., "Preservation of biomolecules in cancellous bone of *Tyrannosaurus rex*," *Journal of Vertebrate Paleontology*, 17:2, (June 19, 1997); Mary H. Schweitzer, Mark Marshall, Keith Carron, D. Scott Bohle, Scott C. Busse, Ernst V. Arnold, Darlene Barnard, J. R. Horner, and Jean R. Starkey, "Heme compounds in dinosaur trabecular bone," *Proceedings of the National Academy of Sciences* (vol. 94, p. 6291-6296) (June, 1997) *http://www.pnas.org/cgi/content/full/94/12/6291*; Virginia Morell, "Dino DNA: The Hunt and the Hype," Science, 261(5118):160-162, (July 9, 1993), p. 160-162; Jeff Hecht, "Dinosaur Bones Yield Blood Protein," *New Scientist*, vol. 154, Issue 2087 (June 21, 1997). Schweitzer and Horner stood by their description of what they saw in the *T. rex* bone as "***red blood cell-like intravascular structures*** . . ." Mary H. Schweitzer, James G. Schmitt, John R. Horner, "Paleoenvironmental and taphonomic parameters of an exceptionally preserved *Tyrannosaurus rex*." GSA Annual Meeting, Nov. 5-9, 2001, Paper 160-0 (emphasis added).

[2] Wieland. See also Schweitzer, et. al, *Proceedings of the National Academy of Sciences, supra.*

[3] Wieland, citing *Science News*, vol. 148, November 11, 1995 (p. 314). Found at AIG website at: *http://www.answersingenesis.org/docs/2987.asp*

[4] Schweitzer, M. H., J. L. Wittmeyer, J. R. Horner, J. K. Toporski, (2005) "Soft-Tissue Vessels and Cellular Preservation in *Tyrannosaurus rex*," *Science* 307: 1952-1955. Hecht, Jeff, "Blood Vessels Recovered from *T. rex* Bone," *New Scientist* (March 24, 2005); "*T. rex* Fossil has 'Soft Tissues'" BBC News, (March 24, 2005); David N. Menton "Ostrich-osaurus Disocovery?" Answers in Genesis, (March 28, 2005) *http://www.answersingenesis.org/docs2005/0328discovery.asp*; Carl Wieland, "Still Soft and Stretchy" Answers in Genesis, (March 25, 2005) *http://www.answersingenesis.org/docs2005/0325Dino_tissue.asp*. Here, a picture really is worth a thousand words, and the reader is encouraged to find and view the photographs on the Internet.

[5] The *Seventh-day Adventist Bible Commentary*, for example, takes this position: "Everywhere over the surface of the earth we find fossil remains of plants and animals obviously deposited by water. These deposits extend, in certain localities, to depths of at least three miles, but the average depth is somewhat over half a mile. The universal

distribution of these remains and the depth of their burial testify unmistakably to both the worldwide extent and the terrific violence of the Noachian Deluge." *The Seventh-Day Adventist Bible Commentary*, vol. 1 (Washington, D.C.: Review and Herald Publishing Association, 1953), p. 257.

[6] Berthault, Guy, "Experiments in Stratification," *http://www.icr.org/index.php?module=articles&action=view&ID=473*, citing McKee, E.D., Crosby, E.J. & Berryhill, H.L. Jr. 1967, "Flood deposits, Bijou Creek, Colorado, 1965," *Journal of Sedimentary Petrology*, 37, 829-851.

[7] White, Ellen G., *Patriarchs and Prophets*, p. 99.

[8] Akridge, G. Russell, "The Hebrew Flood Even More Devastating than the English Translation Depicts," *Creation Research Society Quarterly*, vol. 17, No. 4, (March, 1981).

[9] See, Marsh, Frank Lewis, *Life, Man, and Time*, (Escondido, CA: Outdoor Pictures, 1967), p. 118 ("Always there was the alternate ebb and flow—expressed by the double infinitive absolute, *halokh washobh*, of Genesis 8:3—which laid down successive strata containing contrasting fossils of land forms and water forms.") William Kirby, a member of the Royal Society of London, wrote on this in 1835. Kirby noted how the biblical narrative describes the waters as,

> "'going and returning,' as it is in the Hebrew, but rendered in our translation by the word 'continually.' Almost all the ancient versions adhere to the literal sense, (going and returning) which seems important, for it indicates a *flux and reflux* of the waters, which would effect the deposition of the matters floating upon or suspended in them.

See, Byron C. Nelson, *The Deluge Story in Stone*, (Minneapolis, MN: Bethany Fellowship, 1931, 1968), pp. 90, 91.

[10] See, e.g., Marsh, at 117; Roth, Ariel, *Origins: Linking Science and Scripture*, (Hagerstown, MD: Review and Herald, 1998), p. 209; Coffin, Harold, *Origin by Design*, (Washington, D.C.: Review and Herald, 1980), p. 75.

[11] Some doubt exists whether these "fossils" are real microorganisms preserved in the rock or whether they are pseudofossils, i.e., whether researchers are seeing patterns that are not really present. Roth, Ariel, *Origins: Linking Science and Scripture*, (Hagerstown, MD: Review and Herald, 1998), pp. 150-153.

[12] Roth, Ariel, *Origins: Linking Science and Scripture*, (Hagerstown, MD: Review and Herald, 1998), p. 166.

[13] See, e.g., Gould, Stephen Jay, *Wonderful Life,* (New York: W.W. Norton & Co., 1989), pp. 58-59, 231, 311-314, citing, Adolf Seilacher, "Vendozoa: Organismic Construction in the Proterozoic Biosphere," *Lethaia,* 22 (1989): 229-239; Norman, David, *Prehistoric Life: The Rise of the Vertebrates,* (New York: Macmillan, 1994), pp. 34-38.

[14] Retallack, Gregory J., "Were the Ediacaran fossils Lichens?" *Paleobiology,* 20 (1994): 523-544.

[15] Norman, David, *Prehistoric Life: The Rise of the Vertebrates,* (New York: Macmillan, 1994), p. 38.

[16] See, e.g., Wells, Jonathan, *Icons of Evolution,* (Washington, D.C.: Regnery, 2000), pp. 39-41.

[17] In early British geological literature, the Devonian rocks below the Carboniferous were called "the old red sandstone," while the Permian rocks above were referred to as "the new red sandstone." Hugh Miller's 1851 work, "The Old Red Sandstone" is a classic study of the Devonian rocks.

[18] Milton, Richard, *Shattering the Myths of Darwinism,* (Rochester, VT: Park Street Press, 1997), pp. 80-82; Nevins, Stuart, "The origin of coal," ICR impact No. 41, Nov. 1976.

[19] Milton, supra, at p. 82, citing Hollingsworth, S.E. "The climatic factor in the geological record," *Quarterly Journal, Geological Society of London,* 118:13 (1962).

[20] Stanley, Steven M., *Earth and Life Through Time* (2d ed.) (New York: W.H. Freeman & Co., 1986, 1989), pp. 415-419.

[21] Woodmorappe, John, "A Diluvian Interpretation of Ancient Cyclic Sedimentation," *Creation Research Society Quarterly* 14(3):189-208, 1978.

[22] Troedsson, Gustaf, 1950, "On rhythmic sedimentation in the Rhaetic-Liassic beds of Sweden": *18th Internat. Geol. Congress Rept.*, pt. 4, p. 64-72; Young, R. G., 1955, "Sedimentary facies and intertonguing in the Upper Cretaceous of the Book Cliffs, Utah, Colorado" GSA Bulletin, v. 66, p. 177-202.

[23] Moore, R. C., 1950, "Late Paleozoic cyclic sedimentation in central United States," *18th Internat. Geol. Congress Rept.*, pt. 4, pp. 5-16.

[24] Stokes, W. Lee, *Essentials of Earth History,* (Englewood Cliffs, NJ: Prentice-Hall, 1982), p. 308. Another interesting problem for Lyellians is that there is no coal in the lower Triassic anywhere in the world. After producing coal in such great abundance in the Carboniferous and Permian, all coal-forming processes completely and universally ceased for a period of seven million years, by Lyellian reckoning.

See Retallack, Gregory J., John J. Veevers, Ric Morante, "Global coal gap between Permian-Triassic extinction and middle Triassic recovery of peat forming plants," GSA Bulletin, Feb. 1996, vol. 108, no. 2, pp. 195-207.

[25] Coffin, Harold, *Origin by Design*, (Washington, D.C.: Review and Herald, 1980), pp. 46, 47.

[26] Snelling, Andrew A., "Forked Seams Sabotage Swamp Theory"—condensed version of article first appearing in *Creation Ex Nihilo*, 16(3):10-12, June-August 1994. Found on Answers in Genesis website: *http://www.answersingenesis.org/home/area/magazines/docs/v16n3_zcoal.asp.*

[27] Balsley, John K., 1980. "Cretaceous wave-dominated delta systems: Book Cliffs, east central Utah: a field guide." Unpublished. Amoco Production Company, Denver, Colorado. See also Parker, L.R. and Rowley, R.L., Jr., "Dinosaur footprints from a coal mine in east-central Utah," in Gilette, D.D. and Lockley, M.G. (eds.), *Dinosaur Tracks and Traces* (Cambridge University Press, 1989) pp. 361-366.

[28] Wise, Kurt P., *Faith, Form and Time* (Nashville: Broadman and Holman, 2002), pp. 170-173.

[29] Roth, at 222-232.

[30] Untermann, G.E., and B.R. Untermann, *A Popular Guide to the Geology of Dinosaur National Monument*, (Dinosaur National Monument, 1969), p. 44.

[31] See, e.g., Price, George McCready, *Evolutionary Geology and the New Catastrophism*, (Pacific Press, 1926), pp. 92-104. Price gives several examples, including one from Banff, Alberta, Canada:

> "East of the main divide the Lower Carboniferous is overlaid in places by beds of Lower Cretaceous age, and here again, although the two formations differ so widely in respect to age [Upper Carboniferous, Permian, Triassic and Jurassic are missing], one overlies the other without any perceptible break, and the separation of one from the other is rendered more difficult by the fact that the upper beds of the Carboniferous are lithologically almost precisely like those of the Cretaceous [above them]. Were it not for fossil evidence, one would naturally suppose that a single formation was being dealt with."

Another example is from the Athabasca region of Alberta, Canada:

> "The Devonian limestone is apparently succeeded conformably by the Cretaceous [Carboniferous, Permian, Triassic, and Jurassic are missing],

> and with the possible exception of a thin bed of conglomerate of limited extent, which occurs below Crooked Rapid on the Athabasca, the age of which is doubtful, the vast interval of time which separated the two formations is, so far as observed, unrepresented either by deposition or erosion."

One geologist gives a general summary of conditions in North America:

> "A break corresponding to several geological periods may be no more clearly marked than the relatively brief interruption of sedimentation between two small formations, or between diastrophically distinguished members of a single formation. . . . It makes little difference whether the compared sequences are lithologically similar or dissimilar; the large hiatuses are, as a rule, no more clearly defined than the small breaks."

[32] Carey, S. Warren, *Theories of the Earth and Universe,* (Stanford University Press, 1998), p. 90; Stokes, at 194.

[33] Wise, at 182-188, 278 ("So plate tectonics theories are rather simple; they are also extremely successful. No other theories have been able to better explain such a huge number of diverse geologic features.")

[34] Wise, at 186.

[35] Wise, at 182-195; Steven A. Austin, John R. Baumgardner, D. Russell Humphreys, Andrew A. Snelling, Larry Vardiman, and Kurt P. Wise, "Catastrophic Plate Tectonics: A Global Food Model of Earth History," in Walsh, *Proceedings of the Third International Conference on Creationism,* (Pittsburgh: Creation Science Fellowship, 1994, pp. 609-621.

[36] Wise, at 190, 191. See, also, Hutko, Lay, Garnero, and Revenaugh, "Seismic detection of folded, subducted lithosphere at the core-mantle boundary," *Nature,* **441**, 333-336 (May 18, 2006); "Slab of sunken ocean floor found deep within Earth" Press release from Arizona State University, May 17, 2006: "Deep within Earth, halfway to its center in an area where Earth's core meets its mantle, lies a massive folded slab of rock that once was the ocean floor, reports a team of researchers . . . in the current issue of Nature. . . . The researchers were able to image the buckling and folding of the subducted oceanic slab at the base of the mantle *because of the temperature difference between the relatively cool slab and the hotter mantle rock surrounding it.* . . . Using the method, the researchers found the subducted slab is composed of essentially the same minerals as the surrounding mantle, *but its temperature is about 700 degrees Celsius cooler.*"; S.P. Grand, "Mantle Shear Structure Beneath the Americas and Surrounding Oceans," *Journal of Geophysical Research,* 1994, 99:11591-11621; J.E. Vidale, "A Snapshot of Whole Mantle Flow," *Nature,* 1994, 370:16-17; S. Vogel, "Anti-matters," *Earth: The Science of Our Planet,* August 1995, pp. 43-49.

[37] See, e.g., Ross O. Barnes, "Thermal Consequences of a Short Time Scale for Sea-Floor Spreading" *Journal of the American Scientific Affiliation*, 32(2):123-125 (1980).

[38] Chadwick, Arthur V., "Preliminary Creation/Flood Model," note 23, at *http://origins.swau.edu/papers/global/chadwick/default.html*.

[39] Clark, Harold W., *Fossil, Flood and Fire*, (Escondido, CA: Outdoor Pictures, 1968), p. 146

[40] Chadwick, Arthur V., "Megatrends in North American Paleocurrents," 2000, found at *http://origins.swau.edu/papers/global/paleocurrents/default.html*; Mitchell, Colin, *Creationism Revisited*, (Alma Park, Grantham, UK: Autumn House Ltd., 1999), pp. 103, 104. See also Roth, at 217, 218; Coffin, at 96.

[41] Scheven, Joachim, "The Flood/Post-Flood Boundary in the Fossil Record," in R.E. Walsh, and C.L. Brooks, editors, *Proceedings of the Second International Conference on Creationism Held July 30-August 4, 1990, Pittsburgh, Pennsylvania*. USA, 1990, volume 2, Creation Science Fellowship, pp. 247-266

[42] Wise, at 286, 287; Leonard Brand, *Faith, Reason, & Earth History*, (Berrien Springs, MI: Andrews University Press, 1997), pp. 217; 278; 296, 297

[43] Clark, at 155-157. ("It would seem that the Tertiary must be recognized, not as a long time after the Flood in which the present rate of erosion and sedimentation was active, but a period of great violence. The early part may well be considered as the closing paroxysms of the Flood, after which came a short period of continued violence.")

[44] Dr. Lee Spencer, personal communication. See also Whitcomb and Morris, *The Genesis Flood*, (Phillipsburg, NJ: Presbyterian and Reformed, 1961), pp. 286, 287; Coffin, at 77.

[45] Clark, at 145.

[46] Even this is an oversimplication. Some geological features, such as marine sediments in the Miocene of the Alps, are hard to explain if you believe the Flood ended at the Cretaceous-Tertiary boundary. There are also paleontological data that point to an earlier end for the Flood, such as the fact that Australian marsupials are already different from marsupials on other continents by the Pliocene—hard to explain unless the Flood ended before the Pliocene.

[47] See Hasel, Gerhard, supra; Younker, Randall W., *God's Creation*, (Nampa, ID: Pacific Press, 1999), pp. 46-49.

[48] Hasel, Gerhard F. "The Meaning of the Chronogenealogies of Genesis 5 and 11" *Origins* 7(2):53-70 (1980), found at the website of the Geoscience Research Institute: *http://www.grisda.org/origins/07053.htm.*

[49] Ussher is often ridiculed for having fixed the date and time of creation as "Sunday, the 23rd of October, 4004 BC, at nine o'clock in the morning." But Dr. John Lightfoot, Hebrew scholar and vice-chancellor of Cambridge University, first posited an exact date and time of creation. See, e.g., John Noble Wilford, *The Riddle of the Dinosaur* (New York: Vintage Books, 1987), pp. 43, 44; *Ice Ages,* Windsor Chorlton, Ed. (Alexandria, VA: Time-Life Books, 1983), p. 85. Moreover, in fixing the date and time of creation, Lightfoot and Ussher were following an ancient Jewish tradition in which creation occurred on the first day of the first month of the Jewish year, what we would call Sunday, October 6, at 6:11 a.m., 3,761 B.C. See, Randall Younker, *God's Creation* (Nampa, ID: Pacific Press Pub., 1999), pp. 37-39.

[50] Coffin, at 287-293 (chapter entitled, "A Biblical Perspective on the Age of the World")

[51] Noorbergen, Rene, *Secrets of the Lost Races,* (Indianapolis, IN: The Bobbs-Merrill Company, 1977), p. 29 ("Genesis tells us that Cain and Seth were born in 4969 BC and 4924 BC, respectively." Ussher's chronology would put the birth of Cain at 3,874 B.C., so Noorbergen puts first events about one thousand years farther back than Ussher.) See also p. 179 ("Materials removed from other holes have been dated between 2200 and 2100 BC, which suggests that Stonehenge may have been constructed almost within the first millennium after the Deluge.") See also his general discussion of dating the flood, at pp. 18-39.

[52] Setterfield, Barry, *Creation and Catastrophe Chronology,* (1999) on the web at *http://www.ldolphin.org/barrychron.html.* Also at *http://www.setterfield.org/ccchron/barrychron.html.* Setterfield argues that the Septuagint numbers should usually be preferred to the Masoretic numbers.

[53] The data in Whiston's Josephus, *Antiquities of the Jews,* book 1, chapter 2, par. 4, agrees more often with the Septuagint, but Josephus gives conflicting, and shorter chronology numbers in book 8, chapter 3, par. 1, and different conflicting data in book 10, chapter 8, paragraphs 4, 5. Thus, Josephus is not much help on the subject of biblical chronology. See Hasel, Gerhard F. "Genesis 5 and 11: Chronogenealogies in the Biblical History of Beginnings" *Origins* 7(1):23-370 (1980),

[54] Setterfield, supra.

[55] Setterfield, citing Siegfried H. Horn, "The Old Testament Text in Antiquity," *Ministry,* (November, 1987), pp. 4-8.

[56] Ray, Paul J. "An Evaluation of the Numerical Variants of the Chronogenealogies of Genesis 5 and 11" *Origins* 12(1):26-37 (1985). Found at the Geoscience Research Institute's website at *http://www.grisda.org/origins/12026.htm.*

[57] Ray, supra.

[58] Hasel, Gerhard F. "Genesis 5 and 11: Chronogenealogies in the Biblical History of Beginnings" *Origins* 7(1):23-370 (1980).

[59] Ellen White writes in several places of the earth being six thousand years old. For example, in *The Great Controversy*, she writes, "For six thousand years that mastermind that once was highest among the angels of God has been wholly bent to the work of deception and ruin," and "The great controversy between Christ and Satan, that has been carried forward for nearly six thousand years, is soon to close; and the wicked one redoubles his efforts to defeat the work of Christ in man's behalf and to fasten souls in his snares." In volume 3 of the *Testimonies*, she writes, "God endowed man with so great vital force that he has withstood the accumulation of disease brought upon the race in consequence of perverted habits, and has continued for six thousand years." She also states that there were four thousand years between creation and the Messiah. For example, in *The Desire of Ages*, she writes, "It would have been an almost infinite humiliation for the Son of God to take man's nature, even when Adam stood in his innocence in Eden. But Jesus accepted humanity when the race had been weakened by four thousand years of sin."

A large number of these types of statements could be adduced. In all of them, Ellen White employed the *approximate* age of the earth for literary purposes, to improve the flow and dramatic impact of the writing. In all likelihood, she was accepting Ussher's chronology, which even after two hundred years was still the best chronology available. Ellen White never attempted to make a definitive statement on biblical chronology as such, and her literary use of Ussher's dates—rounded off, as in the examples above, for greater readability—should not be used to limit or foreclose further study in the field of biblical chronology. While she clearly rejected Lyellian notions about the age of the earth, her writings should not be interpreted as expressing a preference between Masoretic and Septuagint chronological data.

CHAPTER 09

# Ice Age— The Flood's Aftermath

"I trust to outlive this mania," wrote Louis Agassiz, referring to Darwinism. Agassiz (1807-1873) was one of the last naturalists of undisputed eminence who never assented to Darwin's theory of evolution. Agassiz was born in Switzerland in 1807, the son of a Protestant minister. He studied at four European universities, earning doctorates of medicine and philosophy. He then went to Paris to study comparative anatomy under Cuvier, where he specialized in ichthyology—the study of fish. Cuvier had been planning to write a treatise on fossil fish, but he was so impressed with Agassiz that he gave Agassiz his notes and drawings on the topic. After Cuvier died, Agassiz accepted a position at the University of Neuchâtel, in Switzerland, where he taught for fourteen years and published a treatise on fossil fish that established his reputation as a naturalist.

At Neuchâtel, Agassiz began studying something completely unrelated to fish: glaciers. His fellow Swiss scientist, Jean de Charpentier, suggested that glaciers had once been much more extensive than they now are. Agassiz was skeptical, but when he summered at de Charpentier's hometown of Bex, Switzerland, his colleague showed him the evidence of extensive past glaciation. Agassiz passionately threw himself into the study of glaciers. To save time, he set up shop right on top of a glacier, building a hut that came to be called "*Hôtel des Neuchâtelois,*" from which he and his students studied the structure and movements of the ice.[1]

In 1840, Agassiz published *Études sur les Glaciers* (Studies on Glaciers), in which he argued that not only Switzerland, but also many other regions, had once been covered by vast ice sheets. "Great sheets of ice, resembling those now existing in Greenland," wrote Agassiz, "once covered all the countries in which unstratified gravel (boulder drift) is found."

**Figure 9.1: Louis Agassiz**

Agassiz showed that glaciers leave distinctive geological traces. They carve out U-shaped valleys. They carry along boulders, called "erratics," which come to rest far from their point of origin. They carry along dirt and gravel, and pile them in unstratified ridges called *moraines.* Lateral moraines form along the sides of the glaciers; terminal or end moraines at the farthest extent of the glacier's advance. As glaciers move over the bedrock, they "polish" it, leaving scratches and grooves called *striations.* There are many other geological traces of glaciation, known by exotic names such as "cirque," "esker," and "drumlin."

Later in 1840, Agassiz visited Scotland and lectured on the Ice Age at a scientific convention attended by Britain's leading geologists, including Buckland and Lyell. His lecture met with a cold reception; Lyell was particularly strident in his denunciation of it. On a previous visit to Switzerland, William and Mary Buckland had become friends with Agassiz, so, after the meeting, Buckland invited Agassiz and Roderick Murchison on a field trip to northern England, Scotland, and Ireland. In these countries, Agassiz pointed out the same evidence of glaciation he had found in Switzerland: moraines, erratics, grooved and polished rocks. Upon approaching a valley he had never seen before, Agassiz correctly predicted exactly where the group would find a moraine.

The geological features that Agassiz was attributing to the action of glaciers, Buckland had attributed to the Genesis Flood in his 1823 book, *Reliquiae Diluvianae* (Relics of the Flood). After publishing *Reliquiae,* however, Buckland had accepted the "tranquil theory"—the notion that the Genesis Flood left no geological traces. But until Agassiz, Buckland still believed that the unstratified gravel and erratic boulders were waterborne deposits. Lyell also believed this, theorizing that the erratic boulders had lodged in drifting icebergs, eventually dropping where they are now found.

Convinced that Agassiz was right, Buckland took it upon himself to convince Lyell, and he succeeded. "Lyell has adopted your theory in toto!!!" Buckland later wrote Agassiz. "On my showing him a beautiful cluster of moraines within two miles of his father's house, he instantly accepted it, as solving a host of difficulties which have all his life embarrassed him."[2]

Even with Lyell's backing, however, the glacial theory was slow to gain acceptance among geologists, and was not firmly accepted until the mid-1860s. In the meantime, Agassiz took a teaching job at Harvard, where he founded the Museum of Comparative Zoology. He was also instrumental in founding the National Academy of Sciences and was a regent of the Smithsonian Institution. After the death of his first wife, he married a prominent Boston woman, Elizabeth Cabot Cary. He continued to teach America's brightest science students until, failing to outlive the mania of Darwinism, he died in 1873.

Glaciers once covered all of Canada and a substantial part of the United States. The glaciated areas of the U.S. include New England, as far south as northern New Jersey and northeastern Pennsylvania, and the Midwest, including most of Wisconsin, Illinois, Indiana, Ohio, and Missouri, and all of Minnesota, Michigan, and Iowa. The ice sheet extended west, covering parts of North and South Dakota and the northern edges of Montana, Idaho, and Washington.

**Figure 9.2: The Ice Age in North America**

Imagine the magnitude of the disaster if the Ice Age were to repeat itself, and the cities of Boston, New York, Cleveland, Detroit, Indianapolis, Chicago, Minneapolis, Milwaukee, Montreal, and Quebec were covered with thick glaciers, as their sites once were. The Ice Age will never repeat itself, however, because it was part of the aftermath of the Genesis Flood.

There is no good uniformitarian explanation for the Ice Age. Over sixty theories have been proposed; none were satisfactory, and most have been discarded.[3] "Pleistocene phenomena have produced an absolute riot of theories ranging from the remotely possible to the mutually contradictory and the palpably inadequate."[4]

Charles Lyell theorized that vertical movements of the earth's crust cause ice ages, noting that a general increase in land elevation would cause temperatures to drop because the atmosphere is colder at higher altitudes.[5] To American Henry Adams, the Ice Age and Lyell's explanation of it appeared closer to catastrophe than uniformity:

> At the very outset Adams struck on Sir Charles's Glacial Theory or theories. He [Adams] was ignorant enough to think that the glacial epoch looked like a chasm between him and a uniformitarian world. If the glacial period were uniformity, what was catastrophe? To him the two or three labored guesses that Sir Charles suggested or borrowed to explain glaciation were proof of nothing, and were quite unsolid as support for so immense a superstructure as geological uniformity. . . . Sir Charles himself seemed shy of them. Adams hinted his heresies in vain. At last he resorted to what he thought the bold experiment of inserting a sentence in the text, intended to provoke correction. "The

> introduction [by Louis Agassiz] of this new geological agent seemed at first sight inconsistent with Sir Charles's argument, obliging him to allow that causes had in fact existed on the earth capable of producing more violent geological changes than would be possible in our own day." The hint produced no effect. Sir Charles said not a word; he let the paragraph stand; and Adams never knew whether the great Uniformitarian was strict or lax in his uniformitarian creed; but he doubted.[6]

Lyell's explanation for the Ice Age has since been discarded.[7]

The explanation currently in vogue is the astronomical theory, proposed in 1864 by James Croll—a janitor at a college in Glasgow, Scotland—and substantially revised by the Serbian scientist Milutin Milankovitch (1879-1958) in the 1920s and 30s. The Milankovitch theory combined three celestial phenomena to explain the cooler weather: 1) cyclical changes in the eccentricity of the earth's orbit around the sun, 2) small variations in the tilt of the earth's axis relative to the plane of its orbit around the sun, and 3) the precession of the equinoxes.[8]

The Milankovitch theory fell into disfavor in the 1950s, in part because it predicted dates for glacial advances and retreats that failed to correlate with Carbon-14 dating. By 1969, the vast majority of scientists had rejected it.[9] Yet the Milankovitch theory was resurrected in the late 1970s thanks to the determined efforts of several influential scientists and to the fact that no one had proposed a better naturalistic explanation.[10]

The Milankovitch theory predicts a slight reduction in solar radiation at higher latitudes. This seems an inadequate explanation for such a profound change in the weather, however, because solar radiation is only one of the ways of heating the higher latitudes. Ocean currents also provide a significant heat source for the higher latitudes. For example, the Atlantic Ocean's Gulf Stream carries warm ocean water from the tropics into the northern hemisphere, where it exchanges heat with the atmosphere and helps warm North America and especially Great Britain and Northern Europe. Some climate scientists believe that the "little Ice Age," a period of lower average temperatures that lasted from the mid-14th to the mid-19th century, was caused by derangement of the Gulf Stream.[11] The Ice Age demands a much better explanation than just slight reductions in direct sunlight on the higher latitudes.[12]

In recent years, creationists have developed an alternative explanation for the Ice Age. Michael J. Oard, a meteorologist formerly with the National Weather Service, has done much work in this area. First, an Ice Age would *not* require colder winters. In fact, colder air works against the formation of glaciers, because Glaciers are made of snow, and colder air cannot hold as much moisture and thus cannot precipitate as much snow. One of the coldest places on earth is Siberia, and it does not have glaciers.

The formation of the vast glaciers of the Ice Age required 1) much more snow; and 2) cooler summers, so that the glaciers would not melt back to their starting points each summer. The creationist model can supply both these conditions.

## A. GREATER PRECIPITATION

The first requirement for an ice age is greater snowfall. This is where the Milankovitch theory is weakest: it has no mechanism to produce the necessary precipitation. In the creationist model, the increased snowfall and generally greater precipitation is a result of greater evaporation from oceans that were substantially warmer during the early post-Flood period.

Creationists have long held that the pre-Flood seas were warmer than today's oceans, and the Flood would have heated the water even more. Genesis 7:11 states that at the beginning of the Flood, "all the fountains of the great deep burst forth," implying that water shot into the air from cracks or fissures in the earth. (Ellen White confirms this: "Jets of water burst from the earth with indescribable force, throwing massive rocks hundreds of feet into the air, and these, in falling, buried themselves deep in the ground."[13]) Assuming that this water came from deep in the earth's crust, it would be hot, because the crust of the earth grows about 20º to 30ºC warmer with each kilometer of depth. The temperature of subterranean water today can be as hot as 350ºC in geothermal vents along the mid-ocean ridges.[14]

Catastrophic plate tectonics theory provides an independent explanation for the heating of the oceans during the Flood. Molten magma rapidly erupted from the submarine ridges, forming new ocean floor as the continents moved apart. The ocean water must have absorbed immense heat in the process of cooling this new ocean floor basalt. We noted that a criticism of catastrophic plate tectonics theory is that the heat generated by such rapid seafloor spreading would have boiled the oceans. Clearly, the rapid seafloor spreading could easily have supplied the heat necessary to warm the oceans.[15] This process would have heated the oceans from the bottom, resulting in a more uniform, and uniformly warm, ocean temperature. Evidence indicates that the ocean floor water temperature was very warm at the end of the Flood and the beginning of the Ice Age.[16]

Warmer water evaporates much more quickly than cold water.

The rate of evaporation depends upon the temperature of the water at the surface. For example, the rate of evaporation when the ocean surface temperature is 30º C is over three times greater than it is at 10º C and over seven times greater than it is at 0º C. (The current average ocean temperature is 4º C, but deep ocean water is just above freezing, whereas surface water averages 17.5º C).

The high rate of evaporation from the warmer post-Flood oceans led to greater precipitation. (Californians are familiar with the affects of "El Niño," the current of warmer water that flows into the eastern Pacific Ocean every three to seven years, creating a much wetter winter.) Oard estimates that the average precipitation over mid- and high-latitude continents would have been at least triple the current rate. Larry Vardiman, a Ph.D. in atmospheric science, designed a computer simulation of the precipitation that would have resulted from the heating of seawater during a rapid formation of new ocean floor. His model predicted a distribution of precipitation that closely tracked the actual distribution of snow and ice during the Ice Age.[17]

## B. COOLER SUMMERS

The second requirement for an Ice Age is much cooler summers. Volcanic activity caused the cooler summers. Plate tectonics theory ties volcanic activity to the movement of plates. Volcanoes tend to be clustered along the margins of plates and along submarine ridges. Most of the world's active volcanoes are located near convergent plate boundaries where subduction is occurring, particularly around the Pacific basin.

In catastrophic plate tectonics theory, plate movement was extremely rapid during the Mesozoic and later stages of the Genesis

Flood. During this time, vast quantities of new ocean floor were being formed, separating the continental plates of North and South America from those of Europe and Africa. We would expect to find evidence of hyperactive volcanism during the Flood and during the first few centuries after the Flood, and that is exactly what we find.

Subduction-zone volcanoes, like Mount St. Helens in Washington and Mount Pinatubo in the Philippines, typically erupt with tremendous explosive force, ejecting volcanic dust and ash high into the atmosphere. The upper Tertiary phase of the Flood witnessed numerous explosive conical volcanic eruptions.[18] In addition to the ejecta from conical volcanoes, enormous cataracts of lava flooded out of fissures in the earth opened by crustal rifting and continental breakup. These fissure eruptions, commonly called flood basalts, spewed aerosols into the upper atmosphere.[19] It is possible that, as the world emerged from the Flood, the skies were dark with volcanic ash and aerosols.[20]

Elevated volcanic activity continued into the Pleistocene—the post-Flood world of the Ice Age. Worldwide studies of volcanic ash layers in deep-sea sediments show that there was heightened volcanic activity during the Pleistocene. Much of the Ice Age volcanic activity was of the explosive type that would have ejected vast quantities of ash and aerosols into the upper atmosphere.[21] In the western U.S. alone, more than sixty-eight ash falls coinciding with the Pleistocene have been counted.[22] An exceptionally large Pleistocene eruption in New Zealand spread a distinct layer of ash over ten million square kilometers of the South Pacific. The amount of dust and aerosols thrown into the atmosphere by the largest of these eruptions would have blocked out most sunlight over the entire world.[23]

Lyellian assumptions space these eruptions by hundreds of thousands of years, over the course of millions of years, minimizing their effect on climate. But if all of these eruptions occurred within a few hundred years, as the creationist model suggests, they would have had drastic effects on the world's climate. The dust and especially the aerosols that volcanoes eject into the upper atmosphere reflect solar radiation away from the earth, lowering temperatures. The effect is the same as the "nuclear winter" scenario frequently discussed during the Cold War years.

The cooling effect of volcanic eruptions is not disputed. Past volcanic eruptions have caused substantial cooling. The 1815 eruption of the Indonesian volcano Tambora lowered worldwide average temperatures by 3ºC (5º Fahrenheit), such that 1816 was known as "the year without a summer." New England and Canada experienced an unprecedented series of cold snaps, with heavy snowfall in June and crop failures caused by frost *in July and August.*[24] The eruption of Krakatoa in 1883 caused a similar, though less severe, worldwide lowering of average temperatures.

Uniformitarian geologists acknowledge that volcanic activity causes cooling and that sufficiently frequent volcanic activity could cause an ice age:

> A large number of volcanic eruptions over a long period could lower global temperatures substantially by injecting huge amounts of volcanic ash and dust into the upper atmosphere. . . . The effect would be to shade the Earth and lower global temperatures. A five percent drop in the amount of solar radiation reaching the surface could cause the temperature to drop by as much as 10 degrees Fahrenheit, which would be enough to bring on an ice age. The long-term cooling could cause glaciers to expand and lower the sea level.[25]

A model in which all Pleistocene volcanic activity is telescoped into the space of a few hundred years predicts substantial summer cooling, so that the glaciers would not melt back to baseline each summer. Thus, the creationist Ice Age model provides both the greater precipitation and cooler summers necessary for the massive glaciers of the Ice Age to form.

Because the oceans have a tremendous capacity to hold their temperature, atmospheric cooling would have lowered ocean temperatures slowly and gradually. The warmer water would always have floated on top, because warmer water rises to the surface and cooler water sinks to the bottom. This would have allowed for greater evaporation and precipitation long into the Ice Age, despite the cooler summers.

Eventually, the factors that led to the Ice Age reversed themselves. The oceans cooled to their present temperatures, reducing evaporation and precipitation to current levels. Meanwhile, the rapid plate movements of the Flood and early post-Flood years decelerated to the present slow pace. Slowed tectonic activity led to less frequent volcanic eruptions. Reduced volcanic activity allowed the skies to clear of dust and aerosols, which allowed the sun to warm up the earth. Warmer summers melted the continental glaciers back to their present locations, but the much colder oceans allowed ice caps to form at the poles.

Oard estimates that five hundred years were necessary to grow the ice sheets to their greatest extent, with an additional two hundred years needed for the ice to melt and retreat. Thus, the entire Ice "Age" could have lasted only about seven hundred years, easily fitting within the biblical time frame.[26]

The greater precipitation predicted by the creationist Ice Age model explains the abundance of lakebeds in areas that are now desert, and the fact that existing lakes once had much higher water levels.[27] For example, the Great Salt Lake is the evaporated remnant of much larger Lake Bonneville. Vast reservoirs existed in the continental interiors, mainly because of greater precipitation in the early post-Flood centuries. Geologist Steven A. Austin suggests that the Grand Canyon was rapidly eroded when one of those reservoirs burst through a weak point in a natural dam.[28]

The creationist Ice Age model explains many other mysteries that cannot otherwise be explained.[29] Suffice it to say that biblical creationists, once reluctant to believe in a glacial age, are now quite enthusiastic about it.[30]

---

[1] Agassiz was not the first to note that glaciation had once been far more extensive than at present. In 1787, a Swiss lawyer named Bernard Friedrich Kuhn had suggested that glaciers had carried the granite erratics of the Jura. Hutton reached the same conclusion in 1795. Swiss guide and mountaineer Jean Pierre Perraudin had written in 1818, "Having long ago observed marks or scars occurring on hard rocks which do not weather, these marks being always in the direction of the valleys, I finally decided, after going near the glaciers, that they had been made by the pressure or weight of these masses . . . I am ready to demonstrate this fact to incredulous people by the obvious proof of comparing these marks with those uncovered by glaciers at present." Agassiz was not even the first to suggest that there had once been at least a local "Ice Age." Ignatz Venetz, a Swiss civil engineer, heard of Perraudin's ideas and conducted his own research. Venetz presented his findings to the Swiss Society of Natural Sciences in 1829, arguing that the glaciers had once extended not only over the Jura but northward onto the European plain. In 1832, a German scientist named Reinhard Bernhardi published a paper claiming that polar ice had once covered Europe as far south as southern Germany. Jean de Charpentier supported Venetz. See Chorlton, Windsor, *Planet Earth: Ice Ages,* (Alexandria, VA: Time-Life Books, 1983), pp. 85, 86. Thus, Agassiz was not the pioneer of the glaciation theory—just as Lyell was not the pioneer of uniformitarianism. Nevertheless, once convinced of the concept, Agassiz became its most enthusiastic and successful supporter, convincing the leading geologists of the day. Agassiz also went further than Charpentier had gone, arguing that the Ice age had been a worldwide event, effecting North America, Northern Europe, and Northern Asia.

[2] Imbrie, John, and Katherine Palmer Imbrie, *Ice Ages: Solving the Mystery,* (Short Hills, NJ: Enslow Pub., 1978), p. 40.

[3] Oard, Michael J., "A Post-flood Ice-age Model can Account for Quaternary Features," *Origins* 17(1):8-26 (1990), citing Eriksson, E. "Air-ocean-icecap interactions in relation to climatic fluctuations and glaciation cycles," in J. M. Mitchell, Jr. (ed.), *Causes of Climatic Change,* Meteorological Monographs 8(30), (American Meteorological Society, Boston, 1968), pp. 68-92. See also Oard, Michael J., *An Ice Age Caused by the Genesis Flood,* (El Cajon, CA: Institute for Creation Research, 1990), p. 13.

[4] Oard, Michael J., *An Ice Age Caused by the Genesis Flood,* (El Cajon, CA: Institute for Creation Research, 1990) (hereinafter "Oard"), p. 13. citing Charlesworth, J. K., The Quaternary Era, (London: Edward Arnold, 1957).

[5] Imbrie and Imbrie, at 65.

[6] Adams, Henry, *The Education of Henry Adams,* (1918) Ch. XV "Darwinism," found at *http://www.bartleby.com/159/15.html*

[7] Imbrie and Imbrie, at 65.

[8] Oard, at 16.

[9] Imbrie and Imbrie, at 141.

[10] Oard, at 16.

[11] See, e.g., Lemley, Brad, "The New Ice Age" *Discover* Vol. 23, No. 9 (September 2002).

[12] "[A] uniformitarian ice age seems meteorologically impossible, and proposed solutions to the problem are inadequate . . ." Oard, at 19. The eminent astronomer Sir Fred Hoyle wrote, "If I were to assert that a glacial condition could be induced in a room liberally supplied during winter with charged night-storage heaters simply by taking an ice cube into the room, the proposition would be no more unlikely than the Milankovitch theory." Oard, at p. 18, citing, Hoyle, Fred, *Ice, the Ultimate Human Catastrophe,* (New York: Continuum, 1981), p. 77.

[13] White, Ellen G., *Patriarchs and Prophets,* (1890), p. 99.

[14] Oard, supra, citing Kerr, R. A., "Ocean Hot Springs Similar Around Globe," *Science* 235:435 (1987). The precipitation of the "vapor canopy," if such a thing existed, also would have generated intense heat, but the vapor canopy theory has many problems that have yet to be worked out, and Oard does not rely on it in his model. See Oard, at 26-28.

[15] Wise, Kurt, *Faith, Form, & Time,* (Nashville, TN: Broadman & Holman, 2002), pp. 210-216.

[16] "The oxygen isotope and other data point to the startling conclusion that the deep ocean was much warmer than now during most of the last 100 million years." Oard, at 71, citing Anonymous, (1978) *Geological Perspectives on Climatic Change,* National Academy of Sciences, Washington, DC. The last 100 million years, in conventional geochronology, correlates to Cretaceous through the Pleistocene or, in a creationist model, the later middle of the Flood through the early post-Flood centuries. Some scientists have estimated that water on the ocean bottoms was as warm as 55º F during the early Tertiary and had been even warmer during the late Cretaceous. Frakes, L.A., *Climates throughout Geological Time,* (New York: Elsevier, 1979), p. 190. While this is significantly warmer than present oceans, Oard believes the Floodwater oceans were warmer still.

[17] Vardiman, Larry, "Numerical simulation of precipitation induced by hot mid-ocean ridges," Presented at the Fourth International Conference on Creationism Pittsburgh, PA, August 3-8, 1998.

[18] Cadet, J.-P., and Fujioka, K., (1980) "Neogene volcanic ashes and explosive volcanism: Japan Trench transect, Leg 57, Deep Sea Drilling Project. In von Huene, R., Nasu, N., et al., *Init. Repts.* DSDP, 56, 57 (Pt. 2): Washington (U.S. Govt. Printing Office), 1027-1041; Cadet, J.-P., Pouclet, A., Thisse, Y., Bardintzeff, J.M., and Azéma, J., (1982). "Middle America Neogene explosive volcanism and ash layers: evidence from the Middle America trench transect, Deep Sea Drilling Project Leg 67" In Aubouin, J., von Huene, R., et al., *Init. Repts.* DSDP, 67: Washington (U.S. Govt. Printing Office), 475-491

[19] The Columbia Plateau of Oregon and Washington is one example of such a continental flood basalt. Individual lava flows on the Columbia Plateau ranged from one hundred to five hundred feet thick, and in places the total accumulation is about three miles deep. It was once thought that flood basalts did not create a large volume of ash and aerosols. But flood basalt magmas are rich in dissolved sulphur, and it is now recognized that sulphuric acid aerosols, formed of sulphur volatiles such as sulphur dioxide, could be injected into the stratosphere, which would cause a long-lasting dry fog. Oard, at 69, 70, citing, Strothers, R.B., et al., (1986) "Basaltic Fissure Eruptions, Plume Heights, and Atmospheric Aerosols," *Geophysical Research Letters*, 13, pp. 725-728. "It has long been asserted that fissural basaltic eruptions could not have a global climatic impact. Recent work has shown that this was not the case. Detailed reconstruction of the history of . . . the 1783 Laki eruption in Iceland . . . one of the largest historical fissure eruptions, . . . has shown that it was responsible for some of the worst climatic change in a century in the Northern Hemisphere, leading to massive pollution, an 'awesome' sulphur-rich dry haze . . ." Courtillot, Thordarson, "Flood basalts appear to be the main cause of biological mass extinction in the Phanerozoic" Geophysical Reseach Abstracts, vol. 7 11196 (2005). The Columbia Plateau (CP) is dated to the Miocene. Other continental flood basalts include the Ethiopia Basalts (Oligocene, four times larger than CP), the Deccan Traps of India (Cretaceous/Tertiary, 8 times larger than CP), and seven other flood basalts dated to the Mesozoic, including the Karoo Dolerites of South Africa (Jurassic, more than 8 times larger than CP) and the Siberian Traps (Permian/Triassic, more than 8 times larger than CP).

[20] Interestingly, there are several Flood legends in which it is recorded that the sky was darkened. See the next chapter.

[21] Kennett, J.P., and Thunnell, R.C., (1975) "Global increase in Quaternary explosive volcanism," Science, 187:497-503; Kennett, J.P., and Thunnell, R.C (1977) "On explosive Cenozoic volcanism and climatic implications," *Science*, 196:1231-1234.

[22] Oard, at 69, citing Izett, G.A, "Volcanic Ash Beds; Recorders of Upper Cenozoic Silicic Pyroclastic Volcanism in the Western United States," *Journal of Geophysical Research*, 86 (B11) (1981), pp. 10200-10222.

[23] Oard, at 69, citing Froggatt, P.C., "An Exceptionally Large Late Quaternary Eruption from New Zealand," *Nature*, 319 (1986), pp. 578-582.

[24]Oard, at 34, citing Hughes, P., "The Year Without a Summer," *Weatherwise*, 32(1979), pp. 108-111.

[25] Erickson, Jon, *Ice Ages: Past and Future*, (Blue Ridge Summit, PA: TAB Books, 1990), pp. 119, 120. (The idea that 10 degrees Fahrenheit would be enough cooling to bring on an ice age is accepted because that is all the Milankovitch theory can supply. More realistic computer simulations indicate that summer cooling would need to be closer to 10-12 degrees ***Centigrade*** (about 18 to 22 degrees Fahrenheit). See Oard, supra, at 6, citing Williams, "An Energy Balance Model of Potential Glacierization of Northern Canada," *Arctic and Alpine Research*, v. 11, n. 4 (1979), pp. 443-456.).

[26] Kurt Wise suggests that the term *Ice Advance* be substituted for *Ice Age* in order to divest it of its Lyellian overtones. Kurt Wise, *Faith, Form, & Time*, (Nashville, TN: Broadman & Holman, 2002), pp. 215, 216, 281.

[27] Beneath the sands of the Sahara Desert is evidence of old river systems and forests, suggesting that the climate was once much wetter than today. Kurt Wise, *Faith, Form, & Time*, supra, at p. 215. Geologist Robert M. Schoch points out that the Sphinx shows evidence of much greater erosion from water than it should if Egypt was very arid at the time it was built. See, e.g., John Anthony West, *Serpent in the Sky* (Wheaton, IL: Quest Books 1993) pp. 186-220, 225-232; Graham Hancock, *Heaven's Mirror* (New York: Crown Publishers, 1998) pp. 92-94. But the creationist Ice Age model, and particularly the higher precipitation that lasted for centuries after the Flood, may provide the solution to this riddle; the archeologists may not be as wrong about when the Sphinx was carved as geologists are wrong about how recently Egypt was wet and rainy. See, e.g., Kurt Wise, *Faith, Form, & Time*, supra, at p. 215.

[28] Austin, Steven A., *Grand Canyon: Monument to Catastrophe* (San Diego: Institute for Creation Research, 1994). See, also, Ed Stiles, "Is the Grand Canyon a Geologic Infant" UANews.org (University of Arizona) July 18, 2002 http://uanews.org/cgi-bin/WebObjects/UANews.woa/wa/MainStoryDetails?ArticleID=5820.

[29] One interesting mystery is that several maps from the sixteenth century show the continent of Antarctica, which is odd because the existence of Antarctica was not confirmed until the 1820s. Some of the maps, particularly the 1531 map of Oronteus Finaeus, show the ice-free coast of Antarctica with amazing accuracy, complete with mountain ranges and rivers going well inland. Now, and certainly in the 16th century, Antarctica is completely covered by thick glaciers, extending well out to sea. This and the fact that several 15th and 16th century maps are so accurate that they would have required the use of spherical trigonometry to produce, and the fact that they reflect the ability to accurately fix longitude (which was not achieved until the mid-18th century), led Charles Hapgood to the conclusion that these maps were copied from earlier maps going all the way back to a sophisticated Ice Age civilization that explored and mapped Antarctica. See Charles H. Hapgood, *Maps of the Ancient Sea Kings,* (Kempton, IL: Adventures Unlimited Press, 1966, 1996). Most have dismissed Hapgood as a crank, but his theory fits with the creationist Ice Age model. The warmer post-Flood oceans would have kept the coastal areas of Antarctica free of ice until the latter stages of the Ice Age, when the oceans had cooled enough to allow the formation of icecaps at the poles. The sophisticated post-Flood people (described in chapter 19) could easily have explored the coastal areas of Antarctica when they were still free of ice. See, e.g., Rene Noorbergen, *Secrets of the Lost Races* (Indianapolis, IN: The Bobbs-Merrill Company, 1977 (reprint: Brushton, NY: TEACH Services), pp. 96-99 ("this survey had to have been made shortly after the Flood (when the land masses were left in their present forms), but before the ice began to accumulate at the poles.")

[30] George McCready Price, for example, was hesitant to admit that there had been an Ice Age, despite the fact that, in his book, *Genesis Vindicated,* (Review and Herald, 1941) pp. 259-284, he correctly identified every factor contributing to the Ice Age:

> "But even more profound climatic results must have followed from the fact that the entire ocean water of the world before the flood must have been warm, and warm not only throughout the Arctic and the Antarctic regions, but warm from the surface to the most profound depths." (p. 267). "There is clear evidence that for a considerable period after the flood the climate of all the world was *much more humid* than at present. This would follow deductively, when we consider the great interior basins which were then full of water, and when we take into account the warm, steaming oceans." (p. 269). "John Tyndall long ago pointed out that, in order to have an extension of glaciers, we need not so much a lowering of temperature, as a more effective system of condensation, with even more evaporation than at present." (p. 278). "Accordingly, we have a right to say that the ocean water, by its *sheer quantity,* would not cool off to its present condition for many

> centuries, perhaps not for a thousand years after the flood. And what warm water remained would always float on top." (p. 280). "But widespread glacial conditions would not prevail suddenly. It would take a long time for these cold conditions to reach their maximum; but finally those very conditions which produced the increase of precipitation over evaporation would *begin to reverse themselves*, and the accumulated snow and ice would begin to disappear. . . . Someone has well said that any adequate explanation of the former glacial conditions must also explain why they ever *came to an end*. Our theory tells the reason; for the chief cause of the oncoming of the period of ice was the original warmth of ocean water; and as this warmth gradually disappeared, the period of accumulated ice and snow also disappeared, leaving us with the climate as it now is." (pp. 280-281). "In all this I am not inadvertently omitting the effects of volcanic dust blown high into the air by the many volcanoes which were produced by the conditions attending the flood. There is plenty of evidence to show that volcanic action was prevalent immediately following the flood, the causes of which are clear enough, but must not detain us now. And W. J. Humphreys has shown how volcanic dust, which will continue to float high in the air for months, will appreciably reduce the temperature of the earth's surface. But I consider this volcanic dust as a minor factor, not the main one, for the oncoming of cold, glacial conditions." ( p. 281). "It is not the mere dry cold which tends to produce the glacier. The coldest region on the globe is northeast Siberia; but it has no glaciers." ( p. 281).

The only differences between Price's explanation and the modern creationist Ice Age model are that Price attributed the warmth of the post-Flood oceans totally to pre-Flood conditions rather than events occurring during the Flood, and he did not place as much weight on volcanic cooling. Ironically, despite having explained the causes of the Ice Age, Price was not willing to admit that the glaciers had extended into continent-wide sheets of ice. He believed (probably relying on Sir Henry Howorth) that the geological evidence of glaciation could also be explained by aqueous causes. Earlier in his career he had written that the Ice Age, *"was invented as a sort of mental buffer, an intellectual shock absorber, to make the transition from the geological world to the present less violent and catastrophic,"* and he was unwilling to retreat from that position even as late as 1941.

Frank Lewis Marsh also described the basic Ice Age mechanism of warm oceans and atmospheric cooling cause by volcanism at pp. 121, 122 of his book, *Life, Man, and Time* (Escondido, CA: Outdoor Pictures, 1957, 1967). Whitcomb and Morris, in *The Genesis Flood* (1961) also mention "probable dense accumulation of volcanic dust particles in the atmosphere" as one of the factors leading to a post-flood Ice Age.

(So defunct was the Milankovitch theory in the late 1950s and early 1960s that Whitcomb and Morris do not even mention it). Harold Coffin also suggests volcanism as a possible cause of the Ice Age, at pp. 203-206 of his book, *Origin by Design* (1969, 1983, 2005).

CHAPTER 10

# The Flood in History

> The one great myth of the world is the story of the Deluge. It is the common heritage of all humanity. The variety of cultures that define one people from another are cut across by something far grander than legend, a planetwide experience of such high drama that it resounds in the folk traditions of each continent. The Deluge was the supremely powerful event that touched every ancestry, so comprehensive in its horror that it lives on, a universal nightmare in our collective unconscious, spanning the millennia from that time to this.[1]

A catastrophe as horrific as the one recorded in Genesis would have left a deep and permanent impression on the survivors. They would never have forgotten how the beautiful world they had known was destroyed and replaced by a world with a different geography, climate, flora, and fauna. We can be sure they would have told their descendants about the Flood, which is why all the ancient nations, tribes, and peoples believed that a flood once destroyed the world.

By one count, scholars had collected fifty-nine flood legends from North American native tribes, forty-six from Central and South America, thirty-one from Europe, seventeen from the middle east, twenty-three from Asia, thirty-seven from the Pacific islands, Australia, and New Zealand, and seven from Africa.[2] These flood legends generally agree on three facts: (1) Water destroyed the human race and most other living things; (2) some type of vessel provided a means of escape for (3) one family that was preserved to perpetuate the human race.

Common sense would indicate that the most accurate flood legends would be those of the oldest civilizations nearest the place of disembarkation, and this is generally true. The Babylonian flood legend stays very close to the Genesis Narrative. Berosus, a Babylonian priest of Bel, lived on a Greek island during the third century before Christ and wrote a three-volume work on the history and culture of Babylon. Berosus' account of the flood parallels the biblical narrative so closely that some skeptics have suggested that the author of Genesis borrowed from Berosus:[3]

> After the death of Ardates, his son, Xisuthros, reigned eighteen sari. In his time happened a great Deluge, the history of which is thus described. The Deity, Cronus, appeared to him in a vision, and warned him that upon the fifteenth day of the month Daesius there would be a flood, by which mankind would be destroyed. He therefore enjoined him to write a history of the beginning, procedure, and conclusion of all things; and to bury it in the City of the Sun at Sippara; and to build a vessel, and take with him into it his friends and relations; and to convey on board everything necessary to sustain life, together with all

the different animals, both birds and quadrupeds, and trust himself fearlessly to the deep. Having asked the Deity whither he was to sail? he was answered, "To the Gods": upon which he offered up a prayer for the good of mankind. He then obeyed the divine admonition; and built a vessel five stadia in length, and two in breadth. Into this he put everything which he had prepared; and last of all conveyed into it his wife, his children, and his friends. After the flood had been upon the earth, and was in time abated, Xisuthros sent out birds from the vessel; which, not finding any food nor any place whereupon they might rest their feet, returned to him again. After an interval of some days, he sent them forth a second time; and they now returned with their feet tinged with mud. He made a trial a third time with these birds; but they returned to him no more: from whence he judged that the surface of the earth had appeared above the waters. He therefore made an opening in the vessel, and upon looking out found that it was stranded upon the side of some mountain; upon which he immediately quitted it with his wife, his daughter, and the pilot. Xisuthros then paid his adoration to the earth, and, having constructed an altar, offered sacrifices to the gods, and with those who had come out of the vessel with him, disappeared. They, who remained within, finding that their companions did not return, quitted the vessel with many lamentations, and called continually on the name of Xisuthros. Him they saw no more; but they could distinguish his voice in the air, and could hear him admonish them

> to pay due regard to religion; and likewise informed them that it was upon account of his piety that he was translated to live with the gods; that his wife and daughter, and the pilot, had obtained the same honor. To this he added that they should return to Babylonia; and, it was ordained, search for the writings at Sippara, which they were to make known to mankind: moreover that the place, wherein they then were, was the land of Armenia. The rest having heard these words, offered sacrifices to the gods; and taking a circuit journeyed toward Babylonia.[4] A portion of Xisuthros' ship, which finally went aground in Armenia, is still found in the Gordyæan Mountains in Armenia, and pilgrims bring away asphaltum, which they have scraped from the fragments; they use it against witchcraft. As to the companions of Xisuthros, they arrived in Babylonia, dug up the writings buried at Sippara, founded a number of cities, built temples, and restored Babylon.[5]

Berosus claimed to have based his history on ancient Babylonian chronicles and inscriptions preserved in the temple of Bel in Babylon. Cuneiform inscriptions found in the libraries and temples of Babylonia and Assyria have verified many of his statements. Berosus does not treat the Deluge narrative as part of Babylon's mythology but rather as part of its history.

Another remarkable Babylonian flood legend turned up during the excavation of the great mound of Kuyûnjik, across the Tigris from the Iraqi city of Môsul. The mound contained the ruins of Nineveh, the capital of ancient Assyria, known to Bible readers as Jonah's mission field. Archeologists found about 25,000 fragments of clay tablets, many of which were in the ruins of the palace of

Ashurbanipal, who ruled Assyria from 669 to 633 B.C. It seems that Ashurbanipal was an avid collector of ancient tradition, including an immense assortment of Sumerian literature.

In 1872, George Smith, a scholar at the British Museum, was given the task of copying these clay tablets to make them available to foreign scholars. As he was copying one of the fragments, Smith came across a sentence that read, "The mountain of Nisir stopped the ship. I sent forth a dove, and it left. The dove went and turned, and a resting place it did not find, and it returned." Upon translating this fragment, Smith reportedly "jumped up and rushed about the room in a great state of excitement, and, to the astonishment of those present, began to undress himself."

When Smith calmed down, he began looking for other fragments of this story. When he had finished translating the rest of the fragments, he had parts of three copies of the *Epic of Gilgamesh*. The London *Daily Telegraph* paid for Smith to return to Nineveh to search for missing fragments of the tablets. In 1873, almost miraculously, Smith recovered a fragment that contained "the greater portion of seventeen lines of inscription belonging to the first column of the Chaldean account of the Deluge, and fitting into the only place where there was a serious blank in the story."

In the *Epic*, Utnapishtim is told to tear down his house and build a ship to save his life. He is told to cause the seed of all living creatures to go into the ship. The ship was shaped like a cube, 120 cubits on each side. It had six decks and was seven stories tall. Utnapishtim waterproofed the ship with thousands of gallons of pitch and asphalt. After loading his family, possessions, and "the seed of all living things," he shut the door. The gods then caused a "destructive rain" to fall. Black clouds and thunder appeared; all light turned to darkness, and the land "broke like a pot." The violence was so terrific that even the gods were terror-stricken and ascended to the heaven of Anu, the sky god. After seven days and nights of storm, the ship came to rest on Mount Nisir. Utnapishtim sent out a dove, which came back because it had no place to land; then he sent a swallow, which also came back. Finally, he sent out a raven, which did not return. After disembarking, Utnapishtim offered sacrifices to the gods.

The discovery of the *Epic of Gilgamesh* created a sensation among scholars. Some had argued that Berosus borrowed his flood account from the Hebrews during their Babylonian captivity, but the reign of Ashurbanipal, in whose palace the *Epic* was found, predated the Babylonian captivity. Moreover, the *Epic* was very ancient even in Ashurbanipal's time. The tablets state that they were copied from an older tablet that was read with difficulty in the seventh century B.C. Archaeologists have found fragments of the *Epic* that were written around 2,000 B.C. during the reign of the fourth successor to Hammurabi. Thus, this flood legend long predates Moses' writing of Genesis. The Babylonian flood traditions are clearly independent of the Old Testament narrative; they are different histories of the same event.[6]

Flavius Josephus, the Jewish historian who became a Roman, treated the Flood as history. Regarding a location in Armenia, he stated, "the Armenians call this place, *Nachidsheuan*, the Place of Descent; for the ark being saved in that place, its remains are shown there by the inhabitants to this day."[7] Josephus also noted the testimony of his fellow historians:

> Now all the writers of barbarian histories make mention of this flood, and of this ark; among whom is Berossus the Chaldean. For when he was describing the circumstances of the flood, he goes on thus: "It is said, there is still some part of this ship in Armenia, at the mountain of the Cordyaeans; and that some people carry off pieces of the bitumen, which they take away,

and use chiefly as amulets, for the averting of mischief." Hieronymus the Egyptian also, who wrote the Phoenician antiquities, and Manases, and a great many more, make mention of the same. Nay, Nicholas of Damascus, in his ninety-sixth book, hath a particular relation about them, where he speaks thus: "There is a great mountain in Armenia, over Minyas, called Baris, upon which it is reported, that many who fled at the time of the deluge were saved; and that one who was carried in an ark, came on shore upon the top of it; and that the remains of the timber were a great while preserved. This might be the man about whom Moses the legislator of the Jews wrote."[8]

**Figure 10.1: The Temple at Edfu**

On the west bank of the Nile in Upper Egypt, sixty-five miles south of Luxor, stands the ancient temple of Edfu. Built during the Ptolemaic dynasty on the site of a more ancient temple, Edfu is the best-preserved Egyptian temple. Hieroglyphics on the walls of the temple tell of a forgotten age long before the first Pharaoh of the First Dynasty sat on the throne of Egypt. In this ancient time, the "builder gods" settled in Egypt and established sacred mounds at various points along the Nile. These mounds were to serve as the foundations of temples that would bring about "the resurrection of the former world of the gods"—a world that had been utterly destroyed. The lost domain, the "homeland of the Primeval Ones," was an island upon which the creation of the world began and upon which "the earliest mansions of the gods were founded." At a certain point during the primeval age, the blessed "former world" was overwhelmed, suddenly and totally, by a great flood, the majority of its "divine inhabitants" were drowned, and the "mansions of the gods were inundated."[9]

The antediluvians and the early post-Flood generations, for reasons we will discuss later, would have seemed like gods to later races. Quite possibly, many of the post-Flood "builder gods" would have wanted to restore the world to its antediluvian beauty and may have devised a technology with the hope of accomplishing that.[10] Thus the Edfu hieroglyphs contain a flood narrative remarkably consistent with biblical history. It is also noteworthy that some of Egypt's most ancient writings, the pyramid texts, state that the god before all gods, the All-Father, was named "Atum."[11]

Figure 10.2: Hieroglyphics at Edfu

The Miao of southern China spoke of the "golden age" that ended when a woman picked some forbidden strawberries. A brother and sister had escaped the floodwaters in a wooden drum, and all post-flood humanity was descended from them. Then there came a time when mankind grew numerous and tried to reach heaven with a ladder. The "Lord of Heaven" struck these dead with lightning. Before this time all people had spoken one language. Now they were given many languages and, not being able to understand one another, separated. The Miao went to a land where the days and nights were six months long. They eventually migrated into Hunan province in

China and were in possession of that land when the Hia or "Chinese" arrived.[12]

China's pictographic writing appears to verify the historicity of the Genesis Flood. Several early civilizations developed pictographic writing. The ancient Egyptians created hieroglyphics, and the Sumerians invented cuneiform. But alphabetic writing, because of its greater versatility and economy, came to predominate and is currently used everywhere except eastern Asia, where pictographic writing persists. Changes and standardizations made since the mid-twentieth century, moreover, have obscured the pictorial roots of the language.

Scholars can demonstrate how the symbols evolved into their present forms. Simple pictographs depict one person or object, such as the sun, a man, a horse, etc. In order to depict a complex concept or idea, it is necessary to combine two or more pictographs into a symbol called an "ideograph." Ideographs combine the simpler concepts that, together, best convey the more complex idea.

When the first known etymological studies were done, around 86 B.C., ancestor worship, Confucianism, Taoism, and Buddhism had become dominant in Chinese religious and philosophical life. Thus, early studies of the origins of the symbols were influenced by the religious views current at that time. But the Chinese were originally monotheistic, and one of the emperor's duties was to offer an annual sacrifice called the "border sacrifice," to "Shang ti," the "emperor above."

In their book, *The Discovery of Genesis*, authors C. H. Kang and Ethel R. Nelson argue that the ideographs of the Chinese language contain parallels to the Genesis narrative from creation to the Babel dispersion. One of the most remarkable ideographs is the symbol for "ship" or boat. It is a combination of the symbol for "vessel," the number eight, and the symbol for "mouth," suggesting Noah's ark with eight human passengers.[13]

A few years later, Ethel Nelson and Richard Broadberry

published another book, *Genesis and the Mystery Confucius Couldn't Solve.*[14] In this volume, the authors explore more of the symbols, generally the oldest and most primitive forms of the language, and show how many of these most ancient pictographs and ideographs reflect the true story of earth's history, as later recorded under inspiration by Moses.

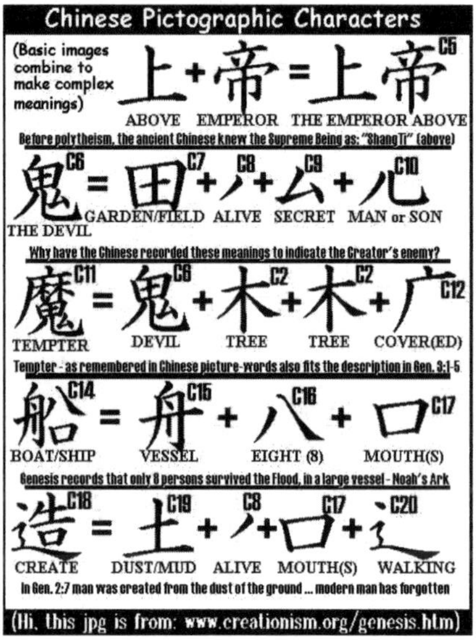


**Figure 10.3: Chinese pictographic writing**

The Greek and Roman flood legends were recorded by the Roman poet Ovid, among others, and are included in Ovid's book, *The Metamorphoses.* Completed in 8 A.D., *The Metamorphoses* is a collection of Greek and Roman myths. In the Roman flood legend, the god who brought about the flood was the Roman god Jove, or Jupiter.[15] The reason for the destruction was the extreme sinfulness of the antediluvian people. Because of its remarkable parallels to scripture, *The Metamorphoses* is quoted at length:

> "Then came the age of iron
> And from it poured the very blood of evil:
> Piety, Faith, Love and Truth changed to Deceit,
> Violence, the Tricks of Trade, Usury, Profit;
>
> * * *
>
> Then land, once like the gift of sunlit air,
> Was cut in properties, estates, and holdings:

Not only crops were hoarded; men invaded
Entrails of earth down deeper than the river
Where death's shades weave in darkness underground;
Where hidden from the sight of men Jove's treasures
Were locked in night. There, in his sacred mines,
All that drives men to avarice and murder
Shone in the dark: the loot was dragged to light
And War, inspired by curse of iron and gold,
Lifted blood-clotted hands and marched the earth.
Men fed on loot and lust; the guest feared host;
Neighbor looked warily with smiles at neighbor;
And fathers had good reasons to distrust
Their eager sons-in-law. If brothers loved
Each other, the sight was rare, and watchful
Husbands prayed for death of wives; stepmothers
Made poison a dessert at dinner—sons
Counted the hours that led to fathers' graves.
Piety was overthrown . . ."[16]

* * *

"Soon it was rumored that earth's taste for blood
Was threatening heaven: giants piled hill on
Mountain to make a stair that reached to the skies,
To clamber to the throne of Jove . . ."

This last passage seems to refer to the Tower of Babel.

Clearly, the *Metamorphoses* parallels Scripture's statement that "the Lord saw how great man's wickedness on the earth had become, and that every inclination of the thoughts of his heart was only evil all the time." There is an even more striking similarity to Ellen White's statements that the antediluvian's lust for gold led to violence, and that the antediluvians sought to contend with heaven:

> The great gifts with which God had endowed man were perverted. There was gold and silver in abundance, and men were constantly seeking to exceed their fellow men in devices. The result was that violence was upon the earth. The Lord was forgotten. This long-lived race were constantly devising how they might contend with the universe of heaven and gain possession of Eden.[17]

According to the *Metamorphoses*, Jove convened a council of lesser gods to warn them of the danger posed by man's impious condition, one of which seems to have been dinosaurs-like monsters:

> "We live in danger greater than the hour
> When lizard-footed giants climbed the hills
> And with a hundred hands clawed at the sky.

Jove concluded that man must be wiped out:

> "All men have joined in Hell's conspiracy
> Since I have said it: all shall pay the toll
> Of early death – and earth an early fall."

Some of the gods at Jove's council disliked the idea of wiping out mankind, because then there would be no one to bring myrrh and sweet herbs to their altars. Jove tells them not to worry:

> "He planned a breed of men of heaven's make,
> Different in spirit, better than the first."

Could there be a better description of the redemptive work that Christ has accomplished for and in us? "Therefore, if anyone is in Christ, he is a new creation; the old has gone, the new has come!" 2 Cor. 5:17.

> "Then Jove raised thunderbolt against the earth—
> And checked the blow. Would heaven break in fire,
> And flames pour over earth from pole to pole?
> He then remembered that the Fates had scored
> A certain distant hour when sea and land,
> Earth and the vault of heaven would be consumed
> In universal fire. He put aside
> The lightning spear Cyclopean hands
> Made as his weapon to assert his will:
> Another doom for man came to his mind
> A death that stormed beneath the waves, and fell
> From air; and then dark rain began to fall."

This is an astounding analog to the scriptural truth that earth's destruction by fire awaits a future time. 2 Peter 3:10, 12 ("The heavens will disappear with a roar the elements will be destroyed by fire, and the earth and everything in it will be laid bare. . . . That day will bring about the destruction of the heavens by fire, and the elements will melt in the heat.").

> "Nor was Jove's rage appeased by pouring heavens.
> Neptune arrived with armies of the waters,
> Rivers assembled at his Ocean's floor
> To hear his orders: 'The hour is all short
> For long orations, open your locks and dykes,
> Your streaming walls, and springs, unleash the horses

> Riding in foam through waterfalls and waves.'
> At his command the mouths of fountains opened
> Racing their mountain waters to the sea.
> Under the blow of Neptune's fork earth trembled,
> And way was open for a sea of waters."

Only one couple, Deucalian—whom Ovid describes as "the best of men"—and his wife Pyrrha, are saved in a boat.

Flood legends are not limited to the ancient civilizations. They are widely dispersed around the world, among all the tribes and peoples. The Hawaiians have a legend in which they say that long after the time of the first man (Kumuhonua), the earth became wicked and failed to worship the gods. Only one man was righteous, Nu-u. He made a large canoe with a house on it and brought food, plants, and animals into it. Then waters overwhelmed the earth and destroyed all mankind except Nu-u and his family. When he landed after the waters had subsided, he looked up and saw the moon, and thought that it was Kane (the great god), and he worshiped the moon. This angered Kane, and he came down on a rainbow and reproved Nu-u but did not punish him, because Nu-u had worshiped the moon by mistake. When Kane returned to the sky, he left the rainbow behind as a token of forgiveness.[18]

The natives of Alaska had a legend in which the father of the tribe, who lived in a land toward the rising sun, was warned in a dream that a deluge would desolate the earth. He built a raft, on which he saved himself, his family, and all the animals. They floated for several months. The animals, who at that time could talk, complained and murmured against him, and when the land appeared again and the animals disembarked, they lost the gift of speech as a punishment for complaining.[19]

Divine warning of the coming catastrophe is also a feature of a legend of the Yakima of the Pacific Northwest. In their legend, one of the good men told the others, "I have heard from the Land Above, the land of the spirits, that a big water is coming—a big water that will cover the land. Make a boat for the good people. Let the bad people be killed by the water."[20]

Some tribes believe that survivors of the flood took refuge on very high mountains. For example, the Mattoles, a tribe of Northern California, say that their forefathers took refuge from the flood on Taylor Peak, a mountain in their area.[21] Another very common theme is that the world was being judged for its wickedness. Among the Southwestern Native Americans, the most common reason for the destruction of past worlds and peoples is that they had been bad.[22]

The natives of the Southeastern United States had numerous deluge stories. In the majority, a man and his wife become aware of the approaching flood due to some form of supernatural warning. Usually, the man and his wife are told to build a raft, make a great earthen jar, or climb inside a huge hollow reed. As the waters subside, various birds are sent to find land (usually the woodpecker or dove) and the couple undertake the task of repopulating the earth, frequently with supernatural assistance.[23]

When missionaries first came to the Choctaw Indians in Mississippi in 1818, the Choctaw related their flood tradition, which incorporates divine judgment for bad behavior, as well as the building of a raft to escape destruction:

> In ancient times after men had lived a long period upon the earth they became very corrupt and wicked, and deluged the earth with so much blood and carnage that the Great Spirit finally decided to

utterly destroy them. He, therefore, sent a prophet among them, who went from tribe to tribe and from village to village proclaiming the fearful tidings that the race was soon to be destroyed. No one paid any attention to him, however, and people went on in their wickedness as carelessly as ever. But one year, with the coming of autumn, mists and clouds gathered over the earth, so that there was no sun shining by day nor did the moon and stars light up the gloom of night. The situation grew gradually worse until the earth became shrouded in total darkness. The air was chill and all animate nature became silent. People too became silent and perplexed, but yet they gave no evidence of repenting for their evil deeds. They were compelled to find their way from place to place by light of torches.

* * *

One day very suddenly there came a crash of thunder much louder than had ever been heard before. The whole earth seemed to shake and tremble with the reverberation. Then, as people looked towards the north, they seemed to see a light—the first they had seen for many a long dark day. But whatever hope may have been aroused in their breasts was dissipated. For what they saw was not the return of the long lost sun, but it was the gleam of a great mountain of water, advancing in great billows from the north, covering the entire earth and destroying everything in its path. With the cry, "Oka Falamah, Oka Falamah" (the returning waters, the returning waters) the

> doomed people turned away in one last vain effort to escape. But there was no escape. The whole earth was soon covered even to the tops of the mountains by the vast flood, and men and animals alike perished, leaving only a desolate wilderness of waters.
>
> Of all mankind, only one remained, and that was the mysterious prophet who had so faithfully yet vainly proclaimed the warnings of the Great Spirit. This prophet had been directed by the Great Spirit to build a raft of sassafras logs, upon which he floated safely above the destroying flood, while he gazed sadly upon the dead bodies of men and beasts as they floated past him in the dark waters.[24]

Some of the American flood legends contain an element analogous to the Tower of Babel. The Native American tribe that lives around Lake Tahoe ascribes the origin of the lake to a great natural convulsion. There was a time when their tribe possessed the whole earth and were strong, numerous, and rich; but a day came in which a people rose up stronger than they and defeated and enslaved them. Afterward, the Great Spirit sent an immense wave across the continent from the sea, and this wave engulfed both the oppressors and the oppressed, all but a very small remnant. Then the taskmasters made the remaining people raise up a great temple, so that they, of the ruling caste, should have a refuge in case of another flood. And on the top of this temple the masters worshiped a column of perpetual fire. Half a month later, however, the Great Spirit caused convulsions, causing the temple to sink into the waters of Lake Tahoe, leaving only the top-most level above water. The Great Spirit took the oppressors out of the temple and tossed them like pebbles into the recesses of a great cavern on the eastern side of the lake. The cavern is still known as Spirit Lodge.[25]

The Apaches also had a flood legend featuring a Tower of Babel:

> The Apaches . . . have a legend that "the first days of the world were happy peaceful days"; then came a great flood, from which Montezuma and the coyote alone escaped. Montezuma became then very wicked, and attempted to build a house that would reach to heaven, but the Great Spirit destroyed it with thunderbolts.[26]

The Pagagos, of the same geographic region as the Apaches, also had a legend involving Montezuma and the coyote. In this legend, the Great Spirit made the earth and all living things, before he made man. A flood that only Montezuma and the coyote escaped destroyed the creation. Even after the flood, Montezuma became puffed up with pride, angering the Great Spirit.

"But Montezuma hardened his heart, and collecting all the tribes to aid him, set about building a house that should reach up to heaven itself. Already it had attained a great height, and contained many apartments lined with gold, silver, and precious stones, the whole threatening soon to make good the boast of its architect, when the Great Spirit launched his thunder, and laid its glory in ruins."[27]

The natives of what is now Nicaragua believed that a flood destroyed the world, wiping out most of mankind.[28] But perhaps the most striking American flood legend, in its similarity to the biblical narrative, is that of the Toltecs of Mexico. Here is the legend as told by historian Don Ferdinand d'Alva Ixtilxochitl (1568-1648):

> It is found in the histories of the Toltecs that this age and the first world, as they call it, lasted 1716 years; that men were destroyed by tremendous rains and lightning from the sky, and even all the land, without the exception of anything, and the highest mountains, were covered up and submerged in water fifteen cubits (caxtolmolatli); and here they added other fables of how men came to multiply from the few who escaped from this destruction in a "toptlipetlocali"; that this word nearly signifies a close chest; and how, after men had multiplied, they erected a very high "zacuali," which is today a tower of great height, in order to take refuge in it should the second world (age) be destroyed. Presently their languages were confused, and, not being able to understand each other, they went to different parts of the earth.
>
> The Toltecs, consisting of seven friends, with their wives, who understood the same language, came to these parts, having first passed great land and seas, having lived in caves, and having endured great hardships in order to reach this land; . . . they wandered 104 years through different parts of the world before they reached Hue Hue Tlapalan, which was in Ce Tecpat, 520 years after the flood.[29]

The duration of the first age, 1716 years, differs by only 60 years from Ussher's calculation of 1656 years for the pre-Flood period. The 15-cubit height of the water above the mountaintops, the tower, the confounding of the languages, and the dispersal of the people are all identical to the Genesis narrative. The uncanny similarity between the Toltec flood legend and the Genesis narrative raises the suspicion that the legend was influenced by the teachings of Spanish priests after Cortes conquered Mexico in 1519. But on closer scrutiny, the legend appears to be genuine.[30]

This is just a sampling of the important flood legends. A quick library or Internet search would turn up many more. The Flood is an almost universal element in the mythology of the tribes and nations. If there never was a global flood, why do so many tribes, peoples and civilizations have a flood legend?

Skeptics argue that one people borrowed from another. While this may be a plausible explanation for any one region, it does not explain the worldwide extent of the flood legends. Another skeptical explanation is that all peoples, being more alike than they are different, invent the same myths and legends. But flooding is not an experience common to all of humanity. Even granting that worldwide destruction might be a common theme of myths and legends, we would expect that such myths would attribute the destruction to many different catastrophes, such as fire, earthquake, volcanoes, famine, drought, disease, pestilence, continuous winter, etc., depending upon local conditions. This is not the case. Of legends that refer to a worldwide calamity, 122 concern a deluge, and only 31 attribute the calamity to various causes other than a flood.[31]

A third skeptical argument is that the native peoples picked up the flood legend from Christian missionaries. But of the scores of flood legends scattered around the world, the overwhelming majority are unquestionably pre-Christian. The compilers of these flood legends were not Christian apologists but secular historians, anthropologists, and ethnographers. The only logical explanation is that the flood legends must have arisen from a real and horrifying event—an event so cataclysmic that it powerfully impressed the founders of each culture and entered the collective memory of each tribe, never to be erased. Thus, these legends constitute both a compelling proof of the accuracy of biblical history and an independent confirmation that a great deluge once destroyed the world.

---

[1] Joseph, Frank, *The Destruction of Atlantis* (Rochester, VT: Bear & Co., 2002), p. 105.

[2] Mitchell, Colin, *The Case for Creationism* (Alma Park, UK: Autumn House Ltd., 1994), p. 160, citing Stickling, J.A., "A Statistical Analysis of Flood Legends," *Creation Research Society Quarterly*, 9(3), 152-155, and Filby, F.A. *The Flood Reconsidered*, (London: Pickering & Inglis, 1970). See also Johannes Riem, *Die Sintflut in Sage und Wissenschaft* (Hamburg: Agentur des Rahen Hauses, 1925); Richard Andree, *Die Flutensagen, ethnologisch betrachtet* (Brunswick, 1891); Byron C. Nelson, *The Deluge Story in Stone* (Minneapolis, Minn.: Bethany Fellowship, 1931, 1968); Sir James George Frazer, *Folklore of the Old Testament* (London: MacMillan, 1918). Even the TalkOrigins website, which defends Darwinism, has a page listing flood legends. Mark Isaak has compiled nine from Europe, nine from the Near East (including the Bible), fifteen from Africa, thirty-seven from the Far East, twenty-seven from Australia and the Pacific, seventy-eight from North and Central America, and twenty-two from South America. See *http://www.talkorigins.org/faqs/flood-myths.html.* While biblical creationists see flood legends as evidence of the historicity of the biblical narrative, others see them as proof that the biblical narrative is not special but just another of the hundreds of (impliedly untrue) flood myths.

[3] See, e.g., de Camp, L. Sprague, *Lost Continents: The Atlantis Theme in History, Science and Literature*, (New York: Dover Publications, 1954), p. 236. De Camp and other like-minded skeptics are wrong. Most Christians and Jews believe that Moses wrote the book of Genesis sometime around 1,500 B.C. Moses was educated in Egypt, not Babylon, and the Egyptian flood legend is not similar to the one related by Berosus. Even if Moses is not credited as the author of Genesis, it is known that the Old Testament was written in paleo-Hebrew around 400 B.C., so it could not have been influenced by anything Berosus wrote around 280 B.C. Moreover, it seems unlikely that the Hebrew scribes would willingly adopt elements of the Babylonian culture and religion when Babylon was a deadly enemy of the Hebrew nation that had conquered Judah and, in 586 B.C., destroyed the most sacred Jewish building, Solomon's Temple. Are we really to believe that the keepers of the Hebrew religion adopted the views of a Chaldean priest of Bel?

[4] Donnelly, Ignatius, *Atlantis: The Antediluvian World*, 1882, Modern Revised Edition, edited by Egerton Sykes (New York: Gramercy Pub. Co., 1949), p. 59. Sykes substituted the translation of I. P. Cory in *Ancient Fragments* (London: 1832) for the translation Donnelly originally used. Unfortunately, no copy of Berosus' entire work has been found; we are left only with fragments other Greek historians, Alexander Polyhistor and Abdynus, quoted in their own works.

[5] *Catholic Encyclopedia.*

[6] Rehwinkel, Alfred M., *The Flood,* (St. Louis, MO: Concordia Publishing House, 1951), pp. 153-164; Byron C. Nelson, *The Deluge Story in Stone,* (Minneapolis, MN: Bethany House, 1931, 1968), pp. 170-173.

[7] Flavius Josephus, *Antiquities of the Jews,* book I, chapter III, paragraph 5, in *Whiston's Josephus,* the Excelsior Edition (London: William P. Nimmo) p. 29.

[8] Josephus, par. 6, p. 29.

[9] Hancock, Graham, *Heaven's Mirror: Quest for the Lost Civilization,* (New York: Crown Publishers, 1998), pp. 63-66, citing E.A.E. Reymond, *The Mythical Origin of the Egyptian Temple,* (Manchester, UK: Manchester University Press, 1969).

[10] This Egyptian story that the "builder gods" mapped out a series of sites in the hope of restoring lost antediluvian glory supports Rene Noorbergen's argument that the early post-Flood peoples mapped out a global ley-line system with idea of harnessing and using mysterious earth forces, perhaps electro-magnetic currents. (See, Rene Noorbergen, *Secrets of the Lost Races* (Indianapolis, IN: The Bobbs-Merrill Company, 1977 [reprint: Brushton, NY: TEACH Services]), pp. 84-119. See, also, John Michell, *The New View Over Atlantis* (London: Thames & Hudson, 1969, 1972, 1983).

[11] Hancock, at 104-106, citing *The Ancient Egyptian Pyramid Texts,* 225, and R.T. Rundle Clark, *Myth and Symbol in Ancient Egypt,* (London: Thames and Hudson, 1991). (The pyramid texts open on the eve of the "First Time"—the golden age before history began, before anger came into being, before strife came into being, before tumult came into being. [*This sounds like the world before the fall.*] "Before the first tick of the First Time, we are told that Atum, the god before all gods, the 'Complete One', the All-Father, lay motionless in the waters of the primordial void—'the infinity, the nothingness, the nowhere and the dark' which the ancient Egyptians called the Nun. Though inert, the god was filled with magic, energized by the power to 'author his own forms,' able to realize endless possibilities of creation and transformation." This would seem to be a corrupted version of the story of the creation of Adam—stripped of references to the Creator God—in which "Atum" essentially creates himself.)

[12] Bernatzek, Hugo, *Akha and Miao* (1970), p. 302, citing F.M.I. Savina, *Histoire de Miao* (Société des Missions étrangères de Paris. Hong Kong, (1930), p. 245.

[13] Kang, C.H., and Ethel R. Nelson, *The Discovery of Genesis: How the Truths of Genesis Were Found Hidden in the Chinese Language* (St. Louis, MO: Concordia Publishing House, 1979), p. 95.

[14] Nelson, Ethel R., and Richard E. Broadberry, *Genesis and the Mystery Confucius Couldn't Solve* (St. Louis, MO: Concordia Publishing House, 1994).

[15] The name *Jupiter* may derive from Noah's son Japheth, who was the father of all the Indo-European peoples. The Greeks worshiped him as *Iapetos* or *Iapetus.* The name was further corrupted in Latin, where he entered the Roman pantheon as *Iupator,* which eventually became "Jupiter." See Cooper, Bill, *After the Flood: The Early Post-Flood History of Europe Traced Back to Noah,* (Chichester, UK: New Wine Press, 1995), p. 199.

[16] Ovid, *The Metamorphoses,* Translated by Horace Gregory, (New York: The Viking Press, 1958), pp. 33-43.

[17] White, Ellen G., Manuscript No. 513, Letter 65, 1898, p. 3. (To Brethren Griggs and Howe, typed August 23, 1898.) {7MR 367.1}

[18] Nelson, Byron C., *The Deluge Story in Stone,* (Minneapolis, MN: Bethany House, 1931, 1968), pp. 190; Rehwinkel, at 137, 138.

[19] Rehwinkel, at 137.

[20] Knight, Christopher, and Robert Lomas, *Uriel's Machine,* (Gloucester, MA: Fair Winds Press, 1999), p. 121, citing E.E. Clark, *Indian Legends of the Pacific Northwest,* (Berkeley, CA: University of California Press, 1963).

[21] Bancroft, Hubert Howe, *Native Races of the Pacific States,* (San Francisco: The History Company, Pub., 1886), vol. III. "Myths and Languages," p. 86.

[22] Natalie Tobert, Fiona Pitt, "The Southwest," *Native American Myths and Legends,* Ed. Colin F. Taylor, (New York: Smithmark Pub., 1994), p. 29.

[23] Carden, Gary, "The Southeast," *Native American Myths and Legends,* Ed. Colin F. Taylor, (New York: Smithmark Pub., 1994), p. 15.

[24] Morrison, W. B., *Ancient Choctaw Legend of the Great Flood.* See also John R. Swanton, *Source Material for the Social and Ceremonial Life of the Choctaw Indians;* Len Green, *The Legend of Oklatabishi and the Oka Falamah.*

[25] Bancroft, Hubert Howe, *Native Races of the Pacific States,* (San Francisco: The History Company, Pub., 1886), vol. III. "Myths and Languages," p. 89.

[26] Bancroft, Hubert Howe, *Native Races of the Pacific States,* (New York, 1874), vol. III, p. 149, as cited in Donnelly, supra, at p. 96.

[27] Bancroft, Hubert Howe, *Native Races of the Pacific States,* (San Francisco: The History Company, Pub., 1886), vol. III. "Myths and Languages," pp. 75-77.

28 Bancroft, Hubert Howe, *Native Races of the Pacific States,* (San Francisco: The History Company, Pub., 1886), vol. III. "Myths and Languages." p. 75.

29 Donnelly, at 87, citing Edward King, Viscount Kingsborough's *Antiquities of Mexico* (London: 1830-1848), vol. 9, pp. 331, 332, in turn citing *Ixtilxochitl's Historia Chichimeca.*

30 "It will of course be said that this account, in those particulars where it agrees with the Bible, was derived from the teachings of the Spanish priests" notes Donnelly, "but it must be remembered that Ixtilxochitl was an Indian, a native of Tezcuco, a son of the queen, and that his 'Relaciones' were drawn from archives of his family and the ancient writing of his nation; he had no motive to falsify documents that were probably in the hands of hundreds at that time." Donnelly, p. 88. The text itself indicates that Ixtilxochitl was skeptical of the legend, as when he states, "and here they added other fables . . ." But even if Ixtilxochitl was not trying to propagate Biblical teachings (and he clearly was not), there are still two ways that the legend might reflect them: 1) Christian or Hebrew teachings had reached the New World prior to the Spanish conquest. There is evidence, compiled by researchers known as "Diffusionists," of pre-Columbian exploration of the Western hemisphere by Old World explorers; and 2) the teachings of Spanish priests in the sixteenth century had already become transmogrified into native myths by the time Ixtilxochitl wrote his history in the early seventeenth century. That the legend reflects Christian teachings cannot be ruled out, but it appears probable that it is genuinely indigenous.

The Toltecs taught that one God, omnipotent and invisible, had made all things, a religion very different from that of the cruel Aztecs. (Bancroft, at 55, 56.) This raises the question: Who were the "Toltecs?" The encyclopedias say they were a warrior nation that controlled central Mexico between A.D. 900 and A.D. 1,200 and were similar to and conquered by, the Aztecs (or Mexica). But until recently, the Toltecs were considered a legendary race of superhuman beings. "Toltec" was more of a claim of cultural affiliation (someone who comes from Tollan "place of reeds" i.e., large city) rather than an actual nation of peoples. (*Code of Kings*, p. 200). The word *Toltec* or "Nahuatl" means "master builders" or "artificers." They were an ancient race of warrior kings from whom the Aztecs claimed descent. "Much of what the Aztecs aspired to can find its roots in the still mysterious epoch of the Toltecs. The Toltec influence over almost all subsequent cultures of Mesoamerica cannot be overestimated. The Mixtecs, the Itza Maya, the Huaxteca, and most importantly the Mexica (Aztecs) claimed descent from the Toltecs, and Toltec blood was considered to be the legitimizing stamp of any ruler. Yet very little is known about these people. There are no contemporary written accounts."

I believe the term *Toltec* refers not—or at least *not only*—to the nation that immediately preceded the Aztecs in Mexico but also to a much more ancient

Central American people: the early post-Flood settlers. This would explain why they left no written records: they had not yet developed a written language. It would explain the eerily accurate flood legend. It would explain why they were called "master builders": they came from the same stock that built the cyclopean and megalithic structures around the world.

It is intriguing that the Toltecs claim to have arrived in Mexico 520 years after the flood. If they wandered for 104 years after Babel before finding Mexico, that places Babel about 416 years after the flood. Many commentators believe the Babel dispersion was around the time of the birth of Peleg, because Genesis 10:25 states that one of Eber's sons "was named Peleg, because in his time the earth was divided." According to Genesis 11 (in the Masoretic text), Shem fathered Arphaxad two years after the flood, who fathered Shelah thirty-five years later, who father Eber thirty years later, who fathered Peleg thirty-four years later. This would place the birth of Peleg only 101 years after the Flood. But if Septuagint ages are substituted for Masoretic ages, the data is 2+135+130+134=401, astonishingly close to the 416 years indicated by the Toltec legend. But it seems unlikely that the Toltec legend would track the Masoretic numbers for the pre-Flood chronogenealogy of Genesis 5—with an age of the Antediluvian world only sixty years different from Ussher—then follow Septuagint numbers in Genesis 11. Also, the Masoretic omits Cainan between Arphaxad and Shelah, whereas some Septuagint manuscripts include Cainan, adding an extra 130 years to the calculation, placing Babel 531 years after the flood, not so uncannily close to the Toltec legend. But then, this argues that the Toltec flood legend is genuinely pre-Columbian and pre-Christian. If the legend merely reflected priestly teaching, the numerical data should reflect only one version of Scripture, not a combination of versions.

[31] Thompson, Stith, *Motif Index of Folk Literature,* (Bloomington, IN: Indiana University Press, 1955), vol. 1, as cited in Ariel A. Roth, supra, p. 307. See also Ariel Roth, "Flood Stories—Can they be Ignored?" *Origins* 17(2):51-55 (1990), online at: *http://www.grisda.org/origins/17051.htm.*

CHAPTER 11

# Where Did the Dinosaurs Come From? The Darwinian View

Because Darwinism is part of the fabric of all historical sciences, most scientists believe that dinosaurs evolved from other creatures, specifically, from the archosaurs, a group of reptiles similar to crocodiles.[1] But if dinosaurs evolved from archosaurs, there should be some evidence of it in the fossil record. There should be a clearly intermediate form—a "missing link" or a "transitional form"—between archosaurs and dinosaurs. In fact, there should be an entire *series* of fossil animals that are intermediate between the sprawling, lizard-like archosaurs and the upright, fully bipedal dinosaurs.

Instead of "transitional forms," a phrase that incorporates a Darwinist interpretation, Kurt Wise suggests the term *stratomorphic intermediate*.[2] This term describes a fossil that is intermediate in two ways. First, it is morphologically intermediate, meaning that the body shape is roughly between the shape of the supposed ancestor and that of the supposed descendant species. Second, the fossil is stratigraphically intermediate, meaning that it appears in the rocks between the putative ancestor species and the putative descendant species. We would expect to find it higher in the geologic column

**Figure 11.1: Subdivisions of Triassic Strata**

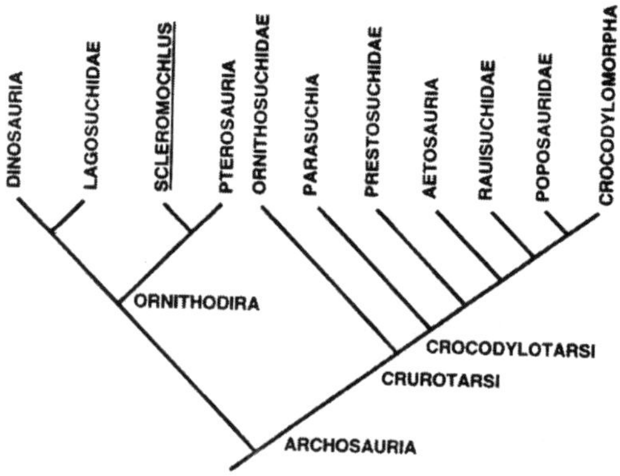


**Figure 11.2: Evolutionary Cladogram of the Archosaurs**[7]

than the lowest representative of its ancestor group and lower than the lowest representative of the group into which it evolved.

The archosaurs appear in the highest Permian strata and continue throughout the Triassic. The lowest dinosaurs in the fossil record, such as *Azendohsaurus, Staurikosaurus, Herrerasaurus, Eoraptor, and Pisanosaurus,* all appear in the Carnian, the lowest subdivision of the Upper Triassic.[3] (See **Figure 11.1**.) Evolutionary theory dictates that we should find a fossil animal with a shape that is between archosaurs and dinosaurs, and we should expect to find it in the lower or middle Triassic, which is above the first appearance of archosaurs in the fossil record, but below the appearance of dinosaurs.

One possible archosaurian ancestor of the dinosaurs is *Euparkeria,* which is found in the lower Triassic of South Africa. This animal was less than three feet long and, although it has often been reconstructed in a bipedal, dinosaurian position, it was most likely quadrupedal. It probably moved like a normal lizard or crocodilian.[4]

Another stratigraphically correct candidate is *Lagosuchus* (also known as *Marasuchus*), an archosaur from the middle Triassic.

Lagosuchids "have pelvic structures, limb proportions, and reduced ankles that clearly link them with the dinosaurs."[5] *Lagosuchus* is also frequently depicted as bipedal, an important characteristic separating dinosaurs from archosaurs. **Figure 11.2** shows the hypothetical evolutionary relationships of the archosaurs and illustrates the current view that the Lagosuchidae are the dinosaurs' most closely related non-dinosaurian sister group. "*Lagosuchus* . . . [was a] long and slender-limbed terrestrial carnivorous or insectivorous reptile of very small size that has many but not all the skeletal features that define primitive dinosaurs."[6]

There were a couple of clues in that last sentence that might have alerted you that something is not quite right with *Lagosuchus.* Did you note the term "insectivorous," meaning that it eats insects, and the phrase "of very small size"? If you know Greek, you're way ahead of me. The name *Lagosuchus* means, "rabbit crocodile." Lagosuchus was about *thirty to forty centimeters long* (about twelve to sixteen inches).[8] By contrast, the lowest and hence "oldest" dinosaur in the fossil record, *Herrerasaurus,* was around three to six meters (ten to nineteen feet long).[9] Thus, although *Lagosuchus* may have been in some senses intermediate, it was anything but intermediate in size. There should be a sequence of fossil intermediates ranging in size from the foot-long *Lagosuchus* to the fourteen-foot-long *Herrerasaurus* and becoming progressively more dinosaur-like as they get larger. There is no such sequence of fossils.

What about the other dinosaurs found in Carnian strata? *Staurikosaurus* and *Azendohsaurus* were over six feet long; *Pisanosaurus* and *Eoraptor* were over three feet long. Thus, even the smallest lower Carnian dinosaurs were more than twice as large as *Lagosuchus.* There is still a large gap between the lizard-like archosaurs and the bipedal dinosaurs. (By the way, an even less likely scenario is buried in Figure 11.2: the supposed evolution of the lizard-like archosaurs into the pterosaurs—the flying reptiles.)

The dinosaur fossil record starts with a large gap between the archosaurs and the dinosaurs, but this is just one of many such gaps. **Plate 15** is a recent "Dinosaur Family Tree" produced by Paul Sereno, a dinosaur specialist at the University of Chicago. Sereno posits a dinosaur family tree consisting mostly of inferred relationships, missing ancestral lineages, and temporal ranges based on fragmentary evidence. **Plate 15** shows that each major group of dinosaurs appears in the fossil record without solid evidence linking it to any other group.

At first, paleontologists thought the dinosaurs had all evolved from one original dinosaur species. Then it was demonstrated that the dinosaurs could be sorted into two orders, the *Saurischians* and the *Ornithischians*, based on whether a species' hipbone configuration resembled the usual configuration of reptiles or that of birds. Based on this insight, scientists began to believe that the two dinosaurian orders had evolved separately. Later, in the twentieth century, it was believed that several sub-orders of dinosaurs within the two main orders had evolved separately. The dinosaurs were seen as "polyphyletic," deriving from some three to six different archosaur ancestors. In the mid-1980s, however, the original view that all dinosaurs had diversified from a single dinosaur species again came into vogue, and is presently the favored view.[10]

Obviously, if a clear trail of "transitional forms" or stratomorphic intermediates existed, there would not be endless controversy and uncertainty about the supposed evolutionary lineage of the dinosaurs. The fact that scientists cannot even agree on a "just so" story as to how the dinosaurs evolved, much less *demonstrate* how they evolved, is evidence that they did not evolve.

Given the current view that all dinosaurs descended from a single species, the *Saurischians* and the *Ornithischians* must have had a common ancestor. Thus, there should be a fossil that embodies characteristics of both the lizard-hipped *Saurischian* and bird-hipped *Ornithischian* dinosaurs, marking where one group split off from the other. No such fossil has been found. The oldest currently known dinosaurs, *Herrerasaurus* and *Eoraptor*, are clearly diagnosable as *Saurischian* theropods on the basis of skull characteristics.[11] "This means that the split into *Saurishcia* and *Ornithischia* must have happened at the latest during the Carnian," writes Michael Benton, "*before the time of the first known dinosaur fossils.*"[12] No, it means that the supposed "monophyletic" evolutionary origin of the dinosaurs remains a hypothesis without supporting fossil evidence.

What about the evolutionary ancestry of the sauropods (**Plates 8, 11, and 20**)? The hoary assumption that they evolved from a group of bipedal dinosaurs called Prosauropods is shaky. Some characteristics of the prosauropods suggest that they were already "too specialized" to have been the ancestors of the sauropods. "The ancestry of the sauropods, *before they burst onto the scene on almost every continent* in the Middle Jurassic, is obscure."[13]

The prosauropods themselves are without plausible ancestors. Commenting on the lack of fossil evidence to support one leading theory as to their ancestry, Jacques VanHeerden writes, "To suppose that light-bodied prosauropods were present but not preserved (or have not yet been found) simply to fit the cladistic hypothesis is a rather tenuous line of reasoning. . . . At the present moment there does not appear to be enough evidence to settle the question of prosauropod origins one way or another."[14]

*Stegosaurus* (**Plate 9**) is another dinosaur that lacks plausible evolutionary antecedents in the fossil record. *Stegosaurus, Kentrosaurus,* and similar Chinese fossil types bear some resemblance to the ankylosaurs and the nodosaurs, but are basically isolated in the fossil record. The fossil record does not reveal the evolutionary development of bony plates; they just suddenly appear. As to the ornithopods, the "bird-footed" dinosaurs, "the evolutionary origins of ornithopods remain *obscure.*"[15] As to the armored dinosaurs, "the origin of the ankylosaurs is not well understood because so few specimens are known from the Early and Middle Jurassic, when ankylosaurs *apparently* first appeared and diversified."[16]

We could go on in this vein, but it would be both redundant and tedious. The fossil record, when measured against Darwinist expectations, is full of "gaps." We might well ask ourselves what kind of theory leads scientists to imply that the foot-long "rabbit crocodile" is the closest non-dinosaur relative of a fourteen foot-long *Herrerasaurus*, to gloss over the absence of any evidence of a forking of the two main dinosaur orders, and to ignore the yawning chasms in the fossil record separating virtually every major type of dinosaur. These problems are real and substantial, yet they are minor nuisances compared to many other problems with the theory of evolution. Let us examine the theory of evolution in greater detail.

---

[1] See, e.g., Benton, Michael J., "Origin and Interrelationships of Dinosaurs," in *The Dinosauria*, David B. Weishampel, Peter Dodson, and Halszka Osmolska, Eds. (Berkeley, CA: University of California Press, 1990), p. 11.

[2] Wise, Kurt, *Faith, Form, and Time*, (Nashville, TN: Broadman & Holman, 2002), pp. 198, 286.

[3] See, e.g., Benton, Michael J. "Origin and Early Evolution of the Dinosaurs," in *The Complete Dinosaur*, James O. Farlow and M.K. Brett-Surman, Eds. (Bloomington, IN: Indiana University Press, 1997), p. 208. Although dinosaur bones appear in the Carnian (upper Triassic), three-toed, dinosaur-like tracks first appear in the lower Triassic. See, e.g., Parrish, J. Michael, "Evolution of the Archosaurs," in *The Complete Dinosaur*, supra, at 198.

[4] See, e.g., Parrish, J. Michael, "Evolution of the Archosaurs," in *The Complete Dinosaur*, supra, at 196.

[5] Parrish, supra, at 197.

[6] Morales, Michael, "Major Groups of Non-Dinosaurian Vertebrates of the Mesozoic Era," in *The Complete Dinosaur*, supra, at pp. 617, 618.

[7] Parrish, at 198. Linnaeus developed the Linnaean system of classification in the 1700s, long before Darwin. After evolution became scientific dogma, scientists began to ask whether the Linnaean classifications represented evolutionary relationships, and, in a discipline known as "evolutionary systematics," they began constructing evolutionary family trees depicting ancestor-descendant relationships. Postulating ancestor-descendant relationships had the unintended consequence of highlighting the almost universal absence of plausible transitional forms. It pinpointed exactly where Darwinists claimed ancestor-descendant relationships existed, and thus where the "missing links" were missing.

The attempt to determine ancestor-descendant relationships has fallen out of favor. The newer trend in classification is "phylogenetic systematics," more commonly called "cladistics," a discipline introduced by German entomologist Willi Hennig (1913-1976) in the 1950s. The cladistics approach assumes that Darwinism is true, eschewing any attempt to prove it. Cladists place species that are assumed to have a common ancestor into a "clade." Cladistics then organizes the clades according to the assumption that for any three groups, two must share a more recent common ancestor than the third. A cladogram consists of a line with shorter lines splitting off at sharp angles and leading to the names of clades. The junctions where the lines join are called nodes.

Although nodes represent a hypothetical common ancestor, the nodes are not typically labeled. Why not? Because cladists are not interested in which group descended from which. "Workers employing 'phylogenetic systematics' are interested in the recency of common ancestry and the interrelationships of clades, without making any statements about which taxa were ancestral to the other." Thomas R. Holtz, Jr., and M.K. Brett-Surman, "The Taxonomy and Systematics of the Dinosaurs," in *The Complete Dinosaur* at p. 100. In other words, workers using cladistics will seldom make a straightforward statement about which group descended from which. Cladistics *assumes* ancestral relationships without ever having to postulate, much less demonstrate, a single such relationship.

Whatever the merits of cladistics, it camouflages the problem of missing links by refusing to posit which species are connected by ancestor-descendant relationships. It could be argued that every unlabeled node is a missing link, but that is not correct. There could be a missing fossil species, or the clade could have evolved from one of the identified members of the closest sister clade. The cladist simply will not hazard a guess.

Cladistics totally ignores the stratigraphic order of fossils and focuses solely on "homologies"—similar morphology—to determine relationships. Cladists construct cladograms without worrying about where fossils appear in the sedimentary pile. "The order in which animals appear in the fossil record also becomes secondary or irrelevant. If evolutionary relationships are inferred solely on the basis of character comparisons, an animal can be the descendant of another even if the supposed ancestor doesn't appear [in the fossil record] until millions of years later." Wells, Jonathan, *Icons of Evolution,* (Washington, D.C.: Regnery, 2000), p. 119. Commenting on the fact that cladists believe that birds descended from dinosaurs that do not appear until much higher in the geologic column than, for example, *Archaeopteryx,* a cladist from the American Museum states, "We don't see time as particularly important. We think the fossil record is incomplete." Wells, at p. 122.

[8] Norman, David, *Prehistoric Life: The Rise of the Vertebrates,* (New York: MacMillan, 1994) p. 134; Keesey, T. Mike, *The Dinosauricon,* found at *http://dinosauricon.com/genera/lagosuchus.html.*

[9] Benton, in *The Complete Dinosaur,* at 208.

[10] Benton, in *The Complete Dinosaur,* at 205; Benton, in *The Dinosauria,* p. 12.

[11] Benton, in *The Complete Dinosaur,* at 209, 210; Currie, Philip J., "Theropods," in *The Complete Dinosaur,* p. 216. Expeditions to Madagascar in the late 1990s yielded jaws and partial skull fragments of two prosauropods the size of small Kangaroos.

The scientists involved believe that the fossils were 230 million years old, as opposed to the 228 million-year-old notional age of *Herrerasaurus.* Even assuming the rocks can be dated that precisely, the problem remains, because the prosauropod fossils are clearly diagnosable as *Saurischian.* "Some of the major divisions in the dinosaurs actually happened early in their history," says Neil Shubin, of the University of Pennsylvania. Shubin suggests that paleontologists might be "missing more of the record than previously thought." If the dinosaurs were monophyletic, an ancestor intermediate between saurischians and ornithischians should exist, but does not. But instead of questioning the idea that dinosaurs evolved, or least rejecting the notion that they were monophyletic, the scientists assume that "we must be missing more of the [fossil] record than previously thought." In this instance, as in so many others, we find contemporary scientists echoing Darwin's claim that the fossil record must be "imperfect" because it does not support Darwinian theory. Kenneth Chang, "Earliest Dinosaurs Uncovered," ABCNews.com.
http://www.geocities.com/CapeCanaveral/Galaxy/8152/earliest.html.
See, also, *Science,* October 21, 1999.

12 Benton, in *The Complete Dinosaur,* at 210 (emphasis added).

13 McIntosh, J.S., Brett-Surman, M.K., and James O. Farlow, "Sauropods," in *The Complete Dinosaur,* supra, at 271-272 (emphasis added). See also McIntosh, J.S. "Sauropoda" in *The Dinosauria,* supra, at p. 399 ("The origins of the sauropods are obscure." . . . "As complicated and uncertain as the problems with the Jurassic phylogeny are, those involving the Cretaceous are far worse.")

14 VanHeerden, Jacques, "Prosauropods," in *The Complete Dinosaur,* supra, at pp. 259, 260.

15 Brett-Surman, M.K., "Ornithopods," in *The Complete Dinosaur,* p. 339 (emphasis added).

16 Carpenter, Kenneth, "Ankylosaurs," in *The Complete Dinosaur,* p. 312 (emphasis added).

CHAPTER 12

# The Copying Error Theory of Evolution

## A. DARWIN'S THEORY OF EVOLUTION

"The nobility of England, my lord, would have snored through the Sermon on the Mount, but you will labor like scholars over a bulldog's pedigree."[1] England's gentry have always excelled at the art of selective breeding, and not only the breeding of dogs, but just about everything else, as well. Charles Darwin, whose wife and first cousin, Emma Wedgwood, was an heiress to the Wedgwood china fortune, was certainly among England's gentry. And Darwin bred pigeons.

It was Darwin's knowledge of selective breeding that led him to formulate his theory of evolution.[2] Darwin discussed selective breeding in the first chapter—indeed, in the very first sentence—of his 1859 book *On the Origin of Species*.[3] Evolutionary theories had been around since the ancient Greeks, at the latest. Darwin's breakthrough was to supply a mechanism for evolutionary change, a mechanism he called "natural selection."

The theory goes like this: Although like produces like, and a child looks like its parents, it doesn't look exactly like its parents. Some offspring have favorable traits—such as being stronger, faster, or more intelligent—that make them more likely to survive (which is the origin of the phrase "survival of the fittest.") Because it survives longer, an individual possessing the favorable trait will mate more often and will leave more offspring than individuals lacking the trait. Some of the offspring will inherit the favorable trait, and those that do will, in turn, survive longer and mate more often than their siblings that lack the trait, thus increasing the incidence of the favorable trait in the third generation. This process of differential survival and breeding continues until the favorable trait permeates the entire breeding population, which now carries a favorable trait that, many generations back, only one or few of its members possessed.

Darwin believed that changes resulting from natural selection would accumulate, and eventually the descendants would be so different from their ancestors that they would be a new species. He believed that such speciation had repeated itself countless times over the course of the millions of years of earth history (provided by Lyellism), such that natural selection had turned a few single-celled organisms into all the life forms on earth today.

Understanding Darwin's high regard for selective breeding is the key to understanding "natural selection," because natural selection is selective breeding, except that nature, rather than man, does the selecting. Darwin was in awe of the remarkable results achieved by the skilled breeders of his day. He thought of selective breeding as a "magician's wand" that could be used to change a species into whatever the breeder wanted:

> Breeders habitually speak of an animal's organization as something quite plastic, which they can model almost as they please. . . . Youatt . . . speaks of the principle of selection as 'that which enables the agriculturalist, not only to modify the character of his flock, but to

> change it altogether. It is the magician's wand, by means of which he may summon into life whatever form and mould he pleases.' Lord Somerville, speaking of what breeders have done for sheep, says: 'It would seem as if they had chalked out upon a wall a form perfect in itself, and then had given it existence.'[4]

Darwin saw no limit to the changes selective breeding could achieve.

The next step in Darwin's chain of reasoning was to argue that selection occurs in wild animals, not just domesticated ones: "Can the principle of selection, which we have seen is so potent in the hands of man, apply in nature? I think we shall see that it can act most effectually."[5] Darwin argued that, in the cruel and remorseless struggle for survival, only the fittest animals survive, breed, and leave offspring. The struggle for survival, argued Darwin, had the same effect of selecting desirable traits and weeding out undesirable traits as if the hand of man had chosen the best specimens for breeding.

He offered the example of wolves that, because of an increase in deer and a decrease in slower prey, had to prey upon deer in order to survive. "I can under such circumstances see no reason to doubt that the swiftest and slimmest wolves would have the best chance of surviving, and so be preserved or selected; . . . I can see no more reason to doubt this, than that man can improve the fleetness of his greyhounds by careful and methodical selection . . ."[6]

Thinking that there was no limit to the change that humans could accomplish by means of artificial selective breeding, Darwin saw no reason to believe that natural selection would be limited, either:

> Slow though the process of selection may be, if feeble man can do much by his powers of artificial selection, *I can see no limit to the amount of change* . . . which may

> be effected in the long course of time by nature's power of selection.[7]

This was Darwin's theory in a nutshell: 1) artificial selection can accomplish limitless change; 2) nature selects for traits of fitness almost as effectively as man can select; therefore 3) natural selection can accomplish limitless change.

Of course, nature cannot select breeding pairs as well as humans beings can. Darwin himself acknowledged the exceptional skill necessary to succeed as a breeder: "Not one man in a thousand," he wrote, "has accuracy of eye and judgement sufficient to become an eminent breeder. If gifted with these qualities, and he studies his subject for years, and devotes his lifetime to it with indomitable perseverance, he will succeed; . . . Few would readily believe in the natural capacity and years of practice requisite to even become a skilful pigeon-fancier."[8] Is it really likely that nature, time, and chance would select breeding pairs as well as such a skilful and persistent human matchmaker?[9]

But a more fundamental flaw in Darwin's reasoning was in his premise that artificial selection can accomplish limitless change. It cannot. Mankind has been selectively breeding animals for millennia without ever having bred a new animal species.[10] Breeders have always known that there is a limit to the amount of change achievable by selective breeding. Plants and animals can be selectively bred for a specific characteristic, but that characteristic can be pushed only so far from the average for that species, and no farther.

One-sided selection, moreover, almost always causes a decline in general fitness. Stockbreeders and horticulturists well know that whenever artificial selection is successful in producing some desirable trait, the improvement is usually gained at the expense of creating some undesirable trait. For example, the modern dairy cow, after uncounted generations of selective breeding for high milk production, must be milked every single day without fail. Stockbreeders don't care about this, however, because the dairy cow is not intended to be able to survive in the wild.[11]

Darwin thought that selection could accomplish limitless change because he misunderstood the relationship between selection and variety. He believed that selective breeding was *generating* variety.[12] If that were true, then Darwin would have been correct that selection could bring about unlimited change, because the very process that was effecting the change—selection—would also have been creating new variations to be selected. Today, it is recognized that selective breeding does not generate variety, but winnows out variety that would otherwise be seen, so that the traits the breeder is aiming for are more frequently, if not uniformly, manifested in the offspring of the animals bred together.

Thus, there was a crucial flaw in Darwin's argument, a flaw that effectively destroyed his theory. In 1859, Darwin's mistake was not obvious, however, because the laws of genetics were a mystery, as Darwin freely admitted:

> The laws governing inheritance are quite unknown. No one can say why the same peculiarity in different individuals of the same species . . . is sometimes inherited and sometimes not so; why the child often reverts in certain characters to its grandfather or grandmother or other much more remote ancestor; why a peculiarity is often transmitted from one sex to both sexes, or to one sex alone, more commonly but not exclusively to the like sex.[13]

Ignorance of the laws of genetics began diminishing rapidly around 1900. In that year, three researchers independently rediscovered the genetic principles previously worked out by Johann Gregor Mendel (1822-1884). In an 1865 paper published in an obscure scientific journal, Mendel had described the results of many years of experimental breeding of garden peas. Mendel described how invisible "characters" determine visible characteristics.[14] Mendel correctly discerned that the "characters," which later came to be called "genes," are paired, with one gene being donated by each parent.

From his garden pea experiments, Mendel correctly inferred some of the laws of genetics, including: 1) The Law of Segregation: Each inherited trait is defined by a pair of genetic "alleles." Each parent randomly donates one allele to its sex cell (sperm or egg), and when the sex cells unite in fertilization, the offspring inherit one genetic allele from each parent; 2) The Law of Independent Assortment: Genes for different traits are frequently inherited independently, so that the inheritance of one trait is not determined by the inheritance of another; 3) The Law of Dominance: Genes are either dominant or recessive. An organism with alternate forms of the gene will express the dominant form, and recessive genes do not manifest themselves unless donated by both parent organisms.

As more was learned about genetics, it became clear that breeders and horticulturists were not generating new variety, as Darwin had thought, but were merely influencing the visible manifestation of genes already found within a given species. In order for breeders to select for a visible trait, it had to be in the "gene pool" of that species to start with. There was now a scientific explanation for the natural barriers that stockbreeders and horticulturists had always encountered. This is why Dutch tulip growers have not been able to breed a black tulip, or rose breeders a blue rose: there is no gene for black coloration in the tulip gene pool, nor blue genes in the rose gene pool.[15]

The hard-won knowledge of stockbreeders and horticulturists has been confirmed and re-confirmed by carefully controlled breeding experiments. Geneticists have extensively studied the fruit fly (*Drosophila melanogaster*) because it produces a new generation every two or three weeks, can be fed and housed inexpensively, is virtually immune from disease, and has only four pairs of chromosomes (compared to twenty-four pairs in many mammal species). There has been a century's worth of experiments, in which thousands of generations of flies have been studied. Many of these experiments have involved selectively breeding the fly for various visible characteristics, such as the number of bristles growing on its body or the size of its wings.

In one such experiment, the researchers set out to vary the number of bristles on the body of the fly, starting from a stock of flies having an average of thirty-six bristles. By selectively breeding flies with fewer bristles, the researchers were able to reduce the average number of bristles to twenty-five. After thirty generations, however, the line became sterile and died out. Another group, with a higher-than-average number of bristles, was selectively bred for twenty generations, and the researchers were able to raise the average bristle count from thirty-six to fifty-six. Again, however, sterility became so common that the experiment was terminated.

"Obviously," wrote Harvard's Ernst Mayr, "any drastic improvement under selection must seriously deplete the store of genetic variability." Moreover, "the most frequently correlated response to one-sided selection is a drop in general fitness. This plagues virtually every breeding experiment."[16] Mayr termed this phenomenon "genetic homeostasis."[17] It is the natural barrier that has been encountered by all plant and animal breeders throughout the ages. There is no evidence that selective breeding can bring about unlimited change; there is an enormous mountain of evidence suggesting that it cannot.

The variability of individuals is limited by the "gene pool" of a species. Sexual reproduction does not add new genes to the gene pool; it merely recombines the genes of the parent organisms and passes on the new combination to the offspring. To borrow a Las Vegas metaphor, sexual reproduction merely shuffles the cards and deals a new hand to the offspring; it does not add new cards to the deck. In order for any truly new traits and characteristics to show up—traits that were not already in the gene pool of a given species—new genetic information has to come from somewhere. New cards have to be added to the deck. But no one knew how this could happen, and around 1900, many scientists were beginning to doubt Darwin's theory of evolution. Especially among geneticists, the opinion was widespread that Darwinism was dead.[18] In order to save the theory, someone had to find a way that the gene pool could be deepened.

## B. THE MODERN THEORY OF EVOLUTION

A Dutch botanist named Hugo de Vries (1848-1935) was working on the problem, not to rescue Darwin's theory, but to replace it with a theory of evolution by large spontaneous genetic changes. While studying the evening primrose, de Vries noticed that some of the flowers were growing wild, and some of these had become dwarfs. Since none of the parent plants had been dwarfs, de Vries assumed that a spontaneous genetic change had occurred—a change he termed a "mutation."[19] A mutation is a new genetic characteristic or trait that does not come from either of the parents, but appears suddenly, in a manner that does not follow the rules of Mendelian genetics. It turned out that dwarfism was already contained within the evening primrose's genetic information; it was not a true mutation in the current sense of the word.[20] Nevertheless, the concept of "mutation" as sudden genetic change entered the scientific mainstream, and geneticists began trying to find and study mutations.

American geneticist Thomas Hunt Morgan (1866-1945) was one of the first to conduct large-scale multi-generation genetic experiments on the fruit fly. He found that inducing mutations was a very effective way of studying how genes work, and he bombarded the flies with toxic chemicals and radiation to induce mutations. *Drosophila* normally has red eyes, but in 1910, Morgan induced a mutation that caused the color of one fly's eyes to be white. Morgan then bred the white-eyed mutant. Although the white-eye gene was both recessive and sex-linked, it could be passed on to offspring. Morgan had discovered an inheritable mutation, just what was needed to rescue the theory of evolution.

The modern theory of evolution, known as the "Neo-Darwinian Synthesis," rests upon the foundation of inheritable genetic mutations. In our Las Vegas metaphor, the mutations are the new cards added to the deck. Once this new genetic information appears in the gene pool, the Darwinian competition to survive ensures that only those mutations that convey a survival advantage will be "selected."

The modern theory of evolution combines genetic mutations with Darwin's concept of natural selection, and also synthesizes the work of several scientific disciplines, hence the name "Neo-Darwinian Synthesis." The theory's architects include Ronald Fisher, J.B.S. Haldane, Julian Huxley, Theodosius Dobzhansky, Ernst Mayr, George Gaylord Simpson, and G.L. Stebbins.

Without the new genetic information supposedly supplied by mutations, neither natural selection nor any type of selective breeding can ever lead to a new species. Mutations are the only way new genetic material gets into the gene pool.[21] "Mutation is the primary source of all genetic variation," wrote Ernst Mayr.[22] "Mutation is ultimately the source of all genetic variation," echoes a current biology textbook, "and therefore the foundation for evolution."[23] "Owing to the inherent stability of the genotype, heredity is a conservative agent," wrote

Dobzhansky. "Evolution is possible only because heredity is counteracted by another process opposite in effect—namely, mutation."[24] The modern theory of evolution *absolutely depends upon genetic mutations.*

So what, exactly, is a mutation?

To answer that question, we must resume our discussion of genetics. We noted that, in 1900, Mendel's laws of biological inheritance were rediscovered. By that time, scientists had found thread-like bodies called chromosomes[25] in the nuclei of cells. Each time a cell divides, it duplicates its chromosomes and passes copies to both cells. Scientists had observed that chromosomes come in pairs, and that in sexually reproducing organisms, parent organisms create sex cells into which one chromosome out of each pair is randomly donated. During fertilization, the sex cells join, and the chromosomes pair up again. This biological process mirrored Mendel's laws of inheritance. Thus, by the time Mendel's laws were rediscovered, it was clear that chromosomes were the anatomical structures that controlled inheritance.

Chromosomes contain nucleic acid, a substance first isolated in 1869 by a Swiss scientist named Friedrich Miescher. Chromosomes also contain proteins, however, so scientists did not know whether the genetic information was carried in the protein or in the nucleic acid. As early as 1884, some had surmised that it was carried in the nucleic acid, but most were betting it was in the protein, because protein molecules are generally more complex than nucleic acid.

In the 1920s, a British researcher named Fred Griffith, while studying the microbes that cause pneumonia, discovered that under some circumstances harmless strains could become virulent. It took another generation of intense research for scientists to discover why. In 1944, Oswald Avery discovered that when deoxyribonucleic acid (DNA) from the virulent strains was transferred to the harmless strains, the latter became just as deadly as the former. Clearly, DNA contained the genetic information. (The white substance adhering to the glass rod in **Plate 16** is DNA that has been released from cells broken open by detergent.)

In 1953, James Watson and Francis Crick were able to elucidate the chemical structure of DNA, building a correct three-dimensional model of it. They determined that DNA was a chain of molecules (a polymer) in the shape of a "double helix"—a double helix looks like a ladder twisted into a corkscrew. The two sides of the ladder are composed of alternating sugar and phosphate groups. Protruding from each sugar group is one of four nitrogenous bases: adenine, guanine, cytosine, and thymine. Each base binds with a base protruding from the other side of the ladder to form the "rungs" of the ladder. Adenine binds with thymine, and guanine binds with cytosine. (See **Plate 17**)

The DNA molecule contains the code—the instructions—for synthesizing the amino acids that form the proteins that build a living organism. The information is contained in the order of the four bases. Just as the information necessary for a computer program, CD, or DVD is expressed digitally as ones and zeros, the information necessary to build a living organism is carried in these four bases. The code (or at least that part of the code that is currently well understood) is written in units called codons, each of which specifies a particular amino acid. A codon consists of a sequence of three bases, for example GAG or TCA. The four bases can be assembled into sixty-four possible codons. Since proteins are constructed from only twenty amino acids, redundancy is built in to the coding system. Each amino acid is coded for by between two and four different codons.

A mutation occurs when a DNA molecule fails to correctly replicate itself. A mutation is essentially a copying error. According to the neo-Darwinian synthesis, the ultimate source of the vast diversity of plants and animals that we observe today is . . . *copying errors!* A mutation may be spontaneous or it may be induced by external factors, such as electromagnetic radiation (e.g., gamma and x-rays) and certain mutation-inducing chemicals (e.g., nitrous acid and nitrogen mustard).

Spontaneous mutations are rare, because cells have special "proofreading" enzymes that correct mistakes occurring during replication. In replication, enzymes split apart, or unzip, the sides of the double helix. An enzyme called *DNA polymerase* moves along the unzipped strands and synthesizes complementary strands by recruiting complementary bases (if the strand has a guanine, the polymerase recruits cytosine; if adenine, thymine).

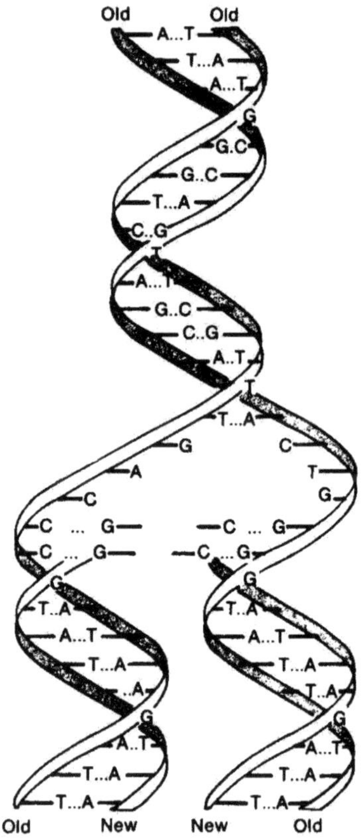


When DNA replicates, the original strands unwind and serve as templates for the building of two new complementary strands. The two new strands of DNA are called *daughter molecules;* the original is called the *parent.* The two daughter molecules are exact copies of the parent because each of the two has one of the unraveled parent strands.

**Figure 12.1: DNA Copies Itself (Replicates)**

In rare cases, a noncomplementary base is incorporated into the new strand. But a second enzymatic system acts as a proofreader. If a noncomplementary base has been bound into the DNA chain, this enzyme will immediately expel it before continuing down the strand. A third enzymatic system corrects errors in the new DNA strand after it has been synthesized. Because of these proofreading enzymes, the mutation rate is extremely low, probably about one in one billion base pairs copied, or even lower.[26] Clearly, our Creator designed the DNA replication system to prevent copying errors, and it is far more successful at this than any humanly devised copying system.

Mutations may be simple "substitution" mutations, in which a single nitrogenous base (or nucleotide) is substituted for another. Substitution mutations affect only one codon, and in most cases the effect is minimal because of the redundancy of the genetic code. But there are also "frame shift" mutations, in which a base is inserted or deleted, rather than substituted. These alter the reading frame of the entire gene, changing every codon from the site of the mutation to the end of the gene. For example, assume that part of a gene reads: TAG GGC ATA ACG ATT. The insertion of an extra adenine in the first codon will cause the sequence to read: TAA GGG CAT AAC GAT T. Because it alters many codons, a frame shift mutation is more likely to have significant consequences.

Mutations in animals that reproduce sexually can occur in two types of cells: *somatic* (or body) cells and *germinal* (or sex) cells. Somatic mutations can produce localized change and frequently cause cancer, but cannot be passed on to offspring. Because they cannot be passed on to offspring, somatic mutations are not significant to the theory of evolution. Germinal mutations, however, affect the sex cells (egg and sperm) and can be transmitted to offspring. Germinal mutations are thought to be the raw material of evolution. As a current biology textbook optimistically puts it, "beneficial mutations, and neutral ones, have been accumulating in different lineages for billions of years. Through all that time, they have been the raw material for evolutionary change—the basis for the staggering range of biological diversity, past and present."[27]

The known facts of genetics do not support this sanguine view of germinal mutations. In real life, germinal mutations, unless a healthy dominant gene donated by the other parent masks them, wreak havoc on the offspring that inherit them. They are usually either harmful or fatal. By some estimates, 90 percent of mutations are fatal to the organism and 99 percent are harmful.[28] Geneticist Hermann J. Muller (1890-1967), who demonstrated that x-rays cause mutations in fruit flies, estimated that 99.9 percent of mutations are deleterious.[29] To say that even one mutation out of every thousand is beneficial may be generous to the theory of evolution.[30] There is no way to calculate the rate of beneficial mutations, however, because mutations that can even optimistically be characterized as "beneficial" are as rare as the unicorn.

This is not surprising. In function, DNA resembles a computer program. Our experience with computer programs is that if one line is botched, the program will not run or, if it runs, it will do something the programmer did not intend. In college, I had the misfortune of having to take a computer programming class using the language COBOL (Common Business Oriented Language). Upon compiling a lengthy program, I found that it would not run. After meticulously checking every line for proper grammar, I gave up and went to the instructor. After taking a quick look at it, he noted one missing period. One dot was missing, and the program was non-functional.

The same thing is true in humans. When our genetic code fails to correctly replicate itself, we have problems. Diseases caused by inheritable genetic mutations include Parkinsonism, spina bifida, muscular dystrophy, cystic fibrosis, diabetes, albinism, dwarfism, hemophilia, Huntington's disease, Marfan Syndrome, lysosomal storage diseases, congenital adrenal hyperplasia, etc. Over 5,700 diseases and disorders are caused by inheritable genetic mutations.[31] Bad things happen when the DNA molecule fails to correctly copy itself. Very bad things. Even Richard Dawkins, an outspoken Darwinist, admits that virtually all the mutations studied in genetics laboratories are bad for the organisms in which they occur.[32]

Mutation is a process of decay and degeneration, and inheritable mutations that are passed along from generation to generation weaken the species. In fact, some scientists wonder, given the high rate of deleterious mutations in the human population, why we are not already extinct.[33] Yet the theory of evolution is based on the hope that some DNA copying errors actually convey a survival advantage. The theory of evolution depends upon good things happening as a result of genetic copying errors, when the only results we observe range from neutral to bad to catastrophic.

Is there even such a thing as a beneficial genetic mutation? Yes and no. The two categories of "beneficial" mutations invariably cited are bacterial resistance to antibiotics and insect resistance to insecticides. Most cases of medically significant antibiotic resistance are not due to mutations, but to complex enzymes that inactivate the poison, which the bacteria either inherit from their "parent" or obtain from other bacteria by the process of lateral gene transfer.[34] There are, however, some cases of antibiotic resistance that are due to spontaneous mutations that alter the bacteria's DNA just enough so that antibiotics can no longer poison them. The same is true of insects; most cases of resistance to insecticides are the result of inactivating enzymes, but some cases are genuine genetic mutations.[35]

But these mutations, it turns out, are just as degenerative as the ones observed in humans. Mycin-type antibiotics, for example, work when the mycin attaches to a matching site on the ribosome[36] of a bacterium and prevents the bacterium from assembling necessary amino acids. As a result, the bacterium makes the wrong proteins and cannot grow, divide, or propagate.[37] If the bacterium has a mutation that alters the site where the mycin would normally bind, the drug can no longer interfere with protein production, and the bacterium has become resistant to mycin. But scientists have found that mutations that make bacteria resistant to mycin degrade their ribosomes. The mutation makes the ribosome less specific and slower in translating RNA codons into protein.[38]

A similar effect has been noticed in insects. The insecticide DDT works when a molecule of DDT binds to a matching site on the membrane of an insect's nerve cells, preventing the nerve cells from functioning properly. Eventually, when enough of the insect's nerve cells have DDT molecules bound to them, its nervous system breaks down and it dies. A mutation that changes the site of the nerve cell at which the DDT molecule would normally bind prevents it from binding and renders the insect resistant to DDT.[39] Here again, however, scientists have noted that resistant insects are less active and slower to respond to stimuli than are nonresistant insects. Their resistance to insecticide is bought at the price of a more sluggish nervous system.[40] Thus, as with other known mutations, the mutations observed in bacteria and insects are degenerative changes, but in these cases the degenerative change saves the organism from the poison.

Another "beneficial mutation" often mentioned is the sickle cell trait's benefit in fighting malaria. Sickle cell anemia is caused by a simple substitution mutation or "point mutation." Out of approximately three billion base pairs of nucleotides present in the DNA of every cell nucleus, only one base pair is different. The result is that instead of producing normal hemoglobin (hemoglobin-A), the body produces *hemoglobin-S.* When the amount of oxygen around the red blood cells drops below a certain point, as during vigorous exercise, the hemoglobin-S comes out of solution and forces the normally flexible red blood cells into rigid crescent or "sickle" shapes. The sickle-shaped cells get stuck in the smaller blood vessels, blocking blood flow and causing tissue damage. Those who inherit the sickle cell allele from *both* parents suffer from sickle cell anemia, which, without medical intervention, is usually fatal.[41] By contrast, those who inherit only one copy of the sickle cell allele, a condition known as "sickle cell trait," do not suffer from sickle cell anemia, because half of their hemoglobin is normal.

In 1946, a physician in Zimbabwe noticed that children with sickle cell trait were better able to resist malaria. Malaria kills over one million people a year. A mosquito bite introduces a parasite known as *Plasmodium falciparum*[42] into the human body, where it spends a short time in the liver, and then emerges into the bloodstream searching for red blood cells. It attaches to proteins on the cell surface, enters the cell, and feeds on hemoglobin, multiplying rapidly. Because *Plasmodium* spends most of its time inside our cells, there are only short periods when our immune system can "see" it and destroy it.

Following up on the discovery in Zimbabwe, researchers learned why people with sickle cell trait are more resistant to malaria. First, *Plasmodium* causes cells with hemoglobin-S to sickle, increasing the likelihood that such cells will be removed by the spleen's "quality control check." As the spleen removes the sickle cells, it also kills the parasites before they can mature. Second, *Plasmodium* does not seem to be able to eat Hemoglobin-S as well as normal hemoglobin, and many of the parasites starve to death, giving the malaria-sufferer a better chance of survival.[43]

Wherever malaria poses a severe threat to life, many carry at least one copy of the sickle cell gene, but races originating in parts of the world where malaria is unknown do not carry it. Darwinists can therefore claim that the sickle cell mutation is an example of a beneficial mutation. They can argue that in malarial regions, the sickle cell mutation was adaptive and beneficial, and thus spread through the population. Yet the sickle cell mutation is obviously a degenerative change leading to sickness and death if inherited from both parents (and even the Darwinist would have to admit that it is not adaptive in areas with no malaria, and hence was apparently selected *against*, or weeded out, if it occurred in those areas). While the sickle cell allele has the accidental side effect of making its owner more resistant to malaria, the overall picture is still one of degeneration.

Accumulated degenerative mutations can never lead to evolution, as it is commonly understood to mean progress toward more complex forms of life. For that type of evolution to occur, there must be many mutations that are not accidentally beneficial within the context of degeneration, but that are always and *merely* beneficial, and not degenerative. *No such mutation has ever been observed.*

Moreover, resistance to insecticides and antibiotics, and the sickle cell mutation, are obviously mutations of the type that do not change the physical appearance of the organism. But the theory of evolution depends upon mutations that change the shape, or morphology, of the organism. Evolution posits innumerable instances of changes in the shape of the bones and overall form of animals, as they evolved from one species into another. Trillions of mutations changing some aspect of animal morphology must have occurred if life on this planet evolved as Darwin believed. *Yet not one single mutation causing a beneficial change to the shape of an organism has ever been observed.*[44]

Scientists have been trying to cause a beneficial morphological mutation for years, without success. German researchers, for example, have used a technique called "saturation mutagenesis" to bring out every possible fruit fly mutation. They have discovered dozens of mutations that affect development and produce a variety of malformations, but they did not turn up a single morphological mutation that would benefit a fruit fly in the wild. Attempts to cause beneficial morphological mutations in zebrafish have also failed.[45] Evidence of beneficial morphological mutations has been extremely difficult to find, despite determined effort. The theory of evolution rests upon a hypothetical event, something more hoped for than real.

## C. THE PROBLEM OF IRREDUCIBLE COMPLEXITY

Plants and animals are composed of fantastically complex organs, organelles, and sub-cellular systems. The eye is the complex organ most often mentioned, and even Darwin acknowledged that it was an organ of "extreme perfection and complication." "To suppose that the eye, with all its inimitable contrivances for adjusting the focus to different distances, for admitting different amounts of light, and for the correction of spherical and chromatic aberration, could have been formed by natural selection," wrote Darwin, "seems, I freely confess, absurd in the highest possible degree."[46]

Darwin nevertheless argued that the eye could have evolved in a stepwise fashion. He did not propose a path of development, but he noted that there are many animals with less complex visual equipment, ranging all the way down to a light sensitive patch of skin. We now know that vision has an extraordinarily complex biochemical component, of which Darwin was not aware, that would also have had to evolve.[47]

The real problem for Darwinists is that much of our biochemical and biological machinery is irreducibly complex. If a biological mechanism consists of five components, and all five are necessary for the mechanism to function, then that mechanism is irreducibly complex. In other words, the machine only works if all five of the component parts are present. Assuming that individual components do not convey any survival advantage, then it does the organism no good whatsoever to have one, two, three, or four of the components. It only helps to have the working mechanism, which needs all five components.

The difficulty irreducible complexity poses for the copying error theory of evolution is obvious. The theory postulates that a mutation that conveys a survival advantage on the organism will be preserved and spread rapidly throughout the breeding population. But how could a mechanism with five necessary components evolve in a stepwise fashion if none of the first four components conveyed any survival advantage? Even if a mutation created one component part of the machine, it would not convey a survival advantage and would not be preserved.

Suppose that an organism has a mutation that conveyed one component of a mechanism with five essential components. Suppose further that, against all odds, this mutation is not deleterious but merely neutral. Because it is not deleterious, it does not harm the organism, but because it does not convey a survival advantage, it does not spread throughout the breeding population. It is inherited within one family line. That line must have four more mutations of a very specific type before that line has a new mechanism that will convey a survival advantage.

Mutations, in the modern theory of evolution, are strictly random copying errors. Orthodox Neo-Darwinists do not believe nature has ways of causing the specific mutation that the organism needs and will find useful in its environmental niche. There are billions of base pairs in the genome of any living thing, and these are separated into several chromosomes. The odds against a copying error occurring exactly where it is needed are astronomical. You are far more likely to win the lottery. Our hypothetical line would die out long before any descendant was the beneficiary of the second mutation needed to build the five-component machine. And it would still need three more very specific mutations before the five-component mechanism could become operational and thus convey a survival advantage.

In Darwin's day, and even in 1930s and 1940s, when the Neo-Darwinian theory was being hammered out, life at the cellular level was largely unknown—a "black box." Rapid progress in biochemical research since the 1970s has shown that life at the sub-cellular level is full of fantastically complex biochemical mechanisms. Michael J. Behe discusses this fact, and the problem it poses for the theory of evolution, in his book, "Darwin's Black Box."

Mechanisms like the blood-clotting cascade, which must work perfectly to prevent death, are composed of many different proteins all working together. None of these proteins can be dispensed with, or the clotting mechanism will fail. It is difficult to imagine, never mind to demonstrate, how these systems could have evolved in a step-wise fashion. Yet to suppose that a single mutation created all components simultaneously is to invoke a miracle.

The larger theory of evolution postulates that many complex organs and biochemical systems evolved not once, but many different times. Most of the phyla are thought to have branched off from each other before these organs evolved. Hence, these complex organs evolved separately in each phylogenetic line, in a process called "convergent" or parallel evolution. The eye, for example, is thought to have evolved separately over forty times.[48] The miracle of flight is supposed to have evolved four different times, in birds, flying reptiles, insects, and bats. In each instance, we are told, these miracles of engineering and design were accomplished by the accumulation of random genetic copying errors.

So that is what the theory of evolution has come to. It holds that all the plants and animals we find around us today are the result of random mistakes in copying the genetic code. It holds that all the evidence of genius and awe-inspiring design we see in the creation is misleading, because there was no intelligence and there was no design—just a series of accidents. It holds that the astonishingly complex biochemical mechanisms we find in nature developed over time by a series of failures in the biochemical copying technology, mistake by mistake, one mistake at a time. The copying error theory of evolution is simply not credible.

As implausible as the copying error theory of evolution seems, is there any way we can test it? Experimentation and observation, the normal methods of science, cannot test the larger claims of the theory of evolution (i.e., microbes into man). There is something, however, that should confirm or refute the theory. The theory of evolution holds that life evolved from "simple" single-celled organisms, over the course of hundreds of millions of years, into all of the diverse varieties of plants and animals that now exist. Over those same hundreds of millions of years, according to Lyellian geology, the sedimentary rocks were gradually forming. Thus, to the Darwinist, the geologic column contains "the history of life." If evolution occurred as the Darwinists claim, we must look to the rocks and the fossils they contain for the evidence.[49]

---

1 So said Sir Thomas More to his friend the Duke of Norfolk, in the play, "A Man for All Seasons," by Robert Bolt.

2 "Wood (1973) and Ruse (1975) have shown that Darwin read widely in the literature of the animal breeders in the year before his theory formed, and that his underlining of crucial sentences in pamphlets by Sebright and Wilkinson (read in the spring of 1838) show how clearly Darwin understood the principles of artificial selection and how important he considered them." Mayr, Ernst, *The Growth of Biological Thought,* (Cambridge MA: Harvard University Press, 1982), p. 486. "Many years after the Malthus episode, *Darwin stated repeatedly that he had arrived at the concept of natural selection by analogy with artificial selection.*" Mayr, at 486 (emphasis added).

3 "When we look to the individuals of the same variety or sub-variety of our older cultivated plants and animals, one of the first points which strikes us is that they generally differ much more from each other than do the individuals of any one species or variety in a state of nature."

4 Darwin, Charles, *The Origin of Species,* 1859 (Reprint: New York: Gramercy Books), p. 90.

5 Darwin, at 130.

6 Darwin, at 138.

7 Darwin, at 153 (emphasis added).

8 Darwin, at 91.

9 See, e.g., Phillip Johnson, *Darwin on Trial,* (Washington, D.C.: Regnery, 1991), pp. 17, 18 ("The analogy to artificial selection is misleading. Plant and animal breeders employ intelligence and specialized knowledge to select breeding stock and to protect their charges from natural dangers. The point of Darwin's theory, however, was to establish that purposeless natural processes can substitute for intelligent design. That he made that point by citing the accomplishments of intelligent designers proves only that the receptive audience for his theory was highly uncritical.")

10 Milton, Richard, *Shattering the Myths of Darwinism,* (Rochester VT: Oak Street Press, 1992), p. 134. Darwin knew of the long history of selective breeding. "The principle of selection I find distinctly given in an ancient Chinese encyclopedia. Explicit rules are laid down by some of the Roman Classical writers." Darwin, at 92. "[Pigeons]

have been domesticated for thousands of years in several quarters of the world; the earliest known record of pigeons is in the fifth Egyptian dynasty, . . . but Mr. Birch informs me that pigeons are given in a bill of fare in the previous dynasty. Darwin, at 87, 88.

[11] Milton, at 137.

[12] "Darwin owed to the breeders good ideas and bad ones. The firm belief that the mere placing of animals or plants under the conditions of domestication would increase their variability was one of the bad ideas." Mayr, supra, at 486. "Sterility is said to be the bane of horticulture," Darwin wrote, "but on this view we owe variability to the same cause which produces sterility; and variability is the source of all the choicest productions of the garden." Darwin at 73.

[13] Darwin, at 76.

[14] For a variety of reasons, Mendel's work did not reach the scientific mainstream during his lifetime. These reasons included: (1) Mendel published his results in 1865 in a small local journal, *The Proceedings of the Brünn Natural Science Society*, in Moravia (now the Czech Republic), only about 140 copies were published, and most were sent reciprocally to other scientific societies where they sat on shelves unread; (2) Mendel was an Augustinian monk—he was not good at self-promotion; (3) Mendel employed more mathematics in his work than biologists were accustomed to seeing; (4) a colleague of Mendel's urged him to repeat his experiments on hawkweed, a plant that—unbeknown to either of them—reproduced both sexually and asexually, with the result that Mendel was unable to reproduce his garden pea results and may have lost confidence in his original results. The three scientists who independently rediscovered Mendel's laws were Hugo de Vries (1848-1935), Carl Correns (1864-1933), and Erich von Tschermak-Seysenegg (1871-1962).

[15] Milton, at 141.

[16] Milton, at 135, citing Ernst Mayr, *Animal Species and Evolution*, (Cambridge, MA: Harvard University Press, 1963).

[17] Milton, at 135, 140.

[18] Mayr, *The Growth of Biological Thought*, at 547.

[19] De Vries, Hugo *Die Mutations-Theorie*, 1901-1903.

[20] Mayr, *The Growth of Biological Thought*, at 743, 744. In 1929, geneticist Alfred H. Sturtevant, a student/colleague of Thomas Hunt Morgan, showed that many of the "mutations" de Vries and others had found in the evening primrose (*Oenothera*) were

not genuine mutations but the expected segregation products from the complex translocations of chromosome arms that are widespread in *Oenothera.*

[21] At least insofar as the Neo-Darwinian theory is concerned. As we will discuss later, however, genetic information can travel from one species to another by means of lateral or horizontal gene transfer. But although horizontal gene transfer can explain how the "gene pool" of a given species might be deepened, it cannot explain the creation of new genetic information in the first place, so Darwinian theory still depends upon mutation and natural selection to explain the creation of new genetic information.

[22] Mayr, Ernst, *The Growth of Biological Thought*, at 56.

[23] Wells, Jonathan, *Icons of Evolution*, (Washington, D.C.: Regnery, 2000), p. 189, citing Guttman, Burton S., Biology, (McGraw Hill, 1999).

[24] Dobzhansky, Theodosius, *Genetics and the Origin of Species*, 3rd ed. (New York: Columbia University Press, 1951), p. 25.

[25] They are called *chromosomes* because they absorbed the colored dyes used to make cell structures visible under the microscope.

[26] Aldridge, Susan, *The Thread of Life, the story of genes and genetic engineering*, (Cambridge and New York: Cambridge University Press, 1996), p. 141.

[27] Wells, at 189, citing Starr, Cecie and Ralph Taggart, *Biology: The Unity and Diversity of Life* (Cole Publishing, 1998).

[28] Mitchell, Colin, *Creationism Revisited*, (Alma Park, Grantham, Lincolnshire, UK: Autumn House, Ltd., 1999), p. 143.

[29] Margulis, Lynn, and Dorion Sagan, *Acquiring Genomes*, (New York: Basic Books, 2002), pp. 11, 12.

[30] Roth, Ariel, *Origins: Linking Science and Scripture*, (Hagerstown, MD: Review and Herald, 1998), p. 97.

[31] Aldridge, at 141, 142.

[32] Johnson, *Darwin on Trial*, (Washington, D.C.: Regnery, 1991), p. 38; Aldridge, at 245.

[33] "Annotations From the Literature," *Origins* 51:31-45 (2001), commenting upon Eyre-Walker, A., Keightley, P.D., "High genomic deleterious mutation rates in hominids," *Nature* 397:344-347 (1999), Crow, J.F. "The odds of losing at genetic roulette," *Nature* 397:293, 294 (1999).

[34] Wells, at 181; see also Spetner, Lee, *Not by Chance!* (Brooklyn, NY: The Judaica Press, 1997), pp. 138, 139.

[35] Wells, at 182.

[36] When our cells produce protein, a process similar to replication produces a modified copy of DNA called "messenger RNA." A *ribosome* then moves along the messenger RNA, translating the three-letter codons into a different alphabet that specifies the twenty amino acids. The specified amino acids are then brought in order, each by a different version of a molecule called "transfer RNA," and each is attached to the last to form a chain in the same order as the codons in the DNA. At the end of this process of concatenating amino acids, a protein has been formed. It is insane to believe that this intricate information processing mechanism assembled itself.

[37] Spetner, Lee, *Not by Chance!* (Brooklyn, NY: The Judaica Press, Inc., 1997), p. 139.

[38] Spetner, at 144, citing Gartner, T.K. and E. Orias "Effects of mutations to streptomycin resistance on the rate of translation of mutant genetic information," *Journal of Bacteriology*, vol. 91, pp. 1021-1028 (1966).

[39] Spetner, at 143.

[40] Spetner, at 144, citing Rowland, M.W., "Fitness of insecticide resistance," *Nature*, vol. 327, p. 194 (1987) (Rowland, who worked for the Crop Protection Division in Great Britain, comments, "Insecticide resistance genes give such an enormous advantage in the presence of insecticide that what would otherwise be quite major disadvantages might count for little.").

[41] Levine, Joseph, and David Suzuki, *The Secret of Life: Redesigning the Living World* (Boston, MA: WGBH, 1993), pp. 38-45.

[42] There are six species of *Plasmodium* that cause malaria, including *Plasmodium vivax, Plasmodium ovale,* and *Plasmodium malariea,* but the deadliest strain, the one that kills over a million people a year, is *falciparum.* Its genome has been sequenced in the hope of furthering drug and vaccine design. (*Nature,* October 3, 2002)

[43] Levine, et al, supra. A similar phenomenon occurs with respect to cystic fibrosis and typhoid. The mutated CFTR gene on chromosome 7 that causes cystic fibrosis protects the body against typhoid, an intestinal disease caused by the *Salmonella* bacterium. People who have inherited only one allele of the mutated CFTR gene do not get cystic fibrosis and are largely resistant to typhoid, with its debilitating dysentery and fever. Again, however, if the mutated CFTR gene is inherited from both parents, cystic fibrosis results. Matt Ridley, *Genome,* (New York: HarperCollins, 2000), p. 142.

[44] Milton, at 157. See also Wells, at 177-193.

[45] Wells, at 189, 190.

[46] Darwin, at 217.

[47] Behe, Michael, *Darwin's Black Box: The Biochemical Challenge to Evolution,* (New York: Free Press, 1996), pp. 18-22.

[48] Johnson, at 35, citing Ernst Mayr, "Toward a New Philosophy of Biology."

[49] "Science is the study of processes," wrote Stephen Jay Gould. "Past processes are, in principle, unobservable; only their frozen results remain as evidence for ancient history—fossils, mountains, lavas, ripple marks." Gould, *Time's Arrow, Time's Cycle,* (Cambridge, MA: Harvard University Press, 1987, p. 105.

CHAPTER 13

# Does the Fossil Record Support Darwinism?

If life on this planet evolved as Darwin imagined, we would expect to find many fossil forms that bridge the gap between species. But we do not. We seldom find them.[1] Charles Darwin was keenly aware of this problem. His theory as to why we do not see *living* links between species was that each such link was driven to extinction by a better-adapted daughter species.[2] But Darwin acknowledged that we should find many transitional forms in the fossil record:

> But just in proportion as this process of extermination has acted on an enormous scale, so must the number of intermediate varieties, which have formerly existed, be truly enormous. Why then is not every geological formation and every stratum full of such intermediate links? Geology assuredly does not reveal any such finely graduated organic chain; and this, perhaps, is the most obvious and gravest objection which can be urged against the theory.[3]

Darwin noted that paleontologists had thoroughly explored only Europe; he argued that more missing links might be found in other parts of the world.[4] That argument cannot be made today. By some estimates, 99 percent of all fossil collecting has been done since 1859, and the lack of transitional forms is as acute as ever.

Darwinists can point to a few fossils as being intermediate between two types. *Archaeopteryx*, which is apparently part bird and part reptile, is a good example. Darwinists can also place eight or nine fossil equine forms from various parts of the world (*Hyracotherium, Orohippus, Epihippus, Mesohippus, Miohippus, . . . Equus*) in a plausible evolutionary sequence; hence, the horse series is almost invariably used in science textbooks.[5] But even the unusual case of the horse does not provide a finely graded series of intermediate forms, and it merely highlights the almost total absence of stratomorphic intermediates (i.e., fossils that are intermediate both stratigraphically and morphologically) that should lead up to each of thousands of genera.

British science journalist Richard Milton, who is not a creationist, has described the failure of the fossil record to yield the type of evidence that Darwinism predicts:

> The case for Darwinism would be made convincingly if someone were to produce a sequence of fossils from a sequence of adjacent strata . . . showing indisputable signs of gradual progressive change of the same basic stock, but above the species level (as opposed to subspecific variation). Ideally this should be demonstrated in a long sequence, ten or twenty or fifty successive fossil species, showing major generic evolution—but a short sequence would be enough.

> But this simple relationship is not what is shown in the sequence of the rocks. Nowhere in the world has anyone met this simple evidential criterion with a straightforward fossil sequence from successive strata. Yet there are so many billions of fossils available from so many thousands of strata that the failure to meet this modest demand is inexplicable if evolution has taken place in the way Darwin and his followers have envisaged. It ought to be relatively easy to assemble not merely a handful but hundreds of species arranged in lineal descent. Schoolchildren should be able to do this on an afternoon's nature study trip to the local quarry, but even the world's foremost paleontologists have failed to do so with the whole Earth to choose from and the resources of the world's greatest universities at their disposal.[6]

Darwinists prefer not to publicize the extreme rarity of transitional forms in the fossil record—Stephen Jay Gould called it "the trade secret of paleontology"[7]—but it is an undisputed fact. "All paleontologists know that the fossil record contains precious little in the way of intermediate forms; transitions between major groups are characteristically abrupt."[8] "It remains true, as every paleontologist knows," wrote George Gaylord Simpson, "that most new species, genera, and families, and that nearly all categories above the level of families, appear in the record suddenly, and are not led up to by gradual, completely continuous transitional sequences."[9]

The fossil record never shows the type of evolutionary transition that Darwinian theory predicts: "The known fossil record fails to document a single example of phyletic evolution accomplishing a major morphologic transition and hence offers no evidence that the gradualistic model [of evolution] can be valid."[10]

The pattern, typical of the overwhelming majority of cases, is that life forms appear in the fossil record, they remain the same for long periods of "geologic time," and then they disappear from the record:

> The history of most fossil species includes two features particularly inconsistent with gradualism: 1. *Stasis.* Most species exhibit no directional change during their tenure on earth. They appear in the fossil record looking pretty much the same as when they disappear; morphological change is usually limited and directionless. 2. *Sudden appearance.* In any local area, a species does not arise gradually by the steady transformation of its ancestors; it appears all at once and "fully formed."[11]

> Paleontologists just were not seeing the expected changes in their fossils as they pursued them up through the rock record. That individual kinds of fossils remain recognizably the same throughout the length of their occurrence in the fossil record had been known to paleontologists long before Darwin published his *Origin.* . . . The observation that species are amazingly conservative and static entities throughout long periods of time has all the qualities of the emperor's new clothes: everyone knew it but preferred to ignore it. Paleontologists, faced with a recalcitrant record obstinately refusing to yield Darwin's predicted pattern, simply looked the other way.[12]

> [S]tasis, or nonchange, of most fossil species during their lengthy geological life spans had been tacitly acknowledged by all paleontologists, but almost never studied explicitly because prevailing theory treated stasis as uninteresting nonevidence for nonevolution. Evolution had been defined as gradual transformation in extended fossil sequences, and the overwhelming prevalence of stasis became an embarrassing feature of the fossil record, best left ignored as a manifestation of nothing (that is, nonevolution).[13]

The fossil record knows nothing of evolution. Fossils appear in the stratigraphical pile, they do not change, and then they disappear.

The fossil record did not support Darwinism in Darwin's day, and it does not today, as candid Darwinists will admit. Mark Ridley, a professor of zoology and a prominent Darwinist, writes of "the false idea that the fossil record provides an important part of the evidence that evolution took place," and notes that "no real evolutionist, whether gradualist or punctuationist, uses the fossil record as evidence in favour of the theory of evolution as opposed to special creation.... No good Darwinian's belief in evolution stands on the fossil evidence for gradual evolution, so nor will his belief fall by it."[14]

Darwin must have known that very few fossil intermediates would ever be found and therefore that the fossil record would never *support* his theory. Because the fossil record clearly did not support his theory, the best Darwin could hope for was to prevent it from being seen as *falsifying* his theory. In order to accomplish that, Darwin had to cast aspersions on the fossil record, he had to argue that the fossil record was spotty, episodic, and incomplete. "The explanation [for the lack of intermediate forms] lies, as I believe, in the extreme imperfection of the geological record," he wrote.[15]

It is important to note what Darwin did. Having acknowledged that the fossil evidence did not support his theory, he asserted that the problem was not with his theory but with the evidence. The Yiddish term for this is "chutzpah." Temerity, audacity, effrontery, nerve, cheek, and gall are other fitting terms that come to mind. If the facts did not support his theory—and the fossil facts clearly did not—then so much the worse for the facts.

Darwin's main argument regarding the "imperfection" of the fossil record was that fossils are formed only when an area is subject to deposition of sediment, usually when the area has subsided below sea level, not when it is above sea level. Thus, he reasoned, there are large time gaps separating fossiliferous formations. The animals must have been evolving during these hiatuses in sedimentation, which is why we do not find fossils of intermediate forms.[16] Even under Darwin's assumptions, however, when one area is not conducive to forming fossils, fossils are being preserved in other places around the world. Erosion and sedimentation cannot cease everywhere throughout the world. By correlating the fossiliferous strata from around the globe, we should be able to find a staggering number of intermediate forms. But we do not.

To this day, the best explanation any Darwinist has found for the absence of fossil intermediates is that the fossil record is incomplete.[17] Maybe Darwin's clever excuse is correct; maybe the fossil record is just a series of snapshots of life taken at long intervals.[18] But there is something disturbing about this kind of excuse-making. Good scientific theories make predictions, not excuses.

For example, Einstein's general theory of relativity holds that gravitational fields bend light beams. Einstein predicted that the presence of the sun between a group of visible stars and the earth would cause an apparent change in the position of the stars, because the sun's gravitational field would bend the light coming from the stars. Einstein suggested that this bending could be measured by photographing a group of stars at night, noting their positions relative to each other and to other stars on the periphery, then photographing the same group during the day when the sun is between them and us. The experiment could be conducted only during a total eclipse of the sun, which, as luck would have it, occurred on May 29, 1919, only four years after Einstein published his theory. The experiment was conducted and Einstein's theory was vindicated.[19]

A good scientific hypothesis must be falsifiable. That means that at least one experiment can be devised that would show that the theory is false. Darwin postulated that an "infinite number" of transitional forms have existed. We should be able to falsify Darwin's theory by showing that the intermediate forms are nowhere to be found in the fossil record. But the extreme rarity of intermediate forms was well established before Darwin published in 1859. Undeterred, Darwin claimed that the fossil record was imperfect, and by doing so, he effectively made the larger claims of his theory unfalsifiable. The scientific world so wanted a naturalistic explanation of origins that it accepted Darwin's *ad hoc* speculations as to why the fossil record did not support evolution.

But the dearth of transitional forms is just the beginning of the problems the fossil record poses for Darwinism. The "Cambrian explosion" strikes a fatal blow to Darwin's theory. The earliest Cambrian animals are already complex and well developed, and could not have been the first animals of their respective phyla, yet there are no transitional forms leading up to the Cambrian. Darwin himself considered the Cambrian explosion a more serious problem for his theory than the lack of transitional forms: "There is another and allied difficulty, which is much graver," wrote Darwin. "I allude to the manner in which numbers of species of the same group, suddenly appear in the lowest known fossiliferous rocks."[20]

Darwin admitted that there simply had to have been evolutionary antecedents to the Cambrian animals:

> Consequently, if my theory be true, it is indisputable that before the lowest [Cambrian][21] stratum was deposited, long periods elapsed, as long as, or probably far longer than, the whole interval from the [Cambrian] age to the present day [over 500 million years]; and that during these vast, yet quite unknown, periods of time, the world swarmed with living creatures.[22]

This is a devastating admission. Darwin was admitting that *more than half* of the fossil record predicted by his theory was *completely missing!* This is far more serious than a mere "imperfection" in the fossil record. There simply is no evidence whatsoever of more than five hundred million years of evolution *necessary* to Darwin's theory.

To his credit, Darwin had no glib excuse for the absence of more than half of the fossil record: "To the question why we do not find record of these vast primordial periods, I can give no satisfactory answer. . . . The case at present must remain inexplicable; and may be truly urged as a valid argument against the views here entertained."[23] He hoped that future exploration and study would solve the problem. He suggested that perhaps the weight of ocean water had metamorphosed all the Precambrian fossiliferous strata,[24] a suggestion that has proven wrong. There are un-metamorphosed sedimentary strata going back 3.5 billion years, in notional "geological" time, and—according to some paleontologists—a line of fossil microbes continuing up through those strata.

The discovery of life below the Cambrian "has not matched Darwin's prediction of a continuous rise in complexity toward Cambrian life," wrote Stephen Jay Gould, "and the problem of the Cambrian explosion has remained as stubborn as ever—if not more so, since our confusion now rests on knowledge, rather than ignorance, about the nature of Precambrian life."[25] Commenting on the single-celled Precambrian forms, Gould wrote, "Such a simple beginning would have pleased Darwin, but the later history of Precambrian life stands strongly against his assumption of a long and gradual rise in complexity toward the products of the Cambrian explosion."[26] Clearly, the Precambrian fossil record was not wiped out by metamorphism, as Darwin thought, but it does not contain intermediate forms leading from single-celled creatures to the amazing diversity of life forms preserved in the lower Cambrian strata. This fact is fatal to Darwin's theory.

The mystery of the Cambrian explosion was deepened by the discovery of the Burgess Shale fauna. High in the Canadian Rockies, in Yoho National Park on the eastern border of British Columbia, are what Gould describes as the "world's most important fossil animals."[27] The Burgess Shale, like the Solnhofen limestone in Bavaria, preserved organisms in exquisite detail, "down to the last filament of a trilobite's gill."

Charles Doolittle Walcott, the head of the Smithsonian Institute, conducted the first major study of the Burgess Shale fauna in 1909. Walcott comprehensively misinterpreted the Burgess fossils, "shoehorning" them into modern animal phyla where they did not belong. In the 1970s, Harry Whittington, a paleontologist at Cambridge University, and two of his students, Derek Briggs and Simon Conway-Morris, took a fresh look at the Burgess fauna. They found that many of the Burgess species were not primitive specimens of modern phyla. Many of the fossils—with such exotic names as *Opabinia, Hallucigenia, Anomalocaris, Wiwaxia, and Sidneyia*—are simply unique. Fifteen or twenty of the Burgess species cannot be placed within any known group and must be classified as separate phyla. Their basic body plans are markedly different from the body plans of any known phylum, modern or fossil.

The creatures of the Burgess Shale "exceed, in anatomical range, the entire spectrum of invertebrate life in today's oceans." For species that can be classified within a known phylum, the Burgess anatomy far exceeds the modern range within the phylum. For example, the Burgess Shale includes representatives of all four major groups of arthropods (trilobites, crustaceans, chelicerates, and uniramians). But the Burgess Shale also contains twenty to thirty kinds of arthropods that cannot be placed in any of these known sub-groups. Thus, "the sweep of anatomical variety reached a maximum" just as the first multi-cellular animals enter the fossil record.[28]

The Cambrian explosion is devastating to the theory of evolution.[29] Darwin predicted that the history of life should reflect a "branching tree" pattern, or a "cone of increasing diversity." In fact, he included a diagram in *Origin of Species*, the only illustration in the

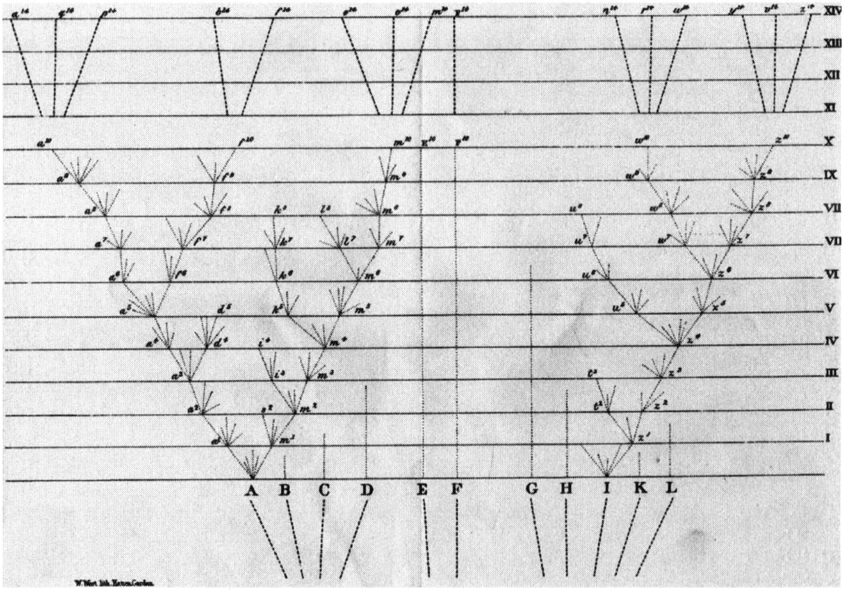


**Figure 13.1: Darwin's Tree of Life**

original edition, to illustrate exactly this point. According to Darwin's theory, life branched out from one or a few single-celled organisms, and these very slowly and gradually evolved into more differentiated forms of life, including separate species, and eventually separate genera.

Over more time, the separate genera then broaden out into families, the families into orders, and the orders diversify further into classes. Finally, after the passage of eons, meaning *much* higher up in the geologic column, we should be able to group the species into several different phyla. In other words, phylum-level differences should emerge in the fossil record only after a very long history of divergence.

But the "branching tree" pattern is not what is observed in the fossil record. Instead, we see many *different phyla* in the lowest layers that contain the fossils of multi-cellular animals. Many of these animals became extinct, and their body plans are not seen in higher levels. "The maximum range of anatomical possibilities arises with the first rush of diversification," wrote Gould.[30] Not only is there no branching tree of life, the tree is upside down. The fossil configuration actually observed is exactly the opposite of that predicted by Darwin and specially illustrated in the first edition of *Origin of Species.*[31]

The fossil record has always been a problem for Darwinists. It certainly was in Darwin's day, as Darwin himself admitted:

> The several difficulties here discussed, namely our not finding in the successive formations infinitely numerous transitional links between the many species which now exist or have existed; the sudden manner in which whole groups of species appear in our European formations; the almost entire absence, as at present known, of fossiliferous formations beneath the [Cambrian] strata, are all undoubtedly of the gravest nature. We

> see this in the plainest manner by the fact that all of the most eminent palaeontologists . . . and all our greatest geologists . . . have unanimously, often vehemently, maintained the immutability of species. . . . I feel how rash it is to differ from these great authorities, to whom, with others, we owe all our knowledge. Those who think the natural geological record in any degree perfect, . . . will undoubtedly at once reject my theory.[32]

The fossil record does not support the Darwinian version of natural history. The near-absence of complex life forms below the Cambrian,[33] the incredible diversity of life forms that suddenly appear in the Cambrian explosion, the fact that fossil forms do not change or evolve during millions of years of geological time, and the extreme rarity of transitional or intermediate forms anywhere in the fossil record are together sufficient to falsify the larger claims of Darwinism. That scientists embraced Darwinism in spite of the fossil record—and continue to subscribe to it in spite of the fossil record—indicates a dogged determination to adhere to naturalism and reject theism, no matter the evidence.

The observed configuration of fossils is, however, exactly what a biblical creationist would expect to find. First, if the fossiliferous layers were formed during the Flood, we would expect that the plants and animals then existing would have been buried and preserved as fossils, and that is exactly what we find.

Second, a biblical creationist who believes that the Genesis Flood began in the oceans with the bursting forth of the "great fountains of the deep" would expect to find a great variety of sea creatures buried by turbidity currents during the early stages of the flood. He would not expect to find intermediate fossil forms leading up to those sea floor creatures; rather, they would all appear suddenly in the fossil record, an "explosion" of diversity and variety at the lowest fossiliferous level. This is exactly what we see in the Cambrian explosion.

Third, instead of the "branching tree" pattern predicted by Darwin, the biblical creationist would expect to find a tremendous range and diversity of life at the very lowest fossiliferous levels. In fact, we find that the "tree" of life is upside down, with almost all the modern phyla and many extinct phyla appearing in the lower Cambrian. This aspect of the "Cambrian explosion" is well illustrated by the Burgess Shale fauna.

Fourth, a biblical creationist would not expect to find fossil evidence of any species changing into a different species. Rather, as each ecological zone was overtaken by the floodwaters, the flora and fauna of that zone were drowned and buried in sediment. After each zone was overtaken, the flora and fauna of that zone were eradicated, and not buried again in higher strata. This is why forms come into the fossil record, they do not change during their tenure of "geological time," and then they disappear from the fossil record, but yield no evidence of having evolved into something else. This pattern of abrupt appearance, stasis, and disappearance is puzzling to Darwinists, but it is exactly what biblical creationists would expect, and exactly what we find.

Finally, a biblical creationist would expect to find many fossil animals in the strata that no longer exist as living animals. This is partly because the ruined post-Flood world cannot support the range, diversity, and type of life that it could support when it came perfect from the hand of its Creator. Seventh-day Adventists have another reason to expect to find fossil animals that no longer exist as living species. We will discuss this reason later. Thus, we find many fossil species that are now extinct and known *only* from fossils.

The fossil record, far from supporting Darwinism, confirms the truthfulness and accuracy of the Word of God. Seventh-day Adventists have been told that the fossil record was intended to confirm our faith in biblical history:

> In the history of the Flood, inspiration has explained that which geology alone could never fathom. In the days of Noah, men, animals, and trees . . . were buried, and thus preserved as an evidence to later generations that the antediluvians perished by a flood. *God designed that the discovery of these things should establish faith in inspired history*; but men, with their vain reasoning, fall into the same error as did the people before the Flood—the things which God gave them as a benefit, they turn into a curse by making a wrong use of them.[34]

We have learned in the last three chapters that all of the dinosaur families suddenly appear in the fossil record, with scarcely a hint of an evolutionary ancestry. We have learned that the modern theory of evolution is an implausible assertion that copying errors in the genetic instructions created the amazing biological mechanisms, organs, and life forms we see around us today. The theory of evolution is radically defective, and the fossil record does not support it. There is no reason to believe that the dinosaurs—or any other animal—ever evolved from a dissimilar animal.

If the dinosaurs did not evolve from other types of animals, where did they come from? Most biblical creationists believe that God created the dinosaurs as part of the original creation.[35] I doubt it. Before exploring why I doubt it, let us first take a closer look at the dinosaurs.

[1] For example, no transitional fossil has ever been found among the marine invertebrates, which comprise 95 percent of all fossils. See, e.g., Kurt Wise, *Faith, Form and Time,* (Nashville, TN: Broadman & Holman, 2002), p. 125 ("Something on the order of 95 percent of the fossil record is of shallow water marine invertebrates, and most of those are found in just a handful of phyla. Yet there seems to be a complete absence of transitional species among these animals—even though these are the best represented animals in the fossil record!") See also oral presentation at Grace Reformed Baptist Church, Mebane, North Carolina, 9/30/2000. Found at *http://www.vor.org/mebane/sermons/2000_origins_conference.html.*

[2] Darwin, Charles, *The Origin of Species,* 1859, (Reprint: New York: Gramercy Books), p. 291 ("The main cause, however, of innumerable intermediate links not now occurring everywhere throughout nature depends on the very process of natural selection, through which new varieties continually take the places of and exterminate their parent-forms.")

[3] Darwin, at pp. 291, 292. (Chapter 10, par. 1.) *http://www.bartleby.com/11/1006.html.* This is one of several places in which Darwin admits that innumerable intermediate or transitional forms are predicted by his theory. "[T]he number of intermediate and transitional links, between all living and extinct species, must have been inconceivably great. But assuredly, if this theory be true, such have lived upon this earth." Darwin, at 293. "Geological research, though it has added numerous species to existing and extinct genera, and has made the intervals between some few groups less wide than they otherwise would have been, yet has done scarcely anything in breaking down the distinction between species, by connecting them together by numerous, fine, intermediate varieties; and this not having been effected, is probably the gravest and most obvious of all the many objections which may be urged against my views." Darwin, at 307. "If then, there be some degree of truth in these remarks, we have no right to expect to find in our geological formations, an infinite number of those fine transitional forms, which on my theory assuredly have connected all the past and present species of the same group into one long and branching chain of life. . . . But I do not pretend that I should ever have suspected how poor a record of the mutations of life, the best preserved geological section presented, had not the difficulty of our not discovering innumerable transitional links pressed so hardly on my theory." Darwin, at 309. It is as though Darwin thought that by repeating the problem over and over, he could make it go away, or at least convince the reader that he had solved it, which, of course, he had not.

[4] "Only a small portion of the surface of the earth has been geologically explored, and no part with sufficient care, as the important discoveries made every year in Europe prove." Darwin, at 298.

[5] Wells, Jonathan, *Icons of Evolution*, (Washington, D.C.: Regnery, 2000), pp. 195-207.

[6] Milton, Richard, *Shattering the Myths of Darwinism*, (Rochester VT: Oak Street Press, 1992), p. 110.

[7] Gould, Stephen Jay, "Evolution's erratic pace," *Natural History*, vol. 86, no. 5 (May 1977). See also Gould, *The Panda's Thumb*, (New York: W.W. Norton, 1980), p. 181.

[8] Gould, *The Panda's Thumb*, p. 189. See also Gould, "The Return of Hopeful Monsters," *Natural History*, 86:22 (1977).

[9] Simpson, George Gaylord, *Major Features of Evolution*, (New York: Simon & Schuster, 1967) p. 360.

[10] Berlinski, David, "The Deniable Darwin," *Commentary*, 101:6 (June 1996).

[11] Gould, Stephen Jay, "Evolution's Erratic Pace," *Natural History*, 86(5) (May, 1977). See also Gould, *The Panda's Thumb*, p. 182.

[12] Niles Eldridge and Ian Tatersall, *The Myths of Human Evolution* (Reprint: New York: Columbia University Press, 1984), pp. 45-48.

[13] Gould, Stephen Jay, "Cordelia's Dilemma," *Natural History*, (Feb. 1993). See, also, *Dinosaur in a Haystack*, (New York: Harmony Books, 1995), pp. 127-128.

[14] Ridley, Mark, "Who doubts evolution?" *New Scientist*, vol. 90, p. 830-832 (1981).

[15] Darwin, at 292.

[16] "But the imperfection in the geological record mainly results from another and more important cause than any of the foregoing; namely, for the several formations being separated from each other by wide intervals of time." Darwin, at 299. "I am convinced that all our ancient formations, which are rich in fossils, have thus been formed during periods of subsidence." Darwin, at 300.

[17] Niles Eldridge and Stephen Jay Gould have promoted a variant of Darwinism called "punctuated equilibrium," which tries to explain why transitional forms are almost never found in the fossil record. Rudely summarized, "punk-eek" holds that animals are in equilibrium and do not evolve for long periods of time. The equilibrium is punctuated by episodes of frantic evolving which occur in isolated valleys that happen not to be conducive to the formation of fossils. Eldridge, Niles,

and S. J. Gould, "Punctuated equilibria: an alternative to phyletic gradualism," in *Models in Paleobiology*, T. J. M. Schopf (ed.) (San Francisco: Freeman, Cooper, 1972) pp. 82-115. Punk-eek is just an elaboration of something Darwin wrote in arguing that the fossil record was "imperfect": "One other consideration is worth notice: with animals and plants that can propagate rapidly and are not highly locomotive, there is reason to suspect, as we have formerly seen, that their varieties are at first local; and that such local varieties do not spread widely and supplant their parent-forms until they have been modified and perfected in some considerable degree. According to this view, the chance of discovering in a formation in any one country all the early stages of transition between any two forms, is small, for the successive changes are supposed to have been local or confined to some one spot." Darwin, Charles, *The Origin of Species*, 1859 (reprint: New York: Gramercy Books, 1979) p. 306.

18 A problem with Darwin's excuse is that it undercuts the very Uniformitarianism that provided the eons of time necessary to Darwin's theory. The view that the fossil record constitutes snapshots of life separated by long ages is Cuvier's catastrophism, not Lyell's uniformitarianism. Snapshots of life implies great cataracts of sedimentation in a short period, followed by long ages of non-deposition during which evolution is occurring unrecorded by fossilization. This is closer to what Cuvier and the catastrophists thought, whereas Lyell envisioned sedimentation as a slow, steady, gradual, continuous process. If deposition was not continuous, how do we know there were long time gaps of non-deposition? Once Lyellism is abandoned, there is no proof that the sedimentary strata took hundreds of millions of years to form, and therefore no proof that there was sufficient time for evolution to have taken place

19 Zukav, Gary, *The Dancing Wu Li Masters*, (New York: Bantam Books, 1979), p. 180-182.

20 Darwin, at 312.

21 Darwin used the term *Silurian*, but changes in geological nomenclature were subsequently accepted, such that the lowest major division of fossiliferous strata is now called the Cambrian. Darwin was referring to the Cambrian explosion.

22 Darwin, at 313.

23 Darwin, at 313, 314.

24 "[I]t might well happen that strata which had subsided some miles nearer to the centre of the earth, and which had been pressed on by an enormous weight of superincumbent water, might have undergone far more metamorphic action than strata which have always remained nearer to the surface. The immense areas in some parts of the world, for instance in South America, of bare metamorphic rocks, which must have been heated under great pressure, have always seemed to me to

require some special explanation; and we may perhaps believe that we see in these large areas, the many formations long anterior to the [Cambrian] epoch in a completely metamorphosed condition." Darwin, at 315.

[25] Gould, Stephen Jay, *Wonderful Life: The Burgess Shale and the Nature of History* (New York: W.W. Norton, 1989), p. 57.

[26] Gould, *Wonderful Life,* at 58.

[27] Gould, *Wonderful Life,* at 23.

[28] Gould, *Wonderful Life,* at 47.

[29] The Cambrian Explosion is now known to have included vertebrates. See, e.g., Shu, D., et al., "Lower Cambrian vertebrates from south China," *Nature* 402(6757):42–46, 1999; Monastersky, R. "Waking Up to the Dawn Vertebrates," *Science News Online* (11/6/99) ("It's important because up to now the vertebrates were absent from the big bang of life, as we call it—that is, the great early Cambrian explosion, where all the major animal groups appeared suddenly in the fossil record," comments Philippe Janvier, a paleontologist at the National Museum of Natural History in Paris.)

[30] Gould, *Wonderful Life,* at 47.

[31] "I know of no greater challenge to the iconography of the cone [of increasing diversity] . . . than the radical reconstructions of Burgess anatomy presented by Whittington and his colleagues. . . . they have inverted the cone." Gould, *Wonderful Life,* at 46, 47.

[32] Darwin, at 315-316. See Chapter 5 for another version of this passage.

[33] The Ediacaran fauna were apparently multi-cellular, but, as previously noted, the current view is that most of these organisms were not related to Cambrian forms, and thus do nothing to ameliorate the problem of the "Cambrian explosion." See, e.g., Gould, *Wonderful Life,* 58, 59, 231, 311-314; Norman, David, *Prehistoric Life: The Rise of the Vertebrates,* (New York: Macmillan, 1994), pp. 34-38.

[34] White, Ellen, *Patriarchs and Prophets,* p. 112, italics mine.

[35] See, e.g., Ham, Ken, *Dinosaurs and the Bible,* Answers in Genisis Ministries.

CHAPTER 14

# What Are Dinosaurs?

The expression "mad as a hatter" entered the language because madness was too often the fate of hat-makers back when dangerous chemicals such as arsenic, lead, and mercury were used in the hat-making trade. Edward Hitchcock's father was a hat-maker. Edward suffered poor health in childhood, possibly because of his father's trade. Edward Hitchcock (1793-1864) studied theology at Yale, becoming a Congregationalist minister. But like Buckland, Conybeare, and Sedgwick, his clergymen contemporaries in England, Hitchcock concentrated on geology, not theology. Hitchcock taught geology at Amherst College in Massachusetts for over 30 years, serving as president of the college for nine of those years.[1]

Hitchcock pioneered the study of fossil tracks and track ways—a science called "ichnology" (from the Greek "ichnos," meaning footprint or track). In 1835, he began collecting and studying fossil footprints from the Connecticut River valley. These tracks had always been assumed to be bird tracks; their discoverer, Pliny Moody, had attributed them to "Noah's raven." When Hitchcock was studying the tracks, it was thought that all dinosaurs were quadrupedal, walking on all four limbs.[2] But the tracks Hitchcock was studying had clearly been made by bipedal animals—animals that walked on two feet. Hence, he never considered that they might have been dinosaur tracks. Hitchcock concluded that large, flightless birds, perhaps extinct relatives of the emu and the ostrich, left the tracks. He knew about the Moa of New Zealand[3], and the Aepyornis, or Elephant bird, of Madagascar, and believed that perhaps these birds, or similar extinct species, made the tracks.[4]

One thing did baffle him, however. One of his track ways clearly contained the drag mark of the animal's tail. Hitchcock was not dealing with any normal bird. "How very strange," he wrote, "must have been the appearance of a lizard, or batrachian [amphibian], with feet and body like those of a bird, yet dragging a veritable tail!"[5]

Another of the track ways was even more obviously the product of a composite creature. Hitchcock labeled this specimen *Anomoepus* ("anomalous feet"). "If a phrase so compound could be used as should imply a participation in the characters of marsupials, birds, batrachians, and even lizards, it would better than any other express my present convictions," he wrote, "for the longer I study these tracks, the stronger my impressions become that some of these ancient animals possessed characteristics now more exclusively belonging to two, three, or even four [vertebrate] classes."[6]

Although Hitchcock was not sure what these creatures were, he knew they were strange and very large. "Whatever doubts we may entertain as to the exact place on the zoological scale which these animals occupied, one feels sure that many of them were peculiar and gigantic: and I have experienced all the excitement of romance, as I have gone back into those immensely remote ages and watched those shores upon which those enormous heteroclitic [deviating from common forms] beings walked."[7]

Over the years, Hitchcock's collection of fossil tracks waxed ever larger; a building was erected at Amherst just to house the collection.

Hitchcock described the strange tracks in his magnum opus, *Ichnology of New England*, published in 1858. He died six years later, but in 1865, his sons produced a *Supplement to the Ichnology of New England*. By then the catalog of track making species included twenty-one species of "ornithoid [bird-like] reptiles."[8] In 1866, Charles Hitchcock, who followed his father into the geology profession, published a description of a track made by a "reptilian bird" from the Triassic of Massachusetts.[9]

Most of the tracks Hitchcock and his sons studied are now known to be dinosaur tracks. An ornithopod ("bird-footed") dinosaur probably made the *Anomoepus* tracks.[10] Hitchcock was quite correct, however, in discerning the mixed nature of the track makers. In fact, Hitchcock's conclusion that *Anomoepus* was a bird-like reptile was prescient in a time when the avian characteristics of dinosaurs were not well understood.[11]

But Hitchcock was not alone in noting the mixed nature of the dinosaurs. Richard Owen (**Plate 7**), who had only the fragmentary remains of a few species with which to work, correctly discerned that the dinosaurs embodied a strange combination of characteristics. In his seminal 1842 paper in which he coined the name "Dinosauria," Owen described how "crocodilian characters of the vertebral column are combined with a Lacertian [lizard] type of the pectoral arch," and how the limb bones "more or less resemble those of the heavy pachydermal Mammals." Owen relied upon "the combination of such characters, some, as the sacral ones, altogether peculiar among Reptiles, others borrowed, as it were, from groups now distinct from each other" to classify the *Dinosauria* as a "distinct tribe."

"When the first dinosaur skeletons were hewn out of gravel quarries in England during the 1820s and 30s," writes dinosaur expert Robert T. Bakker, "the gentleman-naturalist [Owen] immediately recognized their strange combination of characteristics. They combined traits found in lizards, in birds, in mammals, and in crocodiles as well."[12]

The heterodox—mixed, amalgamated—nature of the dinosaurs is their defining characteristic and, other than the enormous size of some of them, the most striking aspect of their physiology. To a skilled anatomist like Richard Owen, it was obvious from only a few bones of a few species. Ever since Owen, the dinosaurs have been considered reptiles, but this traditional classification is now disputed. "No one, either in the nineteenth century or in the twentieth," writes Bakker, "has ever built a persuasive case proving that dinosaurs as a whole were more like reptilian crocodiles than warm-blooded birds. No one has done this because it can't be done."[13]

While Owen thought that the *Dinosauria* were an order or sub-order of the reptiles, Bakker argues that they should be considered an entirely separate class.[14] In the early days, scientists emphasized the dinosaurs' reptilian characteristics; now, largely because dinosaurs are thought to be the ancestors of birds, scientists tend to dwell on the dinosaurs' avian characteristics. But the dinosaurs were neither bird nor reptile. They were, as Edward Hitchcock was able to discern *solely from their tracks*, a strange combination of characteristics now more exclusively belonging to two, three, or even four vertebrate classes. This chapter will survey the anatomical features of several types of dinosaurs, and conclude with a discussion of the dinosaurs' thermal physiology.

## A. TRICERATOPS AND THE HORNED DINOSAURS

The most distinctive thing about *Triceratops* ("three-horned face") (**Plate 18**) is that it had horns. Horns are a mammalian characteristic and are found on several different kinds of hooved mammals (ungulates). Some of the horned ungulates include sheep, cattle, antelope, and bison. Recall that when O. C. Marsh first saw *Triceratops* horn cores, he assumed they had belonged to an extinct bison. The actual horny material, a protein substance called keratin, has not yet been found. That the animal had true horns can be inferred, however, because the horn cores, the underlying bony material, have grooves and channels, suggesting that a blood supply ran under the keratin.[15]

The fact that *Triceratops* had horns like some ungulate mammals raises the question whether he might not also have had hooves. Hooves are made of the same keratin protein that forms the outer covering of horns (and human fingernails and toenails). The animals' unguals (toe bones) closely resemble those of a horse. Peter Dodson, an expert on the horned dinosaurs, notes many similarities between horse unguals and *Triceratops'* unguals, but does not believe that *Triceratops* had an exposed hoof like a horse. He does believe that the toes were covered "in a somewhat hoof-like covering on both the upper and lower surfaces."[16]

After the horns, the next thing you notice about *Triceratops* is his beak. The only modern reptile that has a beak is the turtle, but *Triceratops'* beak was longer and more bird-like than typical turtle beaks. "The beak was extremely narrow, consisting of sharply hooked upper and lower parts, arranged and looking rather like those of a parrot."[17]

*Triceratops* also had teeth, whereas modern birds and turtles lack teeth. *Triceratops* not only had teeth, he had dental batteries, meaning that there were always three to five teeth lined up below the exposed tooth, waiting to take its place when it fell out.[18] Modern reptiles, such as crocodiles, have a system in which socketed teeth are always being replaced as they fall out. In effect, crocodiles have a lifetime supply of teeth. But the dental batteries found in *Triceratops* and the other ceratopsians have no true modern counterpart.

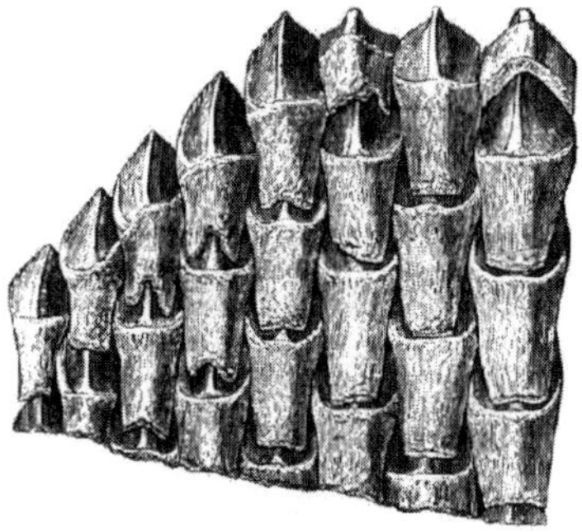

Figure 14.1: Triceratops' Dental Battery

*Triceratops* had reptilian skin. An impression of the skin of the pelvic region of *Chasmosaurus* (another of the horned dinosaurs) was found preserved in contour in rock. From this remarkable fossil, paleontologists determined that the skin consisted of large circular plates up to 5.5 centimeters in diameter set in irregular rows spaced 5 centimeters from each other. Between the large round plates are found smaller polygonal plates.[19]

*Triceratops* had a hipbone structure that has been described as bird-like, which leads us into a discussion of how dinosaurs are divided into two orders. In 1887, Harry Govier Seeley, a professor at King's College in London, noticed that only two patterns of hip bone configuration were present in dinosaurs. One group of dinosaurs had hip bones similar to lizards, with the pubis bone pointing forward and down; the other group had hip bones similar

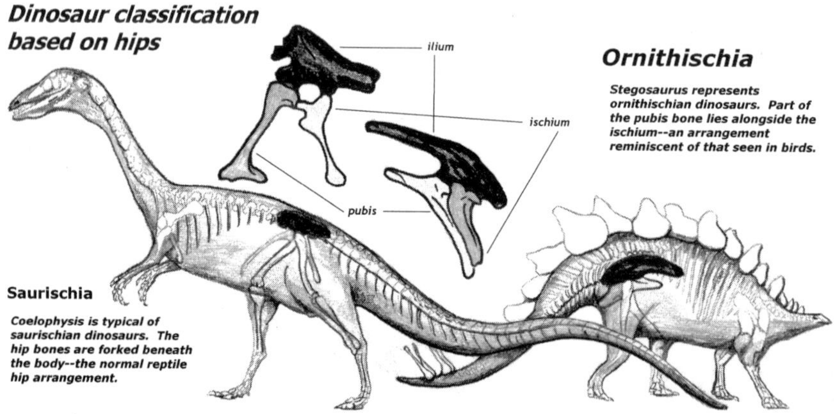


Figure 14.2: Dinosaur Classifications Based Upon Hip Configuration

to birds, with the pubis bone pointing backward and down. Seeley called the dinosaurs with lizard hips *Saurischians*; the bird-hipped dinosaurs he called *Ornithischians.*

It would seem the shape of the hipbones should affect locomotion; intuitively, we would expect the *Saurischians* to be quadrupedal, like lizards and crocodiles, and the *Ornithischians* to be bipedal, like birds. Naturally, we would be wrong. The *Saurischia* includes both bipeds (*T. rex* and other theropods) and quadrupeds (*Brontosaurus* and other sauropods). The *Ornithischia* also includes both quadrupeds (ankylosaurs, nodosaurs, ceratopsians, and stegosaurs) and animals thought to have been primarily bipedal (such ornithopods as the heterodontosaurs, hypsilophodonts, iguanodonts, and hadrosaurs).

Triceratops and the other horned dinosaurs are classified as *Ornithischians*, although the bird-hipped pattern is not as clear in the horned dinosaurs as it is in most other Ornithischians.[20] *Triceratops horridus* is classified as follows:

Class: *Reptilia*
- Subclass: *Archosauria*
  - Order: *Ornithischia*
    - Family: *Ceratopsidae*
      - Sub-family: *Chasmosaurinae*
        - Genus: *Triceratops*
          - Species: *Horridus*

Under this traditional classification scheme, there is no category of "dinosaurs," as such. They are lumped together with archosaurs, which are then subdivided into *Saurischia* and *Ornithischia.*

*Ornithischians* have another avian characteristic in common besides hip structure. All Ornithischians have a small beak-like structure on the tip of the lower jaw, called the predentary bone.

Some paleontologists believe that many of the *Ornithischians*, not just *Triceratops*, had an actual beak.

*Triceratops* and the other Ceratopsians had a body that was much like that of a rhinoceros, the horned mammal found in Africa and tropical Asia.[21] One obvious difference is that a rhino has a cartilaginous, mammalian tail, with fewer and smaller bones, whereas *Triceratops'* tail was the reptilian affair typical of the dinosaurs, in which vertebrae continue down the tail, becoming progressively smaller. There is another major skeletal difference, in that the scapula (shoulder blade) of *Triceratops* is similar to that of a chicken. The arrangement of the scapula, coracoid, and sternum in Triceratops is unlike that found in mammals.[22]

There is also a difference in the shape of the bones of the forelimbs, which has given rise to a heated scientific controversy. A rhinoceros stands with its front legs straight beneath it. Those who have tried to reconstruct horned dinosaurs for mounting, however, have found that the elbow joint is unrealistically configured when the front legs are straightened. They argue that *Triceratops'* forelimbs must have sprawled, like the limbs of a crocodile.[23] But Bob Bakker spent a year researching this very issue as a Yale undergraduate, and he argues vigorously that the forelimbs should be reconstructed to stand directly beneath the animal, just as a rhinoceros stands.[24]

Ichnologists Martin Lockley and Adrian Hunt shed light on this controversy when they identified a *Triceratops* track way near Denver, Colorado. They interpreted the tracks to indicate an upright position of the forelimbs, but Dodson disagrees, noting that *Triceratops'* shoulders were narrower than its hips, and we would therefore expect the forelimb prints (handprints) to fall inside the footprints. But the track way shows the handprints falling outside the footprints, with the hands rotated outward from the direction of travel.[25] This is difficult to explain except as a result of sprawling forelimb posture. If *Triceratops* sprawled, it is a departure from type, because the straight-legged, non-sprawling posture of the dinosaurs is one of the characteristics that separates them as a class from the reptiles of the Permian and lower Triassic, as well as from modern reptiles such as crocodilians and lizards.[26]

To summarize, the horned dinosaurs exhibit characteristics of reptiles, mammals, and birds. The reptilian characteristics include the skin, the teeth, the tail, and the (probable) sprawling posture of the front limbs. Mammalian characteristics include the horns, much of the basic skeletal structure, and the feet. The avian characteristics include the beak, the scapula and structure of the breastbones, and, to a lesser extent, the bones of the hip.

## B. TYRANNOSAURUS REX

The "king of the tyrant reptiles" is easily the most famous of all dinosaurs. But just how much of *T. rex* was reptile? *Tyrannosaurus* was a large, bipedal, carnivorous theropod. All theropod feet resemble the feet of raptors—birds of prey. Rex's feet had three toes ending in sharp claws.[27] *T. rex* remains are found exclusively in the upper Cretaceous, the highest dinosaur strata, and theropods from these strata are the most bird-like of all the dinosaurs. Formerly, *T. rex*, like most large theropods, was classed as a "carnosaur." Lately, however, many have reclassified *T. rex* as a coelurosaur.[28]

When scientists closely examined the nearly complete skeleton of "Sue," they found something that had never been associated with *T. rex* before: a *furcula*, or wishbone. Birds have wishbones, but reptiles generally do not. (Wishbones have also been found in *Velociraptor* and *Allosaurus*.) Rex also had many holes in his skull other than those necessary for the sensory organs. These holes would have reduced the weight of the skull, as they do in modern birds.[29]

**Figure 14.3: Tyrannosaurus "Sue" at the Field Museum in Chicago**

*T. rex* had another weight-saving device characteristic of birds: his vertebrae were perforated with large openings and spaces called *pleurocoels.* In birds, these spaces are part of an air sac system consisting of a series of air-filled passages connected to unidirectional avian lungs.[30] Some scientists believe that the hollow vertebrae in dinosaurs such as *T. rex* were part of a similar air sac system.[31]

The ball joint which connected *T. rex*'s head to his neck was also similar to birds. The ball, a structure called an *occipital condyle,* fits in a hollow in the atlas vertebra—the first neck vertebra. Humans have a double condyle, which limits the motion of our necks. But *T. rex,* like birds, had a single condyle, allowing it greater flexibility and greater ability to look to the rear.[32]

In addition to his avian characteristics, Rex had many reptilian features. He was a lizard-hipped *Saurischian* dinosaur.[33] He lacked the beak-like predentary bone of the *Ornithischians*. Like most dinosaurs, Rex had a self-replacing dental system similar to a crocodile's. For each position, two or three teeth were always developing, getting ready to emerge and replace an older, worn tooth. Although crocodiles replace their teeth about once a year, one study calculated that Rex's teeth would have lasted about two to three years before being replaced.[34] Rex's teeth were frightening. Some were a foot long (including the root and the exposed portion), and they had serrated edges like a steak knife.[35]

Skin impressions from *Albertosaurus*, a dinosaur similar to *T. rex*, show a scaly reptilian hide packed with many small bumps. A skin impression from the tail of another similar species, *Gorgosaurus libratus*, shows small, hexagonal scales.[36] *T. rex* was a combination of avian and reptilian characteristics. The "king of the tyrant reptiles" might better have been named the "king of the tyrant bird-like reptiles."

## C. DUCKBILL DINOSAURS

If it walked like a duck, had a bill like a duck, webbed feet like a duck, and a pelvis like a duck, it wasn't necessarily a duck. It might have been a dinosaur. The *Hadrosaurs* ("bulky reptiles") and several similar genera are typically called the "duckbill dinosaurs;" they are found in Upper Cretaceous sediments around the world. From their abundant fossil record, more is known about the duckbills than any other type of dinosaur. The duckbills were part of a larger group that Professor Marsh labeled *Ornithopods* ("bird-feet"). The *Ornithopods* are, like the horned dinosaurs, the armored dinosaurs, and *Stegosaurus*, a subgroup of the *Ornithischians*, the "bird-hipped" dinosaurs.

Many of the duckbills had wide, flattened bills, which resemble the bill of a mallard duck. The bill was covered in a horny (keratin) sheath, similar to the beaks of birds, turtles, and horned dinosaurs. Fossil molds and mummified remains show a thick, horny sheath bound to the front of the upper jaws.[37] The beak was turned down into a sharp, cookie-cutter edge, with the upper and lower bills closely occluded, making the bill ideal for cropping vegetation. The duckbills had dental batteries similar to those of *Triceratops,* in which new teeth were constantly coming up to replace old. While *Triceratops* had a battery of slicing teeth, the duckbills had batteries of grinding teeth, comparable in function to human molars. The duckbills were obviously herbivorous; their bills and teeth combined to cut and chew the vegetation that composed their diet.

Charles H. Sternberg was a prominent fossil hunter who spent his career working for Cope and later for Osborn and the American Museum. Sternberg involved his sons, George, Charles, and Levi, in the fossil hunting business, and they became renowned dinosaur hunters and paleontologists themselves. The Sternbergs' most celebrated discovery came in 1908, when the four men were digging in southern Wyoming, a two-day ride from the town of Lusk. The elder Charles and his namesake had gone into town for provisions—a round trip that would take five days—leaving George and Levi to pick away at a duckbill dinosaur in the side of a sandstone outcrop. The skeleton lay on its back with its ribs sticking up.

**Figure 14.4: Duckbill dinosaurs drinking at a stream**

"There was nothing unusual about that," wrote George, "but when I removed a rather large piece of sandstone rock from over the breast I found, much to my surprise, a perfect cast of the skin impression beautifully preserved. Imagine the feeling that crept over me when I realized that here for the first time a skeleton of a dinosaur had been discovered wrapped in its skin." The American Museum acquired the specimen and displayed it in a glass case with the label, "Dinosaur mummy." Charles Sternberg called it "the crowning specimen of my life's work."

The Sternbergs' find gave paleontologists their first look at the texture and patterns of dinosaur skin. Duckbill skin was thick and wrinkled, with bony knobs of various sizes embedded throughout.[38] Hadrosaurs appear to have lacked scales; their bumpy, pebbled skin was much like that of an elephant.[39] Another mummified duckbill, a *Brachylophosaurus*, was recently unearthed in Montana; paleontologists hope that the soft-tissue remains preserved with this specimen will expand our knowledge of how these animals lived and moved.[40]

The American Museum displays a third specimen of duckbill mummy, this one found in Alberta, Canada. The skin impressions on this fossil revealed a remarkable fact—the duckbill's front feet were webbed![41] Or so it was long thought. But a recent reinterpretation of the "webbing" holds that the front feet were padded, like the feet of modern camels and other ungulates.[42] The man who believes that the "webbing" was actually padding, Bob Bakker, was visiting South Africa when he saw a dead camel in the desert. He noticed that the desiccated padding on the camel's feet closely resembled webbing. Bakker theorized that what had been interpreted as webbing on the foot of a duckbill mummy was actually a dried-up footpad.

A mysterious aspect of the duckbills concerns the bony crests on the tops of their heads. In *Parasaurolophus* (saurolophus = "crested reptile," **Plate 19**) sometimes called the "trombone-headed" dinosaur, the crest housed a convoluted nasal passage that traveled backward for three feet before doubling back to the throat—a total distance of more than six feet. (**Figure 14.5**) Some believe it was used to enhance the animals' cries, probably their mating calls. Computer analyses have shown that the trombone crest would have made quite an impressive noise.[43]

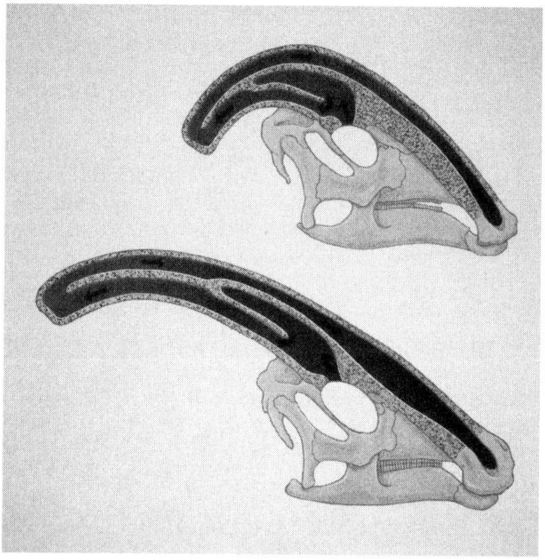

**Figure 14.5: Head of Parasaurolophus (cross-section)**

## D. SAUROPODS

In 1841, when Richard Owen was reviewing the fossil animals he was soon to name the *Dinosauria,* he came across a group of bones that could not be assigned to any known form. The bones had a texture somewhat like whalebone, so Owen christened the animal *Cetiosaurus* ("whale lizard"). Owen thought the remains might be those of a large marine reptile, but they actually belonged to a sauropod dinosaur.[44]

The sauropods were large quadrupedal animals—by far the largest land-dwelling animals that ever lived. Seismosaurus ("Quake Lizard") (**Plate 20**) was so named because when this thirty-ton animal was on the move, the earth shook. Much was learned about them when the competition between Cope and Marsh turned up many sauropod remains in the Jurassic strata of the western United States.

As of this writing, most of the sauropod remains that Barnum Brown quarried at Howe Ranch in the 1930s had never been prepared out of their rock matrices.[45] There must be a surplus of sauropod bones or a lack of interest in these animals, or both.

In the late 1800s and early 1900s, there was a controversy about sauropod posture. Did they walk with their legs directly beneath them, like elephants, or with their limbs sprawled, like crocodiles? Some German paleontologists had reconstructed sauropods as sprawling, but most of the American authorities, including Marsh, displayed them with their legs directly beneath them. Skeletal analysis suggested that the elephant-like posture was correct, and the matter was settled when the American Museum's Roland T. Bird found a long sauropod track way in the lower Cretaceous limestone of the Paluxy River, near Glen Rose, Texas. The narrow track way was not consistent with a sprawling posture.[46]

One of the mysteries about sauropods is how blood reached their brains when their long necks were elevated. The systolic blood pressure required to pump the blood up the twenty-five feet from the chest cavity to the skull would probably have been twice that of giraffes and four or five times greater than human blood pressure.[47] This suggests that the heart must have been extraordinarily large and powerful—and likely four-chambered, like mammals, birds, and crocodilians (other reptiles have three-chambered hearts). It has even been suggested that sauropods had more than one heart.[48]

The sauropods must also have had special soft-tissue structures, found today in giraffes, to keep the blood in one place when the head was being raised and lowered. Giraffes have muscularized arteries with one-way valves to keep the blood from flowing back down their necks, and a structure at the base of their brains to allow blood to pool and prevent them from blacking out when they suddenly raise their heads.[49] Sauropods may also have had a carotid sinus that served as a reservoir for arterial blood to supply the brain when the head was fully elevated.[50]

Although sauropod skeletal structure, posture, and probable soft-tissue anatomy were similar to mammals, the sauropods had the usual suite of reptilian characteristics, such as the extremely long, bony tail. They had self-replacing teeth,[51] they laid eggs,[52] and they had small brain/body-size ratios, the smallest of all the dinosaurs.[53]Some of the sauropods, specifically the Titanosaurs, had dermal armor consisting of embossed plates and small ossicles.[54]

The sauropods were the least bird-like of the dinosaurs. They were *Saurischians,* lacking the avian characteristics of the *Ornithischians.* They also lacked the many bird-like features of the theropods, the other main branch of *Saurischians.* Some paleontologists nevertheless believe that, because of their extremely long necks, sauropods must have been equipped with the one-way avian lung, as opposed to the bellows-type lungs possessed by mammals and crocodilians.[55] In addition, polished stones have been found in association with sauropod remains. Some believe that these were stomach stones ("gastroliths") held in a pouch called a gizzard, where they were used to grind coarse vegetation, aiding in digestion.[56] If this interpretation is correct, it is another avian characteristic. Thus, even the sauropods, which appear to have been primarily a cross between reptiles and mammals, may have had some avian characteristics.

## E. COELUROSAURS

The name *coelurosaur* (pronounced SEEL-uh-ro-SAWR) refers to the hollow bones of these dinosaurs, a characteristic they share with birds. The *coelurosauria* includes such bird-like theropods as *Compsognathus* ("pretty jaw"), *Ornithomimus* ("bird mimic"), *Oviraptor* ("egg stealer"), *Troödon* ("wounding tooth"), *Dromaeosaurus* ("fast running reptile"), *Deinonychus* ("terrible claw") (**Plate 21**), and *Ornitholestes* ("bird robber").[57] Generally, the coelurosauria includes theropod dinosaurs that were relatively small and bird-like, although, as noted, some now classify *Tyrannosaurus* as a coelurosaur.

Darwinists, noting the many anatomical features coelurosaurian dinosaurs share with birds, argue that at least one line of coelurosaurs evolved into birds. We will discuss that claim later, but it cannot be denied that coelurosaurs have much in common with birds. For example, coelurosaurs had hollow, thin-walled bones, elongated foot bones, an S-shaped, curved neck, and an erect stance with the feet positioned directly below the body and the ankles held well off the ground. Coelurosaur eggs have a microstructure similar to bird eggs. Some of the coelurosaurs had collarbones, and some apparently even had a furcula, or wishbone.[58]

Nevertheless, the coelurosaurs partook of the reptilian characteristics common to the dinosaurs: the reptilian tail; the blade-like, serrated, self-replacing, teeth[59]; the configuration of most of the skull; the skeletal bones of the trunk. Moreover, the coelurosaurs were *Saurischians* (lizard-hipped), and most had the classic *Saurischian* hipbone structure, with the pubis bone pointing forward and down. Most coelurosaurs lacked the predentary bone that all *Ornithischian* dinosaurs possessed.

## F. WERE DINOSAURS THERMAL INTERMEDIATES?

One of the most contentious issues in paleontology is whether the dinosaurs were cold-blooded (ectothermic), like modern reptiles, or warm-blooded (endothermic), like mammals. Because dinosaurs have always been classified as reptiles, and modern reptiles are cold-blooded, it was assumed that dinosaurs were cold-blooded. In the late 1960s, however, some paleontologists began to doubt this conventional wisdom, arguing that dinosaurs were more likely to have been warm-blooded. Among the advocates for dinosaurian warm-bloodedness, Bob Bakker has been the most outspoken. His 1986 book, *The Dinosaur Heresies*, is largely a sustained argument for warm-blooded dinosaurs.

Bakker seems to have convinced many of his colleagues, including John R. "Dinosaur Jack" Horner, who served as a technical advisor to Steven Spielberg on the motion picture "Jurassic Park." In that film, fictional paleontologist Alan Grant (who is based on Horner and is played by Sam Neill), is able to declare, after observing a living sauropod for only about forty seconds, "you can just tear up the rulebook on cold-bloodedness, it doesn't apply, they're totally wrong; this is a warm-blooded creature." Later, as they are watching a Velociraptor egg hatch, a technician assures Dr. Grant that the animal is indeed warm-blooded, with a constant temperature of 91 degrees.[60]

In real life, as opposed to celluloid fantasies, there are no living dinosaurs to study; attempting to ascertain the thermal nature of dinosaurs requires careful detective work. One area of study is the growth patterns of periosteal bone, the tissue formed on the side surfaces of a long bone when it grows in thickness. When it grows slowly, this bone is a dense, finely lamellated tissue, with few or none

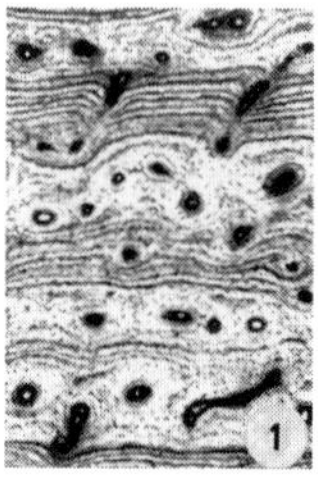

**Figure 14.6: Zonal bone, with lamellated bone forming annuli between successive nonlamellated zones. The large round perforations are cross sections of vascular canals. From the humerus of an Allosaurus.**

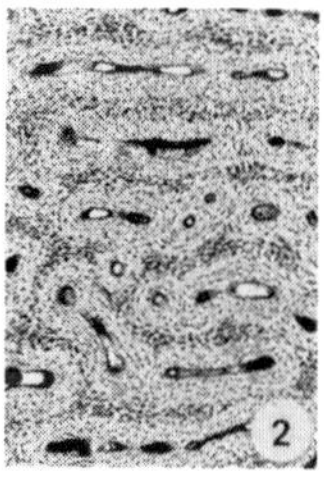

**Figure 14.7: Fribro-lamellar Bone, From the Limb Bones of a Sauropod**

of the channels for blood capillaries known as vascular canals. Modern reptiles exhibit this type of "zonal" bone growth, with annual "growth rings" like the growth rings of trees. When formed relatively quickly, by contrast, periosteal bone tissue is highly vascular and spongy. This second type of bone growth is referred to as *fibro-lamellar*, and it is the characteristic pattern of modern mammals, in which the bone grows continuously, without interruption.[61]

The periosteal bone evidence indicates that dinosaurs were able to grow in the manner of endothermic mammals as well as ectothermic reptiles. In fact, both patterns have been found in the *same animal*. Armand de Ricqles, the University of Paris paleontologist who pioneered this type of research, found uninterrupted fibro-lamellar bone in the limb bones of the large prosauropod *Euskelosaurus* ("good-limbed reptile"), but Robin Reid, of Queen's University, Belfast, found equally good zonal bone in the same animal's ribs. And Reid found a similar contrast between the femur and the pubis bone in *Megalosaurus*. "At present," writes Reid, "these types of bone are seen characteristically in endotherms and ectotherms respectively; but they cannot reflect different types of thermal physiology when both are found in one animal . . ."[62]

The mystery deepens when the zonal growth rings in dinosaur bones are assumed to be annual markers, as they are known to be in modern reptiles. A sauropod whose hipbones indicate that it was at least as large as an elephant showed twenty-three zonal growth rings. So although the bone grew in a reptilian manner, the animal grew at a rate no modern reptile could match. It grew at a mammalian rate similar to modern elephants, which take twenty to twenty-five years to reach full size.[63]

The mystery deepens further when one considers that the sauropod, had it lived, would have grown much larger than an elephant.

Indeed, many of the bones exhibiting the fibro-lamellar growth pattern show no signs of the growth having stopped, or even slowed, before the animal died.[64] This suggests that the dinosaurs combined the uninterrupted but limited growth pattern of warm-blooded animals (a mammal reaches a terminal size and grows no more) with the seasonal but indefinite growth pattern of modern cold-blooded animals (a reptile can grow throughout its life).[65]

Thus, even in such matters as whether they were warm-blooded or cold-blooded, the dinosaurs exhibit a heterodox pattern. They cannot easily be pigeonholed as reptiles, mammals, or birds. In growth patterns and in matters thermal and metabolic, the dinosaurs appear to be a mixture of the characteristics of modern vertebrate classes.

---

[1] Amherst was established to provide a free education to students preparing for the ministry. Between 1845 and 1854, Hitchcock served as Amherst's president. He was also active in a circle of New England intellectuals that included Herman Melville, Emily Dickinson, Henry David Thoreau, and James Wadsworth Longfellow, whose father-in-law helped fund some of Hitchcock's work. Farlow, James O., and Ralph E. Chapman, "The Scientific Study of Dinosaur Footprints," in, *The Complete Dinosaur*, Farlow, James O., and M.K. Brett-Surman, Editors, (Bloomington and Indianapolis: Indiana University Press, 1997), p. 519, 526.

[2] For example, when, in 1854, Benjamin Waterhouse Hawkins, with Sir Richard Owen as his scientific advisor, sculpted life-size models of the then-known dinosaurs for the Crystal Palace exhibition in London, Hawkins depicted all of them as quadrupedal.

[3] Interestingly, the New Zealand Moa remains were excavated and sent to England by Walter Mantell, son of dinosaur discoverer Gideon Mantell

[4] Lockley, Martin, and Christian Meyer, *Dinosaur Tracks and Other Fossil Footprints of Europe*, (New York: Columbia University Press, 2000), p. 105. The remains of seven-foot tall predatory birds, such as the *Diatryma*, are found in Tertiary strata, but it is not clear whether Hitchcock had these in mind when studying the tracks.

[5] Farlow, et al, at 527.

[6] Farlow, et al, at 527, citing Hitchcock, Edward, *Ichnology of New England: A Report on the Sandstone of the Connecticut Valley, Especially its Fossil Footmarks*, (Boston: William White [Reprint, New York: Arno Press] 1858).

[7] Colbert, Edwin H., *The Great Dinosaur Hunters and Their Discoveries*, (New York: Dover Publications, 1984), p. 38.

[8] Spalding, David A.E., *Dinosaur Hunters* (Rocklin, CA: Prima Publishing, 1993), p. 90.

[9] Spalding, at 91.

[10] Lockley and Meyer, at 118.

[11] Lockley, Martin, and Adrian P. Hunt, *Dinosaur Tracks and other Fossil Footprints of the Western United States*, (New York: Columbia University Press, 1995), p. 128.

[12] Bakker, Robert T., *The Dinosaur Heresies*, (New York: Kensington Publishing, 1986), p. 20.

[13] Bakker, at 27. There are lots of in-jokes about Bob Bakker and Jack Horner in the "Jurassic Park" movies. Dr. Grant (the main character, played by Sam Neill) is

modeled on Horner. In "Jurassic Park," Hammond, the entrepreneur who built the park, has two grandchildren, a boy and a girl. The boy, Tim (played by Joseph Mazzello), is a dinosaur enthusiast who has read Dr. Grant's book on dinosaurs. As they are about to leave on the tour of the park, Tim follows Dr. Grant around worshipfully, prattling about all his dinosaur knowledge. " . . . my teacher tells me about this other book, by a guy named Bakker [which Tim pronounces "Backer" although the name is pronounced "Baker"] and he says . ." at which point Grant shuts the car door with Tim inside and Grant outside. Later, after Dr. Grant and the others have left the tour vehicles, Tim catches up to Dr. Grant and says, "like I was saying, there's this other book by a guy named Bakker, and he says the dinosaurs died of a bunch of diseases, . . . and his book was ***a lot*** fatter than yours, it was like this [Tim makes a large bracket with his thumb and forefinger]." This is a humorous reference to the fact that Bob Bakker's 1986 book *The Dinosaur Heresies* is about twice as thick as Jack Horner's 1988 book, *Digging Dinosaurs.* (Universal's "Jurassic Park" screenplay by Michael Crichton and David Koepp, based on the book by Michael Crichton). In the second Jurassic Park movie, "The Lost World," the paleontologist brought to the island by the Ingen expedition (the bad guys) looks exactly like Bob Bakker, complete with long hair, long beard, cowboy hat, and dirty blue jeans.

[14] Bakker, at 459-462.

[15] Dodson, Peter, *The Horned Dinosaurs,* (Princeton NJ: Princeton University Press, 1996), p. 31.

[16] Bakker, at 32, 44.

[17] Norman, David B., *Dinosaur!,* (New York: Prentice Hall General Reference, 1991), p 113.

[18] Dodson, Peter, *The Horned Dinosaurs,* (Princeton NJ: Princeton University Press, 1996), p. 263-264.

[19] Dodson, at 31. (There's a character named "Dodgson" in the movie "Jurassic Park," but the phonetic identity between that name and Peter Dodson's may be a coincidence.)

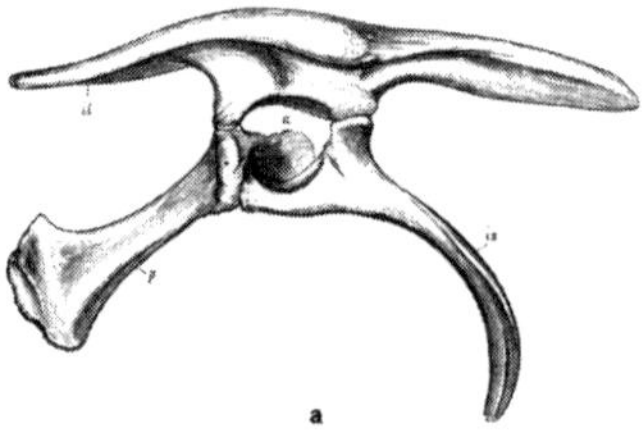

[20] *Triceratops'* pubis bone points forward and down, like a *Saurischian* pubis bone. "This pattern [the *Ornithischian* hip bone structure]," writes David Norman, "becomes a little obscure in some ornithischians, especially among those from the later Cretaceous Period (such as the ceratopsians [horned dinosaurs] and the anky-

losaurs), through shortening of the pubis and growth of a new forwardly pointing part to the bone; but the pattern is still evident." Norman, David B., *Dinosaur!*, supra, at 44. You can see from the illustration below that *Triceratops'* pubis (p) points down and to the front, and only parallels the ischium (is) for a short way near the intersection of all three bones:

[21] Dodson, *The Horned Dinosaurs*, at 15-17, 270.

[22] Dodson, at 39-41.

[23] See, e.g., Lessem, Don, *Kings of Creation*, (New York: Simon & Schuster, 1992), p. 159; Dodson, at 272, 273. Rolf Johnson of the Milwaukee Public Museum excavated and recovered a partial *Torosaurus* skeleton in 1981. "As he prepared the specimen for mounting and exhibition, he discovered anew what everyone who has ever tried to mount a ceratopsid has learned. The forelimb must sprawl." Dodson, at 273.

[24]Dodson, at 273; see, also, Bakker, at 205-225.

[25] Dodson, at 274-276.

[26] Hotton, Nicholas, *Dinosaurs*, (New York: Pyramid Publications, 1963), p. 50. Some of the mammal-like reptiles also appear to have stood with their legs directly below them, like dogs.

[27] Carpenter, Kenneth, *"Tyrannosauridae"* in *The Encyclopedia of Dinosaurs*, Philip J. Currie and Kevin Padian, (eds.) (New York: Academic Press, 1997), p. 767.

[28] See, e.g., Currie, Philip J., "Theropods" in *The Complete Dinosaur*, supra, at p. 229, 230. Cladists seem to enjoy upsetting and reorganizing previously accepted groupings. The sterility of cladistics is far outweighed by the opportunities for professional advancement from this churning of categories.

[29] Horner, John R., and Don Lessem, *The Complete T. rex*, (New York: Simon & Schuster, 1993), p. 106.

[30] Norman, at 102.

[31] See, e.g., Bakker, at 363-365; O'Connor, P. and Claessens, L., "Basic avian pulmonary design and flow-through ventilation in non-avian theropod dinosaurs," *Nature*, 436:253–256, July 14, 2005; Carl Wieland, "Dinosaurs breathed like birds?" Answers in Genesis, *http://www.answersingenesis.org/docs2005/0719dino_lungs.asp.*

[32] Horner, et al, at 117, 118.

[33] Horner, et al, at 119, 120.

[34] Horner, et al, at 110, 111. This is based upon counting tooth rings under a microscope, under the assumption that *T.rex* grew a new tooth ring daily, as modern crocodiles do.

[35] Horner, et al, at 111, 112.

[36] Horner, et al, at pp. 92, 93; Paul, Gregory S., "Restoring the life appearance of dinosaurs," *The Scientific American book of Dinosaurs,* (New York: St. Marten's Press, 2000), p. 91; Carpenter, at p. 768.

[37] McGowan, Christopher, *Dinosaurs, Spitfires, and Sea Dragons,* (Cambridge, MA: Harvard University Press, 1991), p. 71; Forster, Catherine A. "Hadrisauridae," in *Encyclopedia of Dinosaurs,* supra, at 295.

[38] Brett-Surman, supra, at pp. 343, 344.

[39] McGowan, Christopher, *The Successful Dragons,* (Toronto: Samuel Stevens, 1983), pp. 46, 47.

[40] Mayell, Hillary, "Mummified dinosaur discovered in Montana," *National Geographic News,* October 11, 2002. ("Leonardo is one of the most complete *Brachylophosaurus* dinosaur fossils uncovered to date, and the first sub-adult. He is also only the fourth dinosaur fossil in the world to be classified as a 'mummy' because of the soft tissue that is preserved. . . . His fossilized skeleton is covered in soft tissue—skin, scales, muscle, footpads—and even his last meal is in his stomach. The actual tissue has decayed over the millennia, and has been replaced by minerals. What's left for scientists to study is a fossil of a dinosaur mummy.")

[41] Bakker, at 148; McGowan, at 65.

[42] See, e.g., Bakker at 157-159; Brett-Surman, M.K., "Ornithopods," in *The Complete Dinosaur,* supra, at p. 342-344.

[43] Forster, at p. 298; Weishampel, David B., and John R. Horner, "Hadrisauridae" in *The Dinosauria,* Weishampel, Dodson, and Osmolska, Eds. (Berkeley, CA: University of California Press, 1990) p. 561.

[44] McIntosh, J.S., M.K. Brett-Surman, and James O. Farlow, "Sauropods" in *The Complete Dinosaur,* supra, at p. 264; Dodson, Peter, "Sauropod paleoecology" in *The Dinosauria,* David B. Weishampel, Peter Dodson, and Halszka Osmolska, Eds. (Berkeley, CA: University of California Press, 1990), p. 402.

[45] McIntosh, et al, at 268, 269.

[46] McIntosh, et al, supra, at 283.

[47] Fastovsky, David, and David B. Weishampel, *The Evolution and Extinction of the Dinosaurs*, (Cambridge, UK: Cambridge University Press, 1996), p. 249.

[48] McIntosh, et al, at 285.

[49] Fastovsky, et al, supra, at 249; Dodson, supra, at p. 405; McIntosh et al, supra at p. 285.

[50] Dodson, "Sauropod paleoecology," in *The Dinosauria*, supra, at 405.

[51] McIntosh, "Sauropoda", in *The Dinosauria*, supra, at p. 358.

[52] Chiappe, Luis M., and Lowell Dingus, *Walking on Eggs*, (New York: Scribner, 2001), pp. 99-103.

[53] McIntosh, et al, at 269.

[54] McIntosh, et al, at 279.

[55] When animals with bellows-type lungs breathe, the trachea or windpipe creates "anatomical dead space," meaning that some of the inhaled air is exhaled without having reached the lungs. As little as twenty percent (20%) of the oxygen inhaled by bellows-type lungs diffuses into the bloodstream. A longer trachea has more anatomical dead space and makes for less efficient breathing. Giraffes have narrow tracheas, reducing the dead space, but some think that giraffes have the longest necks possible to animals with bellows-type lungs. Because sauropod necks were much longer than giraffe necks are, some believe that sauropods may have solved the problem with unidirectional avian lungs. In birds, the inhaled air travels down the trachea and makes a single pass through a complex system of air sacs and lungs before returning to the trachea to be exhaled. The avian lung doubles—to 40 percent—the percentage of oxygen that reaches the bloodstream. Thus, an avian lung would enable efficient breathing, even with a much longer trachea. Fastovsky, et al, at 253, 254.

[56] McIntosh, et al, supra, at p. 284; Currie, Philip J., "Gastroliths" in *Encyclopedia of Dinosaurs*, supra, at p. 270. Some paleontologists remain skeptical that the polished stones were gizzard stones, calling them "gastromyths."

[57] Norman, at 40, 41.

[58] At least some of the coelurosaurs have the following birdlike characteristics:
1. Pubis in posterior orientation, and bearing a small distal "boot."
2. Elongated arms and forelimbs and clawed manus (hands).
3. Large orbits (eye openings in the skull).
4. Flexible wrist with a semi-lunate carpal (wrist bone).
5. Hollow, thin-walled bones.

6. Three-fingered opposable grasping manus (hand), four-toed pes (foot); but supported by three main toes.
7. Small, stiffened tail.
8. Elongated metatarsals (bones of the feet between the ankle and toes).
9. S-shaped curved neck.
10. Erect, digitgrade (ankle held well off the ground) stance with feet positioned directly below the body.
11. Similar eggshell microstructure.
12. Teeth with a constriction between the root and the crown.
13. Functional basis for wing power stroke present in arms and pectoral girdle (during motion, the arms were swung down and forward, then up and backward, describing a "figure-eight" when viewed laterally).
14. Large pneumatic sinuses in the skull.
15. Five or more vertebrae incorporated into the sacrum (hip).
16. Straplike scapula (shoulder blade).
17. Clavicles (collar bones) fused to form a furcula (wishbone).
18. Hingelike ankle joint, with movement mostly restricted to the fore-aft plane.
19. Secondary bony palate (nostrils open posteriorly in throat).

[59] Although some of the most bird-like coelurosaurs, such as *Ornithomimus*, *Avimimus* ("bird mimic"), and *Oviraptor*, lacked teeth.

[60] Universal Pictures' "Jurassic Park," screenplay by Michael Crichton and David Koepp, based on the novel by Michael Crichton.

[61] R.E.H. Reid, "Dinosaurian Physiology: The Case for 'Intermediate Dinosaurs'," in *The Complete Dinosaur*, supra, at p. 462, 463; See also Norman, at 118, 119.

[62] Reid, at 464.

[63] Reid, at 465, 466.

[64] Reid, at 466.

[65] Reid, at 468-471.

CHAPTER 15

# Other Strange Creatures in Flood-laid Strata

Dinosaurs are not the only large reptiles found in Mesozoic rocks. Reptiles of land, sea, and air dominate the vertebrate fossil record of the Triassic, Jurassic and Cretaceous. Mesozoic rocks contain the remains of marine reptiles, such as ichthyosaurs, plesiosaurs, and pliosaurs, and several varieties of flying reptiles, called pterosaurs. Moreover, below the Mesozoic strata, in Carboniferous and Permian rocks, we find the fossils of very strange creatures called "mammal-like reptiles" which, as their name implies, had some characteristics of mammals and some of reptiles. Everywhere in Flood-laid strata, we find the fossils of animals that were evidently mixtures of reptiles, birds, and mammals.

## A. OTHER MESOZOIC REPTILES

### 1. ICHTHYOSAURS

These "fish lizards" (**Fig. 1.1, Plate 22**) had characteristics of different classes of vertebrates; for a century, naturalists were unable to decide how to classify the ichthyosaurs. They were first thought to be fish, later crocodilians, and still later amphibians. Cuvier was the first to classify *Ichthyosaurus* as a reptile. De la Beche and Conybeare also recognized the reptilian characteristics of the skeleton. By the 1830s, there was general agreement that *Ichthyosaurus* was a reptile, but the animal did not fit comfortably within any existing vertebrate group.[1]

Ichthyosaur remains come in many sizes, varying from five to fifty feet in length, suggesting a reptile's ability to continue to grow throughout life. Large bony rings, called *sclerotic ossicles,* encircled the eyes; these were similar to the plates around the eyes of birds and lizards. The contours of its body were similar to dolphins and porpoises, but with a long, narrow snout, similar to the snouts of river dolphins and swordfish.[2] The snout was stocked with conical teeth, similar to crocodile teeth, but also reminiscent of whale teeth.[3] Although their form was perfectly adapted to their marine environment, ichthyosaurs had no gills. They had to surface to breathe.[4]

Ichthyosaurs were viviparous, meaning that they gave birth to live young, a mode of birth more characteristic of mammals than of reptiles or fish. Numerous fossil specimens reveal baby ichthyosaurs inside mother ichthyosaurs. Paleontologists once interpreted these specimens as evidence that ichthyosaurs ate their own young, just as crocodiles sometimes do today. Remarkably, however, several ichthyosaurs appear to have been buried just at the moment of giving birth. (**See Plate 22**) It is now believed that the other fossils were misinterpreted, and show unborn young, not cannibalized young.[5]

Some interpret fossils such as **Plate 22** as reflecting post-mortem expulsion of the fetus, rather than burial near the moment of birth. They argue that the mothers died shortly before giving birth, and gases generated in their decaying bodies pushed out the babies. Such post-mortem expulsion has been observed among beached whales. But the remarkable preservation of the skeletons indicates that the bodies were buried at death, or soon after, by a significant amount of sediment, otherwise the bones would have been scattered by scavengers. The weight of the sediment would have prevented the expulsion of the unborn ichthyosaur. The better explanation is that these animals were buried suddenly and catastrophically, at the moment of giving birth.

As **Plate 22** shows, Ichthyosaurs were born tail first. Being air-breathing creatures, they would have drowned had they been born headfirst. Clearly, the ichthyosaurs' resemblance to sea-going mammals such as dolphins is more than superficial.

## 2. PTEROSAURS

The pterosaurs (**Plate 4**) were flying reptiles whose remains are found in dinosaur strata. Although they are supposed to have evolved from the same croc-like archosaur stock as the dinosaurs, pterosaurs share many features with birds and bats.[6] The pterosaurs' extraordinarily long fourth finger formed the framework of the leading edge of their wings. The other three fingers terminate about one third of the way to the wingtip. This is why pterosaurs are popularly known as *Pterodactyls,* which means "wing finger."

Some pterosaurs had long, bony tails, tipped with a diamond-shaped vane that may have acted as a rudder. These pterosaurs, of which *Rhamphorhynchus* ("beak snout") was typical, are found in Jurassic strata, and they generally had conical, reptilian teeth. Other pterosaurs, found in Cretaceous strata, had shorter tails or no tail. Some of these pterosaurs had fewer teeth or, like *Pteranodon* ("winged without teeth"), no teeth. Pterosaurs were the largest animals that ever flew—some had a wingspan of over forty feet.

The pterosaurs' skeletal design, including the pectoral girdle and the pelvic region, are similar to modern birds. Pterosaur bones were hollow and tubular, like the bones of birds. As previously noted, bird lungs are connected to several thin-walled air sacs, which are in turn connected to hollow cavities in the bones. Large openings, called *pneumatic foramina*, connect with these cavities. Pterosaurs have similar openings, suggesting that they possessed avian lungs. In this, they differed from bats, which have mammalian bellows-type lungs.[7]

The pterosaurs had a number of mammalian characteristics. Their wings lacked feathers and were formed by a membrane of skin, just as bats' wings are formed.[8] The hind limbs more closely resembled those of a bat than a bird. The hind feet have long toes ending in sharp claws, and it is possible that the animal used its hind limbs for hanging upside down, just as bats do.[9] Interestingly, paleontologists have found in Cretaceous strata an inch-long insect known as *Suarophthirus.* This fossil closely resembles the bat flea—a modern parasite that lives on bats. Paleontologists (assuming that bats did not evolve for several million more years) believe that this Cretaceous bat flea must have lived parasitically on Pterosaurs, the most bat-like animal found in Cretaceous strata.[10]

Russian paleontologist Alexander Sharov discovered that a pterosaur had left an impression of what appeared to be *fur* in the fine-grained limestone of Kazakhstan. The short fur, ranging from about a quarter to a third of an inch, covered most of the body. The fur may not have covered the face, tail, or wing membrane—a pattern similar to modern bats.[11] Some paleontologists had guessed that pterosaurs had fur; its absence would have rendered an ectothermic reptile susceptible to sluggishness caused by rapid cooling in the air.

But some do not believe that pterosaurs were cold-blooded, arguing that they must have been able to generate their own body heat. If so, it "makes pterosaurs comparable to modern mammals and birds, as high-energy, warm-blooded creatures," writes David Norman, "a combination that seems to stretch the definition of 'reptile' to the very limit."[12] Were pterosaurs alive today, "it is very much to be doubted whether biologists would place them in Class Reptilia," argues Bob Bakker, because their physiology puts them, "beyond the traditional limits of the term 'reptile.'"[13] The pterosaurs are yet another example of heterodox creatures combining within themselves characteristics normally found in reptiles, mammals, and birds.

## 3. ARCHAEOPTERYX

The most celebrated case of a heterodox creature is *Archaeopteryx* ("ancient wing") (**Plate 23**). The fine-grained Jurassic limestone near Solnhofen, Bavaria, has been quarried for centuries for various purposes, including roofing tiles and lithographic plates. In 1861, quarry workers found a most unusual fossil. The skeleton was very similar to the chicken-sized dinosaur *Compsognathus*, but around it, the fine-grained limestone had preserved the very clear impression of feathers, identical to a modern bird's flight feathers.[14]

The British Museum paid the handsome sum of seven hundred pounds to procure the fossil. Richard Owen examined it, noting that it had several reptilian features, including a long bony tail, three-clawed fingers at the end of its forelimb or wing, and reptilian ribs and vertebrae. It also had some very bird-like features, including light, hollow bones and a wishbone. Its feathers were flight feathers, with shafts, vanes, hooks, barbs, and barbules. Owen, well aware that the Darwinists—with whom he had no sympathy—would claim that it was a transitional form between reptiles and birds, declared *Archaeopteryx* a bird.[15]

The head was missing from the fossil Owen examined, but a later-discovered specimen was to show that *Archaeopteryx*, while it had a pronounced bird-like beak, also had socketed, conical, reptilian teeth, unlike toothless modern birds. The ten specimens thus far recovered[16] range from about the size of a Blue Jay to the size of a large chicken, suggesting that perhaps *Archaeopteryx* had a reptile's ability to continue to grow throughout life.

All the leading naturalists of the day recognized the significance of *Archaeopteryx* as a possible transitional form between reptiles and birds, and Thomas Henry Huxley, known as "Darwin's bulldog," often cited it as a proof of evolution. But *Archaeopteryx*' role in the supposed evolution of birds remains controversial. What cannot be doubted is that it had some characteristics typical of reptiles and some typical of birds.

## B. ANIMALS FOUND BELOW DINOSAUR STRATA—THE MAMMAL-LIKE REPTILES

In the Upper Carboniferous, Permian, Triassic, and Jurassic strata are found the remains of the Synapsids, more commonly called the "mammal-like reptiles," because they have some characteristics of mammals and some of reptiles.

There is an interesting story behind the name *Synapsid.* Many creatures' skulls have holes, called *fenestrae,* behind their eye sockets. These openings reminded scholars of the arches (in Latin, "apses") of Roman buildings, so the suffix *-apsid* was coined to refer to the various arrangements of the holes.[17] Modern reptiles (except for the turtles, which are anapsids, meaning that they don't have any holes behind the eye sockets) are diapsids—"two-arched"—having two sets of fenestrae behind the eye sockets, whereas mammals have only one. The mammal-like reptiles had only one set of fenestrae, but they were more like reptiles than mammals in other respects, so their monapsid condition puzzled early paleontologists. Someone suggested that these creature's ancestors had possessed two arches that had subsequently fused into one, and coined the name Synapsid, which means "those with fused arches."[18]

The Synapsida are divided into two orders—pelycosaurs and therapsids.

## 1. PELYCOSAURS

Pelycosaurs ("basin reptile" or "bowl reptile," from the shape of the pelvis) were the more reptilian synapsids, having a sprawling crocodilian posture and few mammalian characteristics other than the single set of fenestrae. They are found in Upper Pennsylvanian (Carboniferous) and Permian strata.

The most famous of the pelycosaurs is the "sail-backed" *Dimetrodon* ("teeth of two measures") (**Plate 24**). These are abundant in the Permian, including the Permian Basin of west Texas. It is not clear what the "sail" was for, if not purely ornamental. If the animal was ectothermic, like modern reptiles, the sail might have helped regulate body temperature. *Edaphosaurus* ("ship-lizard") was the only other sail-backed pelycosaur. Interestingly, Darwinists do not believe that *Edaphosaurus* and *Dimetrodon* shared a common sail-backed ancestor (because *Dimetrodon* was carnivorous, whereas *Edaphosaurus* was herbivorous), but that each line evolved the sail separately, another remarkably unlikely case of "convergent" or parallel evolution.[19]

## 2. THERAPSIDS

Of the synapsids, the therapsids ("mammal faced," referring to the mammal-like arch of the cheekbone) were the most mammal-like, but are still classified as reptiles. Therapsid remains are found all over the world, including in Antarctica. They range stratigraphically from the middle Permian to middle Jurassic; they are especially abundant in the Karoo sediments of South Africa and in certain localities in northern Russia.

Therapsid posture varied, ranging from the sprawling posture of crocodiles and lizards to a more mammal-like posture, with the legs held, doglike, directly beneath the body. In most species, the teeth were differentiated into mammal-like nipping incisors, large stabbing canines, and a series of grinding cheek teeth comparable to human molars. The therapsid lower jaw was reptilian, composed of seven bones, whereas a mammal's lower jaw is composed of a single dentary bone. As in mammals, some forms had a bony palate in the roof of the mouth.

These animals were truly bizarre, and many are noteworthy for their extreme ugliness. One such group was the titanosuchians ("Titanic crocodiles"). These were dog to bear-sized animals, with jagged teeth and strong jaws. A specialist with some affection for these creatures described them as "some of the most astonishingly ugly animals ever to have been produced."[20] Titanosuchian posture was intermediate; they had a "three-quarters pushup" posture, as opposed to the "half-pushup" posture of lower reptiles. Another group is called the dinocephalians, meaning "terrible-headed," which is "a genuinely accurate description."[21] Some of these, such as *Estemmenosuchus* ("crowned crocodile") (**Plate 25**), had bizarre bone ornaments projecting from their skulls. Still another group is called the gorgonopsians, which means "terrible-looking." In Greek mythology, the Gorgons were three sisters who were so ugly that the mere sight of them could turn a man into stone.[22]

Another shockingly ugly bunch of animals were the pareiasaurs—an extinct group of anapsid reptiles found in the same strata as the therapsids. Some pareiasaurs, like *Scutosaurus* (**Plate 26**), had bony armor along their backs, not unlike nodosaurs and ankylosaurs, and grotesque bony projections around their faces. So strange looking are these creatures that someone coined the apt name "hallucicrania" to describe the subgroup of anapsids to which they belong.[23] The fact that turtles are also anapsids has led some Darwinists to speculate that turtles evolved from pareiasaurs, or that pareiasaurs are an offshoot or sister group of the turtles. But judging from its looks, *Scutosaurus* would sooner shoot his sister than be related to the turtles.

The dicynodonts ("two dog teeth") were a widespread therapsid group that had two prominent canine teeth, really more like tusks. In most species the remainder of the jaw was toothless, and the animal had a tough, sharp beak, somewhat like that of a turtle. *Lystrosaurus* (**Plate 27**), a very common and widespread type, is typical of the dicynodonts. The fact that *Lystrosaurus* was found in correlating strata in both southern Africa and South America helped the theory of continental drift gain acceptance. Interestingly, large herds of dicynodonts were buried together; many were buried and fossilized standing up.[24]

All the therapsids had a combination of reptilian and mammalian characteristics. The cynodonts ("dog teeth"—not to be confused with the dicynodonts) were the infra-order closest to mammals. There are so many types of cynodonts that have mammalian characteristics that Darwinists have described them as "an infraorder of missing links."[25] *Probelesodon* ("before-dart-tooth") and *Massetognathus* ("chew jaw") are two cynodonts that displayed characteristics of reptiles and mammals. The reptilian features were: (1) the lower jaw was comprised of several bones, (2) the jaw joint was formed between articular and quadrate bones, (3) they had small crania, (4) they had ribs in the neck region, and (5) the number of bones in the fingers and toes. The mammalian features include: (1) teeth specialized for different functions, (2) a lower jaw with prominent coronoid process, (3) a double condyle at back of skull for neck articulation, (4) axis with odontoid process, and (5) the ilium slopes forward.[26]

## C. WHY DO THE VERTEBRATES BURIED BY THE FLOOD HAVE A MIXED NATURE?

The dinosaurs were heterodox, with characteristics of several now-distinct classes of animals. This was also true of other vertebrates found in Mesozoic strata, such as the flying reptiles and the marine reptiles. It was also true of many creatures found below the dinosaur strata, such as the mammal-like reptiles.

Today we rarely find a living species that crosses class boundaries. One such heterodox creature is the duckbill platypus—a mammal native to eastern Australia that lays eggs and has a bill like a duck. When the curators of the British Museum first received the hide of this creature, with its duckbill, dense fur, webbed feet, and flat, beaver-like tail, they thought they were being hoaxed! And that was before they found out it laid eggs. But the platypus is one of only two types of egg-laying mammals of the order *Monotremata.* (The other monotreme is the Echidna, or spiny anteater, also native to Australia and the South Pacific). The vast majority of modern animals are easily sorted into their classes and orders. In sub-Tertiary strata, however, almost the reverse is true. It seems that most of the vertebrates buried by the Genesis Flood were of a mixed, heterodox, or as Edward Hitchcock put it, heteroclitic, nature. Why is this so?

Some suggest that in the earlier stages of life on this planet, evolution had not yet separated the animals into the clearly demarcated classes we observe today. A typical explanation is that of British paleontologist Alan Charig:

> We can . . . take any living four-legged backboned animal and tell very easily whether it is an amphibian, a reptile, a bird or a mammal. This happy situation has come

> about because the main branches of the evolutionary tree separated from each other long ago and are now quite distinct; each class represents all the twigs attached to one major branch of that tree. Indeed, there are very few living vertebrates that cause any problems at all. . . . Fossil vertebrates, however, are quite a different matter. When they occur on the family tree near the point where one class branches off from another they are likely to show a mixture of modern class characters, or characters in a transitional condition. . . . The fact that living members of each class possess a well-defined set of class characters has one unfortunate consequence. We tend to forget that, way back down the family tree, there is a 'blurring' of the classes.[27]

Can the theory of evolution explain the "blurring" of the classes? Phrased another way, is the heterodox nature of the dinosaurs and the other Flood-deposited vertebrates due to the fact that "they occur on the family tree near the point where one class branches off from another"?

The answer is a resounding "no." Typically, there is not even a suggestion of one class branching off from another. No one claims, for example, that mammals either evolved into dinosaurs or descended from dinosaurs. What accounts for the mammalian characteristics observed in many of the dinosaurs? No one claims that bats or birds "branched off" from pterosaurs, or that pterosaurs descended from mammals or birds. What accounts for the bat-like and bird-like characteristics of the pterosaurs? No one claims that dolphins, porpoises, or fish descended from ichthyosaurs, or that ichthyosaurs descended from dolphins or fish. What accounts for the fish-like and dolphin-like characteristics of the ichthyosaurs? Darwinists admit that "Ichthyosaurs . . . may resemble dolphins in some respects, and pterosaurs may resemble birds or bats, but this is merely a convergence, not a real kinship."[28]

It is, of course, claimed that dinosaurs descended from other reptiles, but there is a large gap separating the dinosaurs from their hypothetical archosaur ancestors. It is not claimed that the dinosaurs evolved into any of the modern reptiles. Indeed, except for a very remote, hypothetical common ancestor, dinosaurs are thought to be completely unrelated to living reptiles.

The bird-like characteristics of the dinosaurs present us with a different situation. Many Darwinists do believe that birds descended from dinosaurs, but the facts present a confusing picture. Having learned that dinosaurs are divided into two orders based on hip structure, we would expect that the birds branched off from the "bird-hipped" *Ornithischian* dinosaurs. They did not. No one claims that the birds evolved from the "bird-hipped" dinosaur branch, which included the "bird-footed" *Ornithopod* dinosaurs. Since the *Ornithischian* dinosaurs evolved neither from nor into birds, what accounts for the many bird-like features observed in the *Ornithischian* dinosaurs? The birds supposedly evolved *not* from the "bird-hipped" dinosaurs, but from "lizard-hipped" dinosaurs, which also had a more reptilian configuration of the skull than the *Ornithischian* dinosaurs from which birds *are not* supposed to have descended. Strange.

---

[1] Callaway, Jack M., *Ancient Marine Reptiles*, Callaway, Jack M., and Elizabeth L. Nicholls, Eds. (New York: Academic Press, 1997), pp. 3, 4.

[2] Norman, David, *Prehistoric Life: The Rise of the Vertebrates*, (New York: Macmillan, 1994), p. 128.

[3] Bakker, Robert T., *The Dinosaur Heresies*, (New York: Kensington Publishing, 1986), pp. 320, 321; Motani, R., "Temporal and Spatial distribution of tooth implantations in Ichthyosaurs," in *Ancient Marine Reptiles*, supra, at 96-100; Orndorff, Richard L., Robert W. Wieder, and Harry F. Filkorn, "How the west was Swum," *Natural History*, June, 2001, p. 22.

[4] Orndorff, et al, at 22.

[5] Norman, at 130; see also McGowan, Christopher, *Dinosaurs, Spitfires, and Sea Dragons*, (Cambridge, MA: Harvard University Press, 1991), pp. 228-231.

[6] McGowan, Christopher, *The Successful Dragons*, (Toronto: Samuel Stevens, 1983), p. 191.

[7] McGowan, Christopher, *The Successful Dragons*, at 203; Bakker, at 305.

[8] McGowan, Christopher, *The Successful Dragons*, at 191.

[9] McGowan, Christopher, *The Successful Dragons*, at 202, 203.

[10] Haines, Tim, *Walking with Dinosaurs*, (London: BBC Worldwide, 1999, [1st American Edition: New York: Dorling Kindersley Publishing, 2000]), p. 178.

[11] Norman, at 140, 141; Haines, at 192. Some authorities describe the covering not as fur but rather as, "a dense coat of long, hair like scales." See, e.g., Bakker, at 292, 293. But everyone seems to agree that the pterosaurs had insulation. "Other specimens have confirmed Sharov's discovery. Flying dragons as a group seem to have been insulated." Bakker, at 293.

[12] Norman, at 141.

[13] Bakker, at 293.

[14] Gould, Stephen Jay, *Wonderful Life*, (New York: W.W. Norton & Co., 1989), pp. 63, 64.

[15] Wilford, John Noble, *The Riddle of the Dinosaur*, (New York: Vintage Books, 1985), p. 91.

[16] Specimens found as of this writing include: (1) **a feather** discovered in 1860 near Solnhofen, Germany, described in 1861 by Hermann von Meyer, and currently located at the Humbolt Museum in Berlin. There is no way to be certain whether this feather was from *Archaeopteryx* or some other bird; (2) the **London Specimen** (the type specimen), discovered in 1861 near Langenaltheim, Germany, described in 1863 by Sir Richard Owen, and located in the British Museum; (3) the **Berlin Specimen**, Discovered in 1876 near Blumenberg, Germany, described in 1884 by Wilhelm Dames, and currently Located at the Humbolt Museum. It was the first specimen with a complete head; (4) the **Maxberg Specimen**, discovered near Langenaltheim and described in 1959 by Heller. This specimen was lost after the death of its owner, and is currently missing; (5) the **Haarlem Specimen**, discovered in 1855 near Riedenburg, Germany, described as a *Pterodactylus crassipes* in 1875 by Meyer, and reclassified in 1970 by John Ostrom, and currently located in the Netherlands; (6) the **Eichstätt Specimen**, discovered in the 1950s near Workerszell, Germany, described by Peter Wellnhofer in 1974, and currently located at the Jura Museum in Eichstätt, Germany. It is the smallest specimen, and has the second best head; (7) the **Munich Specimen**, discovered in 1991 near Langenaltheim, described in 1993 by Wellnhofer, and currently located at the Paläontologische Museum in Munich. It is the only specimen with a breastbone; (8) the **Solnhofen Specimen**, discovered in the 1960s near Eichstätt, Germany, described in 1988 by Wellnhofer, and currently located at the Bürgermeister-Müller-Museum in Solnhofen. It was originally classified as a *Compsognathus*; (9) the **Thermopolis Specimen**, long in a private collection in Germany, it was described in 2005 by Mayr, Pohl, and Peters, and is currently located at the Wyoming Dinosaur Center in Thermopolis, Wyoming. It has the best-preserved head and feet; (10) the **Bürgermeister-Müller Specimen** 1997.

[17] McLoughlin, John C., *Synapsida: A New Look into the Origin of Mammals*, (New York: Viking Press, 1980), pp. 28, 29.

[18] The paleontologist who suggested classification based upon fenestrae was A.C. Günther. See Günther, A.C. "Contribution to the anatomy of *Hatteria* (*Rhynchocephalus* Owen)," *Transactions of the Royal Society of London*, 157:595–629 (1867).

[19] McLoughlin, at 34, 37-40.

[20] McLoughlin, at 51.

[21] McLoughlin, at 57.

[22] McLoughlin, at 71.

[23] Haaramo, Mikko, citing Lee, Michael S.Y., "The origin of the turtle body plan: Bridging a famous morphological gap," *Science*, vol. 261, (September 24, 1996) 1716-1720; Lee, M.S.Y., "Correlated progression and the origin of turtles" *Nature*, vol. 379, (February 29, 1996), pp. 812-815.

[24] McLoughlin, at 63.

[25] McLoughlin, at 79.

[26] Vuletic, Mark I., "Frequently Encountered Criticisms in Evolution vs. Creationism," citing Christopher McGowan, *In the Beginning: A Scientists shows why the Creationists are Wrong*, (Gage Distribution Co., 1983 [paperback: Prometheus Books, 1984]).

[27] Charig, Alan, *A New Look at the Dinosaurs*, (New York: Facts on File, Inc., 1983), pp. 78, 79.

[28] Wilford, at 88.

CHAPTER 16

# Did Birds Evolve from the Dinosaurs?

This chapter will examine the claim that the mixed-class physiology of dinosaurs and mammal-like reptiles can be explained as a legacy of evolution. We will explore whether—even granting the long ages assumed in the Lyellian view of geology—birds could have evolved from dinosaurs, and mammals from the mammal-like reptiles.

## A. DID BIRDS EVOLVE FROM THE DINOSAURS?

The notion that birds descended from dinosaurs is not new, but it was out of fashion for a long time, and has only recently regained popularity. When *Archaeopteryx* was discovered in 1861, Huxley argued that it was a transitional form between dinosaurs and birds. Other scientists soon came forward to dispute this, however, arguing that birds and dinosaurs were only distantly related through Triassic reptiles ancestral to both.

In 1926, Danish paleontologist Gerhard Heilmann published an influential book, *The Origin of Birds*, in which he noted the similarities between birds and coelurosaurian dinosaurs, but argued that the dinosaurs could not have been the ancestors of the birds because dinosaurs lacked a very important avian feature: a wishbone (furcula). Not only did the coelurosaurs lack wishbones, Heilmann pointed out, they also lacked what was assumed to be an evolutionary antecedent to a wishbone—clavicles (collarbones). The absence of clavicles in the coelurosaurs, wrote Heilmann, "would in itself be sufficient to prove that these saurians could not possibly be the ancestors of birds."[1]

Heilmann's view was the dominant scientific position until the 1970s, when John Ostrom, a Yale paleontologist who specialized in coelurosaurs, took the position that, because the skeletal anatomy of *Archaeopteryx* was almost entirely that of a coelurosaurian dinosaur, birds must have evolved from dinosaurs. Ostrom personally excavated and named *Deinonychus* (which appeared in the "Jurassic Park" movies under the name of another coelurosaur, *Velociraptor*), and closely studied *Compsognathus* (the chicken-size dinosaur also made famous in the "Jurassic Park" movies). Ostrom noted that one *Archaeopteryx* fossil had for many decades been misidentified as a *Compsognathus* and that some recently discovered coelurosaur species, such as *Velociraptor* and *Segisaurus*, appeared to have collarbones. Ostrom also argued that the hipbones of *Archaeopteryx* had been misinterpreted, and that its pubis bone actually pointed down and forward, as in the *Saurischian* theropod dinosaurs.[2]

Birds do seem to have more in common with dinosaurs than with Triassic archosaurs. Hence, the view that birds evolved from dinosaurs has become the majority view. Not all Darwinists accept it; some bird specialists still favor Heilmann's theory that the birds evolved directly from Triassic archosaurs.[3] "The supposed link between dinosaurs and birds is pretty entrenched in paleontology," states one skeptic, "but it's not as solid as the public has been led to believe."[4] (Some of these dissenters claim to have found the fossil impression of flight feathers attached to *Longisquama insignis*, a small non-dinosaur found in Triassic rocks.[5])

Although the dinosaur-to-bird theory has come back into vogue, most scientists agree that *Archaeopteryx* itself does not connect any particular dinosaur to any surviving line of birds. First, no one can guess what dinosaur, if any, *Archaeopteryx* evolved from.[6] Huxley's candidate was *Compsognathus*, and Ostrom revived the dino-to-bird theory by pointing out similarities between the two species. But whereas *Archaeopteryx* had a wishbone, Compy lacked both a wishbone and collarbones. Moreover, Compy is found in the same Jurassic strata as *Archaeopteryx*, not below it, so Compy would have been a contemporary, not an ancestor, of *Archaeopteryx*. Compy is no longer seriously promoted as the ancestor to *Archaeopteryx*.

Second, *Archaeopteryx* is not an ancestor of modern birds.[7] "*Archaeopteryx* is not ancestral to any group of modern birds," writes paleontologist Larry Martin. Instead, it is "the earliest known member of a totally extinct group of birds."[8] A prominent pro-evolution web-site reports that "no one is claiming that [*Archaeopteryx*] is the transitional species between dinosaurs and birds . . ." Rather, it is "an echo of the actual event" that "represents a grade of organization which the proposed lineage went through to get from dinosaurs to birds."[9] So *Archaeopteryx* is a fascinating mosaic of reptilian and avian characteristics, but it is not a transitional form or "missing link."

For those of us who reject the entire Darwinian paradigm, the idea that any bird evolved from a nonbird is quite a stretch. Consider the supposed evolution of flight feathers—the type of feathers clearly seen in *Archaeopteryx*. The flight feather is marvelously intricate. Each feather consists of a central shaft, called the rachis, from which a series of barbs branch off at sharp angles. Barbules project from the front and back of each barb to form the surface area of the feather. Hooks project downward from anterior barbules and interlock with the ridges on the posterior barbules. In the flight feather of a large bird, about one million barbules are integrated to form an impervious vane. Flight feathers are remarkably light, strong and self-reorganizing.[10]

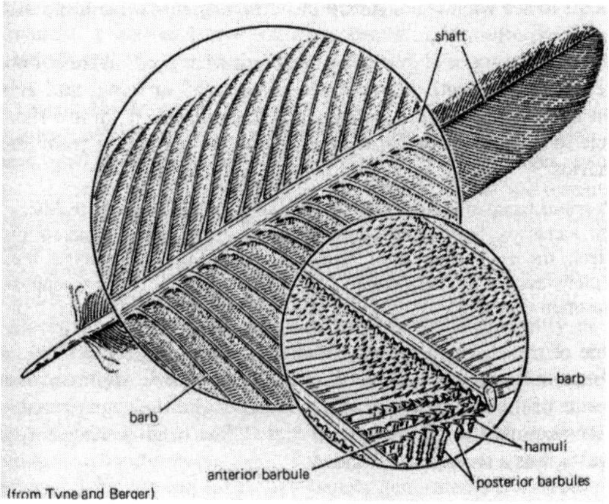

**Figure 16.1: A Flight Feather**

In one of the least plausible "just so" stories that comprise the Darwinian creation myth, feathers evolved from reptilian scales. According to Heilmann, in *The Origin of Birds*, a gliding reptile evolved a set of longish scales along the rear edge of the forearms and the side edges of the flattened tail. "By the friction of the air the outer edges became frayed, the fraying gradually changing into still longer horny processes which in the course of time became more and more feather like."[11]

But even if an animal's scales become frayed during its life, the frayed condition cannot be passed on to its offspring. The offspring will have normal, intact scales. To argue otherwise is like arguing that if a soldier loses a leg during the war, his future son will be born with only one leg. In current Neo-Darwinian theory, acquired changes cannot be passed to offspring; only genetic mutations in the germ line can be passed to offspring.

But let us become Lamarckists[12] for a moment and suppose that the frayed scales could be transmitted to offspring. No conceivable advantage *could* be derived from the frayed scales. In all gliding animals, the wings present a continuous, unbroken surface to the air.[13] Any fraying of the scales would make them pervious to air, decreasing the surface area and lift capacity. This would make flight more difficult, whether the "wings" were used for gliding or were flapped in powered flight. Thus, frayed scales would be disadvantageous, and natural selection would tend to remove them from the gene pool.

Even assuming that frayed scales conveyed some selective advantage and could spread throughout a reptilian population, frayed scales are light years removed from the magnificently designed flight feather:

> Take away the exquisite coadaptation of the components, take away the coadaptation of the hooks and barbules, take away the precisely parallel arrangement of the barbs on the shaft and all that is left is a soft pliable structure utterly unsuitable to form the basis of a stiff impervious airfoil. The stiff impervious property of the feather, which makes it so beautiful an adaptation for flight, depends basically on such a highly involved and unique system of coadapted components that it seems impossible that any transitional feather-like structure could possess even to a slight degree the crucial properties. In the worlds of Barbara Stahl, in *Vertebrate History: Problems in Evolution*, as far as feathers are concerned "how they arose initially, presumably from reptiles scales, defies analysis."[14]

Flight feathers are an example of an irreducibly complex mechanism that could not have evolved in a step-wise fashion. The intermediate steps between a scale and a feather convey no survival advantage; to the contrary, they are disadvantageous.[15]

Current Darwinian theorizing as to how large changes in anatomical shape and structure could occur centers around *homeotic* or *homeobox* genes. As we will discuss again later, these "master genes" control arrays of other genes during the development of an embryo. The most studied master gene is called "eyeless," because it controls the cascade of genes needed to build an eye. Scientists experimenting on fruit flies noticed that when "eyeless" was deleted, the resulting fly did not grow any eyes, but when the gene was inserted into seven different parts of the genome, the fly grew seven sets of eyes, including pairs of eyes on the wings, the legs, and even the antennae.

Applying this information to the problem of the origin of feathers, one Darwinian explanation might be that feathers resulted from mutations to the "master gene" that controls the formation of scales, causing feathers to be formed instead of scales. But master genes do not contain the genetic information needed to form, for example, a scale. The master gene is like an on/off switch that controls the expression of the genes that have the information. If these other genes are not present in the developing organism, no mutation of the master gene could create the necessary information.

Master genes are like a light switch. Unless the switch is wired to a functioning light bulb, flipping the switch will have no effect. Another analogy is to a computer program in which, if a certain condition is met, the program runs a subroutine. If the subroutine is missing from the program, it does not matter that the condition is met. The program will not run a different or modified subroutine. It will crash. Flight feathers are organized so differently from scales, and are so intricately structured, that they must have required new, specific and detailed genetic information to direct their formation.

Recent molecular studies cast doubt on the assumption that feathers evolved from scales. The new data suggests to Darwinists that feathers may be a completely novel structure.[16] Two researchers wrote that, "progress in solving the particularly puzzling origin of feathers has also been hampered by what now appear to be false leads, such as the assumption that the primitive feather evolved by elongation and division of the reptilian scale. . . . The new evidence from developmental biology is particularly damaging to the classical theory that feathers evolved from elongate scales."[17] Thus, Darwinists are abandoning the absurd notion that feathers evolved from scales. But is it any less absurd to argue that feathers evolved from *nothing*?

The dethroning of *Archaeopteryx* as a transitional form leading to modern birds means that it might not have shared a common feathered ancestor with modern birds, meaning that flight feathers must have evolved more than once. Moreover, with some researchers urging that *Longisquama* had feathers, and others urging that recently discovered Chinese dinosaurs had feathers, it appears Darwinists will need to posit that feathers evolved several different times. Not likely. "Feathers are a very complicated structure," noted one paleontologist. The odds of feathers evolving two separate times "would have been astronomically small."[18]

Feathers are a problem for Darwinists. One can appreciate why Darwin wrote that, "the sight of a feather in a peacock's tail, whenever I gaze at it, makes me sick!"[19] But the flight feather is only one of the characteristics that separate birds from reptiles. The avian lung is a unique, complex mechanism, with no workable intermediate forms between it and the bellows-type lung of reptiles.[20] Many other morphological changes must also have taken place, including changes to the skeleton, the muscle configuration, and the nervous system. New genetic information would have been necessary to bring about these changes, all of which was created, Darwinists would have us believe, by random DNA copying errors. There is also the contentious issue of how dinosaurs could have "learned" to fly: did they glide down from the treetops, or did they run along the ground and flap their limbs? Both the "trees down" and "ground up" theories have their partisans, but neither theory is convincing.

There is yet another problem with the purported dinosaur-to-bird evolutionary transition. The fossils are in the wrong order. The most bird-like dinosaurs—*Avimimus* ("bird mimic"), *Oviraptor* ("egg stealer"), and *Ornithomimus* ("bird mimic")—are found only in Upper Cretaceous rocks conventionally dated about eighty million years *younger* than the Jurassic strata in which *Archaeopteryx* was found.[21] Many other species of birds are found in Cretaceous strata, including in the lower Cretaceous levels. Obviously, a dinosaur from the Upper Cretaceous could not have been the ancestor of birds from the Jurassic and Lower Cretaceous—the ancestor would be many millions of years younger than its putative descendant!

This stratigraphical problem has forced paleontologists to conclude that birds must have evolved from dromaeosaurids ("running reptile") or troödontids ("wounding tooth")—even though these types are far less bird-like—because these are found in Lower Cretaceous rocks. But even these theropods are dated about twenty-five million years younger than *Archaeopteryx*, and they are found on the same level as the Cretaceous birds, not below them.[22] When one considers the length of time that would have been necessary for a dinosaur to evolve into a bird, it is hard to believe that any theropod could have been the ancestor of birds.[23]

When tracks are considered, as opposed to just bones, the stratigraphical problem is beyond hope of solution. Bird tracks have been found in the upper Triassic and lower Jurassic. Darwinists are hesitant to admit that the tracks are bird tracks, in the absence of unequivocal skeletal evidence that birds existed during the Triassic—although some claim to have found such evidence.[24] But the tracks clearly are bird tracks,[25] and they falsify the theory that birds evolved from dinosaurs.[26]

Advocates of the dino-to-bird theory seem untroubled by the dire stratigraphical problems confronting their theory. "We don't see time as particularly important," demurs Luis Chiappe, a paleontologist with the American Museum of Natural History. "We think the fossil record is incomplete."[27] Like Chiappe, most contemporary Darwinists seek only morphological intermediates, not fossils that are intermediate both morphologically and stratigraphically. But what is needed to tie dinosaurs to birds is a stratomorphic intermediate, and so far the Darwinists have not found one.

Dinosaur expert Phil Currie argues that the absence of the most bird-like theropod fossils below bird fossils is due to "preservational biases" and that "small theropods are very poorly known in older rocks . . ."[28] But taphonomic (fossil forming) processes could not be more biased against small theropods than against small, hollow-boned birds, yet birds are found in the lower Cretaceous, and *Archaeopteryx* was found in the middle Jurassic. The small theropod *Compsognathus* is also found in the Jurassic. At what point did the strata become biased against preserving small theropods?

The real reason Darwinists suspect "preservational bias" is that certain fossils are not found where a pet theory predicts they should be found. This is an echo of Darwin's claim that the fossil record was "imperfect" because it did not support his theory. The proponents of the dino-to-bird theory will not allow it to be falsified by the fossil record. A good scientific theory should explain the facts; at a minimum, it should conform to the facts. When it comes to evolution, however, if the facts do not fit the theory, it is the facts that are "biased" and "imperfect," not the theory.

Enthusiasts of the dino-to-bird theory are anxious to find fossil evidence that supports their conclusions. In 1999, Stephen Czerkas and the National Geographic Society purchased a Chinese fossil for $80,000 at an Arizona mineral show. Czerkas named it *Archaeoraptor* ("ancient raptor") *liaonengensis*, because it came from the Liaoning province of northeastern China. Czerkas announced that it was "the missing link between terrestrial dinosaurs and birds that could actually fly." The fossil had the forelimbs of a primitive bird and the tail of a dinosaur. *National Geographic* magazine discussed it in an article entitled "Feathers for T. rex?" *Archaeoraptor*, claimed the article, proves that we can now say that birds are dinosaurs "just as confidently as we say that humans are mammals." The article was illustrated by a drawing of a baby *Tyrannosaurus* with feathers.[29]

*Archaeoraptor* was a hoax. Knowing that scientists were expecting to find a feathered dinosaur and that such a fossil might bring big bucks, a clever forger had glued a dinosaur tail to the body of a fossil bird. *National Geographic* fell for the forger's scam because they had already fallen for the dino-to-bird propaganda. Storrs Olson, curator of birds at the Smithsonian Institute, criticized the magazine for being conned by a cadre of "outspoken and highly biased proselytizers of the faith" that birds evolved from dinosaurs. "Truth and careful scientific weighing of evidence have been among the first casualties in their program, which is fast becoming one of the grander scientific hoaxes of our age."[30]

Several species of supposedly feathered dinosaurs have recently been found, all from the same quarry in the Liaoning province of China—where the *Archaeoraptor* hoax originated—and all since 1995. The most well-studied of these species is *Sinosauropteryx prima* ("first Chinese reptile wing"), found in 1996. Fossil impressions seem to show "integument" around the head, trunk, tail, arms, and legs. Several who have seen the "integument" doubt that it is feathers, arguing that it is collagen fibers beneath the skin of the specimen,[31] but even those who believe it is feathers acknowledge that the inch-long structures are not flight feathers.[32]

Other recently discovered, purportedly feathered Chinese fossils include *Protarchaeopteryx* (1997), *Caudipteryx* (1998), *Beipiaosaurus* (1999), *Sinornithosaurus* (1999), and *Microraptor* (2000). Although these creatures could not fly, at least one of them, *Caudipteryx*, appears to have had feathers on its tail and forelimbs. *Microraptor gui* (2003), a small creature with four feathered appendages, even appears to have flight feathers. The scientists who described it speculate that it might have been a gliding animal.[33] As of this writing, there has not been sufficient time for skeptical scientists to analyze these Chinese finds.[34]

But it would not be at all surprising if reports of feathered dinosaurs are substantiated. To the contrary, given how many characteristics some of the dinosaurs shared with birds, it would be surprising if none of the dinosaurs had feathers. Feathered dinosaurs do not prove that birds descended from dinosaurs and do not solve the fundamental problem of how the exquisitely designed flight feather could have evolved in the first place. They do, however, further underscore a key fact: dinosaurs are heterodox creatures, mixtures of several different vertebrate classes, especially including reptiles and birds.

## B. DID MAMMALS EVOLVE FROM THE MAMMAL-LIKE REPTILES?

Darwinists sometimes assert that the reptile-to-mammal transition is the best-documented evolutionary transition in the fossil record—the crown jewel of the fossil evidence for Darwinism.[35] Here, at least, the stratigraphical arrangement of the fossils admits of an evolutionary interpretation. The mammal-like reptiles are found in the Permian, Triassic, and Jurassic strata, and the first mammals are found in the upper Triassic and Jurassic, so the putative ancestors are not younger than their descendants.

But when you dig a little deeper, the evidence does not show an evolutionary transition. We will start a few jumps back, at the supposed transition from amphibians to reptiles. There is a large, unbridged gap in the fossil record between amphibians and reptiles. The early reptiles are so different from amphibians that their specific ancestry has not been established.[36] Gould writes that, "no fossil amphibian seems clearly ancestral to the lineage of fully terrestrial vertebrates (reptiles, birds, and mammals), called Amniota to honor the "amniote" egg . . ."[37]

But the largest difference between amphibians and reptiles is not in their skeletal anatomy but in their mode of reproduction. Amphibians generally lay fragile, jelly-like eggs in water. The eggs are fertilized in the water and develop as waterborne larvae. The young go through a complex metamorphosis before reaching adulthood. Frogs and salamanders go through a tadpole stage in which they have gills and lack limbs. At that stage of their lives, amphibians are more like fish than like the land-dwelling creatures they will become. Indeed, *amphibia* means, "double-lived," reflecting the fact that amphibians go through two very different modes of living.

Reptiles, by contrast, have an egg fertilized in the mother's body in which moisture is enclosed in an amniotic sac, and the sac, embryo, and food supply are all enclosed within a hard outer membrane or shell. Reptiles lay these hard-shelled eggs on land, and the hatchlings emerge from their shells as tiny replicas of the adults. The amniotic egg is much more complex than the eggs of amphibians.

Amphibians and reptiles have very different modes of reproduction and development, and it is difficult to imagine workable intermediate stages between the two modes. It is as though God created the amphibians—and other animals with complex life cycles, such as butterflies—to falsify the evolutionary hypothesis.

Moving on to the mammal-like reptiles, Darwinists believe therapsids evolved from the pelycosaurs. But there are no fossils evidencing this alleged transition.[38] Some have suggested that the sphenacodonts ("wedge teeth"—the sail-backed *Dimetrodon* is in this group) were ancestors of the therapsids, but cladistic analysis indicates that the sphenacodonts were already "too specialized" to have given rise to the therapsids.[39] "The bottom line is that when therapsids first appear they differ significantly from pelycosaurs, and there are no intermediates plausibly connecting any known species from the two orders."[40]

And then there is the problem of the sudden appearance of many different kinds of therapsids. In a sort of Permian explosion, six suborders of therapsids, including ten infraorders, 42 families, and scores of genera, appear virtually simultaneously in the fossil record.[41]

> The sudden appearance of new higher taxa, families, and even orders, . . . with all the features more or less developed, implies a very rapid evolution. . . . It is possible that this is an artifact, and that the new taxa had long histories before they appeared in the fossil record, during which they gradually acquired their characteristic features. However, *in no case is such a long history known by even a single specimen* . . .[42]

"The abrupt appearance of the therapsids is made all the more startling by their amazing diversity."[43] McLoughlin describes the sudden appearance of the therapsids as "an adaptive radiation of unprecedented magnitude, quickly afflicting the Permian world with a host of weird animals . . ."[44] Thus, a staggeringly large variety of therapsid reptiles suddenly appears in the fossil record, and there are no earlier therapsids from which any of them could have evolved.

And what of the transition from mammal-like reptiles to mammals—the crown jewel itself? As previously noted, the cynodonts ("dog teeth") are the therapsids that Darwinists believe gave rise to the mammals. There are three families and seven genera of cynodonts from the Upper Triassic, with various combinations of mammal-like and reptile-like characteristics.[45] But large numbers of eligible candidates are helpful only to the extent that they can be placed in a line of descent leading from a particular reptile species to a particular mammal species.[46]

And no such line of descent can be traced. Darwinists "cannot yet recognize the specific lineage that led to mammals."[47] "As yet," wrote Michael A. Cluver, longtime director of the South African Museum in Cape Town and a specialist in mammal-like reptiles, "no specific cynodont can be confidently pointed to as being directly ancestral to the first mammal."[48]

> It is not known which cynodont family was ancestral to mammals, or whether all the mammals originated from the same group (family) of cynodonts. In the vast literature concerning mammalian origins, it is easier to find suggestions that one or the other therapsid or cynodont family cannot be ancestral to the Mammalia, rather than to find a positive answer.[49]

"The transition to the first mammal, which, probably happened in just one or, at most, two lineages, is still an enigma."[50]

Why the mystery? The reason has to do with character trait reversals. The mammal-like reptiles can be placed in an order leading from reptiles to mammals based upon an increasing number of mammal-like traits. But if a line of creatures had slowly evolved from reptiles to mammals, we would expect that once a reptilian trait was replaced by a mammalian trait, that characteristic would not revert back to a reptilian condition in a later, more mammal-like, descendant.

# Plates

Plate 1: Georges Cuvier

Plate 2: Mosasaurus missouriensis skull

**Plate 3: Mosasaurus**

**Plate 4: Pteranodon**

Plate 5: Mary Anning

Plate 6: Dr. Gideon Mantell

**Plate 7:**
**Sir Richard Owen**

**Plate 8: Camarasaurus**

**Plate 9: Stegosaurus**

**Plate 10: Allosaurus**

Plate 11: Brontosaurus

Plate 12:
Tyrannosaurus rex

Plate 13: Tyrannosaurus rex, "Sue"

Plate 14: Grand Canyon

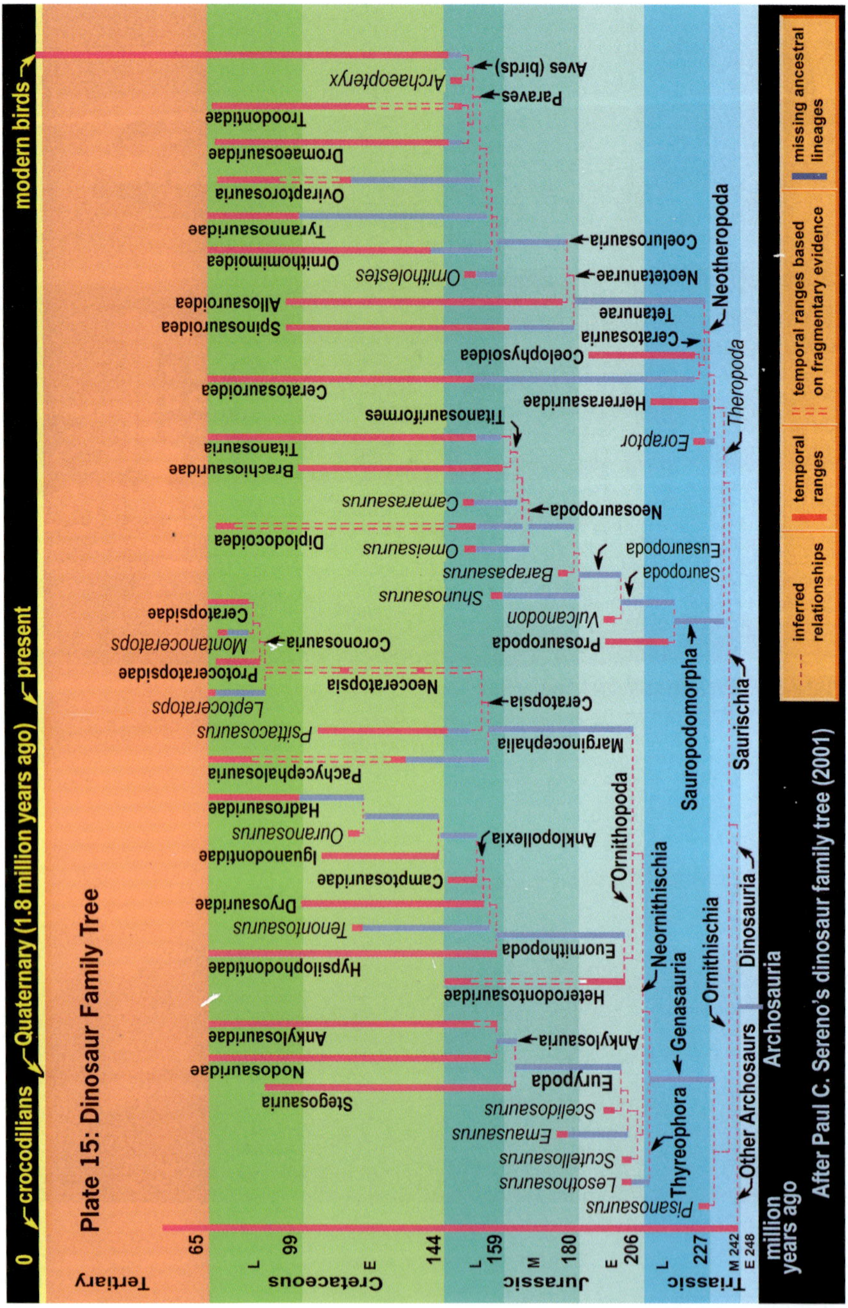
Plate 15: Dinosaur Family Tree
0
crocodilians
Quaternary (1.8 million years ago)
present
modern birds
Tertiary
65
Cretaceous
L
99
E
144
Jurassic
L
159
M
180
E
206
Triassic
L
227
M 242
E 248
million years ago
Pisanosaurus
Lesothosaurus
Scutellosaurus
Emausaurus
Scelidosaurus
Stegosauria
Nodosauridae
Eurypoda
Ankylosauridae
Ankylosauria
Heterodontosauridae
Hypsilophodontidae
Euornithopoda
Tenontosaurus
Dryosauridae
Camptosauridae
Iguanodontidae
Ankylopollexia
Ouranosaurus
Hadrosauridae
Pachycephalosauria
Marginocephalia
Psittacosaurus
Leptoceratops
Ceratopsia
Neoceratopsia
Protoceratopsidae
Montanoceratops
Coronosauria
Ceratopsidae
Prosauropoda
Vulcanodon
Shunosaurus
Sauropoda
Barapasaurus
Eusauropoda
Omeisaurus
Diplodocoidea
Neosauropoda
Camarasaurus
Brachiosauridae
Titanosauria
Titanosauriformes
Eoraptor
Herrerasauridae
Ceratosauroidea
Coelophysoidea
Ceratosauria
Spinosauroidea
Tetanurae
Allosauroidea
Neotetanurae
Ornithomimoidea
Ornitholestes
Coelurosauria
Tyrannosauridae
Oviraptorosauria
Dromaeosauridae
Troodontidae
Paraves
Archaeopteryx
Aves (birds)
Thyreophora
Genasauria
Neornithischia
Ornithopoda
Ornithischia
Other Archosaurs
Archosauria
Dinosauria
Sauropodomorpha
Saurischia
Theropoda
Neotheropoda
inferred relationships
temporal ranges
temporal ranges based on fragmentary evidence
missing ancestral lineages
After Paul C. Sereno's dinosaur family tree (2001)

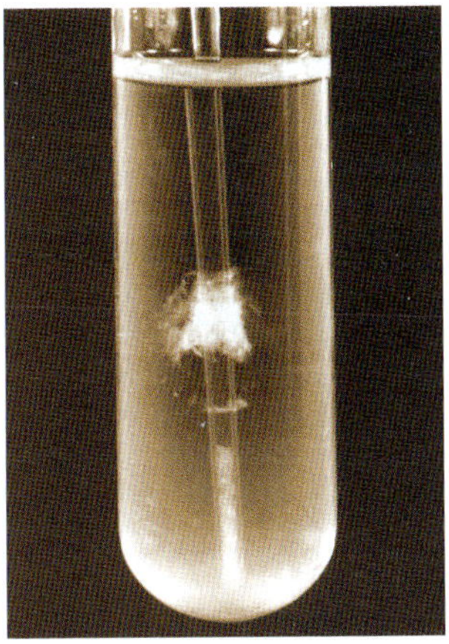

Plate 16: DNA in test tube

**Plate 17: Structure of DNA**

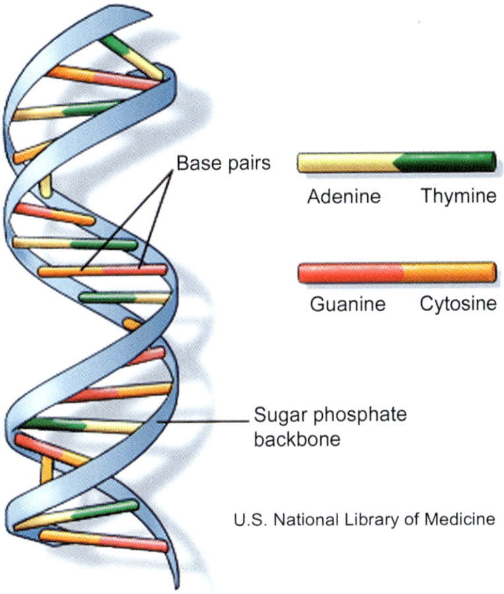


**Plate 18: Triceratops**

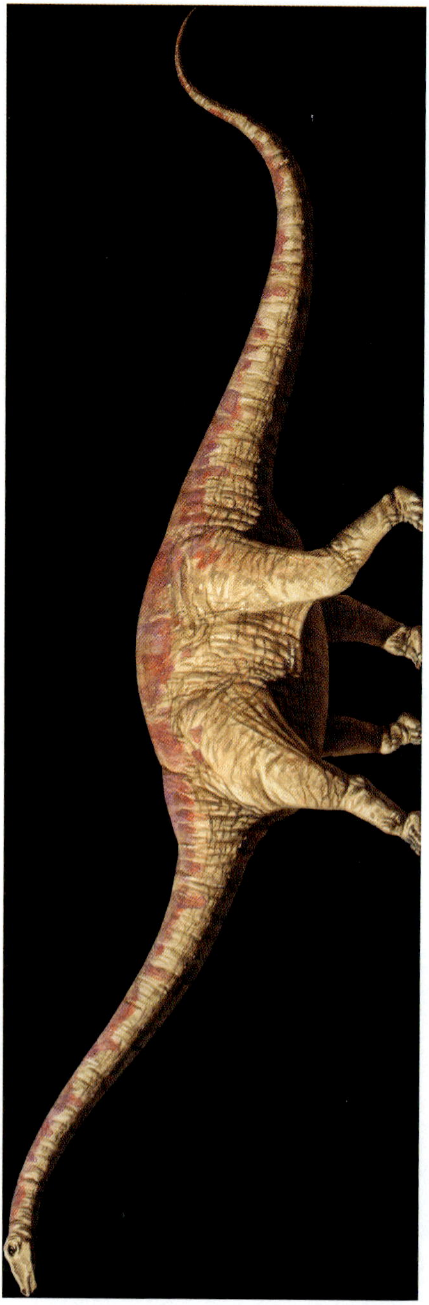

Plate 20: Seismosaurus

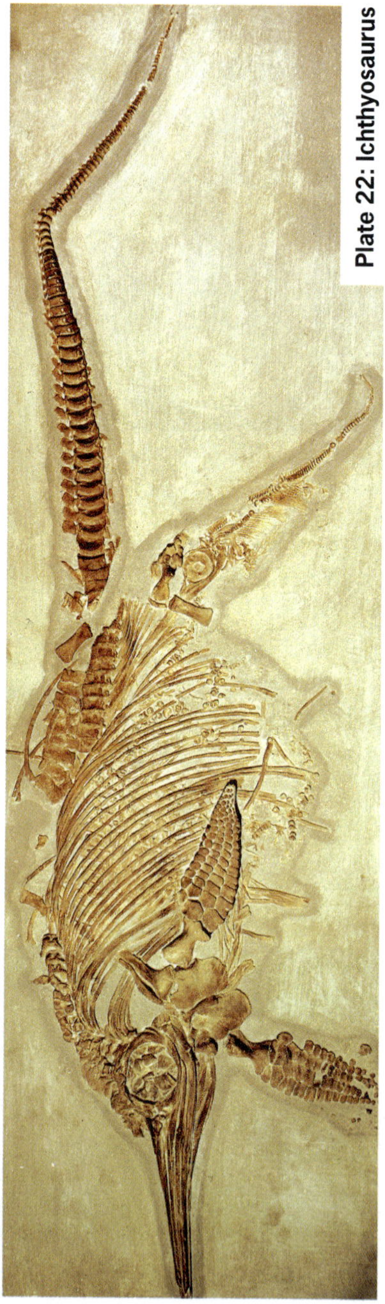

Plate 22: Ichthyosaurus

Plate 19: Parasaurolophus

Plate 21: Deinonychus

**Plate 23: Archaeopteryx**

**Plate 24: Dimetrodon**

Plate 25: Estemmenosuchus

Plate 26: A Pair of Inostrancevia alexandri assaulting Scutosaurus karpinski

Plate 27: Lystrosaurus

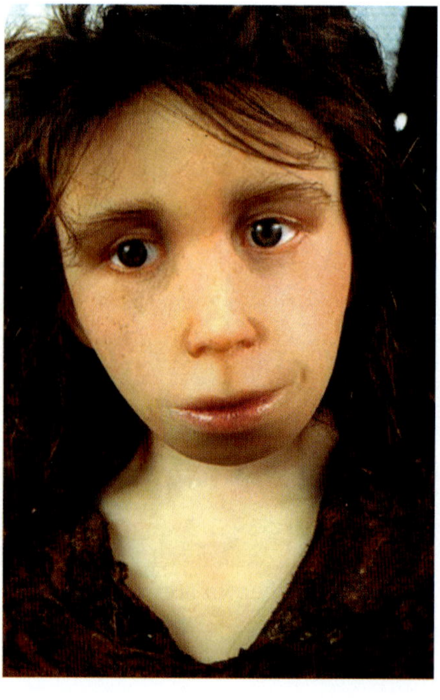

Plate 28: Neanderthal Child of Gibraltar

**Plate 29: Three phases of construction at Stonehenge**

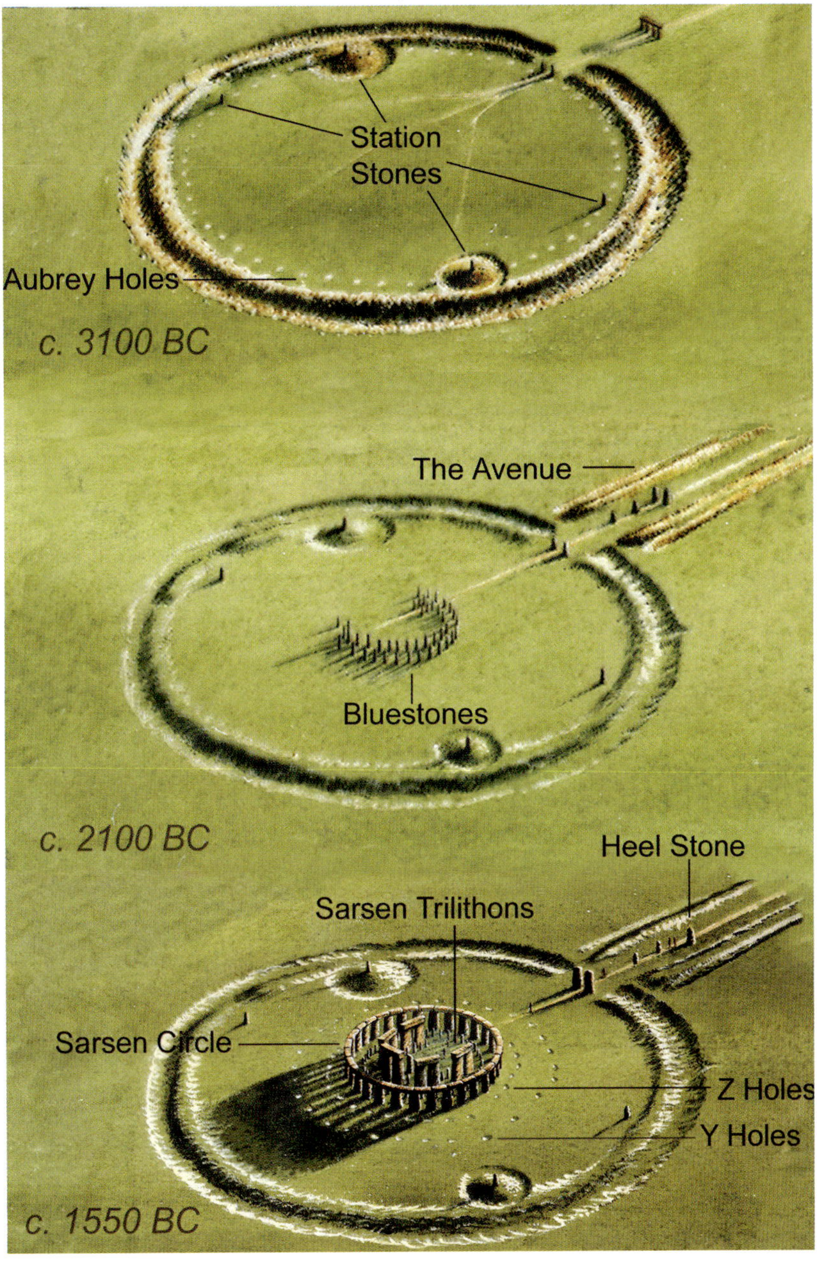

**Plate 30: Central vault at Maeshowe**

**Plate 31: Coelophysis**

But that is not what we find. We frequently find that a mammal-like trait in one creature has reverted to a reptilian condition in what would otherwise be the next link in the evolutionary chain.[51]

> The abundance of reversing traits means that the mammal-like reptiles cannot, by any stretch of the imagination, be portrayed as some sort of quasi-lineage (even a crude one) culminating in mammals. . . . When all of the characters used for the conventional constructions of cladograms are considered, the majority of mammal-like reptile characters do not consistently progress towards the mammalian condition. Instead, within the 'evolutionary' chain of mammal-like reptiles, there are many 'reversals' away from mammalian characteristics.[52]

The fossil record reveals not an evolutionary transition, but a motley group of astonishingly ugly animals with various combinations of reptilian and mammalian characteristics.

Three different orders of mammals (or at least their teeth) are found in the Upper Triassic—the Multituberculata ("many-cusped-teeth"), the Triconodonta ("triple-cone-teeth"), and the Symmetrodonta ("symmetrical teeth"—as if you needed me to explain that one!). These diminutive creatures were similar to modern mice or shrews; they were no more than four inches long and fragile. In most cases, there is nothing left of them but teeth, which is why they are named for their dental characteristics.[53]

All three orders appear in the fossil record virtually simultaneously, without a common ancestor.[54] Two additional orders are found in the Lower Jurassic—Docodonta and Pantotheria—and they do not appear to have evolved from any of the lower three orders. This suggests that the evolution of reptiles into mammals "most probably took place along a broad front."[55] There should be at least three, and maybe *five*, distinct lines of transitional fossil forms, rather than just one—and there is not even one. The fossil evidence of the purported reptile to mammal evolutionary transition is not as clear as Darwinists would have us believe.

Beyond the fossil problems, mammals are different from reptiles in ways that cannot plausibly be attributed to accumulated DNA copying errors. Reptiles grow throughout their lives; mammals reach a terminal size and grow no more. Reptiles are cold-blooded (ectothermic); mammals are warm-blooded (endothermic). Reptiles replace their teeth several times during their lives ("polyphyodonty"); mammals have temporary juvenile teeth and permanent adult teeth ("diphyodonty"). Reptiles have scales or scaly hides; mammals have hair. Reptiles are on their own from the moment they hatch; mammals breastfeed their young.

The breastfeeding mechanism is remarkably complex, when all its anatomical and chemical components are considered. In fact, we are still not completely certain how this intricate bio-chemical system works. It is believed that the baby's suckling stimulates nerves in the nipple that signal the pituitary gland to secrete a hormone called *prolactin*, which encourages continued milk production. As the baby continues suckling, sensors in the nipple cause the pituitary gland to secrete another hormone, *oxytocin*. Oxytocin causes the elastic tissue around each of many milk glands to contract, squeezing a supply of milk through the milk ducts into the sinuses and out the nipple.

Milk is produced on an "as needed" basis; the more the offspring consumes, the more the mother produces. (The mother can even produce two or three times the normal amount, in case there are twins or triplets.) Milk production then automatically turns itself off when breast-feeding has ceased. Darwinists would have us believe that this cleverly designed system is the product of random DNA copying errors.[56]

Another mammalian trait is that mammals (except for the monotremes) give birth to live young. It is true that some reptiles bear live young, but in a way very different from mammals. The common garter snake, for example, retains her eggs within her body until they hatch, and then the tiny snakes crawl out able to fend for themselves. But with mammals, there is no shell and no yolk sac full of food. Instead, mammals have a placenta, "an incredibly complex network of blood vessels joining mother to young," allowing the direct exchange of nutrients and wastes between mother and embryo.[57] You can imagine how different the genetic instruction for an egg with a hard shell must be from the genetic instructions for the placental/live birth method of reproduction, and how unlikely it is that random DNA copying errors effected the change.

So much for the "crown jewel." Closer inspection reveals it to be paste—cheap costume jewelry. Neither the purported transition from dinosaurs to birds, nor from reptiles to mammals, can withstand close scrutiny. The evolution hypothesis fails. There must be some other explanation for the mixed-class characteristics of the animals buried in Flood-laid strata. But before we discuss what that might be, we need to explore the topic of ancient man.

---

[1] Wilford, John Noble, *The Riddle of the Dinosaur,* (New York: Vintage Books, 1985), pp. 217-219.

[2] Wilford, at 103, 220.

[3] See, e.g., Feduccia, Alan, *The Origin and Evolution of Birds,* (New Haven, CT: Yale University Press, 1996).

[4] John Ruben, Professor of paleontology at Oregon State University, as quoted in BBC News, "Earliest Feathers Fan Controversy," June 22, 2000, *http://news.bbc.co.uk/1/hi/sci/tech/802009.stm.*

[5] BBC News, "Earliest Feathers Fan Controversy," supra. See also BBC News, "Bones Make Feathers Fly," August 14, 2000, *http://news.bbc.co.uk/hi/english/sci/tech/newsid_879000/879956.stm. Longisquama insignis* was discovered in Turkestan in the late 1960s by Russian paleontologist Alexander Sharov, who published in *Paleontological Journal* in 1970. Sharov described a row of long narrow appendages down the animal's back, interpreting them as a frill of extremely long scales. Thus, for over twenty years, scientists viewed the fossil as having elongated scales that only *looked* like feathers. In the 1990s, the fossil came to America as part of a traveling Russian dinosaur exhibit. After looking at the fossil, paleontologist Terry Jones stated, "We can identify certain structures in these fossils that you only find in feathers and just don't see anywhere else. So we're quite sure we're looking at the earliest feather. But beyond that, this animal looks like an ancestral bird even if you ignore the feathers. The teeth, pectoral structure, neck, and skull are just like those of birds." The Americans who went to inspect the *Longisquama* fossil were Terry Jones, Larry Martin, Alan Feduccia, Willem Hillennius, and John Ruben. See *Science,* June 23, 2000. Other scientists who have seen the fossil do not believe that feathers made the imprints. See, e.g., Reisz, Robert R., and Hans-Dieter Sues, "Paleontology: The feathers of Longisquama," *Nature,* 408, 428 (November 23, 2000).

[6] Wells, Jonathan, *Icons of Evolution,* (Washington, DC: Regnery, 2000), pp. 112, 116-122, 135.

[7] Wells, at 116, citing Mark Norell, a paleontologist with the American Museum of Natural History.

[8] Wells, at 116, citing Larry D. Martin, "The Relationship of *Archaeopteryx* to other Birds," pp. 177-183 in M.K. Hecht, J.H. Ostrom, G. Viohl, and P. Wellnhofer, Eds., *The Beginning of Birds* (Eichstatt: Freunde des Jura-Museums, 1985), p. 182.

[9] Nedin, Chris, "All About Archaeopteryx," Talk Origins website: *http://www.talkorigins.org/faqs/archaeopteryx/info.html*

[10] Denton, Michael, *Evolution: A Theory in Crisis,* (Bethesda, MD: Adler & Adler, 1985), p. 202.

[11] Denton, at 208, citing Gerhard Heilmann, *The Origin of Birds* (London: Witherby, 1926), p. 199-201.

[12] There is a theory of evolution based upon inheritance of acquired characteristics; it is known as Lamarckism, after its popularizer, Frenchman Jean Baptiste Lamarck (1744-1829). German zoologist August Weissman (1834-1914) conducted the classic experiment to disprove Lamarck. He cut off the tails of several succeeding generations of mice to see if disuse of the tail would eventually lead to shorter tails in succeeding generations. It did not. The offspring always had intact tails of normal length and size. As applied to the feather problem, no matter how badly frayed a reptile's scales became, the saurian's offspring would always have intact scales.

[13] Denton, at 208.

[14] Denton, at 209, citing, Stahl, B.J. *Vertebrate History: Problems in Evolution,* (New York: McGraw-Hill, 1974), p. 349.

[15] Feathers can be divided into two types: "pennaceous" and "plumulaceous." Flight feathers are a subdivision of pennaceous feathers that, in addition to the other qualities discussed, are asymmetrical: their leading edge is shorter. Plumulaceous feathers are downy feathers suitable only for warmth; they do not have aerodynamic properties. Darwinists might argue that scales did not evolve directly into flight feathers, but first evolved into plumulaceous or downy feathers. There are two problems with this idea. First, the transformation from scales to downy feathers is almost as miraculous as a transformation into flight feathers. Second, a downy feather is just as unsuitable for use as an airfoil as a frayed scale, and for all the same reasons. So a transition from downy feathers to flight feathers would be almost as unlikely as from a frayed scale to a flight feather.

[16] Quinn, Thom, Transcript of a paper presented April 18, 1998 at Dinofest. Sponsored by the National Academy of Sciences, Philadelphia, Pennsylvania.

[17] Prum, Richard O. and Alan H. Brush, "Which came first, the feather or the bird?" *Scientific American,* (March 2003)

[18] John Ruben, Professor of Paleontology at Oregon State University, as quoted in BBC News, "Earliest Feathers Fan Controversy," supra.

[19] Charles Darwin, in an 1860 letter to American botanist Asa Gray.

[20] Interestingly, when John Ruben examined a *Sinosauropteryx*, he found that the dinosaur's internal organs were preserved in outline in the rock slab, and resembled the internal organs of dissected crocodiles and other reptiles. Apparently, theropods had the type of compartmentalization of lungs, liver, and intestines you would find in a crocodile, *but not in a bird.* Ruben argues not only that dinosaurs were incapable of the high rates of gas exchange needed for warm-bloodedness, but also that their bellows-like lungs *could not have evolved into the high-performance lungs of modern birds.* See Ann Gibbons, "Lung Fossils Suggest Dinos Breathed in Cold Blood," *Science* 278(5341):1229, 1230 (Nov. 1997) *http://cas.bellarmine.edu/tietjen/images/lung_fossils_suggest_dinos_breat.htm.*

[21] Hutchinson, John R. and Kevin Padian "Coelurosauria" in *Encyclopedia of Dinosaurs,* Philip J. Currie and Kevin Padian, (eds.) (New York: Academic Press, 1997), p. 132; Kevin Padian and Luis Chiappe, "Bird Origins," in *Encyclopedia of Dinosaurs,* supra, p. 78; Currie, Philip J., "Theropods," in *The Complete Dinosaur,* Farlow, James O., and M.K. Brett-Surman, Eds. (Bloomington and Indianapolis: Indiana University Press, 1997), p. 230; Wells, Jonathan, *Icons of Evolution,* (Washington, DC: Regnery, 2000), p. 119-122.

[22] Currie, "Theropods," at pp. 230, 231.

[23] "First, the time line is all wrong. These alleged dinosaurian ancestors of birds occur 25 million to 80 million years after Archaeopteryx, the earliest known bird. Second, by the time you get to dinosaurs, you are dealing with fairly large, earthbound creatures, which means they would have had to evolve flight from the ground up, rather than from the trees down. Evolving flight from the ground up is biophysically implausible. Third, many of the features of birds and dinosaurs—the hands and teeth for example—don't match. The theropod dinosaur hand consists of the thumb and the next two fingers. The bird hand is made up of the middle three fingers. You can't just flip a switch to go from one type of hand to the other." Alan Feduccia, as quoted in *Discover,* vol. 24 no. 2 (February 2003). See also Spivey, Angela, "A Big Bang View of Birds," *Endeavors,* (Spring 1997) *http://research.unc.edu/endeavors/spr97/bird.html.*

[24]Paleontologist Sankar Chatterjee claims to have collected two bird specimens from the upper Triassic of Texas, but Chatterjee's finds are controversial, especially to enthusiasts of the dino-to-bird theory. See, Chatterjee, Sankar (1991) "Cranial anatomy and relationships of a new Triassic bird from Texas," *Philosophical Transactions of the Royal Society B* 332: 277-346; Chatterjee, Sankar, (1995) "The Triassic Bird Protoavis," *Archaeopteryx* 13:15-31; Chatterjee, *The Rise of Birds* (Baltimore, MD:

Johns Hopkins University Press, 1997); Czerkas, Stephen and Sylvia, *Dinosaurs: A Global View* (Limpsfield, Surrey, UK: Dragon's Press, 1990), p. 91-92.

[25] Melchor, Ricardo N., Silvina de Valais, and Jorge F. Genise, (2002) "Bird-like fossil footprints from the Late Triassic," *Nature* 417: 936-938 (The *Nature* article describes: "well-preserved and abundant footprints with clearly avian characters from a Late Triassic redbed sequence of Argentina, at least 55 M.y.r. before the first known skeletal record of birds." Similarity to bird tracks include: 1) an overall similarity to modern bird footprints; (2) footprints that are wider than they are long (not considering the hallux) and of small size; (3) slender digit impressions; (4) a wide angle between digits II and IV; (5) a posterior or posteromedial hallux impression, visible both in shallow and deep tracks; (6) slender claws showing distal curvature of lateral and medial claws away from the foot axis; and (7) a sole or metatarsal-phalangeal impression is visible in some footprints, where digits II to IV converge). See, also, Monani, Salma, "Footprints Step Back Bird Origins?" *Geotimes* (June 27, 2002) ("'The first and most striking feature of these fossil footprints is the overall resemblance with modern bird footprints,' Melchor says. In many birds, the hallux marks the innermost toe. In human beings this would correspond to the big toe, but unlike in human beings or other non-avian animals, the hallux points backwards in birds. . . . The imprint of the hallux in the late Triassic tracks is a distinct bird-like feature, Melchor explains. Other features such as slender digits and claws and wide angles between the second and fourth digits also point to bird-like morphology." Martin Lockley comments, "It's important . . . not to dismiss the track evidence as invalid until the skeletons are found . . . the Argentineans have presented their material well and these footprints seem to be distinctly bird-like and different from any known dinosaurs walking around at the time."); http://www.geotimes.org/june02/WebExtra0627.html. Camp, Ashby L., On the alleged dinosaurian ancestry of birds, True Origins archive: http://www.trueorigin.org/birdevo.asp#101, citing, Chiappe, Luis M., (1995) "The first 85 million years of avian evolution," *Nature* 378: 349-355; Ellenberger, P., (1974) *Paleovertebrata* (Mem. Extraor.) 1-141; Lockley, M. G., S. Y. Yang, M. Matsukawa, F. Fleming, and S. K. Lim., (1992) "The track record of Mesozoic birds: evidence and implications," *Philosophical Transactions of the Royal Society of London B* 336:113-134; Weems, R. E., and P. G. Kimmel, (1993) "Upper Triassic reptile footprints and coelacanth fish scales from the Culpepper Basin, Virginia," *Proc. Biol. Soc.* Wash. 106: 390-401. Ariel Roth supplies a creationist interpretation of these tracks: "We would expect [birds] to escape gradually to higher ground during the months of the flood, leaving only tracks in soft sediments. This could explain the apparent relative abundance of bird tracks in the Triassic below any good fossil bones." Roth, Ariel, *Origins: Linking Science and Scripture* (Hagerstown, MD: Review and Herald Pub. Assoc., 1998), pp. 168-169.

[26] Of course, dinosaur tracks also precede their makers into the fossil record, as we will further discuss in Chapter 24, so it could be argued that the dinosaur tracks evolved into the bird tracks. But Darwinists are not likely to want to call attention to the general phenomenon of tracks preceding the bodies of their makers into the fossil record by huge chunks of "geological time," since it tends to support the creationist/ catastrophist version of earth history, and falsify the Darwinian/Lyellian version.

[27] Wells, at 122.

[28] Currie, "Theropods," at 230.

[29] Wells, at 123-126, citing Christopher P. Sloan, "Feathers for T. rex?" *National Geographic* 196 (November, 1999), pp. 98-107; Rex Dalton, "Feathers fly over Chinese fossil bird's legality and authenticity," *Nature* 403 (2000); Jeff Hecht, "Piltdown bird," *New Scientist* 165 (January 19, 2000). Fossil skin impressions of *Albertosaurus* and *Gorgosaurus*, two species similar to *Tyrannosaurus*, show a scaly reptilian hide, not feathers, and there is no reason to believe *T. rex* had feathers.

[30] Storrs Olson, as quoted by Wells, supra, at 125.

[31] *Science News* reported that an international team of researchers that examined *Sinosauropteryx* concluded that the fibrous structures were not feathers. *Science News*, May 3, 1997, p. 271. See, also, Gibbons, Ann, "Plucking the Feathered Dinosaur," *Science*, 278(5341) (Nov. 14, 1997), pp. 1229, 1230; John Ruben, as reported by Gibbons in, "Lung Fossils Suggest Dinos Breathed in Cold Blood," *Science* 278(5341):1229, 1230.

[32] The structures are about 3 centimeters (1.2 inches). Currie, Philip J. "'Feathered' Dinosaurs," in *Encyclopedia of Dinosaurs*, supra, at 241.

[33] See, e.g., Xing Xu, Zhonghe Zhou, Xiaolin Wang, Xuewen Kuang, Fucheng Zhang & Xiangke Du, "Four-winged dinosaurs from China," *Nature* 421:335-340 (January 23, 2003); Gee, Henry, "Fossil boosts trees-down start for flight," *Nature Science Update* (Jan. 23, 2003); BBC News, "Four Winged Dinos from China," (January 23, 2003) *http://news.bbc.co.uk/1/hi/sci/tech/2684927.stm.*

[34] Alan Feduccia, for example, is skeptical of the Chinese fossils, as indicated by his interview with *Discover* magazine:

***Discover*: So far, only one feathered dinosaur, *Archaeoraptor*, has been publicly acknowledged as a forgery. You think there are others?**
Feduccia: *Archaeoraptor* is just the tip of the iceberg. There are scores of fake fossils out there, and they have cast a dark shadow over the whole field. When you go to these fossil shows, it's difficult to tell which ones are faked and which ones are not.

I have heard that there is a fake-fossil factory in northeastern China, in Liaoning Province, near the deposits where many of these recent alleged feathered dinosaurs were found. Journals like Nature don't require specimens to be authenticated, and the specimens immediately end up back in China, so nobody can examine them. They may be miraculous discoveries, they may be missing links as they are claimed, but there is no way to authenticate any of this stuff.

***Discover*: Why would anyone fake a fossil?**
Feduccia: Money. The Chinese fossil trade has become a big business. These fossil forgeries have been sold on the black market for years now, for huge sums of money. Anyone who can produce a good fake stands to profit.

See "Discover Dialogue: Ornithologist and Evolutionary Biologist Alan Feduccia Plucking apart the Dino-Birds," *Discover*, vol. 24 no. 2 (February 2003)

[35] Johnson, Phillip, *Darwin on Trial*, (Washington, D.C.: Regnery, 1991) p. 75.

[36] Camp, Ashby L., (J.D., Duke, M.Div., Harvard), "Reappraising the Crown Jewel" *Creation Matters*, (September/October 1998) a newsletter of the Creation Research Society, citing, Robert L. Carroll, *Vertebrate Paleontology and Evolution*, (New York: W. H. Freeman, 1988), p. 198.

See *http://www.creationresearch.org/creation_matters/98/cm9809.html.*

[37] Gould, Stephen Jay, *Eight Little Piggies* (New York: W.W. Norton, 1993) pp. 67, 68.

[38] "The transition between pelycosaurs and therapsids has not been documented," Robert L. Carroll, at p. 397, as quoted by Camp, supra.

[39] Carroll, at p. 369, as quoted by Camp, supra. Interestingly, one Clifford Cuffey posted an Internet article criticizing creationist Duane Gish for stating: "It is apparent that the argument for linking pelycosaurs to the therapsids is extremely weak and is based solely on certain similarities. There are no transitional forms that would provide actual evolutionary links between pelycosaurs and therapsids." Cuffey counters that, "In fact, the fossil evidence contradicts Gish's claim: the oldest therapsids and sphenacodonts are very similar in nearly every aspect of their morphology, especially the skulls. [citations omitted] This was demonstrated 90 years ago!"
See *http://www.gcssepm.org/special/cuffey_08.htm.* Yet Carroll points out that the sphenacodonts are too specialized to have been ancestors of therapsids. "However, the more primitive genus *Haptodus* could have filled this role." So Darwinists cannot agree on what *family* might have given rise to therapsids, yet Gish is criticized for pointing out that there is no transitional form.

[40] Camp.

[41] Camp.

[42] Kemp, T.S., *Mammal-like Reptiles and the Origin of Mammals,* (New York: Academic Press, 1982), p. 327, as quoted by Camp, supra (emphasis added).

[43] Czerkas, Sylvia J. and Stephen A. Czerkas, *Dinosaurs, a Global View,* (Limpsfield, Surrey, Great Britain: Dragon's World Ltd, 1990), p. 54.

[44] McLoughlin, John C., *Synapsida: A New Look into the Origin of Mammals* (New York: Viking Press, 1980) p. 52.

[45] Camp.

[46] Johnson, at 76.

[47] Carroll, at p. 398, as quoted by Camp.

[48] Cluver, Michael A., "Fossil Reptiles of the South African Karoo" at South African Museum website.

[49] Kielan-Jaworowska, Z., "Interrelationships of Mesozoic mammals," *Historical Biology* 6(3):195 (1992), as quoted by Woodmorappe, John, "Mammal-like reptiles: major trait reversals and discontinuities," *Creation ex Nihilo Technical Journal* 15(1):44-52, (2001).

[50] Lewin, Roger, "Bones of Mammals' Ancestors Fleshed Out," *Science,* vol. 212, no. 6 p. 1492, as quoted by Camp.

[51] Woodmorappe, John, "Mammal-like reptiles: major trait reversals and discontinuities," *Creation ex Nihilo Technical Journal* 15(1):44-52, (2001). See *http://www.answersingenesis.org/home/area/Magazines/tj/docs/tj_v15n1_mammal.asp.*

[52] Woodmorappe, supra.

[53] McLoughlin, at 120.

[54] "The multituberculates and their shrewlike symmetrodont and triconodont cousins were each independently derived from insectivorous therapsids during the Triassic." McLoughlin, at 125.

[55] Colbert, Edwin H., *Evolution of the Vertebrates,* 3rd ed. (New York: John Wiley & Sons, 1955-1980), p. 250.

[56] Darwinists point to the monotremes, the living mammals that lay eggs, as having an intermediate feeding system. The monotremes have glands that secrete a thick yellow liquid containing fats and proteins. The thick liquid clings to long, stiff hairs and is licked off by the offspring. But no one believes that the monotremes were transitional between reptiles and other modern mammals. McLoughlin writes that, "they are probably derived from a therapsid group different from that which was to produce the rest of the mammals." McLoughlin, at 87, 88. So the monotremes, like *Archaeopteryx*, cannot be a missing link, but at best "a grade of organization which the proposed lineage went through" to get from reptiles to mammals with fully functioning nipples and mammary glands. Moreover, only two jaw fragments identified as being from monotremes are known from the Lower Cretaceous. Otherwise, monotremes do not appear until the Tertiary, which is far too late to be a transition on the way to other mammals.

[57] McLoughlin, at 127.

CHAPTER 17

# Ancient Man—the Darwinian View

Joachem Neander loved to go hiking. As he hiked, the seventeenth-century Lutheran theologian would compose hymns and sing them. One of the products of Neander's creative nature walks was "Praise to the Lord, the Almighty, the King of Creation," a much-beloved Christian hymn, and the first hymn in the *Seventh-day Adventist Hymnal.*

Neander's favorite hiking territory was about ten miles east of Dusseldorf, Germany, in a valley through which flowed the Dussel River. Because of Neander's affection for the place, it became known as Neanderthal. (*Thal* was the Old German word for "valley." The *h* is silent and has since been dropped.) A couple of centuries later, a man named Beckersdorf owned land in the Neanderthal and exploited it as a limestone quarry. In 1856, his workmen discovered bones on the floor of a cave in the side of the gorge. These bones were the first fossils given the name "Neanderthal man." Neanderthals have since been found in western, southern, and Eastern Europe, as well as in the Middle East.[1]

The term *Neanderthal* is synonymous with "caveman" and implies a person who is primitive, backward, brutish, and generally not much better than an ape. Where did this conception of Neanderthal man come from?

In 1908, a Neanderthal skeleton was discovered in a cave in southern France, near the village of La Chapelle-aux-Saints (the chapel of saints). A French anthropologist named Marcellin Boule drew the task of reconstructing and mounting the skeleton, which came to be called the "old man" of La Chapelle. Boule's examination of the specimen's spine indicated that the old man could not walk upright, so Boule placed him in a stooped-over posture. Boule concluded that the old man could not fully extend his legs and must have walked with a bent-kneed gait. Boule envisioned the old man walking pigeon-toed, on the outer edges of his feet. He spread the big toes apart from the other toes, in an ape-like manner. Boule positioned the specimen's head jutting forward, in an ape-like attitude.

In short, Boule reconstructed the old man of La Chapelle as ape-like in every important detail.[2] He published his conclusions in a French scientific journal in 1911. It would be more than forty years before Boule's view of the Neanderthals was seriously questioned; his interpretation dominated scientific thought throughout the first half of the twentieth century.

**Figure 17.1: Boule's erroneous reconstruction of Neanderthal Man**

In 1955, two anatomists—William L. Straus and A.J.E. Cave—visited Boule's Neanderthal exhibit while attending a conference in Paris. Upon examining the La Chapelle Neanderthal, they determined that the "old man" had indeed been very old, as well as arthritic and crippled. It seems that Boule had failed to diagnose the old man's "gross deforming osteoarthritis."

Having lost faith in Boule's competence, Straus and Cave undertook to reexamine the Neanderthals. After examining several sets of remains, they concluded that healthy Neanderthals stood erect and walked normally, just as modern humans do:

> There is no trace of evidence that Neanderthalers had exceptionally divergent great toes or that they were forced to walk orang-like on the outer edge of their feet; there is no evidence that they were unable fully to extend their knee joints; there is no evidence that their spinal columns lacked the convexities necessary for fully erect posture; there is no evidence that the head was slung forward on a peculiarly short and thick neck.[3]

Neanderthals were different from modern humans—they were physically much *stronger* and they had larger brains. But the difference was not so great that if Neanderthal man "could be reincarnated and placed in a New York subway—provided that he were bathed, shaved, and dressed in modern clothing—it is doubtful whether he would attract any more attention than some of its other denizens."[4]

Although no one contradicts Straus and Cave, and most recognize that Neanderthals were only slightly different from modern humans, Boule's view of Neanderthals as primitive and ape-like has been slow to exit the museums, the textbooks, and the public's imagination. The popular conception of Neanderthal man was created by a gross and comprehensive, but "scientific," misinterpretation.[5]

The Neanderthals' basic humanity is beyond question. There is evidence that their burials were conducted with tender ritual.[6] The Neanderthal at La Chapelle was buried with "beautiful pieces of jasper" arranged around his head.[7] A flute has been found at a

Neanderthal campsite, showing that Neanderthals could produce and enjoy music.[8] **Plate 28** is a recent reconstruction of a four-year old Neanderthal child, showing her to be only trivially different from a modern human being.

Boule's blunder was not the only *faux pas* to mar the record of paleoanthropology. Alas, the scramble to find the elusive "missing link" has produced some interesting "science." Another example of weird science involves our friend Henry Fairfield Osborn, of the American Museum of Natural History. In 1922, a Nebraska rancher and fossil hunter named Harold Cook sent Osborn a worn molar tooth from Pliocene (upper Tertiary) strata near Snake Creek, Nebraska, along with his opinion that the tooth "very closely approaches the human type." Osborn examined the tooth, concluded it was from a higher primate, and within a couple of weeks announced that the tooth was the first fossil primate discovered on American soil. Osborn chose

**Figure 17.2: Hesperopithecus haroldcookii, as depicted, for the London Illustrated News, by Amédée Forestier (1861-1930) under the supervision of Sir Grafton Elliot Smith (1871-1937), based upon one cheek tooth.**

the genus name *Hesperopithecus,* meaning, "ape of the western world."

Osborn published an article entitled, "*Hesperopithecus,* the First Anthropoid [man-like] Primate Found in America," in which he asserted that he had found "a new genus and species of anthropoid." Osborn described it as being intermediate between humans and apes and closer to humans than to apes: "On the whole, we think its nearest resemblances are with . . . men rather than with apes."[9] Osborn's colleagues in England produced an artist's conception of Hesperopithecus, showing a couple of anthropoids surrounded by Pliocene fauna such as camels and horses.

By coincidence, when Osborn announced the discovery of *Hesperopithecus,* he was engaged in a newspaper debate over origins with William Jennings Bryan, the three-time presidential candidate who in later life had taken up the creation/evolution controversy. The *New York Times* had published Bryan's lengthy article criticizing evolution. In Osborn's response, he urged Bryan, in the words of Job 12:8, to "speak to the earth, and it shall teach thee."[10] In one of his articles announcing the discovery of *Hesperopithecus,* Osborn wrote:

> It has been suggested humorously that the animal should be named *Bryopithecus* after the most distinguished Primate which the State of Nebraska has thus far produced. It is certainly singular that this discovery is announced within six weeks of the day (March 5, 1922) that the author advised William Jennings Bryan to consult a certain passage in the Book of Job, "Speak to the earth and it shall teach thee," and it is a remarkable coincidence that the first earth to speak on this subject is the sandy earth of the Middle Pliocene Snake Creek deposits of western Nebraska.[11]

Three years later, on the eve of the famous Scopes trial, at which Bryan represented the anti-evolution side, Osborn reissued his *New York Times* essay, expanded to book form, dedicated to John Scopes, and bearing the snide title, "The Earth Speaks to Bryan."

Before gloating so publicly, Osborn should have consulted another passage of Scripture: "Pride goeth before destruction, and a haughty spirit before a fall." (Prov. 16:18) The fall of *Hesperopithecus* was swift and total. The American Museum sent expeditions to Nebraska in 1925 and 1926, and several more teeth were collected from the area where the first was discovered. When all the teeth, including the original find, were closely examined, it was determined that none was from a primate. In 1927, Osborn's colleague, William King Gregory, published an article entitled, "*Hesperopithecus* Apparently not an Ape nor a Man." The teeth had belonged to . . . a pig. Osborn's anthropoid ape and possible missing link was an extinct peccary, *Prosthennops*.[12]

"Nebraska Man" is a minor and almost forgotten piece of Darwinian foolishness compared with *Eoanthropus dawsoni*, better known as "Piltdown Man." *Eoanthropus dawsoni* ("Dawson's dawn man") was named after Charles Dawson, a lawyer and amateur fossil hunter. In 1908, workmen at a gravel pit in Piltdown, Sussex, England, gave Dawson part of a human-looking skull they had found in the gravel pit. Over the next four years, Dawson found more of the skull, and eventually took what he had to Arthur Smith Woodward, a geologist on staff at the British Museum.[13]

A young Frenchman named Marie-Joseph Pierre Teilhard de Chardin, a novice of the Jesuit order, began helping Dawson with the dig. Dawson, Teilhard, and Smith Woodward worked together to excavate the gravel pit, and each found fossils. Dawson found half of an ape-like lower jaw; Teilhard found a canine tooth that fit into that jaw. The gravel pit also yielded animal fossils typical of the Pleistocene, including mastodon, rhinoceros, hippopotamus, beaver, and dear. There were even primitive stone tools.[14] The cranial and mandible fragments appeared ancient, and the animal fossils found with them corroborated that they were of the right geological "age" to be an ancestor of modern humans.

Anthropologists had a ready-made explanation for the fact that the skull looked "modern," whereas the jaw appeared ape-like. An Oxford anthropologist named William J. Sollas had argued that evolution might work at different rates in different parts of the body. It was commonly believed that the most important difference between apes and humans was brain size, and that the enlargement of the brain must have led out in the evolutionary transition. Other characteristics, like the shape of the jaw, lagged behind. Sollas even argued that a human ancestor with a large brain and an apelike jaw was "an almost necessary stage in the course of human development."[15]

Everything seemed to fit. Smith Woodward concluded that they had found a missing link between humans and apes, a conclusion the press reported with alacrity. "PALEOLITHIC SKULL IS A MISSING LINK, DARWIN'S THEORY IS PROVED TRUE," trumpeted *The New York Times.*[16] Almost all of the leading authorities of the day were eventually convinced that Piltdown Man was a genuine primitive human ancestor. Henry Fairfield Osborn was at first skeptical, but eventually became an enthusiastic supporter, making a pilgrimage to England to examine the fossils, and including an exhibit on Piltdown Man in the "Hall of Man" of the American Museum of Natural History. Piltdown Man was in the back of his mind when he declared a Nebraskan pig tooth to be a missing link.[17]

The years between 1915 and 1953 produced many scholarly and popular books expounding on Piltdown Man, including William Sollas' *Ancient Hunters* (1924), Earnest Hooton's *Up From the Ape* (1931), Sir Arthur Keith's *The Antiquity of Man* (1925), Sir Grafton

Elliot Smith's *The Evolution of Man* (1924), *Search for Man's Ancestor's* (1931), and *Discovery of Primitive Man in China* (1931), M.C. Burkitt's *Prehistory* (1921), E. Cecil Curwen's *Prehistoric Sussex* (1929) and *The Archaeology of Sussex* (1937), J. Reid Moir's *The Antiquity of Man in East Anglia* (1927) and Harris Wilder's *The Pedigree of the Human Race* (1926).[18]

Sir Arthur Keith even developed a geographical theory of human evolution (complete with subtle racist implications) based upon Piltdown Man:

> We now know that when the Piltdown type was being evolved in England—or at the Western end of the Old World—a totally different type had come into being in the Eastern Lands of the Old World. The Eastern types had low receding foreheads, modeled as in the gorilla and chimpanzee. The Western or Piltdown type differed; it had a relatively upright and high forehead, modeled not on gorilla lines but rather on those of an orang. While the Eastern forms retained in their shape of head the low squat type of the chimpanzee and gorilla, the Western or Piltdown type tended to assume the higher vaulted skull seen in modern races.[19]

Keith's theory was based entirely upon a few teeth and fragments of bone. The evidence did not warrant the phrase "we now know," but even today scientists constantly speak of theory as though it were fact.

What we *now* know is that Piltdown Man was a hoax. In 1947, a staff geologist at the British Museum tested the Piltdown fossils for fluorine content. Teeth and bones absorb fluorine from the environment. A fluorine test cannot give an absolute age for a fossil, but it is designed to detect whether two fossils have been in the same ground for the same length of time. If two pieces from the same site contain significantly different amounts of fluorine, they probably arrived at the site at different times. Over the course of several years, multiple fluorine tests were done on the Piltdown material. Eventually it became apparent that the skull and mandible had radically different fluorine contents.[20] Something was not right.

Close examination revealed that the teeth had been filed. The canine tooth Teilhard had found was "fake from top to bottom and side to side." The crown looked old because it had been filed down; suddenly, a crisscross pattern of file marks was visible that no one had noticed during the previous forty years. The hoaxer had filed the canine tooth too vigorously, exposing the pulp cavity. He then filled in the cavity with sand and a material similar to chewing gum. The molar teeth had also been abraded, filed, and smoothed out with acid.[21]

Later, after most experts acknowledged that Piltdown Man was a hoax, scientists dated the skull and jaw fragments using Carbon-14, and it was discovered that the skull, which was fully human, was about 630 years old, give or take a century, and the jaw bone, which was from an orangutan, was about 500 years old. In November 1953, the British Museum finally announced that Piltdown Man was indeed a hoax.[22]

No one knows who planned and executed the most famous hoax in scientific history. Dawson has always been the prime suspect; he "found" most of the fossils, including the earliest ones "found." In its article reporting the exposure of the hoax, the *Times* of London suggested that Dawson had chosen Teilhard and Smith Woodward as his helpers because their honesty was unimpeachable.[23] But Stephen Jay Gould has forcefully argued that Teilhard was in on the hoax.[24]

The hoaxer, whomever he was, must have had extensive knowledge of anatomy, chemistry, paleontology, geology, and anthropology. He had to know to file down the ape teeth so they would look more human and fit with the human upper jaw. He had to know to break off the symphysis and condyle so no one could articulate the jawbone with the skull and see that they did not match. He had to know what chemicals would give the fossils a patina of age. He had to know how to seed the gravel pit with index fossils, so that his "missing link" would be dated as expected. He had to know paleolithic stone tools and how to make them or where to find them. Obviously, the hoaxer was either a professional paleontologist or a very well trained amateur.

Because many have suspected that Dawson did not have the requisite scientific knowledge to pull it off—at least not alone—the list of suspects reads like a who's who of British paleoanthropology. William J. Sollas, Sir Grafton Elliot Smith, and Sir Arthur Keith have all been seriously suspected of perpetrating the fraud. Although he was not a professional paleontologist, a good case can be made against Sir Arthur Conan Doyle, the creator of Sherlock Holmes.[25]

It is perhaps more interesting to analyze the hoaxer's motive than to try to guess his identity. If the hoaxer had been a creationist bent on embarrassing the Darwinists, he would have exposed the hoax within a couple of years after the scientific establishment had committed to Piltdown Man's authenticity. But he did not. The hoax remained undisclosed for forty years and was never exposed by the hoaxer. We can only conclude that the hoaxer had no intention of embarrassing Darwinian science.

Most likely, the hoaxer was a Darwinist who wanted people to believe in human evolution. Kenneth Oakley, the British Museum geologist who ran the fluorine tests and thereby played a key role in exposing the hoax, has written:

> The [Java man] material was tantalizingly incomplete, and for many scientists it was inadequate as confirmation of Darwin's view of human evolution. I have

> sometimes wondered whether it was a misguided impatience for the discovery of a more acceptable 'missing link' that formed one of the tangled skein of motives behind the Piltdown forgery.[26]

J.S. Weiner, another of the British Museum authorities who played a role in exposing the hoax, also concluded that this must have been the forger's motive:

> Behind it all we sense, therefore, a strong and impelling motive. . . . There could have been a mad desire to assist the doctrine of human evolution by furnishing the 'requisite' 'missing link.' . . . Piltdown might have offered irresistible attraction to some fanatical biologist to make good what Nature had created but omitted to preserve.[27]

Author Jessica North gives a psychological profile of the Piltdown forger in her novel, *Mask of the Jaguar* (1981):

> The little man who must have perpetrated it—a respected, scholarly gentleman who had nothing – absolutely nothing!—to gain. The hoax brought him no money, no promotions, yet he spent months and maybe years perfecting his forgery. And at a terrible risk of disgrace!
>
> He wanted fame, to be sure, but also to make the world agree with a theory he believed in. He concocted Piltdown Man because he was sure that somewhere,

> unfound, a genuine Piltdown Man *had* to exist. In a warped mind, fraud in the service of truth seems hardly fraud at all.[28]

Charles Blinderman, author of *The Piltdown Inquest*, reckons this one of the best comments made about the hoaxer's motives.

Naturally, paleoanthropology would rather we forget about the Piltdown incident. But "however earnestly scientists may now dissociate themselves and their theory from Piltdown Man, they cannot wipe out the memory of forty years of labor expended on a deliberate and not particularly subtle fraud."[29]

I have recounted the stories of Boule's blunder, "Nebraska man," and "Piltdown Man" not merely to embarrass Darwinists—although that is a gratifying pastime—but because these stories illustrate how science operates.

## 1. THE THEORETICAL MODEL IS MORE IMPORTANT THAN THE "FACTS"

The theoretical model is more important than the facts, because the theory creates the facts. The naïve lay person believes that theory is built upon facts, but in reality, the theory comes first and determines what the "facts" are. The theory creates the data by 1) controlling the research program, i.e., determining what scientists are looking for, and 2) heavily influencing how the primary scientific data are interpreted.

Theory is necessary to scientific inquiry. The universe confronts us with a bewildering, infinite array of natural phenomena. It would be impossible even to try to explain a given phenomenon without at least a working hypothesis as to the explanation. In any given discipline, and with regard to any scientific problem, a consensus emerges among the scientists as to the best theory to apply to the problem. That theory becomes the dominant paradigm.

Thomas S. Kuhn, in his 1962 masterpiece, *The Structure of Scientific Revolutions*, noted that the day-to-day business of science is to elucidate the dominant paradigm. A dominant paradigm acts like a magnet, orienting data in the direction it wants. Typical research programs—which Kuhn calls "normal science"—are designed to show how the data fit within the dominant paradigm and are explained by it. The enterprise of "normal science" is an attempt to force nature into the preformed and relatively inflexible box that the paradigm supplies. As long as the dominant paradigm is sound, the program of normal science is productive, and will tend to expand the frontiers of useful knowledge.

But normal science has a couple of inherent weaknesses. First, scientists are searching for data that fit the paradigm, phenomena for which the paradigm can provide an explanation. Scientists tend not to look for data that do not fit the paradigm. Second, scientists tend to be hostile to anyone who suggests a new and different paradigm, fearing that their work on the established paradigm will have been wasted. As Kuhn explains it:

> No part of the aim of normal science is to call forth new sorts of phenomena; indeed, those that will not fit the box are often not seen at all. Nor do scientists normally aim to invent new theories, and they are often intolerant of those invented by others. Instead, normal-scientific research is directed to the articulation of those phenomena and theories that the paradigm already supplies.[30]

Data that do not fit the paradigm "are often not seen at all" by workers in the discipline, which is why new paradigms are often suggested by outsiders. One example would be *meteorologist* Alfred Wegener's promotion of a new paradigm in *geology*—plate tectonics.

The history of the search for transitional forms between apes and humans illustrates how the paradigm creates the accepted body of scientific "facts." In the 1860s, when Darwin's theory that humans evolved from apes became dominant, *there was no fossil evidence of any such transition.* It simply was not the case that scientists had found strange fossils that appeared to be intermediate between apes and humans, and, having found such fossils, were forced to conclude that humans had evolved from apes. Exactly the opposite was true. Scientists first adopted the Darwinian model and then went in search of evidence to support that model. The Darwinian model determined the type of fossils the scientific establishment deemed significant: fossils that could be interpreted as supporting an ape-to-human transition.

After many long decades of determined, indefatigable search for some evidence of what they had already decided was true, dribs and drabs of "evidence" started to come in. They found Java man, Piltdown Man, Peking man, the Taung skull and the other australopithecines. In fact, every single fossil that Darwinists now point to as evidencing an ape-to-human transition was found long after Darwinism had become the dominant paradigm.

Another way that theory determines the "facts" is by determining how the raw data are interpreted. The theory, model, or paradigm is the lens through which scientists view the raw materials of their science. Whatever the raw facts or data are—in paleontology, the raw materials are the fossils—the model determines how they are *interpreted.* If there are several possible interpretations, and there usually are, scientists will choose the interpretation that best fits the dominant paradigm.

Boule's interpretation of the Neanderthals is a good example. Boule's blunder was not the result of incompetence. Boule was world renowned, one of the top men in his field. But he was a Darwinist. He believed that men had evolved from apes, and when he examined ancient human remains, he expected to see ape-like qualities. When he studied the remains of the "old man" of La Chapelle-aux-Saints, Darwinian expectations blinded him to the possibility that what he was seeing was the result of disease and did not reflect the normal Neanderthal condition. His reconstruction of Neanderthal was more the result of his Darwinian beliefs than of anything in the bones themselves.

English science writer Roger Lewin, commenting on Boule's blunder, writes that,

> It is, in fact, a common fantasy, promulgated mostly by the scientific profession itself, that in the search for objective truth, data dictate conclusions. . . . Data are just as often molded to fit preferred conclusions. And the interesting question then becomes "What shapes the preference of an individual or group of researchers?" not "What is truth?" It is clear that Boule went beyond the evidence of his eyes—perhaps to press more persuasively his version of the Truth.[31]

And Boule's interpretation of Neanderthal became part of the body of scientific knowledge, as reflected in museum displays and textbooks, for half a century.

Likewise, after examining a single, badly worn molar, Osborn might have concluded, "This could be from a number of different species; we need more information before we publish a notice." Instead,

Osborn rushed into print with the assertion that he had discovered a new man-like ape, hinting that it was a missing link. Darwinian theory told him that a missing link between apes and humans existed, so he interpreted the pig tooth as the missing link he expected to find. He then used the result of his subjective interpretation to publicly abuse William Jennings Bryan for doubting Darwinism.

Fortunately, the "ape of the western world" formed part of the body of scientific knowledge for less than five years. Stephen Jay Gould, noting that pig teeth and human teeth are easily confused for one another, argued that the real moral of the "Nebraska man" story is that science is self-correcting.[32] It is true that Nebraska man had a happy ending: arrogance rather swiftly received its comeuppance. But given that both the Piltdown fraud and Boule's misinterpretation of Neanderthal remained part of the body of scientific knowledge for over forty years, there is little solace in the notion that science is self-correcting.

Piltdown Man provides the most dramatic illustration of how theory controls the interpretation of the raw data. Although the Piltdown skull was from a modern human, virtually every British anthropologist, and many others as well, imagined that it had apelike features. And although the jaw was orangutan, the trained eyes of these men of science discerned important human characteristics. "The Piltdown skull, when properly reconstructed, is found to possess strongly simian peculiarities," noted Sir Grafton Elliot Smith. "In respect of these features it harmonizes completely with the jaw, the simian form of which has not only been admitted, but also exaggerated by most writers."[33]

Isn't it interesting that a trained anthropologist could discern "strongly simian peculiarities" in a modern human skullcap? Isn't it revealing that he could accuse others of exaggerating the ape-like characteristics of a fully orangutan jawbone? "How is it that trained men, the greatest experts of their day, could look at a set of modern human bones—the cranial fragments—and 'see' a clear simian signature in them; and 'see' in an ape's jaw the unmistakable signs of humanity? The answers, inevitably, have to do with the scientists' expectations and their effects on the interpretation of the data."[34]

One could hardly ask for a better illustration of the truism, "I wouldn't have seen it if I hadn't believed it." The scientific establishment believes that humans evolved from apes, which is why anthropologists "see" things in the fossils that confirm their belief. Clearly, the model is more important than the data, because the model determines how the data are interpreted.

Kuhn points out that if enough data accumulate that fail to fit within the dominant paradigm, a new paradigm can be substituted. This has happened many times, in many scientific disciplines. But no biblical paradigm will ever be substituted for Darwinism. Science has an *a priori* philosophical commitment to naturalism. Any scientist who proposes a supernatural paradigm of human origins will immediately become *persona non grata*—ridiculed, scorned, and isolated. No mainstream scientific journal will publish a paper that is explicitly creationist. Thus, most scientists will never see any data that is not interpreted according to the Darwinian paradigm and will see very little data that does not seem to fit within it.

The Darwinian paradigm still exerts a controlling effect upon how the data are interpreted. A fragment of an upper arm bone found in Kanapoi, Kenya, provides a glaring example. This fossil, cataloged as KP-271, is thought to be around 4.5 million years old, yet detailed morphological and metrical comparisons with humans, chimpanzees, and an Australopithecine indicated that it was human. "In these diagnostic measurements [KP-271] is strikingly close to the means [averages] of the human sample," wrote Patterson and Howells. "There are individuals in our sample of man on whom measurements . . . of [KP-271] can be duplicated almost exactly."[35]

So scientists have found a human bone that is dated almost five million years old, but theory holds that no humans were around five million years ago. So despite its human morphology, no scientist has been willing to state that KP-271 is human. Howells admits this:

> The humeral fragment from Kanapoi, with a date of about 4.4 million, could not be distinguished from *Homo sapiens* morphologically or by multivariate analysis by Patterson and myself in 1967 (or by much more searching analysis by others since then). We suggested that it might represent *Australopithecus* because at that time allocation to Homo seemed preposterous, *although it would be the correct one without the time element.*[36]

KP-271 is all the more striking because it is not a case, like Piltdown, where theory influenced researchers to "see" ape-like characteristics that objectively were not present in the bone. The researchers' eyes were wide open to the fact that this bone was not ape-like; the measurements were well within the range of human morphology, in fact, close to the averages of the human sample. The specimen was deemed australopithecine in spite of its morphology, based *solely* upon theoretical considerations. Obviously, theory is more important than fact, because the theory *creates* the "facts."

If KP-271 is human, no "younger" fossil could be the ancestor of man. Since all of the putative missing links—*Australopithecus, Homo habilis,* and *Homo erectus*—are younger, the Darwinist theory of human origins would be stripped of every scrap of fossil support. So the Darwinist, despite what his eyes are telling him, simply declares that the bone is not human. Anyone searching the literature will not find a report of a human fossil that is out-of-order stratigraphically and thus poses a challenge to Darwinian theory. He will find a report of an australopithecine fossil, which is just what the theory predicts should be found at that stratigraphic level.

Because only a few specialists are qualified to interpret the raw materials, everyone else tends to accept what those experts say. As scientific knowledge filters down from the original researchers, expert interpretations become part of the "facts." If the expert's interpretation is influenced by his theory—and everyone acknowledges that it is, and must be—that theory becomes part of the "facts."

Considering how theory creates "facts," creationists should be cautious of constructing theories of human origins based only upon the scientifically accepted evidence. Why should we limit ourselves to a data set that Darwinists have developed while searching for fossils that support their theory? If we limit ourselves to the "scientifically accepted evidence," we will be playing with a deck that is heavily stacked against us. We will also be ignoring much evidence that is not significant to Darwinists but should be highly interesting to creationists.

## 2. SCIENTIFIC FRAUD IS A PROBLEM

Piltdown Man was a deliberate fraud. The purpose of the fraud was to provide evidence of human evolution—evidence that did not otherwise exist. Piltdown Man is far from the only example of faked evidence in support of evolution. Scientific deception is a persistent problem.[37]

Ernst Haeckel, a nineteenth-century German Darwinist, concocted the theory that "ontogeny recapitulates phylogeny," meaning that while developing as an embryo, an animal goes through stages in which its form reflects its supposed evolutionary ancestry. "During its own rapid development," wrote Haeckel, " . . . an individual repeats the most important changes in form evolved by its ancestors during their long and slow paleontological development."[38] Haeckel drew numerous vertebrate embryos to illustrate his idea of recapitulation.

Haeckel's drawings were faked. Haeckel distorted every aspect of his drawings, from the embryos he chose to draw, to the stages of development he chose to depict, to the way he depicted the embryos.[39] Stephen Jay Gould wrote that Haeckel "exaggerated the similarities by idealizations and omissions. He also—in a procedure that can only be called fraudulent—simply copied the same figure over and over again."

Haeckel's drawings never fooled experts in embryology, who immediately noticed that he was taking artistic license to support his pet theory.[40] Moreover, Haeckel's idea that "ontogeny recapitulates phylogeny" had by 1910 been "conclusively disproved and abandoned."[41] Yet, despite the fact that it has long been known that Haeckel faked his drawings, and science has long since repudiated his theory, variations of Haeckel's drawings still appear in recent textbooks.[42]

The peppered moths of southern England provide another example of fraud in support of Darwinian science. Peppered moths come in light and dark shades. In the early nineteenth century, most of the moths were of the lighter shade, but over the course of the century, as industrial pollution worsened, the moth population became predominantly dark colored. In the 1950s, a researcher named Bernard Kettlewell conducted a famous study of the peppered moths in which he released moths onto tree trunks and observed predatory birds picking them off.

Kettlewell noticed that pollution had killed the light-colored lichens on tree trunks; he concluded that the lighter moths were too conspicuous against the darker tree trunks. As a result of their camouflage advantage on the darker tree trunks, the dark moths flourished and came to predominate. Peppered moths have been the favorite textbook example of natural selection for over forty years. Usually, the text is accompanied by pictures of moths on tree trunks, illustrating how the lighter colored moths are almost invisible against the lichen-covered trunk but dangerously exposed on the dark tree trunk devoid of lichens.[43]

The photographs were staged. Dead moths were glued or pinned onto tree trunks before the pictures were snapped. Sometimes photographers placed live peppered moths, which are torpid in daylight, on the tree trunks.[44] This would have been a legitimate illustrative technique if peppered moths typically lighted on tree trunks, but scientists now know that they usually do not. First, a series of experiments showed that there was no correlation between lichens and the distribution of light and dark colored moths.[45] Then, a series of experiments demonstrated that most of the time, peppered moths rest in the canopies of trees, usually on the underside of horizontal branches.[46]

The later experiments seem to show a correlation between industrial pollution and moth melanism, but Kettlewell's theory about lichens, camouflage, and selective predation by birds—the part of the story used to illustrate natural selection—has been debunked.[47] Although scientists have known since the 1980s that peppered moths do not normally rest on tree trunks and that the photographs of moths on tree trunks did not reflect reality, the illustrations keep showing up in textbooks, including textbooks published in 2000.[48] Like Boule's blunder, Kettlewell's lichen theory has provided Darwinists with forty years of propaganda, but it is based upon bad science. To keep publishing pictures of moths on tree trunks at this late date is simply fraud.

The manufacturing of false evidence is not common. But scientific cheating also occurs when a researcher fails to report facts that do not accord with his desired conclusions. Christian orthodontist Jack Cuozzo may have come across an instance of this problem while he was researching and studying the Neanderthals. F. M. Bergounioux, describing the original excavation of the Neanderthal at La Chappelle, and the artifacts found with him, noted that the skull was surrounded by beautiful pieces of jasper (*"belles pieces de jaspe"*). The discovery of a semi-precious stone at a burial site is something that anyone studying the Neanderthals would want to know about; it could well affect one's conclusions about the type of people the Neanderthals were. Yet no mention of the jasper appears in any of the scholarly tomes about Neanderthals, including one English-language description of the La Chappelle find by the leading Neanderthal experts.[49] In fact, Neanderthal experts have written: "No ornaments or necklaces have been found with Neanderthal fossils."[50] But if Bergounioux's description is correct, that statement is misleading. The Neanderthals appreciated semi-precious stones.[51]

## 3. ALL SCIENCE IS ODD, BUT PALEOANTHROPOLOGY IS THE ODDEST

One expert on the subject, Lord Solly Zuckerman, compared paleoanthropology to research on extrasensory perception and noted that "students of fossil primates have not been distinguished for caution when working within the logical constraints of their subject. *The record is so astonishing that it is legitimate to ask whether much science is yet to be found in this field at all.* The story of the Piltdown Man hoax provides a pretty good answer."[52]

Part of the problem is that there is not much to work on. Darwinists have little to show for a century and a half of intensive searching for the missing link. "If you brought in a smart scientist from another discipline and showed him the meager evidence we've got he'd surely say, 'forget it; there isn't enough to go on,'" notes Richard Leakey, quoting a colleague, David Pilbeam.[53]

There are a lot of similar quotes: "Scientists concede that even their most cherished theories are based on embarrassingly few fossil fragments."[54] "The human fossil record is short and scant."[55] "The entire hominid collection known today would barely cover a billiard table."[56] "The primary scientific evidence is a pitifully small array of bones from which to construct man's evolutionary history. One anthropologist has compared the task to reconstructing the plot of *War and Peace* from thirteen randomly selected pages."[57]

Because of the scarcity of fossils, paleoanthropology has "more practitioners than objects for study, thus breeding a high level of acquisitiveness and territoriality."[58] Researchers who have access to actual fossils, as opposed to casts and reproductions, hog those fossils and restrict other scientists' access to them. With a "pitifully small inventory of fossils to analyze," writes Lewin, "research access has always been a sensitive issue."[59] Only those in the inner circle—those who agree with the interpretation of the person who controls access—are allowed to see the original fossil. "Virtually every anthropologist has a tale or two to tell about a rival professional improperly preventing others from working on fossils."[60]

This phenomenon has characterized paleoanthropology from its earliest days. The first candidate for missing link status was "Java Man"—a skullcap and femur found in 1891 by Dutch physician Eugene Dubois. After an initial period of access, Dubois locked up the fossils and, for over twenty years, refused to allow anyone to see them.[61] Why? Scientific jealousy:

> In 1899, Schwalbe produced a significant publication on the skullcap that far overshadowed anything Dubois had done. Dubois must have sensed that if he allowed continued access to the fossils, others would

> do the work and get the recognition that rightfully should be his. Hence, the embargo on allowing others to see the fossils.[62]

The problem has not abated since Dubois' day, and paleoanthropology has become "a science in which many of its workers do not have general access to the material upon which their science is based."[63] Casts are more readily available, but a plaster cast is no substitute for the original. The file marks on the orangutan tooth of Piltdown Man were visible on the original; they were not visible on the casts. Thus, there is a general feeling that no scientific paper should be based upon analysis of a cast.[64] In summary, "all sciences are odd in some way, but paleoanthropology is one of the oddest."[65]

One wonders why Darwinists even bother looking for a "missing link." In terms of physical morphology, the gap between apes and humans is no larger than any other gap in the fossil record, and smaller than most. If a morphological gap of that size were deemed sufficient to falsify the theory of evolution, the theory would have been discarded long ago.

Darwinists search for evidence of a physical transition from ape to man, but the physical transition is not what most urgently calls for an explanation.[66] What really needs explaining is the mental and spiritual gap between apes and humans. That gap is huge. The only other discontinuity in nature that even compares is the gaping chasm separating the "simplest" living organism from any non-living thing.

Why do humans have art, architecture, literature, theater, music, fashion, religion, commerce, law, politics, history, paleontology, science, technology, medicine, engineering, manufacturing, agriculture, animal husbandry, sports, cuisine, formal education, marriage ceremonies, funerals, charity benefits, and all the other innumerable things that separate humans from animals? Creationists have an explanation. We believe that we were created in the image of God to commune with God and to be conscious and self-conscious creatures with free will. The animals were not created for this purpose. That is why there is such an enormous gap between humans and animals.

Evolutionary theory has colored the modern conception of ancient man. Museums, books, motion pictures, and even comic strips have depicted early man as a primitive, hairy, ape-like being who worked only with crude tools such as clubs and sharp stones. He lived in caves because he lacked the skill to build a more advanced dwelling. Hunting was his only livelihood, and he wore the furs of the animals he killed. He was ignorant, superstitious, and terrified of the world around him. He had no knowledge of agriculture and no ability to work metals; it took him many thousands of years to develop even these basic technologies. He mated by whacking an unwary female on the head with a club and dragging her back to his cave. He was physically, intellectually, and morally inferior to modern man.

This view of early man is derived entirely from Darwinism. Notwithstanding the failure of the search for mankind's supposed evolutionary antecedents, many Christians have imbibed the Darwinist philosophy of history. Because of the stunningly rapid advances in science and technology that Western civilization has made during the past two hundred years, there is a broadly held belief that the history of mankind has been one of continual advance. Perhaps unconsciously, even Christians have adopted this view. But it is not the biblical view.

---

[1] Lubenow, Marvin L., *Bones of Contention*, (Grand Rapids, MI: Baker Book House, 1992), pp. 59-61.

[2] Lubenow, pp. 36-39.

[3] Lewin, Roger, *Bones of Contention* (New York: Simon & Schuster, 1987), p. 67, quoting C. Loring Brace.

[4] Straus, William L., and A.J.E. Cave, "Pathology and the posture of Neanderthal Man," *Quarterly Review of Biology*, vol. 32, pp. 348-363 (December, 1957).

[5] Scientists have long debated whether Neanderthals interbred with anatomically modern humans, and whether they should be considered the same species or a separate one. This is still a live controversy. Some skeletal remains show a mixture of Neanderthal and more modern characteristics. "This skeleton demonstrates that early modern humans and Neanderthals are not all that different," commented Dr. Erik Trinkaus, on Neanderthal remains found in Portugul. "They intermixed, interbred and produced offspring." Dr. Milford Wolpoff of the University of Michigan states that the find, "shows that fundamentally, Neanderthals are the same species we are and they contributed their genes to European ancestry." See John Noble Wilford, "Discovery Suggests Humans are a bit Neanderthal," *The New York Times*, April 25, 1999.

The Neanderthal story has a dark side. A Neanderthal site in Krapina, Croatia, discovered in 1899, seemed to show signs of cannibalism. Research conducted one hundred years later at a Neanderthal site in southeastern France confirmed that Neanderthals sometimes practiced cannibalism. The Neanderthal bones had marks made by stone cutting tools, indicative of the flesh having been stripped off the bones. All of the large bones had been smashed with stones, presumably to extract marrow.

What is one to make of the fact that some Neanderthal sites show tender, ritualistic burial, while others show cannibalism? "It's a mixed picture," says anthropologist Clark Larsen. "It's a complex picture. It was probably regional, just as you see in living humans. Human social behavior is highly variable across the globe. I wouldn't expect Neanderthals to be any different." (ABC News, September 30, 1999). "When you see some Neanderthals practicing intentional burial and others practicing cannibalism," notes Tim White, "that is a clear indication of behavior that is multidimensional—a pattern that mirrors the behavior of more modern people." (BBC News, Friday, October 1, 1999.) "It shows that Neanderthals are a lot like we are . . . even including things that are sort of the dark side of humanity," said anthropologist Fred Smith. (*The Dallas Morning News*, October 1, 1999.)

[6] At Shanidar Cave in Iraq, the body of a Neanderthal was carefully placed in the fetal position on a bed of woven woody horsetail and surrounded with several different species of flowers. "From the orderly distribution of [pollen] grains around the fossil remains," researchers reported, "there is no question that the flowers were arranged deliberately . . ." Leaky, Richard and Roger Lewin, *Origins*, (New York: E.P. Dutton, 1977), p. 125. The flower specimens, including yarrow, cornflowers, St. Banaby's thistle, groundsel, grape hyacinths, woody horsetail, and mallow, have medicinal qualities which "range from relief from toothache and inflammation to uses as poultices and for spasm." Solecki, Ralph, *Shanidar: The First Flower People* (New York: Knopf, 1971), p. 249. The Shanidar Neanderthals were probably aware of at least some of the medicinal properties of the flowers. Leaky and Lewin, supra, at 125. See also Belfer-Cohen, A., Hovers, E. (1992) "In the eye of the beholder: Mousterian and Natufian burial in the Levant." *Current Anthropology*. 33:463-471; Hayden, B. (1993) "The cultural capacities of Neanderthals: A review and reevaluation." *Journal of Human Evolution*. 24:113-146; Marshack, A. (1990) "Early hominid symbolism and evolution of the human capacity." In (P. Mellars ed.) *The Emergence of Modern Humans*, pp. 457-498. Ithica, NY: Cornell University Press.

[7] Cuozzo, Jack, *Buried Alive: the Startling Truth about Neanderthal Man*, (Green Forest, AK: Master Books, 1998), pp. 121, 122.

[8] Wilford, John Noble, "Playing of Flute May Have Graced Neanderthal Fire," *The New York Times*, October 29, 1996. Archaeologist Ivan Turk found the flute at a site in Slovenia. It had been made from a juvenile bear femur and had two intact pierced holes and two other punctured holes that had partially eroded or had been chewed away. The straight alignment of the holes suggests that a Neanderthal had intentionally manufactured them. Musicologist Bob Fink analyzed this flute and showed that it would have produced four notes and that it had half and full tones. Neanderthals must have used the seven-note scale, the basic formula of western music. Fink states that "the distance between the second and third holes on the old flute is double that between the third and fourth." This means that the first distance represents a full note, and the distance next to it a half note. Fink says, "These three notes ... are inescapably diatonic and will sound like a near-perfect fit within any kind of standard diatonic scale, modern or antique. (See "Neanderthals Lived Harmoniously," *Academic Press Daily Insight*, (April 4, 1997).

[9] Gould, Stephen Jay, *Bully for Brontosaurus*, (New York: W.W. Norton & Co., 1991), pp. 432-447. See also Richard Ritland, *A Search for Meaning in Nature*, (Mountain View, CA: Pacific Press, 1970), pp. 216, 217.

[10] Osborn took the verse out of context. When Job said "speak to the earth, and it will teach thee," Job meant that even the inanimate objects know that God is in control of the Earth and its inhabitants:

> But ask the animals, and they will teach you, or the birds of the air, and they will tell you; or speak to the earth, and it will teach you, or let the fish of the sea inform you. Which of all these does not know that the hand of the LORD has done this? In his hand is the life of every creature and the breath of all mankind.

The obvious retort to Osborn is also from the book of Job, chapter 38, verses 4-7, in which God asks Job:

> Where were you when I laid the earth's foundation? Tell me, if you understand. Who marked off its dimensions? Surely you know! Who stretched a measuring line across it? On what were its footings set, or who laid its cornerstone—while the morning stars sang together and all the angels shouted for joy?

[11] Gould, at 436.

[12] Gould, at 436.

[13] Blinderman, Charles, *The Piltdown Inquest* (Buffalo, NY: Prometheus Books, 1986), pp. 5, 12-14; Lubenow, at 39, 40.

[14] Blinderman, at 13, 14; Lubenow, at 40; Gould, Stephen Jay, *Hen's Teeth and Horse's Toes,* (New York: W.W. Norton & Co., 1983), p. 221.

[15] Lewin, at 71, 72; Richard Ritland, *A Search for Meaning in Nature,* (Mountain View, CA: Pacific Press, 1970), p. 228

[16] Blinderman, at 21.

[17] Blinderman, at 45, 46, 51, 52, 55.

[18] Blinderman, at 54, 58, 60.

[19] Blinderman, at 61, citing Sir Arthur Keith, "The Piltdown Man Discovery," *Nature,* July 30, 1938.

[20] Blinderman, at 65-72.

[21] Blinderman, at 74, 75.

[22] Blinderman, at 76, 77.

[23] Blinderman, at 77.

[24] Gould, *Hen's Teeth and Horse's Toes,* at 201-240. Gould cites the ape's tooth as evidence that Teilhard was in on the hoax. "A tooth in a gravel pit is about as conspicuous as the proverbial needle in a haystack. The hoaxer worked hard on his Piltdown material. He filed the canine and painted it to simulate age. Apes' teeth are not easy to come by. If I had but one precious item, I would not stick it into a large gravel heap and then hope that some innocent companion would find it. It would probably be lost forever, not triumphantly recovered. I doubt that I would ever find it again myself after someone else had mucked about extensively in the pile." Ronald Millar, in *The Piltdown Men* (New York: St. Martin's Press, 1972), at p. 231, states that, "the other successful searcher at Piltdown was Pere Teilhard de Chardin. The evidence against the priest is as black, if not blacker, than that against Dawson. One has merely to recall the incredulity of Dawson and Woodward when Teilhard de Chardin discovered the missing canine tooth in a stretch of gravel which had just been thoroughly searched." But Millar did not think that Teilhard planted the tooth.

[25] See, generally, Charles Blinderman, *The Piltdown Inquest* (Buffalo, NY: Prometheus Books, 1986); Ronald Millar, in *The Piltdown Men* (New York: St. Martin's Press, 1972); John Evangelist Walsh, *Unraveling Piltdown* (New York: Random House, 1996).

[26] Cremo, Michael A., and Richard L Thompson, *Forbidden Archeology: The Hidden History of the Human Race,* (Los Angeles: Bhaktivedanta Book Pub. Inc., 1993), p 523, citing Marcellin Boule and H.V. Vallois *Fossil Men* (London: Thames and Hudson, 1957).

[27] Cremo, et al, at 523, quoting, Weiner, J.S., *The Piltdown Forgery* (Oxford University Press, 1955), pp. 117, 118.

[28] Blinderman, at 101, 102.

[29] Blinderman, at 239, quoting Gertrude Himmelfarb, *The Darwinian Revolution* (1959, reprint 1996, Ivan R. Dee, Inc).

[30] Kuhn, Thomas S., *The Structure of Scientific Revolutions,* (2nd ed.) (Chicago: University of Chicago Press, 1962, 1970), p. 24.

[31] Lewin, at 68.

[32] Gould, at 443-447 ("Anyone who has studied the dental anatomy of mammals knows immediately that this seemingly implausible mix-up of pig and primate is not only easy to understand but represents one of the classic and recurring confusions

of the profession. The cheek teeth of pigs and humans are astonishingly and uncannily similar.").

[33] Lewin, at 61, 73; See also Richard Ritland, *A Search for Meaning in Nature,* (Mountain View, CA: Pacific Press, 1970), p. 219.

[34] Lewin, at 61.

[35] Patterson, Brian, and W.W. Howells, "Hominid humeral fragment from Early Pleistocene of Northeastern Kenya," *Science,* 156: 64-66 (1967), as cited by Cremo, et al, at 685. See also H.M. McHenry and R.S. Corruccini, "Distal humerus in hominoid evolution," *Folia Primatologica,* 23:227-244 (1975); H.M. McHenry, "Postcranial skeleton of Early Pleistocene hominids," Ph.D. dissertation, Harvard University (1972); McHenry, "Early hominid humerus from East Rudolf, Kenya," *Science,* 180:739-741 (1973) ("A humeral fragment has been found at Kanapoi that is almost five million years old yet almost indistinguishable in shape from many modern humeri. Geologically much younger australopithecine humeri at one or two million years are vastly different from those of modern man."); C.E. Oxnard, *Uniqueness and Diversity in Human Evolution,* (Chicago: University of Chicago Press, 1975) p. 97; M.H. Day, "Functional interpretations of the morphology of postcranial remains of early African hominids," in *Early Hominids of Africa,* C.J. Jolly ed. (London: Duckworth, 1978), pp. 311-345 ("It is hard to point to a single anatomical feature or group of features that is not well known in modern man. Functionally it must be nearly identical with the modern human condition.")

[36] Lubenow, at 56, 57, citing, William W. Howells, "*Homo erectus* in human descent: ideas and problems," in *Homo erectus: Papers in Honor of Davidson Black,* Becky A. Sigmon and Jerome S. Cybulski, eds. (Toronto: University of Toronto Press, 1981), pp. 79, 80 (italics in Lubenow).

[37] See, e.g., Ariel Roth, Origins: *Linking Science and Scripture,* (Hagerstown, MD: Review and Herald, 1998), pp. 293, 294, citing William Broad and Nicholas Wade, *Betrayers of Truth: Fraud and Deceit in the Halls of Science,* (Oxford University Press, 1982 [paperback: New York: Simon & Schuster, 1983]).

[38] Ernst Haeckel, as quoted by Stephen Jay Gould, *Ontogeny and Phylogeny,* (Cambridge, MA: The Belknap Press of Harvard University, 1977), pp. 76, 77.

[39] Wells, Jonathan, *Icons of Evolution,* (Washington, D.C.: Regnery Publishing Co., 2000), pp. 81-109; Cremo, et al, at 7, 8.

[40] In the early 1980s, Gould came across Louis Agassiz's personal copy of Haeckel's 1868 work, *Naturliche Schopfungsgeshichte* (Natural History of Creation) in the open

stacks of a Harvard library. Agassiz, who was not a Darwinist, had some experience in embryology. While reading Haeckel's work, Agassiz had freely annotated the margins. Beside Haeckel's embryo drawings, Agassiz had written such comments as "Where were these copied from? Artistically crafted similarities mixed with inaccuracies." One of Haeckel's illustrations shows the embryos of a dog, a chicken, and a tortoise. Haeckel had written, "If you take the young embryos of a dog, a chicken and a tortoise, you cannot discover a single difference among them." "Naturally" retorted Agassiz in the margin, "because these figures were not drawn from nature, but rather copied one from the other! *Abscheulich* (Atrocious)." Gould, Stephen Jay, *I Have Landed,* (New York: Harmony Books, 2002), pp. 305-320.

41 Gould, Stephen Jay, *I Have Landed,* (New York: Harmony Books, 2002) p. 316.

42 Wells, at 107-109. Creationists have been complaining about Haeckel's faked drawings for generations, yet they keep showing up in high school science textbooks, where Darwinists use them to indoctrinate unsuspecting students. See, e.g., Wheeler, Gerald, *The Two-Taled Dinosaur,* (Nashville, TN: Southern Publishing, 1975), p. 139; Coffin, Harold, *Origin by Design,* (Hagerstown, MD: Review and Herald, 1983), pp. 356-360; Ritland, Richard M., *A Search for Meaning in Nature,* (Pacific Press, 1970) pp. 294-298.

43 Wells, at 137-157.

44 Wells, at 150, 151.

45 Wells, Jonathan, "Second thoughts about peppered moths," *The Scientist,* 13(11):13, (May 24, 1999)

46 Wells, Jonathan, "Second thoughts about peppered moths," *The Scientist,* 13(11):13, (May 24, 1999) (Since 1980, however, it has become clear that peppered moths do not normally rest [on tree trunks]. K. Mikkola observed that "the normal resting place of the Peppered Moth is beneath small, more or less horizontal branches (but not on narrow twigs), probably high up in the canopies, and the species probably only exceptionally rests on tree trunks." He noted that "night-active moths, released in an illumination bright enough for the human eye, may well choose their resting sites as soon as possible and most probably atypically." Thus "In 25 years of fieldwork, C.A. Clarke and his colleagues found only one peppered moth on a tree trunk, and admitted that they knew primarily "where the moths do not spend the day." . . . When Howlett and M.E.N. Majerus studied the natural resting sites of peppered moths in various parts of England, they found that Mikkola's observations on caged moths were valid for wild moths as well and concluded: "[I]t seems certain that most *B. betularia* rest where they are hidden . . . [and] that

exposed areas of tree trunks are not an important resting site for any form of *B. betularia*." In a separate study, T.G. Liebert and P.M. Brakefield confirmed Mikkola's observations that "the species rests predominantly on branches . . . Many moths will rest underneath, or on the side of, narrow branches in the canopy.").

[47] Wells, *Icons of Evolution*, at 148-150, citing, among several other sources, Jeremy Cherfas, "Exploding the myth of the melanic moth," *New Scientist* (December 25, 1986-January 1, 1987), p. 25.

[48] Wells, *Icons of Evolution*, at 155-157. See also Jonathan Wells, "Second thoughts about peppered moths," *The Scientist*, 13(11):13, (May 24, 1999).

[49] Cuozzo, at 122, referring to Erik Trinkaus and Pat Shipman, *The Neanderthals: Changing the Image of Mankind*, (New York: Alfred Knopf, 1992), at p. 186.

[50] Cuozzo at 122, citing Christopher Stringer and Clive Gamble, *In Search of the Neanderthals*, (New York: Thames and Hudson, 1993), p. 18.

[51] It is increasingly recognized that Stringer and Gamble's claim is wrong. Neanderthals did practice personal adornment, producing various pendants or beads out of material such as bones and animal teeth. Neanderthals even used exotic sea floor creatures for adornment.

[52] Lubenow, at 25, citing, Solly Zuckerman, *Beyond the Ivory Tower* (New York: Taplinger Pub. Co., 1971), pp. 19, 64 (my italics).

[53] Milton, Richard, *Shattering the Myths of Darwinism*, (Rochester, VT: Park Street Press, 1997), p. 201, citing Richard E. Leakey, The Making of Mankind, (London: Michael Joseph, 1981).

[54] Lubenow, at 31, citing Peter Stoler, "Puzzling Out Man's Ascent," *Time*, (November 7, 1977).

[55] Lubenow, at 31, citing Vaughn M. Bryant, Jr., et al, "The Coprolites of Man," *Scientific American*, (January, 1975).

[56] Lubenow, at 31, citing John Reader, "Whatever Happened to Zinjanthropus?" New Scientist (March 26, 1981).

[57] Lubenow, at 31, citing Constance Holden, "The Politics of Paleoanthropology," *Science*, 213:737 (August 14, 1981).

[58] Gould, Stephen Jay, *Leonardo's Mountain of Clams and the Diet of Worms*, (New York: Harmony Books, 1998), p. 162.

[59] Lewin, at 23.

[60] Lewin, at 24.

[61] Lubenow, at 91, 92; See also Ritland, supra, at 226, 227.

[62] Lubenow, at 91.

[63] Lubenow, at 16.

[64] Lubenow, at 16, 17.

[65] Lewin, at 20, 319, quoting Matt Cartmill.

[66] Darwinism does not have an adequate explanation for man's *inhumanity*, either. Most would acknowledge that man is frequently crueler, more warlike, and more destructive than any animal. Many instances of man's inhumanity are gratuitous and cannot be attributed to the struggle for survival. But Christianity has an explanation for this. Man was created in God's image but fell into sin and thus sometimes exhibits satanic cruelty.

CHAPTER 18

# Ancient Man—The Biblical View

It was late afternoon—Sunday, December 18, 1994. Jean-Marie Chauvet, Éliette Brunel and Christian Hillaire were exploring a cave in the side of a cliff in southern France. The three amateur spelunkers had explored more caves than they could remember. This particular cave seemed small. It was also well known, because it was close to a popular hiking trail. But Chauvet had noticed a gentle current of air emanating from a small opening behind some fallen rocks at the back of the cave. He wondered where this air was coming from, so they started digging, trying to widen the opening. Eventually, they crawled through the small passage they had widened. Sure enough, there was much more cave than anyone had known. The spelunkers found themselves on the edge of a deep shaft.

They did not have the right caving equipment with them, so they hiked back to their automobiles to retrieve it. Night had fallen, and they considered returning to the cave some other day, but they were too excited to postpone their adventure. They gathered the necessary equipment and returned to the cave. One by one, they descended the shaft on a spelunking ladder, finding themselves in a vast chamber with a high ceiling. They made their way through the first gallery and found that it opened onto another chamber as large as the first. The cave was comprised of an extensive network of chambers and galleries. The three admired the natural beauty and noticed animal bones scattered on the floor.

On the way back out, Éliette trained her flashlight on a rocky concretion hanging from the ceiling of one of the galleries. There she saw the image of a mammoth, drawn in red ochre by a human hand. "They were here!" she shouted. All three began intently searching the walls of the cave. They discovered hundreds of drawings and engravings—a veritable ancient art gallery.

The "they" to whom Éliette Brunel referred were the Cro-Magnon cave artists who left such a magnificent legacy of Ice Age art. About 350 sites have been discovered, scattered from southern Spain to the Ural Mountains of Russia. Almost half of the sites are in France, including most of the more important caves. Before the discovery of the cave now known as Chauvet-Pont-d'Arc, the most important caves were Altamira, in Spain, and Lascaux, in France. The artwork at Chauvet-Pont-d'Arc turned out to be the best of all.

"In my opinion," writes American sculptor Bruce Beasley, "Chauvet is compositionally more complex than any of the other caves. In addition, there is a greater effort to represent pictorial depth of space than in any other cave. The compositional complexity can be seen through intentionally complex overlapping and interspersing of animals, as well as groups of animals represented as receding in space."[1]

Ice Age art has long baffled the experts. For decades after the discovery of wall art in Altamira in 1879, scientists refused to believe that the artwork was genuine. But the discovery of more such sites in the 1880s and 1890s finally convinced the skeptics that the work was genuine prehistoric art, not a hoax or a fraud. An article in the 1911 edition of *Encyclopedia Britannica* notes that the artwork is so remarkable that only the existence of multiple sites forced scientists to admit its genuineness:

> [The cave drawings] bring before us a race of artists of first-rate capacity, who, for accuracy of observation and for skill in indicating the character and peculiarities of the animals around them, have never been surpassed.
>
> * * *
>
> If we are forced to marvel at the graphic skill of the cave men, their sculptures in the round are on a still higher plane, as may be seen in the figures of reindeer in ivory in the British Museum. While they are not highly finished, they show a complete understanding of the animal's peculiar forms and contours, which are rendered in a direct, unhesitating way that should betoken a long period of artistic training and an executive power uncommon at any time. These drawings and sculptures have always been appreciated and even regarded as being of a much more advanced style than was to be expected among men who are always classed in the lower grades of culture. But enough stress has not hitherto been laid on the artistic quality of the work, which would be considered fine at any time in the world's history. . . . There are many astonishing problems in archaeology, but none so badly in need of solution. Had the discovery been confined to a single drawing, or even a single site,

> fraud or misreading of the conditions might have been alleged, but the case is very different. The drawings demonstrate that such artistic power was fairly common, while the question of the authenticity and the period of the discoveries has long since been satisfactorily settled.[2]

The scientific establishment continues to be amazed at the artistic capacity of the prehistoric artists. Stephen Jay Gould wrote that "one subject—and only one—elicits absolute unanimity of judgment among students of ancient life . . . Every last mother's son and daughter among us stands in reverent awe and amazement before the great cave paintings done by our ancestors in southern and central Europe . . ."[3]

Why are the scientists so amazed by these finds? The quality of the artwork is good but certainly not beyond the ability of many of today's artists. The real reason for the "reverent awe" is that these artists lived between 10,000 and 30,000 years ago, according to Lyellian chronological notions. "We are, in short, surprised, even stunned, to discover that something so old could be so sophisticated," wrote Gould. That such an ancient people could be so artistically accomplished flies in the face of a deeply held prejudice: the idea that older equals more primitive and newer equals more sophisticated.

What is the source of this prejudice? To the extent that it has any intellectual foundation beyond simple chauvinism, it is Darwinism. Darwinism teaches that we have continually advanced, from microbes to man. We have always been getting better and better, more and more advanced. "As we travel in time down our own evolutionary tree," wrote Gould, "we should encounter ever-older ancestors of ever-decreasing mental capacity. The first known expressions of representational art should therefore be crude and primitive. Instead, we see the work of a primal Picasso—and we are dumbstruck."[4]

Because many of the sites are in France, it is not surprising that the two most noteworthy students of cave paintings, Abbé Henri Breuil and Andre Leroi-Gourhan, were French. Abbé Breuil spent nearly sixty years copying figures from cave walls. Breuil speculated that cave art was "hunting magic" intended to make game plentiful, the idea being that if you draw it, it will come. Leroi-Gourhan approached the cave paintings from the point of view of French structuralism, arguing that the cave art fit into a system of duality of male and female.

But Breuil and Leroi-Gourhan agreed on one thing: cave art had progressed from a more primitive to a more advanced and sophisticated level. "The theory . . . is logical and rational," wrote Leroi-Gourhan. "Art apparently began with simple outlines, then developed more elaborate forms to achieve modeling, and then developed polychrome and bichrome painting before it eventually fell into decadence."[5] This theory won over most of the scholars.

When Breuil and Leroi-Gourhan were studying the cave art, there was no way to date it. Radiocarbon dating was developed in the 1950s, but the amount of charcoal needed to run the test would have necessitated the destruction of whole drawings, an unacceptable sacrifice. In the late 1980s, however, scientists developed a technique called accelerator mass spectrometry, which allows testing with less than one milligram of charcoal. Some of the charcoal drawings from Chauvet were tested and showed dates around 30,000 years before the present.[6] Chauvet was the oldest cave with wall art, yet its wall art was the most sophisticated.

The Chauvet wall art debunked Breuil and Leroi-Gourhan's theory on the progressive development of cave art: "The discovery of Chauvet has certainly dispelled any lingering idea that visual sophistication in art is progressive through time."[7]

> . . . the notion of a linear evolution of art from simple to complex forms is, quite simply, wrong. That notion derived from post-Darwinian perspectives that foregrounded the idea of evolution from simple to complex and that had been applied in almost every research area—zoological and botanical evolution, social evolution, the evolution of language, and so forth. Now, directly contradicting the Darwinian paradigm, the Chauvet Cave shows that "sophisticated" art was being made at the beginning of the Upper Palaeolithic . . .[8]

Breuil and Leroi-Gourhan were wrong. The Darwinian paradigm led them into error.

Wrong premises lead to wrong conclusions. If Darwinism is false, all branches of learning that incorporate the Darwinian view—not just biology, geology, and paleontology, but also history, archaeology, anthropology, and their sub-disciplines—will inevitably come to wrong conclusions on a wide range of issues. We must not let our God-given love of learning blind us to the truth that the workers in these disciplines began with wrong assumptions, and arrived at wrong conclusions.

The biblical view of early man is the polar opposite of the Darwinian view. According to the Bible, God created man in God's own image. He created man with reason and intelligence, with moral faculties, with abundant creative abilities, with self-knowledge, and with knowledge of the Creator. Let us explore in greater detail the biblical view of early man, so that we can begin with right assumptions, and arrive at right conclusions.

## A. MAN WAS CREATED PERFECT

"Then God said, 'Let us make man in our image, in our likeness, and let them rule over the fish of the sea and the birds of the air, over the livestock, over all the earth, and over all the creatures that move along the ground.' So God created man in His own image, in the image of God He created him; male and female he created them." Gen. 1:26, 27. Far from being frightened of the animals, man was created to rule over the animals. Gen. 1:28; Psa. 8:6-8.

The Bible allows no room for the belief that man was created in an inferior condition and left to struggle toward a higher plane of existence. On the contrary, "God saw all that he had made, and it was very good." Gen. 1:31. God planned to make mankind in His own image, carried out that plan, and pronounced His creation "very good." The Bible admits of no evolutionary progression from a lower form to a higher one. This is confirmed by the inspired testimony of Ellen White:

> God created man in His own image. Here is no mystery. There is no ground for the supposition that man was evolved by slow degrees of development from the lower forms of animal or vegetable life. Such teaching lowers the great work of the Creator to the level of man's narrow, earthly conceptions. Men are so intent upon excluding God from the sovereignty of the universe that they degrade man and defraud him of the dignity of his origin.[9]

On the day he was created, man was as perfect as he would ever be. His physical stature and health were perfect, better than they have ever been since that time:

> As man came forth from the hand of his Creator, he was of lofty stature and perfect symmetry. His countenance bore the ruddy tint of health and glowed with the light of life and joy.[10]

## B. ANTEDILUVIAN[i] LIFESPANS

Man was not created immortal. In order to live forever, he had to continue partaking of the fruit of the tree of life. Gen. 2:9; 3:22-24. Fresh from the hand of the Creator, however, man's vital force—his longevity, his ability to keep living—was strong. The Bible records that the typical antediluvian[i] lived almost a millennium, over ten times longer than modern men live. The fifth chapter of Genesis tells us that Adam lived 930 years, Seth lived 912 years, Enoch 905 years, Kenan 910, Mahalalel 895, Jared 962, etc. Methuselah lived 969 years, the longest lifespan recorded in the Bible—hence the expression "old as Methuselah." Clearly, the race that lived before the Flood possessed longevity vastly superior to today's feeble humanity.

---

[i] Antediluvian means "before the deluge" and refers to people who lived before the Genesis Flood, unless it modifies another word, e.g., "antediluvian trees," "antediluvian climate," etc.

## C. THE PHYSICAL STATURE OF THE ANTEDILUVIANS

The sixth chapter of Genesis tells us about the physical stature of the people who lived before the Flood. "There were giants in the earth in those days; and also after that, when the sons of God came in unto the daughters of men, and they bare children to them, the same became mighty men which were of old, men of renown." Some read Genesis 6:4 to mean that the *only* giants were the offspring of the sons of God and the daughters of men.[11] A careful reading of a faithful translation, however, will show that giants lived on the earth even *before* the inter-marriage between the sons of God and the daughters of men.[12]

The Hebrew word translated as "giant" in Genesis 6:4, is *Nephilim*, meaning "fallen ones," or "violent ones." One commentator has written, "In men's eyes they were 'the heroes of old, men of renown,' but in God's eyes they were sinners, fallen ones, ripe for judgment." The word *Nephilim* is used only three times in Scripture—once in Genesis 6:4 and twice in Numbers 13. It is always used to describe people of great size and strength.

When the Israelites were preparing to invade Canaan, they sent out spies to reconnoiter the land. The spies were awestruck by the size of the Canaanites. They called them *Nephilim*. "They said, 'the land we explored devours those living in it. All the people we saw there are of great size. We saw the *Nephilim* there (the descendants of Anak come from the *Nephilim*). We seemed like grasshoppers in our own eyes, and we looked the same to them." Num. 13:31-33. "The people are taller and stronger than we are; the cities are large, with walls up to the sky. We even saw the Anakites there." Deut. 1:28.

The Bible records that the Anakites were people of tremendous size and strength. "The people are strong and tall—Anakites! You know about them and have heard it said: 'Who can stand up against the Anakites?'" Deut. 9:2. The Anakites came from Kiriath Arba, ("City of Arba") later called Hebron. Josh. 21:11. There were three groups of Anakites—the Sheshai, the Ahiman, and the Talmai, all of whom descended from Anak, who was descended from Arba. Josh. 15:13, 14.

In addition to the Anakites, Canaan was also inhabited by the Emites, Zamzummites, and Rephaites, all of whom were giants. "The Emites used to live there—a people strong and numerous, and as tall as the Anakites. Like the Anakites, they too were considered Rephaites, but the Moabites called them Emites." Deut. 2:10, 11. The Zamzummites were, "a people strong and numerous, and as tall as the Anakites." Deut. 2:20, 21. The Rephaites were a giant people that had lived in the area since before Abraham's time. Gen. 14:5, 6. One of the Rephaites was Og, King of Bashan. Scripture records that his bed was made of iron and was more than thirteen feet long and six feet wide. Deut. 3:11.

Of course, the Anakites, and the other tall, strong Canaanites, were mortal and could be defeated in battle. The Moabites, descendants of Lot, defeated and displaced the Emites, and the Ammonites, descended from Lot by a different son, defeated and displaced the Zamzummites. Israel's victory over King Og was celebrated in song and story, and is mentioned at seders to this day. Psalm 135:10, 11; 136:17-22; Neh. 9:22. The land of the Anakites was given to Caleb, one of the spies who was undaunted by the great size and strength of the Canaanites. In fact, Caleb and Joshua almost wiped out the Anakites. They drove the survivors out of the Hebron area and into Philistine cities of Gaza, Gath, and Ashdod. Joshua 11:21, 22; 15:13, 14.

Later, we find the remnant of the Anakites joining the Philistines and waging war against Israel. Goliath, young David's adversary, was from Gath and was apparently descended from the Anakites who fled to Philistia. Goliath was six cubits and a span tall. 1 Sam. 17:4. If a cubit was eighteen inches, Goliath was over nine feet tall; a twenty-one-inch cubit would make him over eleven feet tall.

David's combat with Goliath was not Israel's only encounter with the giants of Gath. Four other Philistine fighters are identified as "sons of Rapha," indicating that they were giants:

> Once again there was a battle between the Philistines and Israel. David went down with his men to fight against the Philistines, and he became exhausted. And Ishbi-Benob, one of the descendants of Rapha, whose bronze spearhead weighed three hundred shekels and who was armed with a new sword, said he would kill David. But Abishai son of Zeruiah came to David's rescue; he struck the Philistine down and killed him. . . . In the course of time, there was another battle with the Philistines, at Gob. At that time Sibbecai the Hushathite killed Saph, one of the descendants of Rapha. In another battle with the Philistines at Gob, Elhanan son of Jaare-Oregim the Bethlehemite killed Goliath the Gittite, who had a spear with a shaft like a weaver's rod. In still another battle, which took place at Gath, there was a huge man with six fingers on each hand and six toes on each foot—twenty-four in all. He also was descended from Rapha. When he taunted Israel, Jonathan son of Shimeah, David's brother, killed him. These four were descendants of Rapha in Gath, and they fell at the hands of David and his men. 2 Sam. 21:16-21.

A parallel passage states that Sibbecai the Hushathite killed Sippai, "one of the descendants of the Rephaites," indicating that the expression "sons of Rapha" does not mean the sons of a man named Rapha but rather the descendants of the Rephaites, meaning giants. 1 Chron. 20:4-8.

In summary, Moses taught that giants lived before the Flood; he used the term *Nephilim* to describe them. When the Israelites found nine-foot-tall men living in Canaan, they associated them with the *Nephilim* that Moses had written about. The Israelites defeated these people and drove the survivors to Philistine cities, but had, occasionally, to fight these giants' descendants several generations later. A reasonable conclusion from the biblical witness is that people who lived before the Flood were of extremely large stature, and very tall people persisted after the Flood in various places for more than a thousand years.

The inspired writings of Ellen G. White confirm that the antediluvians were much larger than modern humans:

> Adam's height was much greater than that of men who now inhabit the earth. Eve was somewhat less in stature; yet her form was noble, and full of beauty.[13]

> As Adam came forth from the hand of his Creator, he was of noble height, and of beautiful symmetry. He was more than twice as tall as men now living upon the earth, and was well proportioned. . . . Eve was not quite as tall as Adam. Her head reached a little above his shoulders. She, too, was noble—perfect in symmetry, and very beautiful.[14]

> The race of men then living were of very great stature, and possessed wonderful strength.[15]

> At the first resurrection [of the saved] all come forth in immortal bloom, but at the second [the resurrection of the unsaved], the marks of the curse are visible upon all. All come up as they went down into their graves. Those who lived before the flood come forth with their giant-like stature, more than twice as tall as men now living upon the earth, and well proportioned. The generations after the flood were less in stature.[16]

Thus, Ellen White left no doubt that the antediluvians were of extremely large stature.

But what are we to make of Genesis 6:4, which states that, "the sons of God came in unto the daughters of men, and they bare children to them, the same became mighty men which were of old, men of renown"? Some non-Christians believe this text indicates that a race of space aliens bred with human woman. Many Christians believe that angels mated with human women, creating a super race. The notion that aliens or angels mated with humans in the distant past has given rise to an extensive and imaginative literature, but it lacks biblical context. It is not sound exegesis.

Seventh-day Adventists, believe that the "sons of God" were the sons of those antediluvians who were faithful followers of God. Romans 8:14 states that those who are led by the Spirit of God are sons of God.[17] By contrast, the "daughters of men" were daughters of antediluvians who were in open rebellion against God. Adventists, among many others, interpret Genesis 6:4 to mean that the godly line inter-married with the ungodly line, as a result of which the godly line eventually died out, with the exception of Noah and his family.[18] A review of pertinent arguments is in order.

First, the rebellion of Cain and his descendants was an open rebellion against a known God. Adam and Eve had seen and conversed with God. Gen. 2:16, 17; 3:8-19. Adam lived to the age of 930 years, more than half of the history of the antediluvian world,[19] and he witnessed to many of his descendants regarding how he was created, how he and Eve sinned, and how they were cast out of the Garden of Eden. The Garden of Eden was still on earth to verify Adam's story, although people could not enter it because cherubim (angels) with flaming swords guarded the gate. Gen. 3:23, 24. Cain had also conversed with God and was specially marked by God as the first murderer. Gen. 4:6-16. Thus, in pre-Flood times, there was not as much room for atheism and agnosticism as there is today.[20] The antediluvians had a choice to make regarding God: obedience or rebellion. This clear decision made for two distinct groups of people.

Second, the Bible is not a general-purpose history of the human race. It is primarily about the relationship between mankind and mankind's Creator and Redeemer. Bible history is the history of the godly line – Adam, Seth, Enoch, Noah, Abraham, Isaac, Jacob, Joseph, Moses, the Hebrew nation, Jesus Christ, the apostles, and the Christians. Other nations and people are discussed only insofar as they touch upon and intersect the godly line. Bearing this in mind, it is logical that a reference to "the sons of God" is a reference to the antediluvian godly line, those few people who were faithful to God.

Third, the Bible teaches that believers should not marry non-believers, because these unions lead to the apostasy of the believer far more often than they lead to the conversion of the non-believer. This theme runs throughout Scripture.[21] It is likely that Genesis 6 relates the first example of the tragic results when believers are joined in marriage with those in rebellion against God.

Fourth, the context of the passage is the increasing sinfulness of the antediluvians, and God's resulting judgment in the form of a deluge. Mankind would not, however, be answerable for the actions of higher beings over which the human race had no control. It would be unjust for God to punish mankind for the actions of non-human beings; we can therefore be certain that God was judging human wickedness. God judged the antediluvians for their own open and notorious rebellion against Him. That rebellion became nearly universal because the godly fell away through intermarriage with the ungodly.

Finally, Numbers 13:33 states that the Anakites were descended from the *Nephilim.* But if the antediluvian *Nephilim* of Gen. 6:4 had been the product of unions between spirit beings and human women, they would have been destroyed in the Flood. Gen. 7:21-23. Obviously, the post-Flood Anakites could not have descended from a race that was destroyed in the Flood. Moreover, Canaanite tribes of giants such as the Anakites were not some sort of super-race. The Israelites nearly destroyed them, apparently without great difficulty. Even before the Israelites arrived, the Moabites and the Ammonites—tribes *not* especially favored by God—had displaced some of the Canaanite tribes of giants. Thus, the scriptural witness, taken as a whole, indicates that the *Nephilim,* although strong and of large stature, were fully human. They were not a hybrid between humans and higher beings.

It could be argued that the "sons of God" were neither men nor angels, but "sons of the gods," as the New English Bible renders them. But this notion is utterly devoid of biblical context. In the Bible, "gods" with a small g are mere idols, and idols are nothing, having no power whatsoever.[22] How can a non-entity have sons, and how can those sons mate with human women?

Among those who differ with our interpretation, the most popular interpretation of the identity of the "sons of God" is that they were angels. There is a plausible argument that the phrase "sons of God" refers to angels when it is used in Job 1:6, 2:1, and 38:7.[23] Although unfallen angels—those still loyal to God—certainly would *not* mate with humans and thus incur God's wrath, Scripture teaches that there are fallen angels, those that Satan took with him when he fell. These are also called demons.[24] Thus it could be argued that fallen angels—demons—bred with humans, thus producing the *Nephilim*. Again, however, there are sound reasons to reject this interpretation.

First, Scripture teaches that angels are much more powerful than mortal human beings.[25] If they bred with humans, we would expect the offspring to be a hybrid race of supermen. But as noted above, scriptural history regarding the post-Flood *Nephilim* of Numbers 13:33 indicates that they were not a race of supermen. They were human and quite conquerable. Since Numbers 13 contains the only other use of the term *Nephilim*, there is reason to believe that the pre-Flood *Nephilim* were not supermen, either.

Second, Christ taught that angels are not sexual beings. Luke 20:34-36; Mark 12:25. In Matthew 22:30, Christ says, "At the resurrection people will neither marry nor be given in marriage; they will be like the angels in heaven." The Bible also indicates that angels are spirit beings. Heb. 1:14. Obviously, non-sexual spirit beings would not have been sexually attracted to human women.

Third, as noted above, angels are much more powerful than humans. If angels chose to mate with human women, humans could do nothing to stop it. Why, then, would God judge mankind for the misbehavior of fallen angels?

By a process of elimination, the best view—or the least objectionable view—is that Genesis 6:4 refers to the corruption of the godly line through intermarriage with the ungodly. This is the correct interpretation based purely upon scriptural considerations.[26]

We also have the prophetic gift of Ellen White confirming that Genesis 6:4 indicates that the godly line of Seth inter-married with the ungodly line of Cain:

> Notwithstanding the prevailing iniquity, there was a line of holy men who, elevated and ennobled by communion with God, lived as in the companionship of heaven. They were men of massive intellect, of wonderful attainments. They had a great and holy mission—to develop a character of righteousness, to teach a lesson of godliness, not only to the men of their time, but for future generations.
>
> * * *
>
> For some time the two classes remained separate. The race of Cain, spreading from the place of their first settlement, dispersed over the plains and valleys where the children of Seth had dwelt; and the latter, in order to escape from their contaminating influence, withdrew to the mountains, and there made their home. So long as this separation continued, they maintained the worship of God in its purity. But in the lapse of time they ventured, little by little, to mingle with the inhabitants of the valleys. This association was productive of the worst results. "The sons of God saw the daughters of men that they were fair." The children of Seth, attracted by the beauty of the daughters of Cain's descendants, displeased the Lord by intermarrying with them. Many of the worshipers

> of God were beguiled into sin by the allurements that were now constantly before them, and they lost their peculiar, holy character. Mingling with the depraved, they became like them in spirit and in deeds; the restrictions of the seventh commandment were disregarded, "and they took them wives of all which they chose." . . . Sin spread abroad in the earth like a deadly leprosy.[27]

In another commentary on this issue, White indicates that the physical size of the godly line diminished less than that of the ungodly line. This explains why the offspring of the godly line were giants even by the standards of those days:

> Those who honored and feared to offend God, at first felt the curse but lightly; while those who turned from God and trampled upon His authority, felt the effects of the curse more heavily, *especially in stature and nobleness of form.* The descendants of Seth were called the sons of God—the descendants of Cain, the sons of men. As the sons of God mingled with the sons of men, they became corrupt, and by intermarriage with them, lost, through the influence of their wives, their peculiar, holy character, and united with the sons of Cain in idolatry. Many cast aside the fear of God, and trampled upon his commandments. But there were a few who did righteousness, who feared and honored their Creator. Noah and his family were among the righteous few.[28]

## D. THE INTELLIGENCE AND TECHNOLOGY OF THE ANTEDILUVIANS

The Darwinian assumption that man evolved from lower animals dictates that it must have taken man many thousands of years to develop technology. Anthropologists, archeologists, and historians have built this assumption into their theories, and it controls how they view and interpret the available evidence. Even among Christians, who should know better, it is widely assumed that the antediluvians were intellectually and technologically primitive. Here again, the Bible differs radically from Darwinian assumptions.

Cain and Abel, only the second generation of humans to live on the earth, farmed (Gen. 4:2, 3), ranched (Gen. 4:3, 4), and built a city (Gen. 4:17). The Bible records that within a few generations, the antediluvians had developed musical instruments (Gen. 4:21) and were working with metals (Gen. 4:22). It did not take thousands of years to develop these skills.

The Bible teaches that mankind has degenerated in terms of physical size and length of life. The intellectual ability of the antediluvians was in proportion to their size and vital force. Their minds were far more powerful than the minds of modern men. Ellen White elaborates on the intellectual ability and technological skills of the antediluvians:

> Notwithstanding the wickedness of the antediluvian world, that age was not, as has often been supposed, an era of ignorance and barbarism. The people . . . possessed great physical and mental strength, and their advantages for acquiring both religious and scientific knowledge were unrivaled. It is a mistake to suppose

> that because they lived to a great age their minds matured late; their mental powers were early developed, and those who cherished the fear of God and lived in harmony with His will continued to increase in knowledge and wisdom throughout their life. Could illustrious scholars of our time be placed in contrast with men of the same age who lived before the Flood, they would appear as greatly inferior in mental as in physical strength. As the years of man have decreased, and his physical strength has diminished, so his mental capacities have lessened. There are men who now apply themselves to study during a period of from twenty to fifty years, and the world is filled with admiration of their attainments. But how limited are these acquirements in comparison with those of men whose mental and physical powers were developing for centuries![29]

Amazingly, because of the power and accuracy of their memories, the antediluvians were able to function without written books and records:

> It is true that the people of modern times have the benefit of the attainments of their predecessors. The men of masterly minds, who planned and studied and wrote, have left their work for those who follow. But even in this respect, and so far as merely human knowledge is concerned, how much greater the advantages of the men of that olden time! They had among them for hundreds of years him who was formed in God's image, whom the Creator Himself

pronounced "good"—the man whom God had instructed in all the wisdom pertaining to the material world. Adam had learned from the Creator the history of creation; he himself witnessed the events of nine centuries; and he imparted his knowledge to his descendants. The antediluvians were without books, they had no written records; but with their great physical and mental vigor, they had strong memories, able to grasp and to retain that which was communicated to them, and in turn to transmit it unimpaired to their posterity. And for hundreds of years there were seven generations living upon the earth contemporaneously, having the opportunity of consulting together and profiting each by the knowledge and experience of all.[30]

Those who are deceived and flattered on in the delusion that the present is an age of real progress, and that the human race has been in ages past progressing in true knowledge, are under the influence of the father of lies, whose work has ever been to turn the truth of God into a lie.[31]

In the antediluvian world there were many wonderful works of art and science. Fresh from the hand of the Creator, these descendants of Adam possessed capabilities that we do not now see.[32]

There perished in the flood greater inventions of art and human skill than the world knows of today. The arts destroyed were more than the boasted arts of today.[33]

> In strength of intellect, men who now live can bear no comparison to the ancients. There have been more ancient arts lost than the present generation now possess. For skill and art those living in this degenerate age will not compare with the knowledge possessed by strong men who lived near 1,000 years. Men before the flood lived many hundreds of years, and when one hundred years old were considered but youths. Those long-lived men had sound minds in sound bodies. Their mental and physical strength was so great that the present feeble generation can bear no comparison to them.[34]

> True knowledge has decreased with every successive generation. God is infinite, and the first people upon the earth received their instructions from the infinite God who created the world. Those who received their knowledge direct from infinite wisdom were not deficient in knowledge.[35]

It would be a mistake to suppose that the technology of the antediluvians was identical to modern technology. We are told that the antediluvians did not have all of the machinery that we have:

> There are many inventions and improvements, and labor-saving machines now that the ancients did not have. They did not need them. The land has felt the curse, more and more heavily. . . . The greater the length of time the earth has lain under the curse, the more difficult has it been for man to cultivate it, and make it productive. As the soil has become barren, and double

> labor has had to be expended upon it, God has raised up men with inventive faculties to construct implements to lighten labor on the land groaning under the curse.[36]

We tend to assume that current technologies are the only technologies, but that is a fallacy. There are usually several different ways to accomplish the same task. Often, two completely different types of technologies can be used to accomplish the same task. Eastern medicine utilizes theoretical principles entirely different from those of Western medicine, yet both methods can effectively cure maladies. Digital technology is entirely different from film, videotape and audiotape, yet these different types of technologies can be use to record, store and replay sounds and images.

In the late 1980s, when the Capitol of Texas was being refurbished, a statue of the "Goddess of Liberty" that adorns the top of the dome was taken down and refitted. In taking it down and putting it back atop the dome, it was necessary to bring in a large helicopter. When the Capitol building was first built in the 1880s, the helicopter had not been invented. It is assumed that the workmen devised a system of pulleys to raise the statue along the outside of the dome, but the exact method used has been lost. Because we have helicopters, we use them. Old technologies are forgotten as new ones take their place.

## E. SUMMARY

The biblical picture of the earliest human beings is the polar opposite of the Darwinian paradigm we are all familiar with, and it takes some getting used to. Man was created perfect. Antediluvian man was physically and mentally superior to modern man and lived many times longer than modern men live. The antediluvian race was possessed of highly advanced arts and sciences. There is no reason to suppose that they had the same types of technology we have, but they were massively intelligent and technically capable.

Although the Flood destroyed the antediluvians and all their works, we would expect that Noah and the first few generations after the Flood would have retained some of the knowledge that was amassed before the Flood. Did the people who lived shortly after the Flood leave us any evidence of advanced technology? They did.

[1] Bruce Beasley, American sculptor, quoted at the web site for the Cave of Chauvet-Pont-d'arc: *http://www.culture.gouv.fr/culture/arcnat/chauvet/en/index.html.*

[2] *Encyclopedia Britannica*, Cambridge University Edition, vol. 2, p. 347, as quoted by George McCready Price, *Evolutionary Geology and the New Catastrophism*, (Pacific Press, 1926), pp. 308, 309.

[3] Gould, Stephen Jay, *Leonardo's Mountain of Clams and the Diet of Worms*, (New York: Harmony Books, 1998), p. 162.

[4] Gould, at pp. 162, 163.

[5] Gould, at p. 173.

[6] The drawings were not done 30,000 years ago. I am assuming, however, that radiocarbon will yield correct ordinal comparisons. In other words, wall art with a radiocarbon date of 30,000 years should be older than art with a radiocarbon date of 12,000 years, even if both were done around 4,000 to 5,000 years ago.

[7] Bruce Beasley, supra.

[8] David Lewis Williams, professor of anthropology, University of Witwatersrand, and Johannesburg, Rock Art Research Institute. At the Chauvet-Pont-d'arc web site. See, also, Jean-Marie Chauvet, Elliet Brunel Deschamps, Christian Hillaire, Jean Clottes, and Paul G. Bahn, *Dawn of Art: The Chauvet Cave*, (Harry N. Abrams, 1996) Jean-Marie Chauvet, et al, *Chauvet Cave: The Discovery of the World's Oldest Paintings* (Thames & Hudson, 1996). Chauvet is not the only support for the "earliest is best" view of prehistoric art. Figurines discovered in Hohle Fels Cave in Germany's Arch Valley, dated around 30,000 years ago, also support it. Dr Anthony Sinclair, an expert in Palaeolithic archaeology at Liverpool University, said the German haul represented "the oldest body of figurative art in the world - pieces that show a coherent set of manufacturing techniques and themes for representation." He said the objects rivaled in age and sophistication the remarkable cave paintings at Chauvet in southern France. "They are as good as anything you will see thousands of years later . . . The quality of the work is very good indeed." "We have always assumed that skills would take some time to evolve - we thought the simplest charcoal drawings were the oldest and the paintings were the most recent. And yet our carbon dates now show that some of the most beautiful and elaborate paintings are among the oldest things we have." Jonathan Amos, "Ivory Bird Displays Ancient Skill," BBC News, December 17, 2003. *http://news.bbc.co.uk/1/hi/sci/tech/3328229.stm.* See, also, Conard, N. J. (2003) "Palaeolithic ivory sculptures from southwestern Germany and the origins of figurative art," *Nature*, 426, 830-832

[9] White, Ellen G., *Patriarchs and Prophets,* (Washington, D.C.: Review and Herald Pub., 1890), p. 45.

[10] White, *Patriarchs and Prophets,* p. 45.

[11] Some paraphrasers write this incorrect interpretation into the text of the Bible. See, e.g., The Living Bible ("In those days, and even afterwards, when the evil beings from the spirit world were sexually involved with human women, their children became giants, of whom so many legends are told.") and the Contemporary English Version ("The children of the supernatural beings who had married these women became famous heroes and warriors. They were called Nephilim and lived on the earth at that time and even later.").

[12] The passage is translated correctly in the King James Version ("There were giants in the earth in those days; and also after that, when . . ."), the Revised Standard Version ("The Nephilim were on the earth in those days, and also afterward, when . . ."), and the Modern Language Version ("There were giants on the earth in those days, and later, too, when . . .").

[13] White, *Patriarchs and Prophets,* p. 45.

[14] White, *Spiritual Gifts,* vol. III, p. 34, (1SP 25.)

[15] White, *Spiritual Gifts,* vol. III, p. 61.

[16] White, *Spiritual Gifts,* vol. III, p. 84. The Bible depicts death as an unconscious, dreamless sleep. See Psalm 146:3, 4; Ecclesiastes 9:5, 6, 10; John 11:11-14. There will be a resurrection of the saved dead at the second coming of Christ. See John 11:23, 24; 1 Cor. 15; 1 Thes. 4:13-18. There will also be a second resurrection, of the unsaved, after the 1,000 years (the millennium). John 5:28, 29; Rev. 20.

[17] See, also, Deut. 14:1; 32:4-5; Psalm 73:15; Hosea 1:10; John 1:12; Philippians 2:15 1 John 3:1-2.

[18] Other Christians who have espoused the theory of intermarriage of the godly line with the ungodly include Chrysostom, Augustine (in *The City of God*), Martin Luther, John Calvin, Matthew Henry, and Francis Schaeffer. See also C.F. Keil and F. Delitzsch, *Biblical Commentary on the Old Testament,* vol. 1.

[19] According to Masoretic numbers—see the discussion of biblical chronology in chapter 8.

[20] Ellen White states, "Skepticism could not deny the existence of Eden while it stood just in sight, its entrance barred by watching angels. The order of creation,

the object of the garden, the history of its two trees so closely connected with man's destiny, were undisputed facts. And the existence and supreme authority of God, the obligation of His law, were truths which men were slow to question while Adam was alive."—*Patriarchs and Prophets*, p. 84.

21 See, e.g., Gen. 24:3, 4; 28:1, 2; Ex. 34:15, 16; Num. 25; Deut. 7:1-4; 17:17; Judges 3:5-8; 1 Kings 11:1-13; Ezra 9, 10; Neh. 10:30; 13:23-27; Mal. 2:11; 1 Cor. 7:12-16; 2 Cor. 6:14-18.

22 See, e.g., Psa. 96:5; 115:4-8; Isa. 44:15-20; Hab. 2:18-20; 1 Cor. 8:4-5; 10:19-20; Rev. 9:20.

23 "One day the sons of God came to present themselves before the Lord, and Satan also came with them." Job 1:6. "On another day the sons of God came to present themselves before the Lord, and Satan also came with them to present himself before him." Job 2:1. "On what were its footings set, or who laid its cornerstone—while the morning stars sang together and all the sons of God shouted for Joy?" Job 38:6, 7. The New International Version translates the phrase *sons of God* in these texts as "angels," dropping a footnote that the Hebrew is literally "sons of God,"—thus admitting that they have paraphrased rather than translated. But is the paraphrase accurate? Ellen White taught that there are other created worlds, worlds whose inhabitants have not fallen into sin. It is possible that the "sons of God" in Job are the inhabitants of these unfallen worlds.

24 See, e.g., Rev. 12:7-9; Luke 10:18; Isaiah 14:12-20; Ezek. 28:11-17; Rev. 12:3-4; Matt. 25:41; 9:32-34; 12:24-28; Luke 11:14-20; Rom. 8:38; James 2:19; 2 Peter 2:4; Jude 6.

25 2 Sam. 24:16; Judges 6:20-22; 13:20, 21; Matt. 13:39, 41, 49; 24:31; Luke 1:19, 20; 2:9, 10; 20:36; Acts 5:19; 12:8-12; 1 Cor. 10:10; 2 Peter 2:11.

26 Proponents of the theory of breeding between angels and humans draw support from the *pseudepigraphal* book of 1 *Enoch* and from an interpretation of verses in Peter (1 Peter 3:18-20, 2 Peter 2:4, 5) and Jude (verses 6 and 7) that is influenced by 1 *Enoch*. Although Jude quotes from 1 Enoch, Enoch contains a mixture of truth and error. *Pseudepigrapha* means "falsely superscribed," meaning falsely titled or attributed. The book of Enoch was written between 200 B.C. and A.D. 50. Enoch, the seventh from Adam, had nothing to do with writing the book. Moreover, the Book of Enoch makes bizarre and incredible assertions. For example, it claims that the Nephilim were 300 cubits tall (450 feet tall), that the antediluvians exhausted themselves trying to feed these outsized giants, that the giants turned on the antediluvian men and started eating them, and that then they started eating each other and drinking blood. (1 Enoch, chapter 7.) The Book of Enoch was never included in the Hebrew canon, and although the book was well known to the early Christians, they did not include it in their canon, either—the Ethiopian church being a notable exception. After the Council of Laodicea in A.D. 364, the church banned

1 Enoch and suppressed it so vigorously that most copies were destroyed. It was thought to be a lost book until the Scottish explorer James Bruce found a copy in Ethiopia, written in Ethiopian, in the early 1770s. Copies were also found among the Dead Sea Scrolls in 1947. For the views of supporters of the theory of breeding between angels and humans, see, e.g., James Montgomery Boice, *Genesis,* (Grand Rapids, MI: Baker Books, 1998); Henry J. Morris, *The Genesis Record,* (Grand Rapids, MI: Baker Books, 1976); Chuck Missler, *Textual Controversy: Mischievous Angels or Sethites?*; Ray C. Stedman, *Signs of Collapse*; Merrill F. Unger, *Biblical Demonology,* (Chicago: Scripture Press, 1952); John Fleming, *The Fallen Angels and the Heroes of Mythology,* (Dublin: Hodges, Foster & Figgis, 1879).

[27] White, *Patriarchs and Prophets,* pp. 84, 81, 82.

[28] White, *Spiritual Gifts,* vol. III, pp. 60, 61 (emphasis added).

[29] White, *Patriarchs and Prophets,* pp. 82, 83.

[30] White, *Patriarchs and Prophets,* pp. 83.

[31] White, *Spiritual Gifts,* vol. IV, p. 156.

[32] White, *Signs of the Times,* Feb. 1, 1889.

[33] White, Letter 65, 1898.

[34] White, *Spiritual Gifts,* vol. IV, pp. 155, 156.

[35] White, *Spiritual Gifts,* vol. IV, p. 154.

[36] White, *Spiritual Gifts,* vol. IV, p. 155.

CHAPTER 19

# How Advanced Were the Ancients?

*There is no remembrance of men of old,*
*and even those who are yet to come*
*will not be remembered by those who follow.*
Ecclesiastes 1:11

"The achievements of civilized man in preliterate times, prior to the building of the first cities in Neolithic antiquity, include a high development of the exact sciences and technologies."[1]

"Mesopotamia cherished a tradition that at the dawn of civilization, long before any period of history known to us, science stood at a level from which historic man has fallen."[2]

Instead of surveying a variety of evidence supporting the view that prehistoric man was highly sophisticated—an approach that would yield superficial and unsatisfying results given our space constraints—we will focus on the British megaliths (*megalith* means "big stone"). There are more megalithic structures in the British Isles than there are modern towns and villages.[3] The megaliths constitute some of the world's most compelling evidence of prehistoric sophistication.

## A. STONEHENGE AND THE BRITISH MEGALITHS

Excluding the pyramids, the most famous megalithic site in the world is Stonehenge, on the Salisbury plain in southern England. Archeological evidence suggests that Stonehenge was built in three phases, the first of which is thought to have begun around 3,100 B.C. (**Plate 29**) During the first phase of construction, a circular mound was created; fifty-six "Aubrey Holes"[4] were dug, forming a circle just inside the mound; two stones were erected at the entrance to the site; and four "station stones" were erected in the Aubrey circle, forming a rectangle.[5]

During the second phase of construction, thought to have begun between 2,500 and 2,100 B.C., about eighty "bluestone" pillars, weighing four tons each, were brought from the Preseli Mountains of Wales, 240 miles (385 km) away, and erected in two concentric semicircles.[6] The entranceway, called the "Avenue," was widened and lined with border mounds, and a pair of "heel stones" was erected.

During the third phase of construction, starting sometime between 2,500 and 2,000 B.C.,[7] the bluestones were removed to make way for a circle of gigantic stones called "sarsens," which were hauled from Marlborough Downs, about twenty miles (30 km) to the north. The sarsen circle consisted of thirty upright stones capped by a ring of horizontal stones called "lintels." Within the sarsen ring, the builders erected five large stone structures called "trilithons" (three stones), arranged in a horseshoe formation. Each of the five trilithons consisted of a pair of gigantic stone uprights, supporting a mammoth stone lintel. Inside the trilithon horseshoe was placed a semicircle of nineteen bluestones. Between the trilithons and the sarsen circle, a circle of perhaps fifty-nine bluestones was erected. Later, perhaps around 1,500 B.C., a circle of thirty "Y" holes and a circle of twenty-nine "Z" holes were dug outside the sarsen circle, in an irregularly spaced fashion.

The sarsen stones are as long as thirty feet and weighed up to forty tons (80,000 pounds). Although the sarsens were formed from extremely hard, silica-cemented sandstone—a stone harder than granite—the builders worked them as if they were wood, dressing and smoothing them with stone hammers. The builders carved mortise-and-tenon joints (convex half-globes on the top of the upright sarsens and matching concave half-globes on the undersides of the lintels) to hold the lintels on the upright sarsens. They curved the lintels so that when joined, they would produce a circle; they fit the lintels to each other with tongue-and-groove joints. Although the monument was built on sloping ground, the builders adjusted the height of the sarsens to compensate, so that the lintel ring was almost perfectly level. Still awe-inspiring after 4,000 years, the monument is an astonishing engineering and construction achievement.[8]

But what is more remarkable about Stonehenge is the celestial knowledge it incorporates. In 1740, a British antiquarian named Dr. William Stukeley (1687-1765) pointed out that "the principle line of the whole work [points to] the northeast where abouts the sun rises when the days are longest."[9] Stukeley noticed that if one stands in the center of Stonehenge on June 21, the summer solstice, the sun rises over the "heel stone," an upright 35-ton sarsen standing some two hundred feet from the center of the monument. Later research by Sir Norman Lockyer

**Figure 19.1: Sunrise Over the "Heel Stone" on the Summer Solstice**

(1836-1920), a scholar with an excellent reputation, confirmed the solstitial alignment of the "avenue" and the "heelstone."[10]

Early investigators, such as John Aubrey and William Stukeley, attributed Stonehenge to the Druids, a Celtic priesthood of the seventh century B.C. But the monument predates the Celts and the Druids by more than a thousand years. Nevertheless, modern-day "Druids" have held summer solstice ceremonies at Stonehenge. Indeed, the solstitial convergence of "Druids" and assorted other enthusiasts became so riotous that the British government was forced to close Stonehenge every summer and winter solstice.

**Figure 19.2: Stonehenge viewed from the Heel Stone**

But Stonehenge was not aligned *only* with the summer solstice. Far more celestial knowledge is built into its design. A major breakthrough in decoding its builders' purposes came after Gerald S. Hawkins visited the site in the early 1960s. Hawkins was a British astronomer who had moved to the United States and become associated with Boston University and the Smithsonian Observatory. When it was still legal to do so, Hawkins and his wife visited Stonehenge and witnessed the dawn on the morning of the summer solstice.

Hawkins felt that the phenomenon of the sun rising over the heel stone must have been carefully planned. He also noticed that the narrow openings through the large trilithons forced him to view specific spots on the horizon:

> The trilithon archways are astonishingly narrow. The space between the gigantic pillars is so small that you can hardly poke your head through (I tried). The average width of the three standing archways is 12 inches, and the average thickness of the bordering uprights is 2 feet, so that when you look through two aligned archways your view is restricted to a very small angle. I felt that my field of observation was being tightly controlled, as by sighting instruments, so that I couldn't avoid seeing something. What was I supposed to see?[11]

Hawkins realized that the significance of Stonehenge's viewing angles could never be comprehended in a lifetime of personal observation. "We need the machine," he thought. Hawkins took a map of Stonehenge, used a mechanical plotter to plot all the viewing lines, determined the compass directions, or azimuths, of the lines, and fed the data into an IBM 7090 mainframe computer. He then checked for matches with the declinations of some of the major stars–Sirius, Canopus, Alpha Centauri, Vega, Capella, and Arcturus—but came up empty.

He next ran the data against the positions of the extreme declinations of the sun and moon. Jackpot! With an average error of less than one degree (out of 360), twelve of the Stonehenge alignments pointed to extreme positions of the sun; with an average error of 1.5 degrees, twelve of the alignments pointed to extremes position of the moon.[12]

What are the extreme positions of the moon? The earth is tilted about 23.5 degrees relative to the plane of its annual orbit around the sun. This is called the angle of the ecliptic, and it causes the seasons. A direct line from the sun to the earth hits the earth at about 23.5° north of the equator (the tropic of Cancer) on June 21st of each year (the summer solstice, for us in the northern hemisphere), and about 23.5° south of the equator (the tropic of Capricorn) each December 21st (the winter solstice). The moonrise and moonset go through this 57° variation monthly, rather than annually, except that, unlike the sun, the moon's minimum and maximum declination are not constant. This is because the moon's orbit around the earth is inclined by a little over five degrees to the angle of the ecliptic. Thus, moonrise and moonset vary between extreme positions at a major standstill, when it is about 28.5° from the angle of the ecliptic, to a minor standstill when it is about 18.5° from it. One complete cycle of variations takes 18.61 years.[13] Astonishingly, the builders of Stonehenge observed the moon over a long-enough period of time to fix the extremes of its 18.61-year cycle.

After Hawkins published his findings in the prestigious journal *Nature,* he learned that another researcher, a retired engineer named C. A. "Peter" Newham, had independently discovered celestial alignments in the station stones.[14] Newham had discovered that certain holes that had once held poles or stones, when viewed from the station stones, created lines that pointed to the equinoxes of the sun and moon. Because these holes lay close to the line between the center of the monument and the "heel stone," Hawkins assumed that they had merely been additional markers of the alignment of the summer solstice sunrise, and he had ignored them.[15]

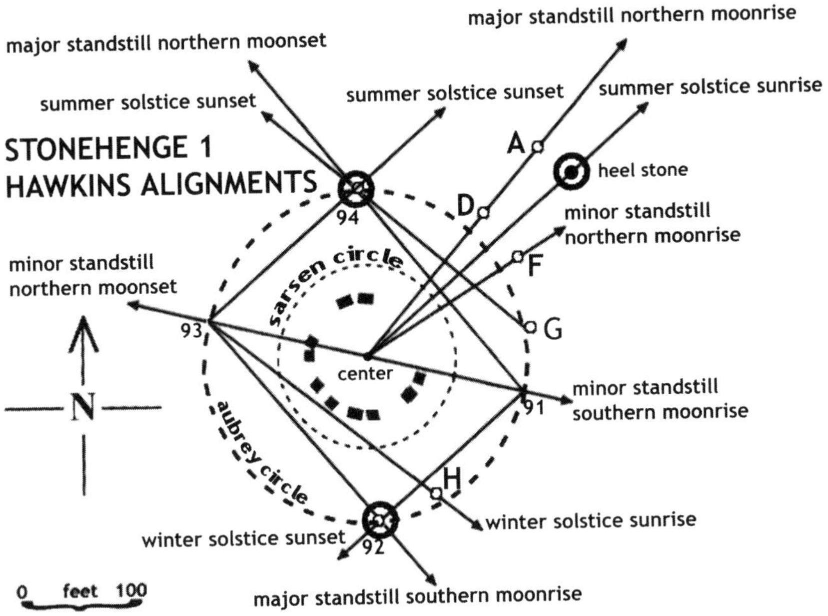


**Figure 19.3: Alignments of the "station stones" of "Stonehenge I"**

Armed with Newham's information, Hawkins went back to the IBM 7090 and entered the new azimuth directions. He found eight more matches—all equinox sunrises or moonrises. In his *Nature* article, Hawkins had calculated the odds that all the alignments were the result of chance as one in a million.[16] After adding Newham's equinox lines, Hawkins recalculated the odds, using Bernoulli's law, as one in ten million.[17] Clearly, the alignments were not accidental.

Another fruitful response to Hawkins' *Nature* article was from R.S. Newall, an archeologist who had rediscovered the circle of holes originally discovered by John Aubrey. Newall had written the official guidebook on Stonehenge.[18] Newall pointed out that Diodorus Siculus (of Sicily), a Greek historian of the first century B.C., had written that beyond the land of the Celts was an island country with a temple "spherical in shape" dedicated to Apollo, the sun god. The sun god returns every nineteen years, "the period in which the return of the stars to the same place in the heavens is accomplished . . ." Diodorus also noted that the people of the island were careful observers of the moon. Newall wondered whether the circular temple of which Diodorus wrote could have been Stonehenge. "Could the full moon," Newall wondered, "do something spectacular once every 19 years at Stonehenge?"[19]

What was the meaning of the nineteen years, and what was the "spectacular" lunar event? Hawkins began to wonder whether the full

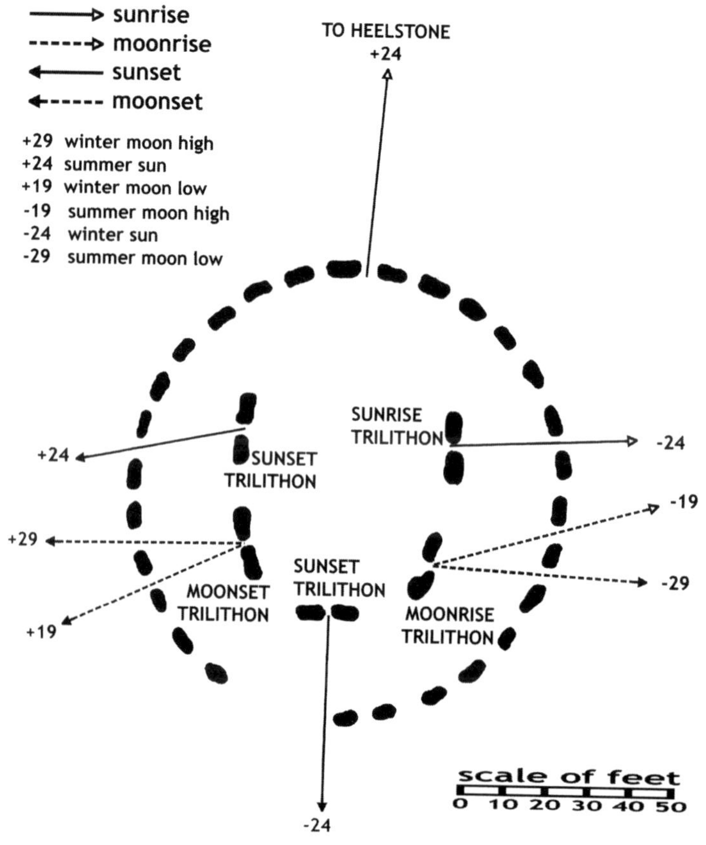


**Figure 19.4: The Megalithic alignments of "Stonehenge III"**

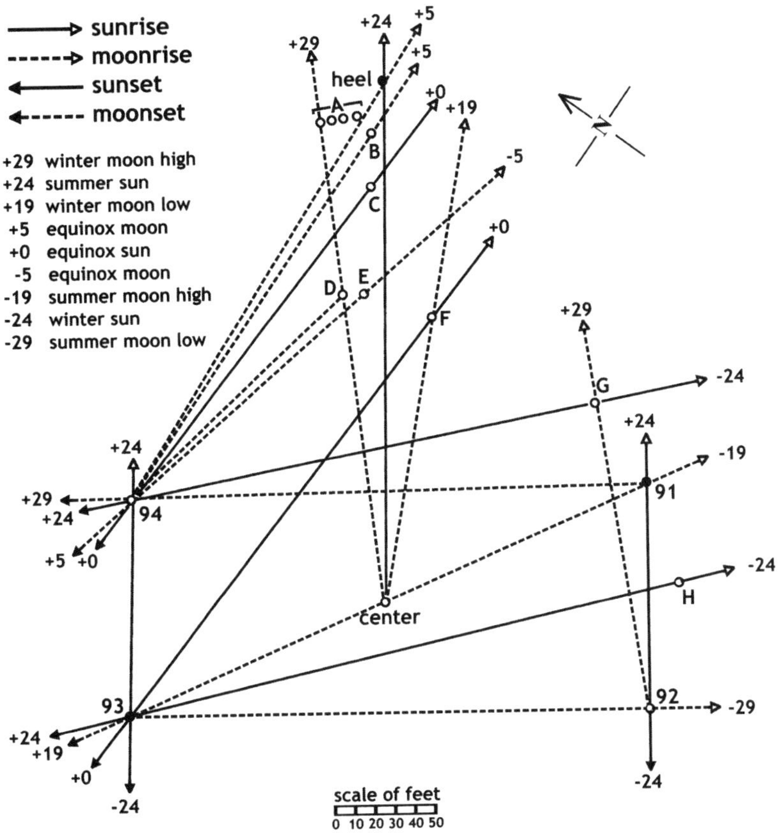


**Figure 19.5: Alignments of Stonehenge I, including equinox lines**

moon might have been eclipsed every nineteen years. Consulting a standard work on historical eclipses, Hawkins discovered that an eclipse of the moon or the sun always occurred when the full moon nearest the winter solstice rose over the heel stone. Moreover, when the winter moon, viewed from the center of the monument, rose over marker D ("winter moon high" in Figure 19.5), or marker F ("winter moon low" in Figure 19.5), the harvest moon was eclipsed that year.

Being able to predict these extremes of the moon would enable the astronomers of Stonehenge to predict when these harvest moons would be eclipsed. But since the moon takes 18.61 years to cycle, how would they keep track? If they used a nineteen-year cycle, it would soon become inaccurate; by the third cycle, it would be off by a year. The smallest number that would have been accurate for multiple cycles was 56, or 19 + 19 + 18. The number 56 rang a bell in Hawkins' mind: "The figure 56 seemed familiar. It was familiar—it was one of the oldest, most puzzling mysteries of Stonehenge. It was the number of Aubrey holes."[20]

No one had ever proposed a satisfactory explanation for the Aubrey holes. They were very carefully spaced and deeply dug, but they had apparently never served any structural purpose. Some of them showed evidence of having been dug and refilled several times.

Hawkins began thinking about ways to use the Aubrey hole circle to predict eclipses. In a 1964 article in *Nature*, Hawkins argued that by moving a series of six markers around the Aubrey holes—counter-clockwise, one hole per year—the operators of Stonehenge could have predicted every important lunar event for hundreds of years.[21] The number 56 is the smallest number that measures the swing of the moon with an overall accuracy of better than three days, and lunar cycles provide the only method of long-range eclipse prediction related to the seasons of the year.[22]

Mainstream archeologists, convinced that the builders of Stonehenge were primitive, were quick to attack Hawkins' work. They were not prepared to entertain the suggestion that the Aubrey holes were used in aid of predicting eclipses:

> Although the whole of the monument had now been given an astronomical basis, archaeologists such as Atkinson and Jacquetta Hawkes were not the least bit convinced by the evidence put forward—believing that the whole idea advanced by Hawkins seemed much too

> fanciful and totally beyond the intellectual capabilities of the so-called barbarian races who had inhabited the British Isles during the prehistoric period.[23]

> But the new implication that Neolithic man had the intellectual capabilities to conceive and then operate such a computer-like device raised further serious doubts among archaeologists. . . . Most archaeologists were not entirely negative to the idea of seeing certain astronomical alignments built into Stonehenge . . . But the evoking of the Aubrey holes at Stonehenge as a prehistoric eclipse predictor was totally unacceptable to all but a tiny minority of archaeologists no matter how favourably swayed towards the idea some astronomers and the general public following appeared to be.[24]

The criticism of Hawkins' work was motivated entirely by the fact that his discoveries clashed with Darwinian presuppositions. His findings did not fit the Darwinian paradigm that man has continually advanced from an initially primitive condition. Despite the fact that the engineering aspects of the monument amply demonstrate the high intellect of its designers and builders, Darwinist archeologists would not believe that a people so ancient–*who left no written records and apparently had no written language*—were so sophisticated. Only the inspired insight that the superiority of the ancients' minds allowed them to function without a written language can explain the seeming contradiction between engineering and astronomical sophistication, on one hand, and pre-literacy, on the other.

If Stonehenge were the only megalithic site that seemed to show that its builders were observers of the heavens, we would have reason to doubt Hawkins' conclusions. But it is not. There are many megalithic sites in Great Britain, Ireland, and France that have celestial orientations. For example, Callanish is a group of standing stones situated on the Isle of Lewis, the northernmost island of Scotland's Outer Hebrides. It had long been observed that Callanish pointed to extreme declinations of the sun and moon, and Hawkins came to the same conclusion. What is most interesting about Callanish, however, is its location. It is near the Arctic Circle, at a latitude where the moon, at its extreme declination, remains hidden just below the southern horizon. Callanish is located one degree three minutes south of this critical latitude, and the full moon at midsummer stands about one degree above the southern horizon once every 18.61 years. A row of stones points to its rising, transit and setting at this time.[25]

A Scottish engineer named Alexander Thom had for many years been meticulously investigating and measuring English and Scottish megalithic sites, including their celestial alignments, but his work did not become well known until after the publicity and controversy created by Hawkins' Stonehenge articles in *Nature*. Based upon his careful examination of many megalithic sites, Thom concluded that the megalithic builders knew of the moon's minor variation.

What is the minor variation? As noted above, the moon's orbit around the earth is inclined by about five degrees, eight minutes, to the angle of the ecliptic. It also has a minor variation of up to 9 minutes, caused by gravity. Thus, the moon's deviation from the angle of the ecliptic ranges from a maximum of five degrees, 17.5 minutes to a minimum of four degrees, 58.5 minutes. The maximum is reached when the moon is between the earth and the sun—which coincides with solar eclipses, making them predictable. The minimum is reached when the earth and the moon are at right angles to the sun. The minor variation was unknown to the Greeks. Tycho Brahe, in the sixteenth century, was the first to measure the minor variation, although Arab astronomers knew of or suspected its existence in the tenth century.

Thom argued that the megalithic builders not only knew of the moon's minor variation but also were able to determine the exact time of its maximum. Thom based this claim upon two Scottish sites, Temple Wood in Argyllshire and Mid Clyth in Caithness. At these sites, stone markers point to natural "notches" in the hills. At Temple Wood, for example, the standing stones point to a well-defined notch in the hills about 1.25 miles (2 km) away. By using an extremely distant natural foresight, Thom argued, the megalithic builders were able to measure the sun and moon's declinations accurately enough to calculate the value of the minor variation.[26] If Thom was correct, the megalithic builders had knowledge that was soon lost and that has been rediscovered only rarely in all subsequent human history. Clearly, Stonehenge is not alone as a megalithic monument oriented to points of celestial significance. Far from it.

Sir Fred Hoyle, an astronomer of immense reputation, jumped into the debate on Stonehenge with a 1966 article in *Nature* entitled "Stonehenge—An Eclipse Predictor." There could be absolutely no doubt, Hoyle wrote, that the Aubrey holes had operated as an ingenious eclipse cycle computer. He criticized Hawkins' interpretation of *how* the Aubrey holes were used, however, and suggested a different method, one that imputed even greater sophistication to the astronomers of Stonehenge.

Hoyle suggested that the Aubrey circle represented the ecliptic itself, and markers representing the sun, the moon, and the position of the observer could be moved around the circle so as to represent the actual motions of the Sun and Moon. With this method, the Aubrey circle could be used to predict almost every eclipse, even though only roughly half of them would be visible from the position of an observer at Stonehenge.[27]

Hoyle also disagreed with Hawkins regarding supposed azimuth or sight-line errors. Ten out of twelve lines that Hawkins cited as being slightly in error (because he assumed that the builders had intended to point exactly to the solar and lunar extremes), Hoyle believed were not errors. Hoyle suggested that the builders had not intended to mark the extremes exactly. In most cases, the sight line was slightly ahead of the turning point, which, Hoyle argued, would have allowed the observer to determine the *date of turning* more accurately.[28]

An article by Peter Newham followed Hoyle's article in the same issue of *Nature*. Newham pointed out that one of the thirty sarsens (stone 11) making up the sarsen circle was smaller than the other twenty-nine. Newham believed that this was intentional, and that the smaller sarsen was intended to represent half of a day, such that the sarsen circle would signify the 29.5 day lunar month. Newham also noted that the circles of thirty "Y" and twenty-nine "Z" holes allowed the builders to track the lunar month by moving a marker alternately through the Y and Z circles.

And then, there was also the circle of possibly fifty-nine bluestones inside the area that would later be encircled by the thirty sarsens. If there really had been fifty-nine stones, and many archeologists agree that there were, this bluestone ring would have provided an even better method of tracking the lunar months. Newham also pointed out that the nineteen-year phase, or Metonic, cycle could have been signified by the nineteen bluestones located inside the horseshoe of the five massive trilithons.

Newham and a French architect had discovered something extraordinary about the location of Stonehenge. The latitude fifty-one degrees seventeen minutes north of the equator very nearly makes a perfect 90-degree right angle between a line pointing toward the extreme solar horizon points and a line pointing toward the extreme lunar horizon points. Thus, the "station stones" form a rectangle. If the site were moved thirty miles north or south, it would change the celestial geometry. "Outside this zone the rectangle would be noticeably distorted. . . . perhaps these people were aware that the angles of the quadrangle formed by the station stones would change as one moved north or south."[29] The right angle would only repeat at the same latitude *south* of the equator. Stonehenge was placed at the only latitude in the Northern Hemisphere that made its unique geometry possible.[30]

**Figure 19.6: Part of the Sarsen circle at Stonehenge**

Stonehenge was designed and built as an observatory and eclipse predictor. Every detail of the monument, including its location, furthers this purpose, and the significance of many of its features can only be understood in light of this purpose.[31] But this purpose implies that its builders had intellectual attainments far beyond those previously attributed to "stone age" people. Sir Fred Hoyle thought that perhaps there existed, in prehistoric times, groups whose intellectual norm was considerably higher than that of the present day. "A veritable Newton or Einstein must have been at work," wrote Hoyle, "but then why not?"[32]

Why not? Because of the myth of perpetual progress, because early man was supposed to have been bridging the intellectual gap from apes to humans. The "why not" is the Darwinian paradigm. But Hoyle's statement has a very familiar ring to the Seventh-day Adventist ear. "In strength of intellect," we have been told, "men who now live can bear no comparison to the ancients." The Darwinian paradigm has it exactly backward. "True knowledge has decreased with every successive generation."[33]

Stonehenge itself evidences a history of intellectual decline. The use of Stonehenge as an observatory is already apparent in Stonehenge I, the oldest part of the site.[34] Even harsh critics of the astronomical interpretation of Stonehenge have admitted the celestial significance of Stonehenge I. Richard Atkinson, an archeologist who worked at Stonehenge during the 1950s, said that, "most of what has been written about Stonehenge is nonsense or speculation. No one will ever have a clue what its significance was."[35] But Atkinson admitted that, "we must accept, I think, that the positions of at least the heel stone and the station stones, and indeed, the latitude of Stonehenge itself, are astronomically determined . . ."[36] Of course, these features of the site were part of the original construction, as were the Aubrey holes. If the Aubrey holes were used to predict eclipses, as both Hawkins and Hoyle believed, the true genius of the earliest Stonehenge builders is apparent.

The later stages of the building, though far more elaborate, and an amazing display of engineering and building skills, do not represent any advance in astronomical science. In fact, they represent a decline:

> Very few of the major sight-lines of the first two periods were blocked by the imposing architecture of the final temple. But Stonehenge III was not in any sense an accurate or sensitive observing instrument, in the way that the first phase of the monument had been. The intriguing possibility exists, therefore, that over the 1,000-year history of the monument there was actually a decline in the level of astronomical investigation—a decline from the systematic and practical layout of Stonehenge I to the monumental and symbolic architecture of Stonehenge III.[37]

Sir Fred Hoyle came to the same conclusion: "I was convinced that the inner part, which was built around 1,500 BCE, was really mostly a matter of simply religious construction," Hoyle said. "I thought the people who built the first structures there, approaching 3,000 BCE, were the cleverest and that the later people didn't know what they were doing."[38] This may explain why the megalithic portion of the monument was built. It is as though the megalithic builders realized that their culture was slipping into darkness and decided to build a massive, permanent structure at the site to ensure that the genius of their ancestors would never be forgotten.

The graves of the British megalithic builders document the same pattern of decline. The tombs of 3,000 B.C., including Newgrange and Maes Howe, were monuments. These tombs are called "passage graves," because each had a long, megalith-lined passage leading to a central burial chamber, all covered by a large earthen mound.

The tomb at Maes Howe, in the Orkney Islands of Scotland, was one of the great achievements of prehistoric architecture.[39] The passage is 50 feet long, three feet wide, and 4 feet tall; 20 feet of the passage is formed by three megaliths, each 7 inches thick, 4 feet wide, and 18 to 20 feet long. The central vault (**Plate 30**) was 18 feet tall. Four megalithic buttresses at the corners support the central vault. The most striking feature of the central vault is the quality of the masonry. The wall faces are flush, with fine joints between regular, rectangular blocks. Many of the masonry courses consist of a single stone slab, about ten feet long, running from buttress to buttress.[40]

Newgrange on the River Boyne, County Meath, Ireland, is another example of a splendid megalithic tomb dating from around 3,000 B.C. The mound is 260 feet in diameter and 30 feet high, ringed at its base with megaliths. The passage is 62 feet long, almost 12 feet tall at the entrance, and roofed with large stone slabs. It is lined with

43 megaliths, many of them more than six feet tall and weighing ten to twelve tons each. The central vaulted chamber rises to a height of 20 feet. Newgrange has a unique feature, a "roof box" above the roof of the passage, set back a few feet from the entrance.[41]

Like Stonehenge, these tombs demonstrate celestial awareness. The passage at Maes Howe is oriented toward the west, toward sunset on the winter solstice. For about six weeks, centered on December 21, the rays of the setting sun shine through the passage and illuminate the inner vault. At Newgrange, by contrast, the passage is oriented toward the east, toward *sunrise* on the winter solstice. Michael and Claire O'Kelley, who excavated Newgrange in the 1960s, were told of a folk legend that the tomb was illuminated at dawn on the winter solstice. They positioned themselves in the vault at dawn on December 21, 1969, and, sure enough, "a pencil of direct sunlight shown through the roof-box and along the passage to reach across the tomb chamber floor as far as the front edge of the basin stone in the end recess."[42]

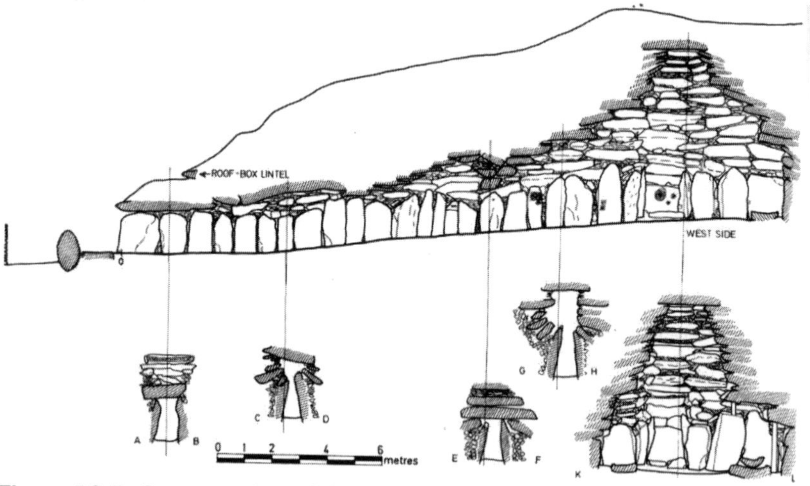


**Figure 19.7: Cross-section of the passage grave at Newgrange (viewed from the north)**

This effect was not accidental. The passage runs uphill to the central vault, and sunlight cannot reach the vault through the doorway of the passage. The roof box is above the doorway, however, about six inches above the height of the floor of the vault, so that sunlight streaming through the roof box will reach the vault. The O'Kelleys consulted an astronomer to make certain that the effect was present when Newgrange was built. He reported that the winter solstice sunrise, "has shown [into] the chamber ever since the date of its construction and will probably continue to do so forever, regardless of secular changes in the obliquity of the ecliptic."[43]

Later burials were more modest, but still attest to a very high cultural level. There were multiple burial chambers called "long barrows," which consisted of long earthen mounds anywhere from two feet to twenty-two feet high, and always more than twice as long as they were wide. These mounds typically contain several burial chambers, usually constructed of megalithic slabs, but sometimes of massive timbers. England's famous long barrows bear such names as "Belas Knap," "Hetty Peglar's Tump," "Wayland's Smithy" and "the Toots."[44]

Still later, toward the end of the "Stone Age" and the beginning of the "Bronze Age," there was a transition to "round barrows."[45] These typically consisted of a small, round earthen mound, not containing any stone or wooden burial chamber. There are over three hundred round barrows within a two-mile radius of Stonehenge. These early Bronze Age barrows often contained grave goods, some of which are quite spectacular. In the early 1800s, a barrow near Stonehenge was excavated, revealing the skeleton of a "tall and robust" man,[46] along with a number of gold objects, including a breastplate and belt buckle of pure gold, finely hammered and etched. Other barrows revealed gold earrings and buttons, and bronze daggers and spear points.[47]

The earliest Bronze Age interments were non-cremation "inhumation" burials, but within a couple of hundred years, this practice had largely been abandoned in favor of cremation. Earlier interments tended to be accompanied by a larger variety of artifacts, such as jewelry and ceremonial weapons. Bronze Age cremations were usually placed in ceramic urns; the form and decoration of these urns was more complex during the early Bronze Age.

By the middle Bronze Age, barrow building had virtually ceased. Old barrows were reused. Where a typical early barrow would contain only one or two interments, reused barrows might contain up to one hundred cremations placed at the outer margins of the barrow. As the form and decoration of the urns became simpler, other grave goods disappeared altogether. By the later Bronze Age (1,000-700 B.C.) ceramic urns were not typically interred with the cremated remains.[48]

Thus, the burials of prehistoric Britain and Ireland show extreme brilliance, skill, and celestial awareness early in the third millennium before Christ. Gradually, the tombs shift from monumental mounds with extraordinary internal megalithic chambers, to long barrows containing modest megalithic structures, to round barrows with no megalithic structures, to the re-use of existing barrows. Burials shift from remarkable monuments, to inhumations with elaborate grave goods, to cremations in decorated urns, to cremations without urns. This is a remarkable chronicle of the decline of a culture, a culture going in the opposite direction of that predicted by the Darwinian model.

But it fits the Genesis model perfectly. The first people to reach the British Isles after the Flood were the most important and august personages. They were few in number, brilliant in intellect, long-lived, physically strong, and able to build astonishing monuments to house their dead. Gradually, lives became shorter, death more common, people more numerous, and individuals less important. The funerary architecture reflects these trends, until finally it is abandoned altogether; the dead are cremated and their ashes spaded into the nearest barrow.

## B. A LEGACY OF ATLANTIS?

The megalithic culture of the British Isles is not the only one to show a pattern of early brilliance followed by decline. The same pattern is observed throughout the world. Of course, academic historians, archeologists, and anthropologists do not acknowledge this pattern, because it does not fit the Darwinian paradigm that forms the framework for their endeavors. While grudgingly acknowledging that "stone age" man accomplished some remarkable things, mainstream academics always insist that these must have been led up to by a long period of development.

Often, however, there is no evidence of any such developmental period. Consider the case of Egypt:

> Every aspect of Egyptian knowledge seems to have been complete at the very beginning. The sciences, artistic and architectural techniques and the hieroglyphic system show virtually no signs of a period of "development"; indeed, many of the achievements of the earliest dynasties were never surpassed, or even equaled later on. This astonishing fact is readily admitted by orthodox Egyptologists, but the magnitude of the mystery it poses is skillfully understated, while its many implications go unmentioned.[49]

Although mainstream academics are loath to mention the advanced capabilities of the ancients, there is a vast and enthusiastic non-academic literature on the subject. The authors of this literature have concluded that the ancients must have inherited their technology from an even older source. To continue quoting the writer quoted above:

> The answer to the mystery is of course obvious, but because it is repellent to the prevailing cast of modern thinking, it is seldom seriously considered. *Egyptian civilization was not a "development," it was a legacy.*[50]

But a legacy from where? From Atlantis, argue many authors. Ignatius Donnelly, an eccentric but erudite congressman from Minnesota, wrote the standard work on Atlantis back in 1882. Hundreds have written on the Atlantis theme before and since, but few have approached Donnelly's originality or his level of scholarship.[51] The Atlantis authors argue that Plato's "Atlantis" really existed, and after it was destroyed in a deluge, the survivors scattered to different parts of the globe. These writers usually cite the same evidence of an advanced ancient civilization—Egyptian sites (the pyramids and the Sphinx), European megaliths (Stonehenge), South American sites (The Nazca lines, Tiahuanaco, Sacsayhuaman, Ollantaytambo, Machu Picchu), Central American sites (Teotihuacan), Islands of the Pacific, such as Easter Island, etc. They also point to the many similarities between the Egyptians and other early civilizations.[52]

With the advent of the "space age" in the 1960s, some authors put a new slant on this material, theorizing that aliens from other planets visited the earth in ancient times, bestowing their advanced technology upon the earthlings.[53] The outlandish nature of this theory and the poor scholarship of some of its proponents have tended to marginalize the entire "ancient technology" literature and the underlying facts. Darwinian science now feels safe in assuming that there is nothing particularly extraordinary about any of these sites. The pendulum of opinion has swung from invoking space aliens to explain the evidence of ancient sophistication, to the view that there is nothing remarkable to be explained.

But these sites are remarkable. They *do* demonstrate that ancient man possessed extraordinary abilities. They do not fit within the Darwinian paradigm of continual progress from apes to modern technology. They demonstrate that human history has been one of peaks and valleys, with stunning achievements being followed by decay and dark ages. They show that some of the earliest civilizations were some of the most advanced. They imply a "lost civilization"—a sophisticated people who left their mark in the third millennium B.C., then vanished or degenerated to a lower level of civilization.

This is exactly what Adventists expect to find. The lost civilization was not Plato's "Atlantis," but the antediluvian world that was destroyed by the Genesis Flood. Noah and his nearer descendants retained at least some of the technology of the antediluvian world. That the early post-Flood race had remarkable capabilities is confirmed in the biblical story of the tower of Babel: "If as one people speaking the same language they have begun to do this, then nothing they plan to do will be impossible for them." Genesis 11:6. The confusion of tongues at Babel dispersed the people around the world, where they planted civilizations showing remarkable sophistication, as well as some remarkable similarities to each other.

Because the early post-Flood peoples were of taller stature than modern men, there are legends that prehistoric monuments were built by giants. Because the antediluvians did not need or use a written language, early post-Flood monuments such as Stonehenge reflect a people with high intellectual capabilities but no written language.

After the Flood, there was a gradual decline in physical strength, intellectual ability, and length of life. (See Genesis 11:10-26.) The decline in intellect and length of life necessitated the creation of written languages, but much knowledge was lost before this was accomplished. It is probable that, in some cases, no written language was ever created, and entire nations' histories were lost, leaving only durable artifacts as mute witnesses to their achievements.

What we find, preserved in stone and legend, is exactly what Seventh-day Adventists expect to find. Lacking prophetic insight to the contrary, other Christians have tended to accept the myth of perpetual progress. This is why a Seventh-day Adventist named Rene Noorbergen was the only author to place the ancient technology material in a biblical—rather than an Atlantis or ancient astronaut—perspective. Of Stonehenge and the other megalithic monuments, Noorbergen wrote:

> Rather than corroborating the concept of slow, gradual development of the arts and knowledge, a concept in line with evolutionary theory, the monuments left by our early ancestors point decidedly to an advanced technology in the remote past, which either matches or surpasses our own. . . . Early man was no ape. He certainly had a well-developed knowledge of mathematics and architecture, and he must have belonged to a social order that combined efforts for the construction of cities and for the organization of civilizations. Considering the astonishing accomplishments achieved by the first generations that survived the Flood . . . we may well wonder what advanced forms of structural and technological feats the antediluvians were responsible for prior to their being swallowed up by the waters of the Deluge.[54]

Indeed, what might have been accomplished by a people to whom "men who now live can bear no comparison"?

---

[1] Gaddis, Vincent H., *American Indian Myths & Mysteries* (Radnor, PA: Chilton Book Co., 1977), p. 18, citing William J. Perry, *The Children of the Sun: A study in the Early History of Civilization* (UK: Methuen & Co., 1923 [reprint: Grosse Pointe, MI: Scholarly Press, 1968]).

[2] Gaddis, at p. 18, citing Cyrus H. Gordon, *Before Columbus* (New York: Crown Pub. Inc., 1971).

[3] Knight, Christopher and Robert Lomas, *Uriel's Machine* (Gloucester, MA: Fair Winds Press, 1999), p. 213.

[4] After antiquarian John Aubrey (1626-1697), who investigated the site in the 1660s.

[5] "Stonehenge," *Encyclopedia Britannica.*

[6] Aubrey Burl suggests that Ice Age glaciers moved the bluestones as erratic boulders to within a few miles of Stonehenge. See, e.g., Aubrey Burl, "How Did the Stones Get There?" *History Today*, 51:3, 19-25 (March, 2001). The argument seems driven by skepticism that "Stone Age" people could have transported the bluestones from Wales. Others, like Tim Darvill, claim to have found the exact place in Wales where the bluestones were quarred. "Stonehenge Quarry Site 'Revealed'" BBC News, June 24, 2005. http://news.bbc.co.uk/2/hi/uk_news/wales/south_west/4123764.stm.

[7] As with any site of extreme antiquity, the dates are estimates. Some scholars now believe that that the monumental sarsens were erected between 2,500 B.C. and 2,250 B.C. Some believe that the entire monument was completed by 2,500 B.C. See NOVA, *Secrets of Lost Empires: Stonehenge* (Boston: WGBH, 1997).

[8] In 1997, the PBS television show "NOVA" commissioned archeologist Julian Richards, engineer Mark Whitby, and stonemason Roger Hopkins to move, raise, and cap a Stonehenge-like trilithon, armed only with stone-age stools and hundreds of volunteers. They succeeded in dragging a forty-ton poured concrete "sarsen" a few feet (just to show it could be done), toppling it into a hole, and raising it to vertical, all with only ropes, lumber, and muscle power. They then planted another stone by its side, but when it came time to mount the lintel on the two upright sarsens, their plan hit a snag. One way to accomplish this task would have been to lever up the lintel stone and put lumber beneath it, and keep repeating that process until the height of the lumber platform reached sixteen feet, the height of the upright sarsens. Then they could slide it over onto the sarsens. Another method would have been to build an earthen ramp up to the top of the sarsens. British safety officials would allow neither method, however, so the modern henge builders were required to construct an elaborate steel ramp up to the top of the sarsens.

Thus, NOVA recorded the amusing spectacle of 120 people dragging the huge concrete lintel up a very modern steel framed ramp, in support of the notion that "primitive" Neolithic people could have built Stonehenge. The modern henge builders did not even attempt to carve mortise and tenon joints with stone hammers. NOVA, *Secrets of Lost Empires: Stonehenge* (Boston: WGBH, 1997).

[9] Hawkins, Gerald S., with John B. White, *Stonehenge Decoded* (Garden City, NY: Doubleday, 1965), p. 20.

[10] Apparently, the sun was to rise in the middle of the Avenue, rather than over the heel stone, which is placed slightly right of center in the Avenue. The exact spot of the solstice sunrise shifts by 2/100th of a degree every century. See Peter Lancaster-Brown, *Megaliths, Myths, and Men* (New York: Taplinger Pub. Co., 1976; Reprint: Mineola, NY: Dover, 2000), pp. 61-63, 68-70.

[11] Hawkins, et al., at pp. 93, 94.

[12] Hawkins, et al., at pp. 104-118. Hawkins did, of course, account for the precession of the equinoxes, calculating the extreme positions of the sun and moon as of 1,500 B.C. Hawkins, et al., at pp. 106, 107. Since the mid-1960s, many archeologists have come to believe that entire monument was completed by 2,500 B.C. or 2,000 B.C. at the latest.

[13] Mitchell, Colin, *Creationism Revisited* (Alma Park, Grantham, Lincolnshire, UK: Autumn House Ltd, 1999), p. 196. See also Lancaster-Brown, *Megaliths, Myths, and Men* (New York: Taplinger Pub. Co., 1976), pp. 118-121.

[14] Although Newham had submitted his results to the journal *Antiquity*, that journal chose not to publish Newham's article. (At the time, *Antiquity* was edited by Glyn Daniel, an archeologist decidedly hostile to the idea that megalithic constructions had celestial orientations. See note 43, below.) A Yorkshire newspaper did publish Newham's claims, establishing his priority, but his scholarly article did not appear until after Hawkins' article in *Nature* had already created an avalanche of publicity. Lancaster-Brown, Peter, *Megaliths, Myths, and Men* (New York: Taplinger Pub. Co., 1976), pp. 101-110; Hawkins, et al., at pp. 132-137.

[15] Hawkins, et al., at pp. 132, 133.

[16] Hawkins, Gerald S. "Stonehenge Decoded," *Nature*, October 26, 1963.

[17] Hawkins, et al., at pp. 135, 136.

[18] Hawkins, et al., at p. 127; Lancaster-Brown, at p. 125.

[19] Hawkins, et al., at pp. 127-131. Diodorus does not identify Britain by name, but refers to the land of the "Hyperboreans," which means the people who live "beyond the north wind." It is not clear that the Hyperboreans are the British, but based upon Diodorus' description, many scholars think so. Assuming that the ancient British were Diodorus' Hyperboreans, it does not prove that the "spherical temple" was Stonehenge. Other sources report a Temple of Apollo beside a cliff, and another Temple of Apollo in what is now London, on the present site of Westminster Abbey. See, e.g., Bill Cooper, *After the Flood* (Chichester, UK: New Wine Press, 1995), pp. 64, 65 online at *http://www.ldolphin.org/cooper/ch4.html*; Charles Dickens, *A Child's History of England*, ch. 2, online at: *http://www.underthesun.cc/Classics/Dickens/achoe/achoe2.html*. The point is not whether Diodorus' "spherical Temple" was Stonehenge, but that Diodorus' report made Gerald Hawkins wonder whether there was a nineteen-year "Metonic cycle" connected with Stonehenge—and indeed there was.

[20] Hawkins, et al., at p. 140.

[21] Hawkins, Gerald S., "Stonehenge: A Neolithic Computer," *Nature*, June 26, 1964. See also Hawkins, et al., at pp. 174-181.

[22] Hawkins, et al., at pp. 138-148; Lancaster-Brown, at pp. 123-127.

[23] Lancaster-Brown, at p. 111.

[24] Lancaster-Brown, at pp. 126, 127.

[25] Hawkins, et al., at pp. 182-190; Gerald S. Hawkins, "Callanish, A Scottish Stonehenge," *Science*, 147:3654 (January 8, 1965); Mitchell, Colin, *Creationism Revisited* (Alma Park, Grantham, Lincolnshire, UK: Autumn House Ltd, 1999), p. 197; *Feats and Wisdom of the Ancients* (Alexandria, VA: Time-Life Books, 1990), pp. 88, 89.

[26] Lancaster-Brown, at pp. 194-201; Mitchell, Colin, *Creationism Revisited* (Alma Park, Grantham, Lincolnshire, UK: Autumn House Ltd, 1999), pp. 196, 197, citing Alexander Thom, *Megalithic Lunar Observatories,* (Oxford, UK: Clarendon Press, 1973), p. 18; Alexander Thom, *Megalithic Sites in Britain* (Oxford: Clarendon Press, 1967); Alexander Thom and Archibald S. Thom, *Megalithic Remains in Britain and Brittany* (Oxford: Clarendon Press, 1978).

[27] Lancaster-Brown, at pp. 130-133; Sir Fred Hoyle, "Stonehenge—An Eclipse Predictor," *Nature*, 211:454 (July 30, 1966).

[28] Lancaster-Brown, at pp. 134, 143-147; Gerald Hawkins, *Beyond Stonehenge* (New York: Harper & Row, 1973 (Reprint: Hubert Allen & Assoc., 2001)), pp. 77, 78; Sir Fred Hoyle, "Stonehenge—An Eclipse Predictor," *Nature*, 211;454 (July, 1966); Hoyle, "Speculations on Stonehenge," *Antiquity*, 40:262-276.

[29] Hawkins, et al., at p. 190.

[30] Lancaster-Brown, at p. 141; Mitchell, at p. 197.

[31] In the 1960s and 1970s, science was going through a relatively open-minded phase, and astro-archeological interpretations were published in prestigious scientific journals (such as *Nature*). In recent years, the Darwinian paradigm has reasserted itself, and this applies to Stonehenge, as well. But even writers skeptical of astro-archeology have had to admit its usefulness in interpreting Stonehenge. "It is now abundantly clear that astronomy does have some considerable part to play in the way we interpret the monument. . . . given the sophistication of so many other features of this and other monuments, recognition of the 18.6-year moon cycle can surely not have been beyond the builders' capabilities. . . . In the end, after the flurry of activity during the 1960s, it was clear that Stonehenge *could* have been used as some sort of astronomical observatory and calendar device, but there was no clear evidence that it *had been* so used." David Souden, *Stonehenge Revealed* (New York: Facts on File, 1997), pp. 124, 125. This is like saying, "I found a metal object in the garage with four wheels, an internal combustion engine, a passenger compartment, and a plate with the word 'Ford' on the front. It could have been used as an automobile, but there is no clear evidence that it had been so used." Such astonishing obtuseness can only be explained as slavish adherence to the Darwinian paradigm.

[32] Lancaster-Brown, at pp. 147, 148, citing Hoyle in *Antiquity*. See, also John Michell, *A Little History of Astro-Archaeology* (New York: Thames & Hudson, 1977, 1989), p. 68. Hoyle is also reported to have said that Stonehenge "demands a level of intellectual attainment orders of magnitude higher than the standard to be expected from a community of farmers." Quoted in Rodney Castleden, *The Making of Stonehenge* (London & New York: Routledge, 1993), p. 26. For further reading on Stonehenge directed toward a general audience, see Peter and Connie Roop, *Great Mysteries: Stonehenge: Opposing Viewpoints* (San Diego, CA: Greenhaven Press, 1989); Robert Wernick, *The Monument Builders* (New York: Time-Life Books, 1973); Rene Noorbergen, *Secrets of the Lost Races* (Indianapolis, IN: The Bobbs-Merrill Co., Inc., 1978), pp. 177-186; David Hatcher Childress, *Technology of the Gods: The Incredible Sciences of the Ancients* (Kempton, IL: Adventures Unlimited Press, 2000) pp. 57-60. On the subject of Astro-archaeology, see John Michell, *A Little History of Astro-Archaeology* (New York: Thames & Hudson, 1977, 1989), a very brief treatment, sympathetic to the astronomers and critical of the archeologists. ("The theory of what the Germans call *Urdummheit* (original stupidity) has been the most constant source of opposition to astro-archeology. . . . the main objection to [Lockyer's] thesis was simply that it did not fit the notion of *Urdummheit*. This notion was so confidently held by the British archaeologists, totally conditioned by Darwinian historical theory, that they were mentally incapable of giving serious consideration to contrary evidence." p. 45)

("Not everyone was convinced by Hoyle's expert testimony. The archaeologists fought hard in defence of the primitive view of antiquity. The whole basis of archaeological orthodoxy in respect of ancient history was obviously threatened by the Stonehenge decoders, and those who had become professionally dependent on maintaining that orthodoxy felt threatened too." p. 69)

[33] White, *Spiritual Gifts*, vol. IV, p. 156.

[34] Hadingham, Evan, "The Secrets of Stonehenge," in *The World's Last Mysteries* (Pleasantville, NY: The Reader's Digest Assoc., 1978), pp. 90, 91.

[35] "Stonehenge," *Encyclopedia Britannica.*

[36] Hawkins, *Beyond Stonehenge* (New York: Harper & Row, 1973 (Reprint: Hubert Allen & Assoc., 2001)), p. 78, citing *Antiquity* vol. 41, p. 97 (1967); John Michell, *A Little History of Astro-Archaeology* (New York: Thames & Hudson, 1977, 1989), p. 69. ("Atkinson himself came finally to accept the legitimacy of astro-archaeology and made generous amends for his former skepticism.").

[37] Hadingham, at p. 91.

[38] Dimitrakopoulos, Sandra, "Mystery of Stonehenge Points to the Heavens," October 12, 1999, EXN.ca, on the web at *http://exn.ca/mysticplaces/Enigma.cfm* and at *http://www.exn.ca/hiddenworlds/stonehenge/readstory.asp?storyID=1999101264.* See also Rodney Castleden, *The Making of Stonehenge* (London & New York: Routledge, 1993), p. 22. Of course, more recent radiocarbon testing indicates that Stonehenge was finished earlier, before 2,000 B.C. The process of decay might well have been over the course of as little as 500 years, not the 1,500 years that Hoyle indicates.

[39] Forde-Johnston, J., *Prehistoric Britain and Ireland* (New York: W.W. Norton & Co., 1976), p. 90.

[40] Forde-Johnston, at p. 93.

[41] O'Kelley, Michael J., *Newgrange: Archeology, Art, and Legend* (London: Thames and Hudson, 1982), p. 21. See also *Feats and Wisdom of the Ancients* (Alexandria, VA: Time-Life Books, 1990), pp. 86, 87.

[42] O'Kelley, at pp. 123, 124. See also Peter Harbison, *The Archaeology of Ireland* (New York: Charles Scribner's Sons, 1976), pp. 34-38.

[43] O'Kelley, at p. 124, quoting Dr. Jon Patrick, "Midwinter sunrise at Newgrange," *Nature*, 249:517-519 (1974). An astrophysicist named Tom P. Ray found that in 3,150 B.C., the beam of light would have entered the vault not four minutes after

sunrise, but exactly at sunrise. *Feats and Wisdom of the Ancients*, at p. 87. As with Stonehenge, archeologists were loath to admit the celestial sophistication of the builders of Newgrange. As recently as 1964, two archeologists dismissed the winter solstice sunrise phenomenon as "an example of the jumble of nonsense and wishful thinking indulged in by those who prefer the pleasures of the irrational and the joys of unreason to the hard thinking that archaeology demands." Sean P. Ó Riordáin and Glyn Daniel, *Newgrange and the Bend of the Boyne*, (London 1964), p. 19. Just a few years later, the O'Kelleys proved that the winter solstice sunrise phenomenon was real; it was Ó Riordáin and Daniel who were indulging in "nonsense and wishful thinking."

Newgrange is part of a complex of passage graves that includes Dowth and Knowth, which are each as large as Newgrange. On the winter solstice, the same sun that shines into Newgrange at sunrise shines into the chamber of Dowth at sunset. The third great passage grave, Knowth, has two passages designed to receive sunlight at the equinoxes, the eastern passage from the rising sun, the western passage from the setting sun. See John Michell, at p. 95.

44 See, e.g., "Megalithic Sites and Mounds" *http://witcombe.sbc.edu/earthmysteries/EMMegaSites.html*; English Heritage's "Monument Class Descriptions" at *http://www.eng-h.gov.uk/mpp/mcd/jump3top.htm.*

45 NOVA, *Secrets of Lost Empires: Stonehenge* (Boston: WGBH, 1997).

46 The men of the "Beaker people" of northern Europe were "tall and strongly built." Harrison, Richard J., *The Beaker Folk* (London: Thames and Hudson, 1980) p. 160.

47 NOVA, *Secrets of Lost Empires: Stonehenge* (Boston: WGBH, 1997); Robert Wernick, *The Monument Builders* (New York: Time-Life Books, 1973), pp. 88, 89.

48 "Barrows," *http://www.geocities.com/CollegePark/Theater/1377/page6.html.* Interestingly, Heinrich Schliemann, the famous excavator of Hisarlik Tell in Western Turkey (the site of Troy), found fabulous gold artifacts at a level he assumed to be Homer's Troy. More recent analysis indicates that the gold objects were from Troy II (2600-2250 B.C.), not Homer's Troy, which is probably the higher level of Troy VIIa (1300-1190 B.C.), which Schliemann plowed through in his un-scholarly haste to get to the bottom of the ruins. The archeology of Troy is more proof that older civilizations may be more advanced, not less advanced, than younger ones.

49 West, John Anthony, *Serpent in the Sky, The High Wisdom of Ancient Egypt* (Wheaton, IL: Quest Books, 1993), p. 1.

50 West, at p. 1 (italics in original).

51 Ignatius Donnelly, *Atlantis: The Antediluvian World*, 1882, Modern Revised Edition, ed. by Egerton Sykes (New York: Gramercy Pub. Co., 1949).

52 Typically, these books are written in "travelogue" format, i.e., we stayed here, we ate such and such, we drove up the winding road to there, etc., so that fifty pages of interesting or anomalous material can be padded out into a 300-page book. These writers usually recite flood legends from around the world, some eerily similar to the biblical account, making this literature a convenient place to look for collections of flood legends. Yet these authors never entertain the idea that the Genesis Flood took place as described in the Bible. And it is not as though they oppose catastrophic geological theories. Many are enthusiastic about Charles Hapgood's theory of "earth crust displacement." See Charles Hapgood, *The Path of the Pole*, 1958 (Reprint, Adventures Unlimited, 1999); Rand and Rose Flem-Ath, *When the Sky Fell: In Search of Atlantis* (New York: St. Martins, 1997); Colin Wilson and Rand Flem-Ath, *The Atlantis Blueprint: Unlocking the Ancient Mysteries of a Long Lost Civilization* (New York: Delacorte Press, a Bantam/Dell imprint, 2001). Despite a willingness to embrace unorthodox geological theories, these authors are not willing to consider the truthfulness of biblical history. Plato's obscure and fragmentary account of the destruction of an island metropolis called Atlantis is given more weight and credibility than the Scriptures upon which are based all of the world's monotheistic religions.

53 Erich von Daniken and Zechariah Sitchin are noteworthy proponents of this theory.

54 Noorbergen, Rene, *Secrets of the Lost Races: New Discoveries of Advanced Technology in Ancient* Civilizations (New York: The Bobbs-Merrill Co., Inc., 1977), p. 178. Noorbergen is no longer the only author to treat this material from a Christian perspective. See, Donald E. Chittick, *The Puzzle of Ancient Man: Advanced Technology in Past Civilizations?* (Newberg, OR: Creation Compass, 1998).

CHAPTER 20

# Genetic Engineering

It was the summer of 1971. Dr. Paul Berg of Stanford University was preparing to conduct a revolutionary scientific experiment. He was planning to transfer a virus that causes cancerous tumors in monkeys into the DNA of the common intestinal microbe, *Escherichia coli.* But when geneticist Robert Pollack learned of Dr. Berg's plan, he became very worried. He telephoned Berg, urging him not to carry out the experiment. "Do you really mean to put SV40 into *E. coli*?" he asked. Berg thought Pollack's concerns were unfounded. "Just forget about it," he said. But Pollack would not forget about it. "It's dangerous, very dangerous," he replied. "I don't think you should do it. The issue isn't just *E. coli.* This is an intrinsically dangerous line of work. We're all in over our heads."[1]

Berg finally agreed to host an informal gathering of about one hundred scientists to discuss the issue. Much to his surprise, most of the convened scientists agreed with Pollack that the experiment raised serious safety issues. For the most part, the concerns were not about the larger implications of inter-species DNA transfer but with the specific experiment Berg had planned: transferring a cancer-causing strand of DNA into an organism that resides in the human intestine.

What if the altered gene caused *E. coli* to become tumor factories? And what if the dangerously modified *E. coli* escaped the laboratory and colonized human intestines, causing an epidemic of cancerous tumors? It was a nightmare scenario.

The National Academy of Sciences formed a committee to issue guidelines for dealing with what by then had come to be known as *recombinant DNA*. Seemingly leaving the fox to guard the hen house, the Academy appointed Paul Berg to chair the committee. But the Berg committee took a cautious approach. In a 1974 open letter to several scientific journals, the Berg committee called for an immediate moratorium on certain recombinant DNA experiments.[2] In a remarkable example of science effectively regulating itself, everyone abided by the moratorium.

Eventually, safety guidelines were adopted specifying progressive levels of containment for progressively more risky experiments. For example, a particularly risky experiment might be conducted in an airtight cabinet, with the researcher manipulating the microbes from outside the cabinet, through gloved apertures. Within a few years, however, the scientific community reached a consensus that the dangers of recombinant DNA had been grossly overstated. In 1981, the Recombinant DNA Advisory Committee voted to lift the guidelines.[3]

Recombining the DNA of different species, as Dr. Berg proposed, is possible because the DNA molecule is present in all living things. Of course, different organisms have different genes, different genetic information, but the information is written in the same language in every living thing. That language is DNA.[4] Because DNA works the same way in all living things, genes from one plant, animal, or microbe, can be transferred to any other plant, animal, or microbe. In 1969, a graduate student at Stanford worked out a technology for transferring genes from one organism to another.[5]

Genetic engineering is basically a "cut and paste" operation. A gene is cut out of the DNA strand of one organism and pasted into the DNA of a different organism. One of the breakthroughs leading to genetic engineering was the discovery of restriction enzymes. These are found in bacteria, in which they act as a defense mechanism against viruses. (A virus is just DNA or RNA encased in a protein shell. It inserts itself into a cell, and takes over the cell's machinery to make multiple copies of itself.) When a virus inserts its DNA through a bacterial cell wall, the bacterium's restriction enzymes start chopping up the viral DNA at specific points along the polymer in order to defend the bacterium against the invader.[6] Scientists realized they could use these enzymes as "scissors" to cut DNA. Another enzyme, called *DNA ligase*, bonds complementary ends of DNA together. Scientists realized that they could use DNA ligase to "paste" the DNA back together after splicing in a foreign gene.

The early genetic engineers also needed a "vector," that is, a way to insert the altered DNA into an organism. Scientists discovered that bacteria contain free-floating loops of DNA called *plasmids*. Although not part of the bacterium's regular genome, plasmids are made of functioning DNA that codes for proteins. Plasmids are naturally passed back and forth between bacteria and can thus serve as "vectors" for inserting modified DNA into living bacteria.

With these tools at hand, genetic engineers could isolate the gene they wanted to transfer, cut it out, cut open a plasmid, insert the foreign gene into the plasmid, re-close the plasmid, and allow the plasmid to insert itself into bacteria. In order to identify the genetically modified bacteria, an antibiotic resistance gene was added to the plasmid. The researchers could then colonize the bacteria and treat the colony with an antibiotic—thereby killing any microbes that lacked the engineered plasmid.

Paul Berg never carried out the experiment he planned in 1971. In 1973, however, two other researchers, Stanley Cohen and Herbert Boyer, spliced genes from a staphylococcus bacterium with genes from a toad and inserted them into an *E. coli* bacterium. We had entered the age of genetic engineering. Gene splicing experiments proceeded in the 1970s—and the pace of such experiments accelerated in the 1980s—because scientists were learning from them. Moreover, it was obvious that the technology had enormous commercial potential. Herbert Boyer, for example, established a very successful biotechnology company named Genentech. In 1980, the U.S. Supreme Court put the biotechnology business on a firm intellectual property basis by ruling that genetically engineered organisms could be patented.[7]

The firestorm of controversy created by Paul Berg's monkey virus experiment seems quaint now, the product of a more innocent era. Today, biotechnology is big business. As of this writing, Genentech's market capitalization (the number of shares issued multiplied by the price per share) is over 28 billion dollars. Another leading biotechnology firm, Amgen, has a market capitalization of 65 billion dollars. These two companies and a third, Biogen, have a combined market capitalization of over $100 billion, but there are estimated to be over a thousand companies in the biotechnology sector. Investors have poured an enormous amount of money into gene splicing technology. What has this money bought?

## 1. MEDICAL USES OF GENE SPLICING TECHNOLOGY

An early commercial use of the new gene splicing technology—one that started Genentech on the road to success—was the production of insulin. Insulin was previously available only from the pancreases of pigs and cows. But porcine and bovine insulin are not identical to human insulin, and 5 percent of diabetics are allergic to nonhuman insulin. Bioengineers took the human gene that codes for insulin and spliced it into bacteria, so that the bacteria would produce human insulin. A simple process of fermentation could then produce plentiful and inexpensive insulin. Transgenic insulin has been commercially produced since 1982.[8]

Similar techniques have been used to produce human growth hormone (hGH). In some children, the gene that should code for hGH is defective, causing a condition known as pituitary dwarfism. HGH was previously available only in very small quantities from the brains of cadavers, but hGH from cadavers was at risk of contamination by an infectious agent (a protein crystal called a *prion*) that causes Creutzfeldt-Jakob disease, a horrible disease that ends in dementia and death (Creutzfeldt-Jakob is the human equivalent of Bovine Spongiform Encephalopathy—"Mad Cow" disease). But hGH can now be produced in quantity by fermenting bacteria into which the human gene has been transferred. The hormone is thus more plentiful, cheaper, and carries no risk of Creutzfeldt-Jakob.[9]

The same considerations favor transgenically produced factor VIII, a blood protein that some hemophiliacs lack. Many hemophilia patients treated prior to 1985 with factor VIII from human blood became infected with the AIDS virus. But factor VIII fermented from transgenic bacteria carries no risk of HIV.[10] Transgenically produced medical substances are generally at a lesser risk of contamination from a wide variety of infectious agents.

Other examples of substances that have been transgenically produced include interleukin-2 (for treatment of kidney cancer), hepatitis B vaccine, erythropoietin (for anemia), and whooping cough vaccine. Most hormones and enzymes can be produced in quantity by inserting the gene that codes for that substance into bacteria and allowing the bacteria to ferment. This use of gene splicing has been a triumph of the biotechnology industry.[11]

## 2. AGRICULTURAL USES OF GENE SPLICING TECHNOLOGY

An organism that has been given a gene from a different plant or animal is referred to as "transgenic" or "genetically modified." You have probably been eating genetically modified foods for many years. Farmers began planting genetically modified crops in 1996, when about 4.2 million acres of such crops were planted. By 2002, 145 million acres—22 percent of all land under cultivation worldwide—were planted with genetically modified crops. The acreage planted with genetically modified crops in the U.S. increased from 3.7 million acres in 1996 to 101.5 million acres in 2003.[12] Worldwide, 167 million acres were planted with genetically modified crops in 2003. Worldwide sales of genetically modified foods increased from an estimated $75 million in 1995 to $2.3 billion in 1999.[13] Transgenic food crops include canola, corn, papaya, potatoes, soybeans, squash, sugar beets, and tomatoes.[14]

One clever use of transgenic technology was to engineer crops that produce their own insecticide. A species of moth known as the European corn borer (*Ostrinia nubilalis*) lays its larvae on corn, where the larvae feed on tassels, whorl and leaf sheath tissue, as well as on silk, kernels, and cobs. Corn borer larvae can destroy up to 30 percent of the U.S. corn crop each year.[15] To combat it, farmers used massive amounts of insecticide, but insecticides also kill beneficial insects.

About a century ago, scientists noticed that a bacterium named *Bacillus thuringiensis* (Bt) produces a spore that contains protein crystals that are harmless to humans but toxic when ingested by insect larvae. A Bt toxin insecticide was produced, but why spray Bt toxin onto the plants when you can make the plants produce their own? Bioengineers have done just that, transferring the bacterial gene that codes for the toxin into the corn plant's genome. The resulting transgenic corn, called Bt corn, produces its own insecticide, decreasing the need for spraying or dusting.[16]

By switching to Bt corn, farmers in the United States and Canada reportedly experienced average yield increases of 7 percent in 1997; meanwhile, the use of artificial insecticides declined 30 percent in 1998.[17] Cotton, tobacco, tomatoes, potatoes, apples, rice, broccoli, and walnuts have all been engineered to produce their own Bt insecticide.[18] At least eighteen insecticidal transgenic crops have been field tested, so the future will see more such plants come under cultivation.[19] By 2003, about eighty percent (80%) of the soybeans and about forty percent (40%) of the corn (*maize*) planted in the United States was genetically modified; almost all Americans have eaten genetically modified corn and soybean oil.[20]

Commercial cotton farmers rely on large quantities of insecticides, including calcium arsenate, DDT, and toxaphene, to kill the boll weevil (*Anthonomus grandis*), a beetle whose life-cycle includes a larval stage. Monsanto, a leading biotechnology company, has developed and patented a Bt cotton called "Nucoton." According to a 1996 report, Monsanto licenses Nucoton to farmers for $32 an acre, but the farmers save from $50 to $60 an acre that they would have spent applying pesticides to a nontransgenic cotton crop.[21] By 2003, 73 percent of cotton grown in the U.S. was genetically modified. Several southern states, including Arkansas, Georgia, Louisiana, and Mississippi reported that more than 90 percent of their cotton crop was genetically modified.[22]

In addition to pesticides, farmers spend a lot of money on

fertilizer. They do not want weeds, because weeds divert nutrients away from the cash crop. But eradicating weeds without also killing the crop is a delicate and expensive business. Monsanto has an herbicide called "Roundup," which they market to suburban gardeners as well as commercial farmers. The active ingredient in Roundup is glyphosate. Monsanto has developed genetically modified soybeans with built-in immunity to glyphosate. Farmers can now spray for weeds with Roundup and other glyphosate-based herbicides without hurting their soybean crops.[23]

Monsanto has also engineered glyphosate-tolerant canola, cotton, corn, and sugar beets.[24] Other companies have modified crops to be resistant to a variety of herbicides, including glufosinate[25], bromoxynil, imidazolinone, and sulfonylurea.[26] One of Monsanto's competitors, Novartis, developed a broad-spectrum herbicide and simultaneously developed a gene named "Acuron" to make crops resistant to the new herbicide. Novartis planned to market Acuron-modified corn in the 2003 season and plans to market Acuron beets, wheat, soybeans, rice, canola, cotton, and sorghum at a later date.[27]

Humanitarian concerns motivate some uses of the gene splicing technology. In poorer countries, millions do not get enough Vitamin A in their diet. Around 350,000 children go blind each year as a result. A Swiss scientist named Ingo Potrykus developed transgenic rice that incorporates DNA from daffodils. Because it is full of Vitamin A and beta-carotene, the modified rice is yellowish and is called "golden rice." Golden rice could potentially prevent thousands of children in developing countries from going blind.[28]

Charles Arntzen, a scientist at Arizona State University, is developing a transgenic banana with a built-in vaccine. This would allow vaccine manufacturers in Third World countries to make vaccines with simple food processing technology that they already have, rather than sophisticated pharmaceutical technology that they may not have.[29]

According to reports, a vaccine for Hepatitis B has been built into genetically modified bananas. One dose, in the form of a single banana chip, will cost just one cent, as opposed to $15, and there is no need to refrigerate the banana chips. Arntzen has also developed transgenic potato and tomato vaccines.[30]

Genetically modified crops are viewed with suspicion in Western countries, where food is already plentiful and relatively inexpensive. They have met with a warmer reception in Third World countries, which feed themselves largely through subsistence farming. Third-world farmers cannot afford fertilizers and pesticides. They need agricultural solutions that are self-contained, and do not require extensive training or infrastructure. Transgenic crops meet this need, because the technology is in the seed itself, and the farmers can plant the seeds and raise the crop just as they always have.[31]

Sweet potatoes, for example, are a staple crop in Kenya, but pests such as weevils and the feathery mottle virus can destroy 75% of a crop. In 1992, Florence Wambugu, a Kenyan scientist, began genetically engineering the sweet potato to defeat the feathery mottle virus. She spent three years at Monsanto's laboratories developing a transgenic sweet potato, and field tests have been successful.[32]

On a lighter note, the gene splicing technology can be used to make vegetarian cheese. Cheese is made by curdling milk with an enzyme called *chymosin*, which is extracted from the stomachs of calves. But some strict vegetarians do not want to eat something that has been produced using a calf stomach. Scientists can extract the section of bovine DNA that codes for chymosin and insert it into yeast, ferment the yeast, and create all the chymosin needed to make cheese. For the really squeamish, scientists can synthetically build the relevant section of DNA using a gene machine—a computer-controlled chemical synthesizer with lots of tubes, valves, and pumps—which can assemble the DNA nucleotides in the right order, eliminating the need to go anywhere near a cow.[33]

Any connoisseur of tomatoes knows that vine-ripened tomatoes are delicious, but most store-bought tomatoes taste like cardboard. This is because tomatoes must be picked while still green in order to be marketed before they rot. Just before the tomatoes are put in stores, they are sprayed with ethylene to trigger or accelerate the final ripening process. Bioengineers at Calgene removed the gene that causes tomatoes to become soft, and then reinserted it backward, canceling out the gene for rotting. The goal was to allow the tomato to be picked when ripe, and still make the journey from farm to market to dinner table without bruising or rotting.[34] In 1994, the FDA approved Calgene's "FlavrSavr," tomato, but the taste was not satisfying, and it was removed from the market in 1997. Biotechnology firms are still trying to solve the problem of how to produce and market tasty tomatoes.

Bioengineers have also taken a gene from the arctic flounder that allows that fish to withstand near-freezing waters, and inserted the gene into a tomato. The modified tomato produces it own antifreeze, and is more tolerant of frost.

## 3. TRANSGENIC ANIMALS

Genetic engineers have created hundreds of varieties of transgenic animals.[35] The most widely used technique for making a transgenic animal is called microinjection. The scientist isolates the strand of DNA to be transferred, prepares a solution with multiple copies of the gene, and injects, with an extremely tiny syringe, a "picoliter"—a millionth of a millionth of a liter—of the gene solution into an undivided, fertilized egg. The injected gene diffuses to the chromosomes of the egg and inserts itself into the genome at random positions on the chromosomes.[36]

If the transgene integrates before the first cell division, every cell in the embryo will contain at least one copy. If it integrates or is injected after the first cell division, only some of the cells of the resulting animal will contain a copy of the transgene. When this happens, the animal is called a *mosaic*. The tortoise-shell cat is a naturally occurring mosaic. Some of its cells have the gene for orange fur, and some have the gene for either white or black fur. Another method for creating transgenic animals is to introduce foreign embryonic cells into young embryos. The resulting animal will contain cells (and genomes) from two different organisms. These animals are called *chimeras*.[37]

The first transgenic animal to be patented was a mouse developed for medical research. Scientists at Harvard designed it to be susceptible to cancer-causing chemicals and to succumb to breast cancer.[38] Another early experiment involved linking the gene that codes for human growth hormone to a regulatory sequence controlling a gene for binding metal ions, and inserting this combined genetic mechanism into mice. When the mice were fed zinc, the gene that codes for hGH was switched on, causing the mice to grow twice as large as normal.[39]

It is relatively easy to create transgenic fish, because fish eggs are fertilized and develop outside the fish's body, obviating the need to extract and reinsert the egg into the mother.[40] Transgenic rainbow trout, goldfish, and carp have been produced.[41] As with tomatoes, the antifreeze gene from the arctic flounder was inserted into the Atlantic salmon, allowing the salmon to live in colder waters. The gene that codes for a rainbow trout's growth hormone has been inserted into catfish and carp embryos, causing the transgenic fish to grow 60 percent faster than regular fish.[42]

Increasingly, fish are farmed in large hatcheries rather than caught in the open ocean. Aqua Bounty Technologies has introduced a growth gene from Chinook salmon and a promoter sequence from ocean pout into its farmed Atlantic salmon, causing them to grow four times faster than normal salmon. These "AquAdvantage™" salmon do not grow larger than normal, but mature in one-fourth the normal time span. Obviously, it is far more efficient and profitable to grow and sell four fish crops in the time it previously took to grow one.[43]

As of this writing, Aqua Bounty is seeking the FDA's permission to license transgenic salmon to commercial fish hatcheries, including those that have offshore pens. Every year, millions of fish escape from offshore fish pens, a fact that prompted environmentalists to wonder what would happen if transgenic salmon escaped into the open ocean. One environmentalist conducted a computer simulation. It predicted that the genetically modified male salmon would have a breeding advantage because of larger size at mating, such that the modified gene would soon spread throughout the wild salmon population. It also predicted that the transgenic salmon would be so unfit for life in the wild that the entire species would eventually become extinct.

Officials at Aqua Bounty argue that AquAdvantage™ salmon are not larger at sexual maturity than normal salmon, invalidating the key assumption of the computer simulation.[44] Moreover, Aqua Bounty states that they engineer all of their fish to be both sterile and female—a claim that is less than reassuring, given that a fictional bioengineer at "Jurassic Park" made an identical claim regarding his dinosaurs. Aqua Bounty expected a ruling from the FDA by 2005. Aqua Bounty executive Joseph McGonigle predicted that approval would next be sought for bioengineered trout. Studies are under way involving thirty other commercially sold fish and shellfish species.[45]

Some transgenic animals are created for the same reason that scientists made transgenic yeast and bacteria—to produce medically useful substances. This use of transgenic animals, which is a cross between agriculture and pharmaceuticals, is called *pharming*.[46] (Pharming is also the name of a Dutch company in the industry.) Pharming typically involves transferring a human gene into a bovine (or other mammalian) genome, in the hope that the cow will produce the desired substance in its milk. The story of one pharming project illustrates this technique.

The human body normally produces a protein called alpha-1-antitrypsin (AAT), which is necessary to the proper functioning of our lungs, and the lack of which causes emphysema and cystic fibrosis. Replacement therapy works, but supplies are limited. Scientists at Pharmaceutical Proteins Limited (PPL), a Scottish company based at Roslin Institute, set out to engineer transgenic sheep that could produce AAT in their milk.

First, the scientists isolated the human gene that codes for AAT and joined it with a regulatory sequence in the sheep's DNA that causes the gene to express itself in the mammary glands.[47] The researchers next experimented by inserting the transgene into mice eggs to create a mouse that would produce AAT in its mammary glands. They successfully created mice that produced AAT, but not in large enough quantities to justify trying the technique with larger mammals.

An astonishing fact about the human genome is that most of our DNA, about 95 percent to 97 percent of it, does not code for any protein. Geneticists had assumed that this huge proportion of our genome was simply useless "junk DNA"—perhaps some legacy of evolution. When portions of junk DNA intrude upon sections of DNA that code for proteins, these intruding strands are called *introns*. Assuming that introns would merely interfere with the normal expression of the gene, the researchers at PPL had excised or edited out the introns from the AAT gene. But when the initial experiments with mice failed to produce the protein in sufficient quantities, the researchers started tweaking the gene.

They discovered that the gene actually worked *better* with the introns than without them. "We left some of these random bits of DNA in the gene, essentially as God provided it," stated one of the researchers, "and that produced high yields."[48] This story illustrates that although we may not understand the role "junk DNA" plays in regulating organic processes, it is becoming very clear that non-coding DNA has a role. It is not "junk."

The PPL bioengineers next injected 550 sheep eggs with the gene that had succeeded in mice. Of these 550 eggs, 439 survived and were implanted into ewes. The ewes gave birth to 112 lambs, of which only five had incorporated the transgene, and only three of the five actually produced AAT-enriched milk. But one of these, a sheep dubbed "Tracy," was a winner—she produced thirty grams of AAT per liter of milk. PPL has bred—the old fashioned way—two herds of AAT-producing sheep from Tracy's offspring.[49]

Obviously, it takes a lot of work to get a pharming operation going, but once there is a herd that produces the desired substance, the animals can produce it in large quantities and without the moment-by-moment supervision necessary in fermenting-type operations.

Pharming Group, based in the Netherlands and Wisconsin, has developed a shortcut called "nuclear transfer." Under this method, which is essentially cloning, the transgene is introduced into fetal cow cells. Those fetal cells that incorporate the transgene are isolated and their nuclei placed in cow eggs from which the original nuclei have been removed. The resulting transgenic embryos are then implanted in the wombs of surrogate mother cows and carried to term, making it possible to generate only female transgenic calves modified to produce the biopharmaceutical product.[50]

One of the weirdest transgenic experiments involves tweaking the human gene that codes for hemoglobin to make hemoglobin more efficient at oxygenating tissue. Hemoglobin is a blood protein that carries oxygen from our lungs to our body tissues. But only about 30 percent of the oxygen is released into the body's tissues; the rest returns to the lungs and is exhaled. The crocodile version of hemoglobin is far more efficient than the human version, allowing crocodiles to stay underwater for about an hour before surfacing to breathe. A group of scientists in England is redesigning the human hemoglobin gene to make it more like the crocodile gene. Clinical trials to test the new hemoglobin on humans were underway in 1996.[51]

Scientists have spliced genes that affect visible characteristics. Japanese researchers transferred a gene for bioluminescence from the North American jellyfish into mice. The resulting mice glowed green under ultraviolet light.[52] Scientists have also transferred a firefly gene—called "luciferase"—into a tobacco plant. When watered with a solution of the proper chemicals, the resulting tobacco plant glows in the dark.[53]

Austin, Texas-based Yorktown Technologies is betting that there is a market for genetically engineered aquarium fish. Researchers at the National University of Singapore transferred a gene from a sea anemone into the freshwater zebra fish, causing the zebra fish to take on a bright red hue and glow under ultraviolet or black light. Yorktown Technologies purchased the right to sell these fish in the United States. In December 2003, Yorktown began distributing the ornamental "GloFish,™" marking the first time an entire transgenic animal has been sold as a consumer product.

Environmentalists fear that the GloFish™ will open the door for many more genetically engineered animals. "What makes this GloFish important is it is the first of what could be hundreds or thousands of animals that are genetically engineered," said Andrew Kimbrell of the Center for Food Safety. "We can mix and match the genetic makeup of the whole living world for fun and profit." No

U.S. federal agency has claimed regulatory jurisdiction of these fish, because they are not intended for human consumption, but the state of California has laws specifically governing genetically altered fish and has declined to allow the sale of GloFish™ in California.[54]

## 4. THE FUTURE

This brief survey will be out of date before this book goes to press. What genetic engineers have thus far done is the tip of the iceberg of what they will do:

> Scientists are currently only touching the surface of transgenic organisms. This process will greatly speed up in the twenty-first century, as the genomes of various crops and animals are decoded more or less simultaneously with the human genome. . . . From now to 2020, the pace of creating transgenic animals will vastly accelerate because we will have the complete genome of thousands of life forms on the earth to guide us.[55]

Since any gene can be transferred from any living thing to any other living thing, the possible combinations and uses are almost endless. "It is a fact that this technology, and extensions of it which can be logically foreseen, give mankind the possibility to find out more about the basic processes of life than ever before, and to create life forms in ways that nature never intended."[56]

What is the future of biotechnology? So far, we have discussed only small transgenic modifications—a gene that codes for one protein is transferred from one species to another; a plant is modified to make its own pesticide, or to tolerate herbicides or viruses. What about big changes to the genome—true genetic engineering that radically changes the shape of an organism? In ancient mythology, a creature called a chimera was part eagle, part goat, and part serpent. Will genetic engineers ever be able to produce true chimeras?

In order to make major changes to the shape of an organism, we will need to learn how thousands of genes work together to produce wings, eyes, scales, feathers, teeth, and other body parts. This is where gene mapping and genomics come in. The genomics puzzle is coming together. Scientists have recently discovered master genes that direct thousands of other genes in the work of forming a human or animal body. These are also called "homeobox" genes or "selector" genes. The most famous of these is called "eyeless," because it turns on the cascade of genes necessary to build an eye. Scientists experimenting on fruit flies noticed that when this gene was deleted, the resulting fly had no eyes.[57] In 1995, Swiss scientists placed the "eyeless" gene in different parts of a fly genome, creating a fly that had seven sets of eyes growing on its body, including on the wings, the legs, and even the antennae.

These inappropriately placed eyes were real. They possessed all 800 essential eye units: the red pigment, the bristles between each unit eye, the lens, the light-sensitive cells inside, etc.[58] Scientists estimated that the "eyeless" gene supervises over 2,500 separate genes that are involved in making an eye. To celebrate the discovery, the journal Science published an issue with a picture of a multi-eyed fly on the cover.[59]

Homeobox genes appear not only in fruit flies and mice but in most other animals studied so far, including humans. Researchers have found that they can freely interchange homeobox genes between very different types of species.[60] For example, mice have an "eyeless" gene called Pax-6; researchers found that transferring Pax-6 into the fruit fly genome caused the fly to develop eyes just as its own gene would have. Even the squid version of "eyeless" will, when inserted into the fly genome, cause fly eyes to develop.[61] Master genes from mice and even squid work just fine in the fruit fly genome. The mapping and study of homeobox genes is an important step in unraveling how DNA controls visible morphological features.

Futurist Michio Kaku predicts that between A.D. 2020 and 2050, scientists will probably be able to identify many of the genes that shape crucial body organs, but it will probably be decades after that before we are able to manipulate them with much dexterity. In a book section entitled, *Beyond 2050: Angels in America,* Kaku discusses what would be necessary in order to genetically engineer a human to grow its own wings and fly with them (like an angel, hence "angels in America").[62]

We could start by isolating the homeobox genes in birds that control the development of wings and transfer these to the human genome. But this would not work, because only birds have the cluster of genes needed to make wings. Inserting the master gene for wings into a human genome, if it had any effect, would probably only trigger the development of analogous human appendages, such as arms. Although the genome for birds may be sequenced early in the twenty-first century, identifying the thousands of genes necessary to form wings might take decades after that. Scientists would then have to determine how these genes act together to produce bones, muscles, tendons, feathers, the blood supply, and many other things—which might take further decades.

Aerodynamics would also have to be considered. Bird's bones are hollow, reducing their weight and making flight easier. Human bones are thick and solid, and our wingspans would have to be around twenty feet to achieve flight. But the muscle power necessary to flap wings of this size would exceed anything possible with the human frame. Scientists would have to reengineer the human body, developing massive back muscles and lighter but stronger bones.

Genetically modifying the human body to grow usable wings would involve modifying every other organ and system in the body, and hence most of the genome. And although we completed, in the year 2000, the mapping of the human genome, what we actually have is just a map of the positions of the four bases (A, T, G, and C). In other words, we have a DNA map—we do not have a gene map. We do not yet know where all the genes are, what they all do, and how they work together.

The biotechnological capability that we already have does not translate into creating true chimeras with traits of multiple animals. Such chimeras will require knowledge that will not be available until after 2050, perhaps not even until the twenty-second century. "True chimeras are probably beyond the reach of biotechnology for perhaps a century or more," notes Kaku.[63] But the fact that such feats of genetic engineering are even being discussed shows what biotechnology will be in the future.

## 5. WILL WE EVER BE ABLE TO RESURRECT THE DINOSAURS?

Will we ever be able to genetically engineer a dinosaur? The prospect has tantalized futurists and novelists for decades. The famous science fiction writer Arthur C. Clarke, for example, became hooked on dinosaurs as a boy. "To this day," he writes, "I retain my fascination with dinosaurs and eagerly look forward to the time when the genetic engineers will recreate the *Tyrannosaur.*"[64]

In "Jurassic Park," the best-selling novel by Michael Crichton made into a movie by Steven Spielberg, entrepreneur John Hammond decides to build a theme park around dinosaurs—real dinosaurs brought to life by the wonders of biotechnology. The movie begins at the *Mano de Dios* Amber Mines in the Dominican Republic. Why? Because amber has sometimes trapped, encased, and preserved mosquitoes ancient enough to have bitten a dinosaur. If such a mosquito had bitten a dinosaur, it might have some dinosaur blood in its abdomen. Blood and tissue contain DNA, and DNA is the blueprint of life. With dinosaur DNA, Hammond's genetic engineers will build dinosaurs. But how realistic was Crichton's scenario, and how long before a real entrepreneur will turn "Jurassic Park" into reality?

Crichton's novel was just that—fiction—and there are several technical problems, currently insurmountable, that prevent it becoming non-fiction. First, it may not be possible to recover ancient DNA from amber. Several scientists claim to have done so. Some claim to have extracted DNA from yeast, bacteria, viruses, and plants preserved in amber.[65] Scientists claim to have extracted the DNA of a termite,[66] a 30-million-year-old bee, and a 120-million-year-old weevil, all preserved in amber.[67] Other scientists are skeptical of these claims, however, noting that these results have not been replicated. Samples are easily contaminated, because all kinds of DNA are floating in the air, in dust, dirt, fibers, dead skin cells, germs, etc. Scientists who claim to have found ancient DNA, it is argued, really found modern DNA and did not realize it.[68]

But there is another problem with the mosquito scenario of "Jurassic Park." A mosquito draws blood not to store it for posterity but to feed itself. The mosquito's digestive acids will immediately start breaking the blood down into nutrients, destroying the dinosaur DNA. Even if the mosquito bit a dinosaur, flew immediately to a tree, became trapped by—and engulfed in—resin, and killed, the digestive processes might not cease in time to save the dinosaur DNA.[69]

Instead of collecting mosquitoes preserved in amber, why not get the DNA directly from the dinosaur fossils? Many fresh-looking, lightly mineralized dinosaur bones have been found. This was true of a *T. rex* femur that Mary Schweitzer was studying, and her team successfully isolated parts of a hemoglobin molecule.[70] Others have isolated the delicate protein osteocalcin from a dinosaur bone.[71] So why can't scientists get dinosaur DNA straight from the dinosaur?

Some scientists claim to have done just that. In 1994, Scott Woodward and his team at Brigham Young University removed fragments of dinosaur bone from a seam of coal. They cracked open the bone in ultra-sterile laboratory conditions and extracted DNA that appeared to be equidistant from modern mammalian, avian, and reptilian DNA. Again, however, most scientists were skeptical of Woodward's findings, arguing that modern DNA had contaminated his sample. Attempts to reproduce his results have not been successful.[72] Mary Schweitzer and the MSU researchers could not obtain DNA from the dinosaur bone they were working on.[73]

Assuming that dinosaur DNA is somehow recoverable, it would probably have broken down into fragments of a few hundred base pairs of nucleotides.[74] Any dinosaur genome would probably have at least a billion base pairs, so Hammond's genetic engineers would have to figure out how to stitch millions of fragments of DNA back together. If they managed to get fragments constituting several copies of the entire genome, and the genome had broken at different places on different copies, they might, given enough computing power, figure out how to put the genome back together again. If, that is, the genome consisted of one long strand of DNA.[75]

But, of course, it is not one long strand. An animal's genome is divided into chromosomes, which come in pairs. Hammond's technicians would have no way of knowing how many pairs of chromosomes they were looking for and thus how many strands of DNA they should end up with. The ends of chromosomes are marked by repetitive patterns of DNA called *telomeres*, which might help locate the end of each chromosome. But there are many sections of repetitive DNA in all animals, so that clue might not be enough to solve the problem.[76]

Another complication is that cells contain organelles called *mitochondria*, typically about a thousand of them per cell. Each mitochondrion has its own short complement of DNA, called mitochondrial DNA, or mtDNA. Hammond's bioengineers would have to assume that their recovered DNA contained some mtDNA; they would need to segregate the mtDNA from the nuclear DNA of the genome.[77] And the technicians of Jurassic Park would not have a living dinosaur genome to use as an example. What they would be attempting is like trying to put together a million-piece jigsaw puzzle with thousands of extra pieces, without any picture to illustrate the end product.

At Jurassic Park, an orientation film informs visitors that gaps in the dinosaur DNA have been filled in with frog DNA. This is puzzling. If you've come with me this far, you know that bird, crocodile, or even mammalian DNA would be more appropriate for filling in gaps in dinosaur DNA. (Later in the story, the frog DNA facilitates a plot development that vindicates Ian Malcolm's prediction that life is not easy to contain—that "life . . . finds a way.")

Bridging the gaps with a closely related species seems a clever idea, but if there are any gaps that could not be bridged by comparing one copy of the genome with another copy, the project is probably stalled.[78] Hammond's bioengineers could use another animal's genome for purposes of comparison, say perhaps a bird or a crocodile (after they'd mapped it, that is), but they could know with certainty where and how long the dinosaur DNA gaps are only if they already had a map of the dinosaur genome they were trying to reconstruct.

Assembling the genome is just the beginning of the problem. There are many things about how an egg turns into a chicken that we do not now understand.[79] Life still has its mysteries. Hammond's bioengineers would not know how to insert the DNA into an egg. DNA doesn't just float around in a cell; it is wound up and "packaged" in a certain way, and the way DNA is folded and packaged effects how it functions. If the DNA is to function correctly, it must be packaged so that the egg's biochemical machinery can work with it.[80]

Another problem is that the bioengineers would not know what kind of egg might suffice to gestate a dinosaur. Eggs are very complex, containing RNA, proteins, and mitochondria, and these components probably vary from species to species. How could Hammond's bioengineers modify the egg to make it work with the dinosaur DNA? In order to modify an existing egg to serve as a dinosaur egg, "you would pretty much have to understand how to make the entire dinosaur—which is the problem you're trying to solve in the first place."[81]

So resurrecting the dinosaurs would be vastly more difficult than Michael Crichton and Steven Spielberg made it appear. It is impossible at present, and it may never be possible. Consider the thoughts of Rob DeSalle and David Lindley, who wrote a book on *The Science of Jurassic Park and the Lost World*:

> It may simply be that no one will ever find any dinosaur DNA at all, or never any more than a few scraps of it. And even if sufficient dinosaur DNA is found, there are countless difficulties and complications in creating a dinosaur—enough to make the task essentially impossible. But the sorts of problems that remain perplexing—how to identify and "mend" unknown genes, how to persuade an egg cell to take up a foreign genome, how to manipulate the genetic characteristics of a growing embryo—are those that scientists are working on right now, in many different contexts, in laboratories around the world.[82]

> Fifty years from now, it may be that many of the technical obstacles to dinosaur re-creation—obstacles that seem insurmountable today—will have been overcome through the efforts of scientists pursuing other goals.[83]

If history tells us anything, it is that "insurmountable" problems can be, and frequently are, overcome. The cloning of a sheep from the cell of an adult ewe provides an example of how a seemingly insurmountable technical obstacle can be solved. With the exception of some specialized cells such as red blood cells, every cell in the body has a complete copy of the DNA. Theoretically, then, it should be possible to create a genetically identical duplicate from any body cell. Scientists have known this for decades, and cloning has long been the holy grail of molecular biology. One way to clone an animal would be to extract its DNA, insert that DNA into a fertilized egg from which the DNA had been removed (enucleated), and then insert the modified egg into a surrogate mother. The animal born to the surrogate mother would be a genetic duplicate of the animal that donated the DNA. Scientists attempted this many times, but it never worked. It had become textbook dogma that DNA from a cell of a mature animal could not be used to clone the animal.

But Ian Wilmut and Keith Campbell, of Roslin Institute near Edinburgh, Scotland, solved the problem. By closely studying cell cycles, they determined the exact stage of the cell cycle at which a donor cell can successfully donate DNA for cloning. They also determined how to culture the donor cell—what to put in the serum, what to withhold, and how long to withhold it—in order to stop the donor cell at the crucial stage of its life cycle. In 1996, having mastered these techniques, they successfully cloned a sheep using DNA from an adult, differentiated, epithelial cell. Overnight, the resulting sheep, "Dolly," became the most famous little lamb since Mary's.[84]

Most of the more difficult problems preventing us from turning "Jurassic Park" into reality—how to recover dinosaur DNA and repair the fragmented genome, how to persuade an egg to take up a foreign genome, how to manipulate the genetic characteristics of a growing embryo—have to do with recreating a dinosaur from its DNA alone. What if you started with a living animal, progressively modifying it until you had something that looked like a dinosaur? That seems a more likely scenario.

Ian Wilmut predicts that we will eventually have that sort of god-like mastery of nature. He believes that the confluence of the three high biotechnologies—genetic engineering, genomics, and cloning—make it "possible in principle to build new organisms at will."[85]

> Until the birth of Dolly, scientists were apt to declare that this or that procedure would be "biologically impossible"—but that expression, biologically impossible, seems to have lost all meaning. In the twenty-first century and beyond, human ambition will be bound only by the laws of physics, the rules of logic, and our descendants' own sense of right and wrong. Truly, Dolly has taken us into the age of biological control.[86]

A few years ago, I served on a jury with a young man who was a research assistant in genetics at the University of California at Los Angeles. When I learned what he did, I asked him about recent genetic experiments such as attempts to get tomatoes to ripen without rotting. We also discussed the plot of the film *Jurassic Park.* I asked him if Crichton's proposed method of recreating the dinosaurs would ever be technically feasible. We would not need to try to recover actual dinosaur DNA, he said, because scientists would soon be able to build something that looked like a dinosaur from the DNA of existing animals or genes generated with a gene machine. In fact, he indicated that we were less than a generation away from being able to do this. That startled me. I think it will take much longer; my fellow juror was confusing a clear view with a short distance. But we will inevitably get there. It is just a question of when.

---

[1] John Lear, *Recombinant DNA: The Untold Story* (New York: Crown Pub. 1978), p. 26.

[2] Lampton, Christopher, *DNA and the Creation of New Life* (New York: Arco Pub., 1983), p. 51.

[3] For more on this episode, see Christopher Lampton, *DNA and the Creation of New Life* (New York: Arco Pub., 1983); Sheldon Krimsky, *Genetic Alchemy: The Social History of the Recombinant* DNA Controversy (London and Cambridge: MIT Press, 1982); John Lear, *Recombinant DNA: The Untold Story* (New York: Crown Pub. 1978); James D. Watson, *A Passion for DNA* (Cold Spring Harbor, NY: Cold Spring Harbor Laboratory Press, 2000), pp. 49-79.

[4] This fact has been used to argue that every living thing evolved from a single ancestral cell, but it can as easily be used to argue that everything was created by one God. At the announcement of the completion of mapping the human genome, for example, Bill Clinton stated: "Today we are learning the language in which God created life."

[5] Lear, John, *Recombinant DNA, the Untold Story* (New York: Crown Pub., 1978), pp. 39-52. Peter Lobban was the graduate student who worked out the protocol.

[6] When this happens, the bacterium tags its own DNA (with a methyl compound added to the cytosine bases), essentially telling its restriction enzymes, "This is my DNA—don't cut it."

[7] *Diamond v. Chakrabarty*, 447 U.S. 303; 100 S.Ct. 2204; 65 L.Ed.2d 144 (1980) In 1972, Ananda Chakrabarty, an engineer with General Electric, applied for a patent for a genetically engineered bacterium capable of breaking down crude oil. Chakrabarty collected bacteria from hazardous-waste dumps and found four different strains of the *Pseudomonas* bacteria that lived on various components of oil. He took the genes that allowed these four strains to live on oil (the genes were found in plasmids, rather than in the genome proper) and inserted them into a single bacterium. The idea was to use the genetically engineered bacteria to help clean up oil spills. The Supreme Court held that although a naturally occurring organism is not patentable, a genetically engineered one is: "Here, by contrast, the patentee has produced a new bacterium with markedly different characteristics from any found in nature and one having the potential for significant utility. His discovery is not nature's handiwork, but his own; accordingly it is patentable subject matter under § 101." See, also, Christopher Lampton, supra, at pp. 102-106; Linda Tagliaferro, *Genetic Engineering: Progress or Peril?* (Minneapolis, MN: Lerner Publications Co., 1997), pp. 77-95; Nossal, G.J.V., *Reshaping Life*, (Cambridge, Melbourne: Cambridge University Press, 1985-1989), pp. 126, 127.

[8] See, e.g., Sheldon Krimsky, *Genetic Alchemy: The Social History of the Recombinant DNA Controversy* (London and Cambridge: MIT Press, 1982), p. 288; Lampton, Christopher, *DNA and the Creation of New Life* (New York: Arco Pub., 1983), pp. 82-83; Nossal, G.J.V., *Reshaping Life* (Cambridge, Melbourne: Cambridge University Press, 1985-1989), pp. 43-45; *Genetic Engineering: Opposing Viewpoints* (San Diego, CA: Greenhaven Press, 1990), pp. 22, 23.

[9] Kaku, Michio, *Visions: How Science Will Revolutionize the 21st Century* (New York: Doubleday Anchor Books, 1997), p. 223; Lampton, at p. 85; Nossal, at pp. 42, 43; Caskey, C. Thomas, "The Future of Biotechnology," in *Biotechnology* (Washington, D.C.: Joseph Henry Press, 1996), p. 54; Levine, Joseph, and David Suzuki, *The Secret of Life* (Boston: WGBH, 1993), pp. 161-164; Aldridge, Susan, *The Thread of Life: The Story of Genes and Genetic Engineering* (Cambridge and New York: Cambridge University Press, 1996), p. 45.

[10] Caskey, C. Thomas, "The Future of Biotechnology," in *Biotechnology*, (Washington, D.C.: Joseph Henry Press, 1996), p. 54.

[11] Kaku, at p. 223.

[12] Fact Sheet, "Genetically Modified Crops in the United States," *Pew Initiative on Food and Biotechnology*, August 2003. *http://pewagbiotech.org/resources/factsheets/display.php3?FactsheetID=2.*

[13] Tyson, Peter, "Should we Grow Genetically Modified Foods?" PBS online in-depth coverage of the *Frontline-NOVA* special, "Harvest of Fear: Exploring the Growing Fight over Genetically Modified Foods" (2001) *http://www.pbs.org/wgbh/harvest/exist.*

[14] Fox, Michael W., *Beyond Evolution: The Genetically Altered Future of Plants, Animals, the Earth . . . and Humans* (New York: The Lyons Press, 1999), p. 208.

[15] Levidow, Les, "Regulating Bt Maize in the United States and Europe: A Scientific-Cultural Comparison," *Environment*, Dec. 1999.

[16] Aldridge, Susan, *The Thread of Life: The Story of Genes and Genetic Engineering* (Cambridge and New York: Cambridge University Press, 1996), pp. 209, 210; "Harvest of Fear: Exploring the Growing Fight over Genetically Modified Foods," (2001) *Frontline/NOVA*.

[17] Levidow, supra.

[18] Moses, Vivian and Sheila, *Exploiting Biotechnology* (Switzerland: Harwood Academic Publishers, 1995), p. 172; Fox, at pp. 37, 38; Milius, S. "Bt broccoli test: Refuges cut pest resistance," *Science News*, March 4, 2000.

[19] Obrycki, John J., "Transgenic Insecticidal Corn: Beyond Insecticidal Toxicity to Ecological Complexity," *Bioscience*, May 2001.

[20] Fact Sheet, "Genetically Modified Crops in the United States," *Pew Initiative on Food and Biotechnology*, August 2003.

[21] Feder, Barnaby J., "Out of the lab, a Revolution on the Farm," *The New York Times*, March 3, 1996, in *Genetic Engineering: A Documentary History*, Thomas A. Shannon, ed. (Westport, CT: Greenwood Press, 1999), pp. 82, 83. There was a report that Monsanto's Bt cotton failed to control the boll weevil in field tests. Fox, at p. 148, citing J. Rissler and M. Mellon, eds. "Bt Cotton Fails to Control Boll-worm," *Gene Exchange* (Union of Concerned Scientists), vol. 7, no. 1 (1996). There is always the race against resistance. Insects can become resistant to the Bt toxin just as to any other insecticide. Mutations alter the molecules in the lining of the stomach to which the Bt protein crystals bind, such that the crystals no longer bind. In order to slow such mutations from spreading throughout the pest population, federal regulations mandate that farmers plant a certain percentage of nontransgenic crop to allow the nonresistant pests to survive. The regulations allow growers to choose one of the following three plans: (1) Plant a maximum of 80 percent Bt cotton along with 20 percent non-Bt, with the option of spraying the latter as needed with anything but foliar Bt insecticides; (2) Plant a maximum of 95 percent Bt cotton along with 5 percent non-Bt, but the latter cannot be sprayed with any insecticides; (3) Plant a maximum of 95 percent Bt cotton along with 5 percent non-Bt, with the latter planted in the same field with the Bt and sprayed only if the 95 percent is sprayed with non-Bt foliar insecticides. Weaver, Tara, "Protecting Farmers' Investment in Bt Cotton," *Agricultural Research*, February 2001. See also Mike Holmberg, "Those who need Bt technology most have the most to lose when resistance develops," *Successful Farming*, May-June 1999.

[22] Fact Sheet, "Genetically Modified Crops in the United States," *Pew Initiative on Food and Biotechnology*, August 2003.

[23] Aldridge, at p. 211; Paarlberg, Robert "The Global Food Fight," *Foreign Affairs*, May/June 2000, in *Biotechnology*, Lynn Messina, ed. (New York: H.H. Wilson and Co., 2000), p. 148; Hodgson, John, *Bio-Technology: Changing the Way Nature Works* (London: Cassel, Equinox, 1989), p. 74. Interestingly, bioengineers got the idea from a natural plant parasite called *Agrobacterium tumefaciens*. This parasite infects scratches or cuts in plants, causing a cankerous growth called a crown gall. *Agrobacterium* invades plant cells and releases a free-floating strand of DNA called a *Ti plasmid*. Part of the plasmid inserts itself into the chromosome of the plant. The inserted DNA converts the plant into a factory that produces food for *Agrobacterium*. Genetic engineers have modified the *Agrobacterium* Ti plasmid for use

as a "vector" to transfer foreign genes into plants. They remove the genes that create the canker, thus "disarming" the plasmid, but retain the genes that allow for the insertion of DNA into the plant chromosome. They then splice the genes they want to introduce into the plant into the disarmed plasmid. Hodgson, at p. 72. The state of the art for creating transgenic plants has advanced beyond this elegant but elaborate method for transferring genes. In 1987, scientists showed that a .22 caliber blank cartridge could serve to literally shoot DNA into plant cells. Now, bioengineers use a "gene gun," to shoot genes into a plant. See, e.g., Paarlberg, at pp. 147, 148; Kaku, at p. 224.

24 Fox, at pp. 68-74; Holmberg, Mike, "Transgenic Beets," *Successful Farming*, Feb. 2000.

25 Aventis Cropscience (A Swiss company formerly known as Hoffman-LaRoche) developed the first rice genetically modified to be glufosinate-resistant. The folks at Aventis got ahead of themselves. Three U.S. Government agencies regulate various aspects of genetically modified food. The FDA and the USDA had approved Aventis' glufosinate-resistant rice for human consumption. But the EPA had not approved glufosinate itself for use with rice, although it had approved glufosinate for use with corn, cotton, and rapeseed. Aventis had to bury 2,272 metric tons of rice in a landfill in southeast Texas. "Aventis to dispose of GM rice," *Food & Drink Weekly*, May 28, 2001.

26 Fox, at pp. 68-74.

27 Fairly, Peter, "Novartis Debuts PPO Herbicide and Tolerance Gene," *Chemical Week*, Feb. 24, 1999. The herbicide is called a PPO herbicide because it kills plants by inhibiting an enzyme in the photosynthetic machinery—protoporphyrinogen oxidase (PPO). The "Acuron" gene codes for a variant of *PPO* engineered to resist the inhibitor. One drawback of transgenic herbicide resistance is that some cash crops are closely related to weeds and could donate their herbicide resistance to their weed cousins. Oilseed rape, for example, is a valuable crop in some areas and a nuisance weed in others. Should the herbicide-resistance gene cross from the cash crop to the weed, one farmer's herbicide-resistance cash crop has created another farmer's herbicide-resistant weed. See Wilson Wall, *Sexing the Parrot: Changing the World with DNA* (London: Cassell, 1999), pp. 164, 165.

28 "Harvest of Fear: Exploring the Growing Fight over Genetically Modified Foods," *Frontline/NOVA* (2001); Bailey, Ronald, "Dr. Strangelunch: or Why we should Stop Worrying and Love Genetically Modified Food," *Reason*, Jan. 2001; *Frontline/NOVA* interview with Charles Arntzen, *http://www.pbs.org/wgbh/harvest/interviews/arntzen.html*

29 Tagliaferro, Linda, *Genetic Engineering: Progress or Peril?* (Minneapolis, MN: Lerner Publications Co., 1997), pp. 26, 27; "Harvest of Fear: Exploring the Growing Fight over Genetically Modified Foods," *Frontline/NOVA*; "Edible Vaccines," *Pediatrics*, Jan. 2001.

30 "Hepatitis B," *Environment*, Nov. 2001.

31 Cyrus Ndiritu, a former director of the Kenya Agriculture Research Institution, states. "I'd like to make something clear. It is not multinationals that have a stranglehold on Africa. It is hunger, poverty and deprivation. If Africa is going to get out of that, it has to embrace GM technology." Lynn J. Cook, "Millions Served," *Forbes* (December 23, 2002). http://www.forbes.com/global/2002/1223/064.html.

32 "Harvest of Fear: Exploring the Growing Fight over Genetically Modified Foods," *Frontline/NOVA* (2001). See also Paarlberg, Robert, "Promise or Peril? Genetically Modified Crops in Developing Countries," *Environment*, Jan. 2000; Lynn J. Cook, "Millions Served," *Forbes* (December 23, 2002).

33 Aldridge, at pp. 104-110.

34 Tagliaferro, at p. 36; Wilson Wall, *Sexing the Parrot: Changing the World with DNA* (London: Cassell, 1999), pp. 168-179.

35 Wilmut Ian, Keith Campbell, and Colin Tudge, *The Second Creation: Dolly and the Age of Biological Control* (Cambridge, MA: Harvard University Press, 2000), p. 30; Fox, at p. 93.

36 Aldridge, at p. 114.

37 Aldridge, at pp. 114-116; Wilmut, Ian, et al., at pp. 34, 35

38 Fox, at p. 94.

39 Nossal, G. J. V. and Ross L. Coppel, *Reshaping Life* (Cambridge, Melbourne: Cambridge University Press, 1985-1989), p. 106; Fox, at p. 94.

40 Aldridge, at p. 116.

41 Nossal, et al., at p. 107.

42 Aldrige, at p, 117. See also "The Coming of Biotech Animals," *Nutrition Research Newsletter*, March 2001, citing, C. Lewis, "A New Kind of fish Story: The Coming of Biotech Animals," *FDA Consumer* (Jan.-Feb. 2001).

43 "Harvest of Fear: Exploring the Growing Fight over Genetically Modified Foods," *Frontline/NOVA* (2001); "Biotech Brief: Fast Growing Fish," *Sierra*, July 2001.

44 "5 Myths about Transgenic Salmon" from Aqua Bounty's official website: *http://www.aquabounty.com.*

[45] "Harvest of Fear: Exploring the Growing Fight over Genetically Modified Foods," *Frontline/NOVA* (2001). See also "The Coming of Biotech Animals," *Nutrition Research Newsletter,* March 2001.

[46] Levine, Joseph, and David Suzuki, *The Secret of Life* (Boston: WGBH, 1993), pp. 158-190; Wilmut, Ian, et al., at pp. 32-42; Fox, at pp. 93-122.

[47] Levine, et al., at pp. 167, 168.

[48] Levine, et al., at p. 169.

[49] Levine, et al., at p. 170. See also Wilmut, Ian, et al., at pp. 36-39.

[50] "Pharming Announces Birth of First Female Transgenic Calves through Nuclear Transfer," *PR Newswire,* January 27, 1999.

[51] Aldridge, at pp. 120-122.

[52] Associated Press, "Japanese researchers create glowing mice," June 13, 1997.

[53] Roth, Ariel, *Origins* (Hagerstown, MD: Review and Herald Pub., 1998), pp. 277, 278, citing Ow, Wood, DeLuca, de Wet, Helinski, and Howell, "Transient and stable expression of the firefly luciferase gene in plant cells," *Science,* 234, pp. 856-859 (1986); De Wet, Wood, DeLuca, Helinski, and Subramani, "Firefly luciferase gene: structure and expression in mammalian cells," *Molecular and Cellular Biology,* 7(2), pp. 725-737 (1987).

[54] See, e.g., Don Thompson, "State exception sought for genetically engineered Fish," *Contra Costa Times*/Associated Press, November 30, 2003; Guy Boulton, "GloFish to make waves in court," *The Tampa Tribune,* December 28, 2003

[55] Kaku, at p. 223.

[56] Nossal, G.J.V., *Reshaping Life* (Cambridge, Melbourne: Cambridge University Press, 1985-1989), p. 118.

[57] This is typical of genetic research. Geneticists delete or disable genes, then see what happens to the organism. The method has been compared to someone who, wanting to know how an automobile is assembled but being unable to see inside the plant, kidnaps a worker on the way into the factory each morning. If the automobiles coming out of the plant that day lack seats, then he knows the kidnapped worker's job was to install the seats.

[58] Halder, G., et al., "Induction of ectopic eyes by targeted expression of the eyeless gene in Drosophila," *Science,* 267:1788-1792 (March 24, 1995).

59 Kaku, at pp. 235, 236. ("It's the paper of the year," remarked one observer, "This is Frankensteinian science at its best.")

60 Kaku, at p. 237.

61 Incidentally, how is it that mammals and insects, which are thought to have evolved separately for more than 500 million years, share the same master control gene for eye morphogenesis? It seems extremely unlikely, if the DNA copying error theory of evolution is true, that two creatures that supposedly began evolving (accidentally mis-replicating their DNA) along separate ancestral lines shortly after life appeared would share the same genetic control mechanisms. More likely, one Designer created everything, using the same "off the shelf" design solutions in many different creatures.

62 Kaku, at pp. 238-240.

63 Kaku, at p. 240.

64 Wilford, John Noble, *The Riddle of the Dinosaur* (New York: Alfred A. Knopf, 1985), p. 79.

65 Raul Cano, a microbiologist with Cal-Polytechnic at San Luis Obispo, California. See, e.g., "Ancient DNA and micro-organisms," *http://www.calpoly.edu/~rcano/ancientdna.html.*

66 Rob DeSalle and David Grimaldi, of the American Museum of Natural History, claimed, in 1992, to have extracted the DNA of a termite trapped in amber.

67 Cano, R.J., Poinar, H.N., Pieniazek, N.J., Acram, A. and Poinar, G.O., "Amplification and sequencing of DNA from 120-135 million year old weevil," *Nature*, 363:536-8 (1993).

68 See, e.g., Richard Thomas and Jeremy Austin, of London's Natural History Museum, *New Scientist*, (October 18, 1997) ("It is highly unlikely that geologically ancient DNA survives in any fossil material so far studied.") Michael Braun of the Smithsonian Museum states, "its the end of the story for very ancient DNA." But Raul Cano of California Polytechnic State University, one of the leading amber DNA researchers, denies claims that the DNA he has isolated is the result of contamination. "Just because they failed doesn't mean we did—They are too dogmatic in their conclusions." See also Gibbons, Ann, "Ancient History," *Discover*, Jan. 1998.

69 DeSalle, Rob, and David Lindley, *The Science of Jurassic Park and the Lost World or, How to Build a Dinosaur* (New York: Basic Books, 1997), pp. 41-43.

70 That was Horner and Schweitzer's 1997 report, of a 1990 find. Regarding whether there will be recoverable DNA in the far more spectacular 2005 find, Schweitzer says,

"We don't know yet. We are doing a lot in the lab now that looks promising," "Scientists recover T. rex soft tissue," *Reuters/MSNBC.com*, March 28, 2005.

[71] The osteocalcin was detected by immunological reaction and by direct measurement of Gla/Glu ratios using high-performance chromatography. Gerard Muyzer, *et al.*, "Preservation of the Bone Protein Osteocalcin in Dinosaurs" *Geology*, October, 1992, 20:871-874; J. Collins, A.M. Gernaey, C.M. Nielsen-Marsh, C. Vermeer, and P. Westbroek, "Slow rates of degradation of osteocalcin: Green light for fossil bone protein?" *Geology*, December 1, 2000, 28(12):1139-1142.

[72] DeSalle, et al., at pp. 9-10. See, also, Bada, Jeff, Hendrick N. Poinar, Matthias Höss and Svante Pääbo, *Science*, May 10, 1996.

[73] Gibbons, Ann, "Ancient History," *Discover*, Jan. 1998.

[74] DeSalle, et al., at p. 32.

[75] DeSalle, et al., at pp. 60-64, 67.

[76] DeSalle, et al., at pp. 64-67.

[77] DeSalle, et al., at pp. 96-98.

[78] DeSalle, et al., at p. 68 ("And if there are gaps in the recovered dinosaur DNA, you're stuck. If you find two unmatched ends and can't find any fragment that bridges the gap between them, you come to a dead stop. There's no way of knowing whether 5 bases are missing or five million; if the whole genome is a billion bases long, even a five-million-base gap is a small slice of the whole thing.")

[79] DeSalle, et al., at introduction, p. xxix.

[80] DeSalle, et al., at pp. 91-93.

[81] DeSalle, et al., at p. 96.

[82] DeSalle, et al., at pp. 180-182.

[83] DeSalle, et al., at pp. 182-183. (". . . although any number of these problems might be insoluble now, it's not clear that any of them are insoluble in principle. Genetics and molecular biology are, comparatively speaking, in their infancy." DeSalle, et al., at introduction, p. xxix).

[84] Kaku, at pp. 226, 227. See also Wilmut, Ian, et al., at pp. 312, 313. Of course, some mad scientists will attempt to clone a human being, with all the ethical problems that entails. (UPI, "Italian doc says he will clone human, January 26, 2001; Alex Kirby,

BBC News, "Cloned human planned 'by 2003'", January 30, 2001; BBC News, "Doctors defiant on cloning," March 9, 2001) ("Doctors from Italy and the United States said on Friday they intended to push ahead with their plans to clone human beings, despite the objections and doubts raised by religious and scientific groups. . . . "Dolly is here and we are next," says Dr. Panos Zavos.") But cloning is still far too novel and risky to consider experimenting with human cloning. The efficiency of animal cloning is very low—around 2 percent of the embryos created by cloning make it to term, notes Dr. Harry Griffin of Roslin Institute. Many animal clones die late in pregnancy or soon after birth and show developmental abnormalities. Dolly aged prematurely, got arthritis, contracted lung disease, and finally had to be put to sleep at age 6, about half the normal life expectancy of a healthy sheep (BBC News, "Cloned sheep Dolly has arthritis," January 4, 2002; Jane Wardell, Associated, Press, "Dolly the cloned Sheep put to sleep," February 14, 2003.)

[85] Wilmut, et al., at p. 9.

[86] Wilmut, et al., p. 5.

CHAPTER 21

# An Alternative Theory of the Dinosaurs

*Is there anything of which one can say,*
*"Look! This is something new"?*
*It was here already, long ago;*
*it was here before our time.*
Ecclesiastes 1:10

## A. A DEBATE IN SAN FRANCISCO

It was the morning of September 8, 1947. In a conference room not far from San Francisco, fifteen men were seated behind a long table, facing a podium flanked by two chairs. The men seated at the table were officials of the Seventh-day Adventist Church, including its highest officer, James Lamar McElhany, President of the General Conference. They were there to hear a debate. The two men seated in front of the table, on either side of the podium, were two Adventist scientists, Harold W. Clark and Frank Lewis Marsh. They were there to argue.

Shortly after 9 a.m., Milton E. Kern rose to introduce the issue to be debated. As Chairman of the Board of Trustees of the

Ellen G. White Estate, Kern was an appropriate person to introduce the issue, because it concerned some of Ellen White's most puzzling written statements. In the four-volume work, *Spiritual Gifts, Important Facts of Faith in Connection with the History of Holy Men of Old,* White had written of a concept she called "amalgamation:"

> But if there was one sin above another which called for the destruction of the race by the flood, it was the base crime of amalgamation of man and beast which defaced the image of God, and caused confusion everywhere.[1]

> Every species of animal which God had created were preserved in the ark. The confused species which God did not create, which were the result of amalgamation, were destroyed by the flood.[2]

Adventists were familiar with these passages, but no one was certain what the term *amalgamation* meant, or which species were "the confused species." The passages had remained mysterious and controversial, noted Kern, ever since their first publication in 1864.

Over the years, most commentators had gravitated toward one of two interpretations. One group held that the passages referred to crossbreeding between humans and animals; the other group argued that amalgamation was merely the intermarriage of the godly line of Seth with the ungodly line of Cain (something we discussed at length in chapter 18). Both points of view were ably represented that day in 1947. The men who would argue were the two most prominent Seventh-day Adventist naturalists of their generation. Each published several books over the course of long careers in science.[3]

Kern first called upon Harold Willard Clark (1891-1986). Clark was an Adventist from New England who had taught school and then attended Pacific Union College, an Adventist college in northern California. After graduating from PUC, Clark taught at PUC and took graduate classes part-time at the University of California at Berkeley, eventually earning a master's degree in biology.[4]

Clark had begun to grapple with the meaning of the "amalgamation" passages during the 1920s, because his biology students at PUC kept asking about them. Ellen White had died in 1915, but Clark inquired of some who had been close to her as to the meaning of these passages. The consensus seemed to be that "amalgamation" referred to hybridization among animals and between animals and humans. In 1940, Clark had completed the book *Genes and Genesis*, in which he supported this interpretation and suggested examples of possible crosses in the animal kingdom. He was at a loss to suggest an example of a cross between man and beast.

That morning, Clark began his presentation by complimenting Marsh on his scientific work furthering a biblical understanding of origins. The two men were friends who agreed on the need to rebut the theory of evolution, and, Clark noted, they even agreed on some aspects of the amalgamation statements. By beginning in this gracious manner, Clark helped eased the tension in the room. He then moved on to the substance of his argument.

Clark noted that the first amalgamation statement ("But if there was one sin above another . . .") comes at the end of a chapter dealing with the sins of the antediluvians. One of the issues White had already discussed, earlier in the chapter, was intermarriage between the godly line of Seth and the rebellious line of Cain:

> The descendants of Seth were called the sons of God—the descendants of Cain, the sons of men. As the sons of God mingled with the sons of men, they became corrupt, and by intermarriage with them, lost, through

> the influence of their wives, their peculiar, holy character, and united with the sons of Cain in their idolatry. Many cast aside the fear of God, and trampled upon his commandments.[5]

It seems very unlikely, Clark argued, that the amalgamation passage would refer to an issue she had explicitly addressed just four pages earlier.

Second, Ellen White's contemporaries did not seem to be in doubt about the meaning of the term *amalgamation.* W. C. White, Ellen's son, and D. E. Robinson, her secretary, both believed that the amalgamation statements referred to the crossing of man and beast. Some of White's contemporaries had argued in print that amalgamation referred to the crossing of humans and animals, and White had never attempted to correct those statements.

Third, the passage clearly refers to a sin—the one sin above another that called for the destruction of the race by a flood. History showed, argued Clark, that human cohabitation with animals—bestiality—was one of the greatest sins of antiquity. Clark admitted that there was no evidence that modern humans can mate with nonhumans and produce offspring, but this did not mean that such crosses could not have occurred in the antediluvian past. To say that amalgamation between men and animals had never occurred because it does not occur today, Clark argued, is to fall into the same uniformitarian error that led geologists to the conclusion that the world is many millions of years old.[6] Ellen White must therefore have intended for her readers to interpret the "base crime" as the sexual crossing of man and beast.

At about 9:45, Clark concluded his argument, and Kern called upon Marsh. Frank Lewis Marsh (1899-1992) had aspired to become a physician, but lack of funds steered him into nursing and, later, teaching. While teaching school near Chicago, Marsh earned a master's degree in zoology from Northwestern University and later completed a Ph.D. in botany at the University of Nebraska.[7] Some ten years after that 1947 debate, Marsh was to become the first director of the Geoscience Research Institute, an institute founded by the Adventist denomination to address questions regarding origins and the age of the earth.

Upon publishing a book entitled, *Evolution, Creation, and Science,* Marsh sent copies to Harvard's Ernst Mayr and Columbia's Russian-born geneticist, Theodosius Dobzhansky. Mayr declined to comment on the book, but Dobzhansky carried on an extensive correspondence with Marsh regarding it.[8] Dobzhansky had written that, "among the present generation, no informed person entertains any doubt of the validity of the evolution theory in the sense that evolution has occurred." After reading Marsh's book, Dobzhansky admitted that Marsh had written a sensibly argued defense of special creation. Several years later, in the sixth edition of his *Genetics and the Origin of Species,* Dobzhansky cited Marsh as the exception to the rule that an informed person cannot doubt evolution.

Marsh began as Clark had, by praising his adversary of the moment and noting that he and Clark were good friends who agreed on most things. He began the substance of his argument by reading an 1859 dictionary definition of *amalgamation* that defined it as the mixing of the African and Caucasian races. In Ellen White's time, the "amalgamation of man" would have evoked interbreeding between two races. In the context of this passage, Marsh argued, the "two races" would have been the descendants of Cain and the descendants of Seth.

Marsh then turned to the shelf of Ellen White books that lined the wall of the conference room. He selected a copy of *Fundamentals of Christian Education* and read a statement from page 499: "The enemy rejoiced in his success in effacing the divine image from the minds of the people . . . through intermarriage with idolaters and constant association with them . . ." Ellen White had written that intermarriage with idolaters "effaced the divine image," and that amalgamation "defaced the image of God," basically the same thing. The weight of the literary evidence, Marsh argued, fell on the side of interpreting amalgamation to mean the mixing of the godly line of Seth with the ungodly line of Cain.

But the gravamen of Marsh's argument was that modern science did not support Clark's interpretation of the passage. One of the most clearly demonstrated principles of biology, Marsh noted, was that fundamentally different kinds of animals cannot interbreed. Marsh also pointed out that humans cannot breed with animals today, and there was no reason to believe that humans had ever been able to breed with animals. We do not find, he argued, the fossils of human-animal hybrids.[9]

The Bible itself, argued Marsh, explains why different kinds of animals cannot interbreed, and why humans cannot breed with animals. The Bible states that God created each type of animal "after its kind." Based upon this passage, Marsh had coined the term *baramin* from the Hebrew words *bara,* meaning "created," and *min,* meaning "kind."[10] The Bible implied, Marsh argued, that species from one baramin, or created kind, cannot interbreed with species from another baramin. Humanity is a separate created kind, created in the image of God. (Gen. 1:27; 5:1, 2.) Because mankind was a separate baramin, there was never a time when humans could have bred with animals. And the many different animal baramins could never have interbred. It therefore appeared to Marsh that modern science was merely confirming what Scripture taught.

Clark's view was closer to the opinion held in Ellen White's time, but Marsh's position was more in line with modern scientific findings and was more politically palatable. No vote or official action was taken on that day in 1947, but, for a variety of reasons, the generally sanctioned opinion in the church swung toward the Marsh position, where it has remained. Adventist interpreters such as Francis D. Nichol, long-time editor of the *Review and Herald* (now *Adventist Review*), have argued that "amalgamation" refers to the intermarriage of the godly line of Seth with the ungodly and rebellious descendants of Cain.[11]

Clark and Marsh were both correct on their main points. Clark was correct that the passage does not refer to intermarriage between the godly and ungodly peoples. First, White had adequately treated the intermarriage issue just four pages earlier. She would not have revisited, in a vague and inscrutable fashion, an issue she had just clearly and explicitly addressed. Second, the second amalgamation passage makes clear that "amalgamation" was something that happened primarily to *animals*:

> Every species of *animal* which God had created were preserved in the ark. The confused species which God did not create, which were the result of amalgamation, were destroyed by the flood.

The intermarriage of the line of Seth with the line of Cain has nothing to do with animals.

Clark was also correct that the early interpreters assumed that sexual reproduction was implied in the amalgamation passages. This is not surprising, because in 1864 selective breeding was the only method by which mankind could influence the form and characteristics of living creatures. Perhaps, like Darwin, White's early interpreters thought there were no limits on the extent to which a species could be modified by selective breeding.[12]

Marsh, also, was correct on his main point. While different species of the same basic kind of animal can mate and produce fertile offspring (wolves with coyotes, for example) and others can hybridize to the extent of producing an infertile offspring (horses with donkeys, for example), fundamentally dissimilar types of animals cannot interbreed. This cannot happen today, and there is no reason to believe that it was ever possible. And although Scripture does not explicitly state that one separately created kind, or baramin, cannot breed with another created kind or baramin, Marsh's view is a reasonable inference. Marsh was probably correct in arguing that there was never a time when humans were interfertile with the animals.[13]

In summary, Clark was correct that amalgamation means the mixing of very different kinds of creatures, and Marsh was correct that this cannot be done by means of sexual reproduction. How then could it have been accomplished? This was an insoluble problem in Ellen White's time, and it appeared even farther from a solution in 1947. But the scientific advances we reviewed in the previous chapter suggest a solution. Clark and Marsh both made one assumption that, in retrospect, was unwarranted. They both assumed, along with most other commentators, that "amalgamation" was accomplished by means of *sexual reproduction.* But Ellen White never said that "amalgamation" had anything to do with sexual reproduction.

## B. AN ALTERNATIVE THEORY OF THE DINOSAURS

In previous chapters, we established that it is not necessary to interpret the geologic column as the residue of long ages. The strata give evidence of having been deposited rapidly and catastrophically. The fossils reflect an order of burial that roughly corresponds to the order in which different ecological zones would have been overtaken and buried during the Genesis Flood. The bulk of the sedimentary strata, including the dinosaur strata, were accumulated during the year of the Genesis Flood.

If the Bible is correct that all plant, animal, and human life was created in six days, and that animal predation and death are the result of human sin, then humans and dinosaurs were both on earth at the time of the Flood. Most other biblical creationists have also concluded that humans and dinosaurs must have lived contemporaneously, because this is the only conclusion consistent with a straightforward reading of God's Word.

We rejected the theory of evolution as an explanation for the existence of the dinosaurs, noting that the copying error theory of evolution seems improbable and the rocks do not yield the countless intermediate fossil forms Darwin's theory predicted. We established that typical dinosaur anatomy is a combination of the characteristics of several vertebrate classes, including reptiles, birds, and mammals. The dinosaurs, along with many other extinct vertebrates, such as marine reptiles, flying reptiles, and mammal-like reptiles, appear to be amalgamations of the characteristics of different classes of vertebrates.

The mixed nature of these fossil creatures is not well explained by the theory of evolution. We specifically addressed whether evolutionary "branching off" can explain the heterodox nature of the dinosaurs and concluded that it cannot. Darwinists claim that dinosaurs evolved into birds, but the evidence does not support that claim. They also claim that the mammal-like reptiles evolved into mammals, and although the stratigraphic order of the fossils is roughly consistent with that theory, the fossils do not show a clear transition.

We have also taken a look at early man and seen that the Darwinian version of man's origins determines what data are sought and how those data are interpreted. The biblical view of early man is the polar opposite of the Darwinian view. Pre-Flood humanity was markedly superior to present-day humanity in longevity, physical stature, and intelligence. Sites such as Stonehenge indicate that early post-Flood humanity achieved remarkable sophistication in the centuries immediately after the Flood, but then degenerated to a lower cultural level.

In this chapter, we examined Ellen White's mysterious statement that the worst sin committed by the antediluvians was the sin of "amalgamation," which "caused confusion everywhere." An amalgam is a combination of different elements. Amalgamation is the act of combining disparate elements. It seems that Ellen White was trying to tell us that the people who lived before the Flood combined the characteristics of different species.

Many dinosaurs share several characteristics with birds and several with reptiles, but one cannot mate a bird with a lizard and produce offspring. Modern genetic engineering has demonstrated, however, that sexual reproduction is not necessary in order to transfer genes from one animal to another. Genetic information can be transferred from one organism to another by splicing the DNA. If we can genetically manipulate life, it stands to reason that the vastly superior antediluvians also learned the secrets of DNA.

So we arrive at an alternative theory of the dinosaurs—that the dinosaurs were the product of genetic engineering. The people who lived before the Flood created them by mixing the genetic elements of reptiles, birds, mammals, and probably a little of everything else, as well.[14]

Why did the antediluvians engage in genetic engineering, creating the amalgamated species? Because they could. If history teaches us anything, it is that we always use any technology we invent. The antediluvians were, at first, probably motivated by the same considerations that motivate modern genetic engineers: medical and scientific research. Once they became adept at amalgamation, however, they may have wanted to design animals as beasts of burden, or for their meat,[15] or for the sheer novelty of inventing new forms. We would be unlikely to find any fossils of the experiments, the failures, or the fits and starts. Only when the genetically modified creatures bred in the wild, in large numbers, would we be likely to find fossil remains.

Although some paleontologists and small children have a fondness for Mesozoic creatures, most people find them monstrous—much uglier and more frightening than modern animals. They seemed that way to the early nineteenth-century fossilist Thomas Hawkins, as well:

> The sheer ungodliness of these cannibalistic sea creatures, with their smiling jaws and monster appetites, beggared belief. As the geological enthusiast and amateur collector Thomas Hawkins pointed out, these were more like Satan's creatures: "armed with all the virility of Evil, instant sprung, Satan seems to have been thrice seized, generating Horrors and realizing a teeming Spawn fitted for the lowest abysm of Chaos!" Why would God create such hideous monsters?[16]

Although we do not know exactly what dinosaurs and the other Mesozoic creatures looked like as living animals, their fossils convey an overall impression of ugliness and ungainliness different from modern animals. The carnivorous Mesozoic reptiles typically had outsized jaws full of serrated teeth. Compared to modern carnivores—with the scary exceptions of sharks and crocodiles—the Mesozoic vertebrates seem to have had an excessive killing ability. It is easier for me to view these creatures not as God's creations but as the products of progressively more radical genetic engineering by the hopelessly sinful antediluvians.

There was a class of heavily armored dinosaurs, including the ankylosaurs and nodosaurs, that had thick, bony plates all along the top of their bodies, and were essentially invulnerable except from below. These are referred to as armored dinosaurs, and sometimes "fighting dinosaurs." *Ankylosaurus* had a specially designed bony club on the end of its tail, rather like a medieval mace. Watching animals fight—any animals, including roosters, dogs, and bears—has been a spectator sport at many times and in many cultures. It is an ugly aspect of fallen human nature. Ellen White tells us that the antediluvians, "loved to destroy the lives of animals."[17] It would not be surprising if the fighting dinosaurs were designed to fight each other for the amusement of spectators.

The dinosaurs lived lives of extraordinary violence. Most skeletons have broken bones that healed prior to the animal's ultimate death and burial. The famous *T. rex* "Sue" had many injuries to her legs, tail, ribs, and skull. Sue's skull has a series of perforations, caused by an attack by another large predator, or by a bone disease, or by abscesses that ate through her jawbone. She had multiple broken ribs on both sides, probably from two separate incidents. The ribs had tried to heal, leaving ugly, swollen bone growth. She had an abnormality of the left leg that has been variously interpreted as a healed break, arthritis, or a long-term infection resulting from a major wound. Sue must have lived a life of violence, pain, and suffering.

"Big Al," an *Allosaurus* found in the Morrison Formation of Wyoming, had nineteen skeletal abnormalities, including numerous fractured and healed bones, infections to bones of the hands and feet, and perhaps even the bone marrow infection osteomyelitis.[18] Broken and rehealed bones are common among predatory dinosaurs, but many herbivores show evidence of injury, as well:

> Most dinosaur skeletons seem to have rehealed bone injuries, ranging from broken fingers, toes and ribs to broken leg bones that must have been very painful to the living animals. One hadrosaur had a broken jaw that caused some messy problems in the growing

> teeth, but the animal obviously survived for a long time after it had the injury. Injuries at the backs of young horned-dinosaur skulls could cause neck crests to grow in strange, asymmetrical ways. Shallow wounds sometime got infected, and abscesses would invade the adjacent bones to leave great gaping holes. There are several specimens of plant-eaters that appear to have been attacked and bitten to the bone by carnivores, but escaped and lived long enough for the bones to heal.[19]

Clearly, the dinosaurs lived in a violent world.

A couple of observations are in order. First, God would not have created the dinosaurs in this violent, disease-prone condition and called His creation "good." It should be beyond argument that the dinosaurs lived after predation, suffering, and death had come into the world through sin. Since the dinosaurs clearly lived after the Fall of man, they must been buried during the Genesis Flood. People who say otherwise are envisioning a god very different from the loving God of the Bible.

Second, the violence evidenced by the dinosaur skeletons is a remarkable verification of the Genesis narrative:

> Now the earth was corrupt in God's sight and was full of violence. God saw how corrupt the earth had become, for all the people on earth had corrupted their ways. So God said to Noah, 'I am going to put an end to all people, for the earth is filled with violence because of them. I am surely going to destroy both them and the earth.'" Gen. 6:11-13 (NIV).

As Randall Younker points out, scholars have shown that the Hebrew word translated in this passage as "violence" implies not only human violence but also the shedding of blood by animals.[20] One of God's reasons for destroying the earth with a flood was because the earth was full of violence, including animal violence, which was becoming ever more serious.[21] This violence is clearly evident in the skeletons of the dinosaurs.

Genesis 6 tells us not only that (1) man was wicked, and (2) the earth was filled with violence, but also that (3) the earth was filled with violence, including animal violence, *because of man.* This could merely mean that the animal violence was a direct result of Adam's fall. But what if man did something, beyond falling into sin, to exacerbate the problem of violence in nature? The KJV reads: "And God looked upon the earth and behold it was corrupt . . ." The Hebrew word translated as "corrupt" means "ruined," which conveys the idea that man had ruined the earth. What if man's tinkering with nature through genetic engineering had caused the animals to become more predatory and violent, thus effectively ruining the earth?

## C. THE EXPLANATORY POWER OF THE AMALGAMATION THEORY

Scientists use the phrase *explanatory elegance* to describe a theory's ability to explain a large amount of data in a simple manner.[22] The amalgamation theory of the dinosaurs has tremendous explanatory elegance within the biblical creationist paradigm. The amalgamation theory can explain facts that are difficult to explain outside of the Darwinian paradigm.

The amalgamation theory explains why the dinosaurs are such obviously heterodox creatures, as Edward Hitchcock discerned merely by examining their tracks. It explains why, although the dinosaurs are classified as reptiles, many dinosaurs appear to be closer to birds than to reptiles. It explains why the "bird-hipped" *Triceratops* appears to be mostly mammal and reptile, but with a parrot-like beak. It explains why the dinosaurs appear to be thermal intermediates, rather than following the warm-blooded or cold-blooded pattern of modern animals.

The amalgamation theory explains why the other extinct Mesozoic vertebrates, such as pterosaurs and ichthyosaurs, also appear to incorporate characteristics of several different classes of animals. It explains why *Archaeopteryx* appears to be part bird and part reptile. It explains why the mammal-like reptiles incorporated both mammalian and reptilian characteristics. No honest biblical creationist can deny the mixed nature of *Archaeopteryx* and the mammal-like reptiles, but that mixed nature is difficult to explain outside of the Darwinian paradigm, *except by reference to the amalgamation theory.*

The amalgamation theory explains why all the heterodox creatures of the Mesozoic appear to be at least part reptilian—such that the Mesozoic has been called "the age of reptiles"—and why many also have prominent avian characteristics. Birds and most reptiles lay eggs, and thus gestation and development takes place outside the mother's body. Common sense indicates that an embryo developing outside the mother's body would be easier to genetically modify. This is probably the reason the antediluvians included so much bird and reptile in their creations.

The amalgamation theory also solves a serious problem confronting the biblical creationist model of earth history: there are no remains of dinosaurs, pterosaurs, marine reptiles, or mammal-like reptiles in any layer of rock deposited after the Flood. If every animal that lived before the Flood was brought on board the ark, how is it that no remains of these hundreds of species are found in any post-Flood rocks?

Creationists often assert that the dinosaurs were brought aboard the ark but, because the post-Flood climate was much more severe, failed to survive for long after the Flood, and failed to leave any trace of themselves in post-Flood strata. They argue that although conditions for fossilization were ideal during the Flood, fossilization is relatively rare in the post-Flood world. For example, millions of buffalo once roamed the plains of North America, but they were slaughtered, almost to the point of extinction, during the late nineteenth century. Typically, their carcasses were left where they were killed. Yet we do not find the fossils of these slaughtered millions of buffalo, because conditions were not right for their preservation.

The problem with this argument is that we find numerous fossils of species that apparently survived the Flood and subsequently became extinct. The most famous of these are the woolly mammoth and the saber-tooth cat. Their remains are found in large numbers in Ice Age sediments, and in association with human fossils. The rate of erosion and deposition of sediment, and the burial and fossilization of animals, did not stop suddenly after the Flood, but gradually settled down. Had dinosaurs been taken aboard the ark and reproduced after the Flood, we would expect to find at least some of their remains in post-Flood strata. That no such remains have ever been found is compelling evidence that the dinosaurs were not taken aboard the ark and did not survive the Flood.

It strains credulity to assert that the dinosaurs, mammal-like reptiles, and flying reptiles—hundreds of genera—all lived and died without leaving any remains in post-Flood strata. Biblical creationists have not typically drawn attention to this problem, but it is a problem with our model. Ellen White's inspired insight that "the confused species which God did not create, which were the result of amalgamation, were destroyed by the flood" solves this problem.

The amalgamation theory may also help explain why some sea creatures survived the Flood and some did not. Pairs of land-dwelling animals and birds were taken on the ark, but sea-dwelling creatures were left to survive the Flood as best they could. Thus, the extinct marine reptiles, such as ichthyosaurs and plesiosaurs, must have shared the pre-Flood seas with non-extinct marine mammals such as whales and dolphins.[23] Why is it that marine mammals, such as the cetaceans (whales, dolphins, and porpoises) and pinnipeds (seals, sea lions, and walruses) survived the Flood, but the marine reptiles (ichthyosaurs, plesiosaurs, and mosasaurs) did not survive?[24]

There are several possible explanations. Perhaps the marine reptiles inhabited an area of the antediluvian seas separate from the marine mammals, and during the Flood, the former were caught in catastrophic killing zones that the latter managed to avoid. But it would seem that dolphins and ichthyosaurs inhabited the same environmental niche. They had the same body style, both ate fish, both had to surface to breathe, and both gave birth to live young, tail first. Ichthyosaur fossils are found all over the world, and dolphins are found in all of the world's seas. An ecological zonation theory does not seem an adequate explanation for the differential survival of these two groups.

Another naturalistic suggestion is that changes in the temperature and/or salinity of the seawater during the Flood were fatal to the marine reptiles, but not to the marine mammals. As previously discussed, current creationist modeling suggests that the sea became much warmer during the Flood, and ultimately, after the Ice Age, much colder than the antediluvian seas. Arguably, the endothermic (warm-blooded) marine mammals were able to survive these swings in water temperature, whereas the possibly ectothermic (cold-blooded) marine reptiles were not.

While it is possible to hypothesize naturalistic explanations for why marine mammals survived and the Mesozoic marine reptiles did not, I believe that the marine reptiles were selected for destruction because they were some of the "confused species" that God did not create. God supernaturally intervened to preserve the marine mammals and ensure the destruction of the marine reptiles.

The amalgamation theory may also help explain faunal succession in the rock strata. Somewhat counter-intuitively, we find the fossils of land vertebrates in a fairly orderly succession. Creationists explain this order on the basis that the animals were overtaken and buried by floodwaters according to the location, and primarily the elevation, of their ecological niches. This is known as the "ecological zonation" or "biome succession" theory. It explains why we find seafloor-dwelling creatures buried first, then fish, then amphibians, and then reptiles, including the dinosaurs.

We find very few large mammals in the higher Mesozoic strata, despite the fact that these strata must correspond to pre-Flood ecological niches that were well inland. The fossils of most of the larger mammals are found only in the Cenozoic strata, which are thought to correspond to the mountains and highlands of the pre-Flood world.

This pattern can be explained if the dinosaurs controlled the lowlands, while the mammals inhabited higher ground. The fur covering and higher metabolism of the mammals would have enabled them to thrive in the colder temperatures of higher altitudes. The dinosaurs, on the other hand, with scaly hides and a reptilian metabolism—well, no one knows what kind of metabolism they had, but it probably was not as high as true warm-blooded mammals—controlled the lowlands. The carnivorous dinosaurs like *T. rex* could easily have hunted to extinction any mammals that did not flee to higher altitudes where the dinosaurs preferred not to go.

Ellen White's description of the animals supernaturally being led to the ark provides support for this explanation: "Beasts of every description, the fiercest as well as the most gentle, were seen coming from mountain and forest and quietly making their way toward the ark."[25] They were *not* seen coming from plains, grasslands or lowlands, but rather from "mountain and forest." "Mountain" obviously refers to higher altitudes, but "forest" might also refer to highlands, because the pre-Flood mountains were forested, not bare and jagged like modern mountains.[26] Could it be that the animals to be preserved on the ark were seen coming from "mountain and forest" because that is where the dinosaurs and other amalgamated creatures that controlled the lowlands had driven them?

Even in the Tertiary rocks, however, the fossils of modern forms are rare. The fossil mammals of the Tertiary rocks, especially the lower Tertiary levels of Paleocene, Eocene, and Oligocene, are very strange and mostly extinct. Modern mammals are found mostly in the Quaternary—the Pleistocene and Holocene/recent.

A detailed examination of extinct Tertiary animals is beyond the scope of this book, but it may be that some of the strange Tertiary creatures were also products of amalgamation. We find obviously heterodox creatures in the seas, in the air, and in the lower altitude inland niches of the pre-Flood world. There is no reason to suppose that the antediluvian genetic engineers held mammals sacred. One possible example of amalgamated mammals are the *Amphicyonids* ("in-between dog"), the "bear dogs" of the Tertiary—creatures that appear to be part bear and part dog. Their remains are found in Eocene, Oligocene, and Miocene strata. Anatomically, the bear dogs (or "beardogs" or "bear-dogs") ranged from small and dog-like to large and bear-like, but they are neither dog nor bear, and they are not thought to be the forebears of either dogs or bears.[27]

The fact that we do not find many modern mammals until the upper Tertiary (which may have been deposited after the Flood) and the Quaternary (most of which probably was deposited after the Flood) raises an inference that these animals may have been almost entirely missing from the pre-Flood world. This could be because the amalgamated species had driven some of the created "kinds" almost to extinction. Ellen White's statements imply that the introduction of amalgamated species led to a severe derangement of the created order. They "defaced the image of God and caused confusion everywhere." Amalgamation was the "one sin above another which called for the destruction of the race by the flood."

It is not going too far beyond the light of inspiration to infer that the "confused species" had come to dominate most ecological niches, including the higher-altitude niches corresponding to the Tertiary, especially the lower Tertiary. Had God not brought the Flood when He did, the confused species might have driven the created kinds completely to extinction. Representatives of the originally created kinds did survive until the time of the Flood, however, and God supernaturally guided these animals onto the ark.

## D. WERE HUMANS AMALGAMATED WITH ANIMALS?

Ellen White wrote of "amalgamation of man and beast." Were humans and animals mixed or amalgamated in some way? Probably, the answer is yes. Right from the beginning of modern genetic engineering, human genes have been mixed with animal genes. Human genes have been placed into farm animals and microbes in order to manufacture medically useful substances. There have been efforts to mix crocodile hemoglobin with human hemoglobin. Researchers have inserted a human marker gene into a rhesus monkey.[28] Scientists have also successfully inserted a human gene into a pig embryo.[29] In the future, more cloning, stem cell research, and transgenic alteration of the human genome will take place, as the medical benefits derived from this type of research become manifest and overwhelm the ethical concerns.

Mixed creatures such as chimeras and griffins are prominently featured in ancient mythology. On a trip to the Getty museum in Los Angeles, I was struck by how pervasive is the griffin—a mythical beast that was part eagle and part lion—in ancient Greek art and artifacts. Combinations of humans and animals are also common in ancient mythology, which features centaurs (half horse), harpies (half bird), minotaurs (half bull), satyrs (half goat), and mermaids (half fish).

One wonders why human/animal combinations are so common in mythology. Could it be that the ancients had a dim collective memory of a time when human and animal combinations existed? Perhaps Noah and his children told stories of human/animal combinations that existed before the Flood, and, over the millennia, elements of these stories worked their way into the mythologies of the ancient nations. "With the revolution in recombinant DNA," writes Michio Kaku, "we have to re-analyze many of these ancient myths from an entirely different perspective. The ancient dream of being able to control life is gradually becoming a reality via the bio-molecular revolution."[30]

When we contemplate the possibilities inherent in gene splicing technology, and what might be done to alter the basic nature of humans and animals, we can well appreciate that such a practice would "deface the image of God and cause confusion everywhere."

---

[1] Ellen G. White, *Spiritual Gifts*, vol. III, p. 64. (*The Spirit of Prophecy*, vol. I, p. 69).

[2] Ellen G. White, *Spiritual Gifts*, vol. III, p. 75. (*The Spirit of Prophecy*, vol. I, p. 78).

[3] Clark's books include: *Back to Creationism* (1929), *Genes and Genesis* (1940), *The New Diluvialism* (1946), *Creation Speaks* (1947), *Nature Nuggets* (1955), *Skylines and Detours* (1959), *Wonders of Creation* (1964), *Crusader for Creation* (a biography of George McCready Price) (1966), *Genesis and Science* (1967), *Fossils, Flood, and Fire* (1968), *Battle Over Genesis, and New Creationism* (1980). Marsh's books include: *Fundamental Biology* (1941), *Evolution, Creation, and Science* (1944), *Studies in Creationism* (?), *Evolution or Special Creation?* (?), *Life, Man, and Time* (1967), and *Variation and Fixity in Nature* (1976).

[4] As an undergraduate at PUC, Harold Clark studied with George McCready Price, a Seventh-day Adventist creationist who published some twenty books in his long career. Price was a stalwart soldier in the trenches of early creationism who, virtually alone for over thirty years, pioneered Flood geology. His writing style was acerbic and occasionally sarcastic, and is out of fashion today, although it was similar in tone to the writings of H.L. Mencken and other literary stylists of Price's day. Price, through his own books and through his influence on Byron C. Nelson, Alfred Rehwinkel, John Whitcomb and Henry Morris, was probably the most influential American creationist. Interestingly, at the time of the debate with Marsh, Clark had just broken ranks with his old mentor Price over the issue of order in the sedimentary strata. Based upon many instances of out-of-order strata in mountainous regions, Price argued that the geologic column was an illusion, a theory contradicted by the empirical facts. Clark had come to accept the reality of the order of the strata, based upon the testimony of oil field geologists and the data from thousands of well cores. He came to believe that out-of-order strata were a result of tectonic activity and mountain building. Clark had just published, in 1946, *The New Diluvialism*, in which he explained the order in the fossil record on the basis of ecological zonation.

[5] White, *Spiritual Gifts*, vol. III, p. 60.

[6] Harold Coffin echoed this argument in his book *Origin by Design* (Hagerstown, MD: Review and Herald, 1983), p. 20. ("Since Genesis One does not declare that animals of diverse kinds cannot interbreed or cross, we ought to keep our eyes and minds open to the possibility that in the pre-Flood world crossing between more diverse kinds may have occurred on a greater scale than today. To deny emphatically that it could have happened in the past because it does not take place now is to stand dogmatically on the principle of uniformity.")

[7] Marsh had something in common with Clark besides being an Adventist science teacher: both had studied under George McCready Price. Marsh had taken geology from Price at Emmanuel Missionary College and idolized the man he once called the "old warhorse."

[8] Numbers, Ronald L. ed. *Creationism in Twentieth-Century America* (New York & London: Garland Publishing, 1995), p. xviii (Numbers characterizes the exchange as "more than two months of nearly weekly exchanges").

[9] The story of the 1947 meeting is from Gordon Shigley, "Amalgamation of Man and Beast: What did Ellen White Mean?" *Spectrum* (the journal of the Association of Adventist Forums) 12(4):10-19 (1982). The biographical details of the lives of Harold W. Clark and Frank Lewis Marsh are from Ronald L. Numbers, *The Creationists* (New York: Alfred Knopf, 1992 (paperback: University of California Press, 1993)), pp. 105, 123-133. Marsh set out his views on the "amalgamation" issue in his 1941 book, *Fundamental Biology*. "In the first place stands the fact that in nature today, although often attempted in both directions, man and beast cannot cross and produce offspring. Scientific records state that no such cross ever occurred. . . . In the second place, plants or animals which are sufficiently different to constitute two Genesis kinds are also incapable of crossing. At least, it is very clear that they cannot produce fertile offspring. . . . Recent dictionaries agree that the meaning of amalgamation in the field of biology is the 'the coalescence or blending of races', and the crossing of the white and black races of man is given as an illustration." Marsh devoted most of *Fundamental Biology* to elucidating his interpretation of the amalgamation statements. Frank Lewis Marsh, *Fundamental Biology* (self published, 1941) in *Creationism in Twentieth-Century America, vol. 8: The Early Writings of Harold W. Clark and Frank Lewis Marsh*, Ronald L. Numbers, ed. (New York and London: Garland Pub. Inc., 1995), pp. 399-527.

[10] Creationists are exploring methods for determining which modern species are members of a single created "baramin," and the study of the created kinds has come to be called "baraminology." See, e.g., Wise, Kurt P. (1990) "Baraminology: a young-earth creation biosystematic method," in R.E. Walsh, ed., *Proceedings of the Second International Conference on Creationism. Volume II, Technical Symposium*, pp. 345-360. Creation Science Fellowship, P.O. Box 99303, Pittsburgh, PA 15233-4303; Frair, Wayne, "Baraminology: Classification of Created Organisms," *Creation Research Society Quarterly*, 37:2, p. 82-91 (September, 2000); Wood, Todd Charles, and David P. Cavanaugh, "A Baraminological analysis of subtribe Flaveriinae (Asteraceae: Helenieae) and the Origin of Biological Complexity," *Origins*, no. 52, p. 7 (2001) http://www.grisda.org/origins/52007.htm;

[11] See, e.g., Francis D. Nichol "Amalgamation: Ellen G. White Statements Regarding Conditions at the Time of the Flood," (adapted from his book *Ellen G. White and Her Critics*, pp. 306-322) on the web at *http://www.WhiteEstate.org/issues/amalg.html.*

[12] See chapter 12. Subsequent advances in genetics have shown that selective breeding cannot achieve unlimited modification. Nossal, G.J.V. and Ross L. Coppel, *Reshaping Life* (Cambridge, Melbourne: Cambridge University Press, 1985-1989), p. 105 ("there are inherent limits within a breed beyond which it is not possible to go.")

[13] As I understand the *Baramin* concept, animals that are members of the same created kind may be, and frequently are, no longer inter-fertile. Despite the fact that two species are not currently inter-fertile, other factors may indicate that they descended from a common ancestor and must have been inter-fertile in the past. Taking this aspect in isolation, one could argue that humans and apes, although no longer inter-fertile, may have been in the distant past, and thus that Clark's argument is not so far fetched (if applied only to humans and apes). The apes are definitely animals, however, and we know from Scripture that animals were created separately from humans, and only humans were created in the image of God. Gen. 1:24-26; 2:7. Thus, apes and humans were never part of the same created kind, and it is a fundamental tenet of *Baraminology* that species from different baramins are not inter-fertile. Hence, apes and humans were never inter-fertile.

[14] The amalgamation theory of the dinosaurs is far from original to this work. Probably since the amalgamation statements were written in the 1860s, many Adventists have believed that the dinosaurs were among the results of amalgamation. For example, in his 1969 book, *Creation—Accident or Design?* Harold Coffin stated that one explanation of the dinosaurs were that they were the result of amalgamation. (See, e.g., James L. Hayward, "Noah's Ark or 'Jurassic park'?" *Spectrum*, vol. 23, no. 2 (August 1993).) I remember hearing the amalgamation theory—interpreted as genetic engineering by the advanced antediluvians—discussed by one of my teachers when I was in elementary school in the 1970s. In informally polling Adventists on their views of the dinosaurs, I found that about 75 percent of those who had a theory on the dinosaurs held the amalgamation theory.

[15] "The people who lived before the flood ate animal food, and gratified their lusts until their cup of iniquity was full, and God cleansed the earth of its moral pollution by a flood." White. *Spiritual Gifts*, vol. IV, p 121.

[16] Cadbury, Deborah, *Terrible Lizard* (New York: Henry Holt, 2000 (published in England as *The Dinosaur Hunters*), p. 151.

[17] White, *Spiritual Gifts*, vol. III, pp. 63, 64 ("They loved to destroy the lives of animals.

They used them for food, and this increased their ferocity and violence, and caused them to look upon the blood of human beings with astonishing indifference.")

[18] See Laws, Rebecca R. "Allosaur trauma and infection: Paleopathological analysis as a tool for lifestyle reconstruction." *Society of Vertebrate Paleontology, 57th annual meeting, Chicago, IL, Abstracts of Papers,* v. 17, no. 3, p. 59A-60A.

[19] Currie, Phillip J. and Eva B. Koppelhus, *101 Questions about Dinosaurs* (Mineola, NY: Dover Publications, 1996), p. 36.

[20] Younker, Randall, *God's Creation* (Nampa, ID: Pacific Press, 1999), pp. 62, 73, citing a 1977 article by Professor Tikva Frymer-Kensky in *Biblical Archaeologist.* See also Younker, "Creation, Christ, and the Cross," *Ministry,* May 2003 ("Some old Testament scholars have persuasively argued that the violence (camas) that filled the earth prior to the Flood was concerned as much with animals as humans")

[21] Younker, *God's Creation,* p. 73 ("One of the reasons God had to destroy the earth by a flood was because of the increasing violence of animals. This violence was so great it included the spilling of blood, including human blood.").

[22] Wise, Kurt P., *Faith, Form, and Time* (Nashville, TN: Broadman & Holman, 2002), p. 278.

[23] Some creationists argue that the last Flood-laid level was the Cretaceous, and that the fossils found in the Tertiary and Quaternary reflect post-Flood speciation and diversification. It this view is correct, it could be argued that marine mammals—which do not enter the fossil record until the Tertiary—did not exist before the Flood, but evolved after the Flood. If so, the creative power of evolution is everything the Darwinists claim and much more, because this astonishing evolutionary transformation must have occurred within a few hundred years of the Flood.

[24] Some creationists have argued that the Japanese fishing trawler *Zuiyo-maru* pulled up the remains of a plesiosaur in 1977, but the better argument—as most now acknowledge—is that the carcass was that of a basking shark. Answers-in-Genesis ministry, for example, repudiates the notion that the find was a plesiosaur. *http://www.answersingenesis.org/home/area/magazines/docs/v24n2_moving_forward.asp.* See also Jerlström, Pierre G., "Live plesiosaurs: weighing the evidence," *Creation Ex Nihilo Tech. J.* 12(3):339–346, (1998) ; Jerlström, Pierre G., and Bev Elliot, "Letting Rotting Sharks Lie," *Creation Ex Nihilo Technical Journal,*13(2):83–87, (1999) *http://www.answersingenesis.org/docs/4216cen_m1999.asp#r1*; Kuban, Glen J., "Sea-monster or Shark? An Analysis of a Supposed Plesiosaur Carcass Netted in 1977," *Reports of the National Center for Science Education,* 17:3 (May/June 1997), pp. 16-28. *http://paleo.cc/paluxy/plesios.htm.* But see Goertzen, John, "New Zuiyo Maru

cryptid observations: Strong Indications it was a Marine Tetrapod," *Creation Research Society Quarterly*, 38:1 (June 2001). *http://www.creationresearch.org/crsq/articles/38/38_1/Cryptid.htm.*

[25] White, Ellen, *Patriarchs and Prophets*, p. 97.

[26] White, Ellen, *Patriarchs and Prophets*, p. 44 (". . . the hills and mountains were not abrupt and rugged, abounding in terrific steeps and frightful chasms, as they now do; the sharp ragged edges of earth's rocky framework were buried beneath the fruitful soil, which everywhere produced luxuriant growth of verdure. . . . The heights were crowned with trees more majestic than any that now exist.") *Spiritual Gifts*, vol. III, p. 33 ("The earth was not one extensive plain, but the monotony of the scenery was broken by hills and mountains, not high and ragged as they now are, but regular and beautiful in shape. The bare, high rocks were never seen upon them, but lay beneath the surface, answering as bones to the earth. . . . The hills, mountains, and very beautiful plains, were adorned with plants and flowers, and tall, majestic trees of every description, which were many times larger, and much more beautiful, than trees now are.").

[27] The modern canids are thought to have descended from a fox-like animal called Leptocyon, which is thought to have descended from Hesperocyon

[28] Adam, David, "Biotechnology: First GM [genetically modified] primate bred," Nature/science update, Macmillan Magazines, Ltd., January 12, 2001.

[29] "Cloning teams cross pig and human DNA," Times Newspapers, Ltd., October 8, 2000.

[30] Kaku, Michio, Visions, *How Science will Revolutionize the 21st Century* (New York: Doubleday/Anchor Books, 1997), p. 221.

CHAPTER 22

# Defending the Amalgamation Theory

This chapter will address some of the criticisms that will be leveled at the amalgamation theory. We will begin with some general objections. First, most people will assume that antediluvian man could not possibly have accomplished something that we are probably 100 years away from accomplishing. But how long has it taken us to reach our current level of science? Virtually everything that is known by modern Western civilization has been discovered—or rediscovered from the ancients—since A.D. 1400. It has taken us only 600 years to reach our current level of science and technology.

Moreover, most of what is known in the natural sciences has been learned in the past 300 years. Many scientific disciplines are even younger. Almost everything known about genetics, for example, has been learned since 1850. Let us assume that it will take another century before our ability to manipulate the genome has advanced to the point where we can design and build creatures like the dinosaurs by mixing, matching, and modifying genes from existing creatures. In that case, it will have taken us only 250 years to advance from no theoretical knowledge—*tabula rasa*, a blank slate—to astonishing mastery over nature.

Over 1,650 years elapsed between the Creation and the Genesis Flood. Is it reasonable to insist that the antediluvians could not have learned, in sixteen centuries, what we will have learned in three? To the contrary, it is sheer nonsense. But there are other biblical facts we need to consider.

First, before the Flood, people lived many times longer than we live now—the antediluvians typically lived 800 or 900 years. Imagine that all the great pioneers in the field of genetics[1] were still alive, still relatively young, and would contribute to science for several more centuries, and you can begin to appreciate how much faster scientific discovery might have proceeded in antediluvian times.

Second, we have no indication that there was more than one language before the Flood. According to Scripture, God confused the languages for the first time at Babel, after the Flood. Contemporary scientists must spend time learning foreign languages and translating scientific papers in order to stay abreast of work being done by scientists in other countries. Jack Horner, a famous dinosaur paleontologist, never finished his Ph.D., largely because of such academic requirements as French. But he later found, when he needed Armand de Ricqles' expertise in bone histology, that French would have been useful.[2] The antediluvians did not need to spend time on translation, and could concentrate solely on the substance of their investigations.

These antediluvian advantages—longevity and the unity of language—would have helped their science advance more rapidly than ours. These biblical facts alone indicate that the antediluvians might have taken less time to achieve the mastery of genetics that we will have achieved in a span of 250 years.

When the relevant Ellen White statements are also considered, it is beyond clear that the antediluvians were capable of the scientific facility imputed to them by the amalgamation theory of the dinosaurs:

Notwithstanding the wickedness of the antediluvian world, that age was not, as has often been supposed, an era of ignorance and barbarism. The people were granted the opportunity of reaching a high standard of moral and intellectual attainment. They possessed great physical and mental strength, and their advantages for acquiring both religious *and scientific* knowledge were unrivaled. . . . Could illustrious scholars of our time be placed in contrast with men of the same age who lived before the Flood, they would appear as greatly inferior in mental as in physical strength. As the years of man have decreased, and his physical strength has diminished, so his mental capacities have lessened. There are men who now apply themselves to study during a period of from twenty to fifty years, and the world is filled with admiration of their attainments. But how limited are these acquirements in comparison with those of men whose mental and physical powers were developing for centuries!

It is true that the people of modern times have the benefit of the attainments of their predecessors. The men of masterly minds, who planned and studied and wrote, have left their work for those who follow. But even in this respect, and so far as merely human knowledge is concerned, how much greater the advantages of the men of that olden time! They had among them for hundreds of years him who was formed in God's image, whom the Creator Himself pronounced "good"—*the man whom God had instructed*

> *in all the wisdom pertaining to the material world.* Adam had learned from the Creator the history of creation . . ."[3]

It must have been particularly valuable to antediluvian science to have Adam around for over 900 years, because Adam and Eve had studied "all the wisdom pertaining to the material world" at the feet of the Creator Himself:

> The holy pair were not only children under the fatherly care of God but *students receiving instruction from the all-wise Creator.* They were visited by angels, and were granted communion with their Maker, with no obscuring veil between. They were full of the vigor imparted by the tree of life, and their intellectual power was but little less than that of the angels. *The mysteries of the visible universe*—"the wondrous works of Him which is perfect in knowledge" (Job 37:16)—afforded them an exhaustless source of instruction and delight. The *laws and operations of nature,* which have engaged men's study for six thousand years, were opened to their minds by the infinite Framer and Upholder of all. They held converse with leaf and flower and tree, *gathering from each the secrets of its life.*[4]

> God is infinite, and the first people upon the earth received their instructions from the infinite God who created the world. Those who received their knowledge direct from infinite wisdom were *not deficient in knowledge.*[5]

The result of these inestimable advantages was that the antediluvians developed an extremely advanced civilization:

> In strength of intellect, men who now live can bear no comparison to the ancients. There [have] been more ancient arts lost than the present generation now possess. For skill and art those living in this degenerate age will not compare with the knowledge possessed by strong men who lived near one thousand years.[6]

Our scientific attainments have accumulated rapidly in the century since Ellen White wrote those words, but it may be that only now have we advanced to the point where we can begin to understand what the antediluvians were doing. Stonehenge stood for many centuries a total mystery. It was not until the 1960s that we were able to discern its purpose and fully appreciate the genius of its builders. Likewise, it might be that we only now understand the nature of pre-Flood amalgamation because we have just begun to discover the secrets of life at the molecular level and imagine what will be possible in the future.

But could it really be true that the antediluvians possessed the high technology necessary to do genetic engineering? Surely, it will be argued, the ancients had nothing as sophisticated as an electron microscope. There are two responses. First, how do we know they lacked electron microscopes? Antediluvian humanity and all its architecture, arts and technology were destroyed and buried without a trace by the Genesis Flood. If they had such equipment, we would be unlikely to find it, having found virtually nothing else from their civilization.

We ought to be chastened, moreover, by our experience of rediscovering other technologies of the ancients, which sometimes lay unrecognized in museum storage rooms. Perhaps the most frequently cited example is a set of twelve electric batteries found in the village of Khujut Rabu, near Baghdad, Iraq. The batteries date from no later than A.D. 640 and possibly from as early as 250 B.C. Each battery consisted of a pottery jar with an asphalt stopper. An iron bar protruded through the asphalt stopper into the jar, where it was encased in a copper cylinder. In 1938, a German archeologist named Wilhelm Konig recognized these objects as batteries.

Later, Willard F. M. Gray of the General Electric High Voltage Laboratory in Pittsfield, Massachusetts, built reproductions. Many replicas have since been made and tested. When filled with an electrolyte solution—which could have been made of grape juice—each battery could produce between one and two volts of electric power. If enough of the batteries were produced and wired together, they could have produced a strong current. Several archeologists have suggested that the batteries were used for electroplating—transferring a thin layer of metal onto another metal surface.[7]

British researcher and author Robert Temple has studied the ancients' use of optical technology, including lenses and telescopes. Visiting the museums of Egypt, Europe, and the Middle East, Temple cataloged over 450 ancient optical lenses.[8] When Austen Henry Layard was excavating at Nimrud in the mid-nineteenth century, he found a piece of crystal in the palace of Sargon II, who ruled Assyria from 722 B.C. to 705 B.C. This crystal lens

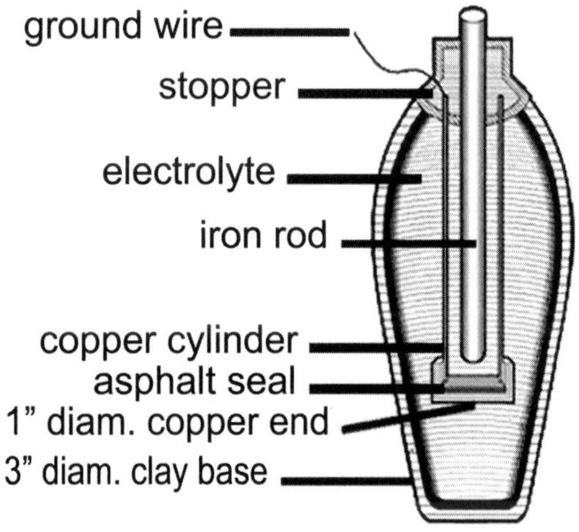


**Figure 22.1: the Baghdad Battery**

may have functioned as a monocle to correct for the king's astigmatism.[9]

Second, how do we know the antediluvians needed an electron microscope? Their genetic engineering techniques may not have mirrored ours. Remember the story of the Capitol Dome and the Goddess of Liberty: we do not know how the builders of the 1880s got the statue to the top of the dome, but they did not use a helicopter. The antediluvians may have discovered a way to segregate genes and transfer them from one species to another without having to see the DNA.

Finally, we do not know how good antediluvian eyesight might have been. One of the artifacts Robert Temple discusses is a carved knife handle from pre-dynastic Egypt, discovered at Abydos and dated to around 3,300 B.C. The figures carved into the handle cannot be seen without magnification.[10] Temple argues that, since the carvings cannot be seen without magnification, they could not have been carved without magnification either. But what if, when the handle was carved, the carvings could be seen without magnification? In other words, what if the eyesight of the antediluvians and the early post-Flood peoples was much better than ours?

The *Popol Vuh,* the sacred book of the ancient Quiche Maya, records that the first men . . .

> . . . were endowed with intelligence; they saw and instantly they could see far, they succeeded in seeing, they succeeded in knowing all that there is in the world. When they looked, instantly they saw all around them, and they contemplated in turn the arch of heaven and the round face of the earth. The things hidden [in the distance] they saw all, without first having to move; at once they saw the world, and so, too, from where they were, they saw it.[11]

In this passage, eyesight may be a metaphor for high intelligence and discernment, but it might literally mean that the first men had better eyesight and could see both far away and up close.

Ellen White made many statements to the effect that modern man is much degenerated compared to Adam and his nearer descendants. Our eyes have probably degenerated along with everything else. Beyond the fact that individual sets of eyes do not focus properly—a problem now routinely corrected by glasses, contacts, and laser surgery—the inherent capabilities of a perfectly functioning modern human eye may not approach those of an antediluvian human eye.

We now need to discuss some issues relating to Ellen White's writings. The amalgamation statements appeared in the work *Spiritual Gifts,* published in 1864—one of White's early works, considering that she lived until 1915 and produced an enormous literary estate. Most of White's early writings on biblical history were incorporated into the books of the "Conflict of the Ages" series, which includes *Patriarchs and Prophets* (1890), *Prophets and Kings* (1917), *The Desire of Ages* (1898), *Acts of the Apostles* (1911), and *The Great Controversy* (1888, revised 1911). These books are White's most mature and polished writings. Many of her earlier writings on Biblical history were re-written for inclusion in this series. For example, *Acts of the Apostles* incorporates most of an earlier book entitled *Sketches From the Life of Paul.* Likewise, *Patriarchs and Prophets* incorporates much of the history of the antediluvians and the Flood covered in volume III of *Spiritual Gifts.*

We would therefore expect the topic of amalgamation to appear in *Patriarchs and Prophets.* But it does not. There is nothing about amalgamation in *Patriarchs and Prophets.* This puzzling omission occasioned sufficient comment to warrant the following editorial statement, found in Appendix C (p. 452) of book 3 of *Selected Messages*:

> Regarding the two paragraphs which are to be found in *Spiritual Gifts* and also in the *Spirit of Prophecy* regarding amalgamation, and the reason why they were left out of the later books, and the question as to who took the responsibility of leaving them out, I can speak with perfect clearness and assurance. They were left out by Ellen G. White. No one connected with her work had any authority over such a question, and I never heard of anyone offering to her counsel regarding this matter. In all questions of this kind, you may set it down as a certainty that Sister White was responsible for leaving out or adding to matters of this sort in the later editions of our books.

One has to wonder whether, and to what extent, White changed her mind regarding amalgamation. We lack any statement from Ellen White on this issue, and no one else can answer it with certainty.

One possibility is that, while she never repudiated the truth revealed to her, she realized that she did not fully understand the concept of amalgamation[12] and that some of her co-religionists were misusing it to support racist theories. In order to appreciate this explanation, we need to consider a sentence in the second amalgamation passage that we have not previously quoted. The second passage reads, in its entirety:

> Every species of animal which God had created was preserved in the ark. The confused species which God did not create, which were the result of amalgamation, were destroyed by the flood. *Since the flood, there has been amalgamation of man and beast, as may be seen in the almost endless varieties of species of animals, and in certain races of men.*

The italicized sentence casts doubt upon our alternative theory of the dinosaurs. It seems unlikely that genetic engineering has taken place in post-Flood times (up until just the past few years). It would seem that by the time Noah's descendants had apostatized to the point where they would dare to dabble in genetic engineering—a sin that above all others caused God to destroy the antediluvian world—their intellectual and technical capabilities would not have been up to the task. Thus, it seems unlikely, although perhaps not impossible, that the early post-Flood peoples experimented with genetic engineering.[13]

More troubling, however, is the last clause of the last sentence, regarding "certain races of men." Ellen White did not identify any race or races as having resulted from amalgamation, but it is still fair to ask, was Ellen White a racist? At this juncture, we must suppress the widespread but grossly unfair impulse to judge an historical figure by the standards of our day, instead of by the standards of his own time. Ellen White lived in a time of deeply entrenched and universally accepted racism. During her adult life, a bloody civil war was fought over the issue of race-based slavery. A whole section of the United States had for years justified its "peculiar institution" on the basis that the African race was inherently inferior and fit only for slavery. Northerners, while generally opposed to slavery, held racial attitudes similar to, and often harsher than, those held by southerners.

Abraham Lincoln, for example, favored colonizing Blacks outside of the United States. In 1862 he asked Congress to pass a constitutional amendment to promote Negro emigration; on his own, he started a Negro colony in what is now Panama even as he was preparing to issue the Emancipation Proclamation.[14] In addressing a group of free Blacks whom he was trying to convince to emigrate to Panama, Lincoln very bluntly described the reality of the United States of the 1860s: "But even when you cease to be slaves, you are yet far removed from being placed on an equality with the white race. You are cut off from many of the advantages which the other race enjoys. The aspiration of men is to enjoy equality with the best when free, but on this broad continent, not a single man of your race is made the equal of a single man of ours. Go where you are treated the best, and the ban is still upon you."[15]

During Ellen White's life and long after her death, racism was frequently clothed with scientific justification. Darwinism gave scientific cover to racism. "Biological arguments for racism," wrote Stephen Jay Gould, "increased by orders of magnitude following the acceptance of evolutionary theory."[16] Darwinism has given aid and comfort to racism because Darwinists see racial diversification as a precursor of speciation. In the theory, some racial varieties become reproductively isolated from the larger gene pool, and eventually evolve into a new species.[17] Hence, one race may be closer than another to becoming a "higher" species. Darwinists viewed the African race as being closer to our putative simian ancestors, whereas they viewed the white race as being more advanced.

Darwin himself held this opinion. He believed that humanity had originated near the baboon, then evolved into something like the gorilla, then into the Negro, and finally into the Caucasian. Darwin even predicted racial genocide: "At some future period, not very distant as measured by centuries, the civilized races of man will almost certainly exterminate, and replace, the savage races throughout the world. At the same time, the anthropomorphous apes . . . will no doubt be exterminated. The break between man and his nearest allies will then be wider, for it will intervene between man in a more civilized state, as we may hope, even than the Caucasian, and some ape as low as a baboon, instead of as now between the Negro or [aborigine] and the gorilla."[18]

Darwinian racial theorizing of this type persisted into the 1920s. Henry Fairfield Osborn wrote in 1926: "The Negroid stock is even more ancient than the Caucasian and Mongolians, as may be proved by an examination not only of the brain, of the hair, of the bodily characteristics . . . but of the instincts, the intelligence. The standard of intelligence of the average adult Negro is similar to that of the eleven-year-old-youth of the species *Homo Sapiens*."[19] A strong case can be made that evolutionary thinking had a profound impact on twentieth-century European philosophy, and, as interpreted by Germans such as Ernst Haeckel, was responsible for the Nazi genocide.[20]

Compared to the learned men of science, Ellen White was far ahead of her time on the issue of race. "The Black man's name is written in the book of life beside the White man's," she wrote. "All are one in Christ. Birth, station, nationality, or color cannot elevate or degrade men."[21] "Those white people who appreciate the ministry of Christ in their behalf cannot cherish prejudice against their colored brethren."[22] She was ashamed of the treatment that Blacks received at the hands of Christians in America: "How little of the spirit of Christ has been manifested in the treatment given to the colored race in this so-called Christian country! . . . They are often treated as if it were a disgrace to sit by their side, or even to worship in the same congregation."[23]

Ellen White wanted to see more done on behalf of Blacks and was critical of the Seventh-day Adventist Church for failing to do more: "For many years I have borne a heavy burden in behalf of the Negro race. My heart has ached as I have seen the feeling against this race growing stronger and still stronger, and as I have seen that many Seventh-day Adventists are apparently unable to understand the necessity for an earnest work being done quickly."[24] She wanted to train blacks for gospel work: "Among the Negro race there are many who have talent and ability. Let us search out these men and women, and teach them how to engage in the work of saving souls."[25]

Considering the times in which she lived, Ellen White was astonishingly sensitive on the issue of race. She was no racist. To the contrary, she was greatly troubled by the racism she saw preventing the education and uplift of the former slaves. She would not have wanted her followers to use the clause, "and in certain races of men," to support a view that some races were inferior to others. There was a very real danger that this would happen. Not long after White published the amalgamation statements, some began to speculate that black African tribes and some Native-American Indian tribes, were the result of hybridization between humans and apes.[26] But Ellen White never endorsed that view, and it would be wrong to impute it to her.

It was probably the race issue that led White to omit any mention of "amalgamation" in later publications. She would not have wanted anything she had written to contribute to the growing problem of racism. Rather than risk fueling racist speculations, she probably thought it better not to mention amalgamation again. Keep in mind that in 1890, when *Patriarchs and Prophets* was published, the climate of racism in the United States was considerably worse than it had been during the civil war,[28] and it did not begin to improve until the 1930s. It was not until the 1960s that African-Americans were afforded full civil rights and, at least on paper, legal equality with whites.

We know that Ellen White was concerned that her writing not offend unnecessarily. "In several places," in the 1911 edition of *The Great Controversy*, "forms of expression have been changed to avoid giving unnecessary offense. An example of this will be found in the change of the word 'Romish' to 'Roman' or 'Roman Catholic.'"[28] Thus, it is likely that the "amalgamation" statements were omitted from later books to avoid racist implications, and not because Ellen White had not been inspired when she wrote of amalgamation having taken place.[29]

But let us return to the issue of what Ellen White could have meant in writing that:

> Since the flood, there has been amalgamation of man and beast, as may be seen in the almost endless varieties of species of animals, and in certain races of men.

We were noting that it seems unlikely that post-Flood man experimented with genetic engineering. What might Ellen White have been referring to?

At the outset, we should note that she used the term *amalgamation* in more than one context, apparently to mean different but related things. In addition to using the term to describe the antediluvians' greatest sin, White also used it to describe Satan's biological modifications of the fallen world:

> Christ never planted the seeds of death in the system. Satan planted these seeds when he tempted Adam to eat of the tree of knowledge, which meant disobedience to God. Not one noxious plant was placed in the Lord's great garden, but after Adam and Eve sinned, poisonous herbs sprang up. In the parable of the sower the question was asked the master, "Didst not thou sow good seed in thy field? From whence then hath it tares?" The master answered, "An enemy hath done this" (Matt. 13:27, 28). All tares are sown by the evil one. Every noxious herb is of his sowing, and by his ingenious methods of amalgamation he has corrupted the earth with tares.[30]

Satan's "amalgamation" that produced noxious herbs and tares is not the amalgamation that was the "one sin above another which called for the destruction of the race by the flood." God would not punish the antediluvian people for Satan's work, over which the human race could have had no control. Based solely upon our knowledge of God's character, we can rule out the possibility that the "amalgamation" that hastened God's judgment in the Flood was Satan's work of making noxious herbs and tares.

So there were at least two separate instances of "amalgamation"—(1) what the antediluvians did and (2) what Satan did. In both cases, amalgamation was a type of biological modification, apparently involving mixing genetic elements.

The amalgamation that occurred after the Flood may have been yet a third instance, slightly different from the other two. Regarding post-Flood amalgamation, Ellen White does not state who did it. In fact, *she does not state that anyone did it.* She wrote only that "there has been" amalgamation after the Flood, and that it led to (1) an "almost endless variety of species of animals," and (2) the racial diversification of humanity. There is no question that these two things happened.

First, racial diversification occurred after the Flood. The Bible gives no indication that there was more than one race or language prior to the Flood, and only one family, Noah's, survived the Flood, so the diversification of humanity into several races must have occurred after the Flood.

Second, something caused an astonishing diversification in the plant and animal kingdoms after the Flood. It is unfortunate that, in times past, Christians taught that the species were fixed and unchanging, and that they were all created just as we see them today. The notion of fixity or immutability of species is not taught in Scripture; it entered Christian thought from Greek philosophy (Aristotle), via the writings of Augustine.[31] The evidence Darwin marshaled in *Origin of Species* demolished the notion of the fixity of species. Darwin showed that species have in fact changed over time. In some cases there has been true speciation, in the sense that animals can no longer interbreed with similar types with whom they apparently share a common ancestor. All responsible creationists admit this.[32]

This process of speciation or diversification has led to an "almost endless variety of species of animals," just as Ellen White stated. For example, more than 800,000 species of insect have been described, and entomologists estimate that there are at between five and ten million living insect species. Among the insects, there are at least 250,000 species of beetles. There are 4,625 species of mammals, distributed among 24 orders and 125 families. There are 8,700 living species of birds, and 6,000 living species of reptiles.

No creationist argues that every one of these species was originally created by God in its present form and brought aboard Noah's ark. There was a period of rapid speciation immediately following the Flood.[33] Some creationists believe that as few as 2,000 pairs of animals were brought on board the ark and that all modern land animals and birds descended from these.[34] There must have been a great deal of post-Flood diversification and speciation of plants and animals.

Creationists do point out that the extent of provable change is limited. There may be 250,000 species of beetles, all of which may have descended from a few ancestral pairs, but they are all still beetles. They did not evolve into some different life form. Moreover, the species concept is not as clear-cut as most people assume. Many different species are inter-fertile and, by analogy to mankind, might be considered mere races rather than distinct species. For example, there are thirteen species of finch on the Galapagos Islands, but they can and do interbreed.[35]

The fact that the species concept does not correspond to the created kinds, or *baramins*, as Frank Lewis Marsh called them, has led creation scientists to develop the discipline of *baraminology*.[36] By studying factors such as the fossil record, inter-fertility, the number of chromosomes, and DNA sequences, creationist scientists hope to determine which species are all descended from a single, originally created kind, or *baramin*. Obviously, modern creationists do not argue for the fixity of species, but believe that a genus or even an entire family of animals could have descended and diversified from a single baramin.

What genetic mechanism facilitated this rapid post-Flood diversification? Most scientists, including many biblical creationists, assume that the mechanism driving this sort of "micro-evolution" is natural selection acting upon genetic mutations, i.e., the copying error theory of evolution. But as previously discussed, the copying error theory can explain only degenerative change. Occasionally, a DNA copying error has a fortuitous side effect for the organism concerned—as when a mutation prevents DDT from binding to a specific site on the membrane of an insect's nerve cells—but the overall picture is one of degeneration. Mutations cannot add new genetic information, such that the organism would acquire new organs or new functionality. Thus, creationists have argued that species change has been entirely degenerative, entailing no increase in biological complexity.

Yet some of the diversification and species change that can be inferred from baraminological analysis has not been strictly degenerative.[37] In fact, much of it has been exquisitely adaptive and beneficial. For example, most creationists do not believe that the pre-Flood world contained arctic environments or experienced harsh winters with large areas covered with snow and ice—and the fossil record supports this belief.[38] Yet there are several types of mammals and birds whose fur or feathers turn white in the winter and brown in the summer—or even dappled, for the patchy spring thaw—an astonishingly intricate adaptation to the harsher post-Flood climate.[39] Some mechanism other than genetic copying errors must be at work. What could that be?

To answer that question, we must go back to the 1940s, when scientists were working to identify Griffith's "transforming principle," which turned harmless bacteria into the virulent kind that causes pneumonia. It was discovered that DNA from the virulent strains could travel to the harmless strains and transform the latter into disease-causing microbes. Thus, at the very moment science established that DNA carried the "genes," it also established that genes could be passed between organisms without sexual or asexual reproduction.

By the 1960s, scientists had discovered that microbes that were resistant to antibiotics were passing the resistance genes not only to their own lineal descendants, but to non-related microbes as well, by horizontal gene transfer. It eventually became obvious that, at least at the microbial level, a lot of DNA was being passed back and forth. During the 1970s, to defuse the controversy created by nascent genetic engineering, scientists argued that, by introducing foreign DNA into a bacterium via a *plasmid*, they were doing nothing different than what nature has always done.

In recent years, as scientists have mapped the genomes of various microbes, they have realized that microbes regularly pick up genes from organisms outside their own species, even from non-related species. In a study of the common intestinal bacteria *E. coli*, researchers estimated that nearly 20 percent of its DNA originated in other microbial species.[40]

The knowledge that microbes have been swapping genes has forced researchers to acknowledge that such lateral DNA transfers have played a large role in evolution. "I see [horizontal gene transfer] as one of the real major forces in evolution," says Gary Olsen of the University of Illinois, "It's totally changed how I think about evolutionary history."[41] Jeffrey Lawrence and Howard Ochman's study of DNA sequences convinced them that horizontal gene transfer paved the way for the creation of new bacterial species by endowing microbes with abilities that allowed them to exploit different environmental niches.

For example, the acquisition of the *lac operon*, a cluster of genes coding for proteins that metabolize lactose, allowed *E. coli* to live in the colons of mammals. "Horizontal gene transfer drives the diversification of bacterial species," concludes Ochman.[42] The pace of gene transfer today is impressive, but scientists believe that it pales in comparison to the pace of gene swapping in life's earliest days.[43]

Although it is easy to see how bacteria and other microbes composed of cells lacking nuclei (called *prokaryotes*) could swap DNA, scientists do not yet understand the mechanism by which higher organisms could horizontally transfer genes. A mechanism such as viral insertion seems workable with regard to prokaryotes, which have free-floating DNA, but less plausible with higher organisms whose genomes are encased within the membrane of a nucleus.[44]

Nevertheless, as researchers map the DNA of various plants and animals, they have learned that many species have sections of DNA that are almost identical to those found in different species. Hence, it seems clear that horizontal gene transfer is not limited to prokaryotes. "Certainly, there can be little doubt that most, if not all, genomes include a number of foreign genes . . ."[45] Plants and insects are known to have acquired genetic material by horizontal transfer. It has been argued that the diversification of the angiosperms (flowering plants) cannot be explained without recourse to horizontal gene transfer.[46]

Humans are not exempt from picking up genes from other species. During the Human Genome Project, scientists encountered 113 human gene sequences that were almost identical to DNA sequences found in bacteria. These appear to be the result of direct transfer from bacteria to humans.[47] (One study indicated that twenty-eight of these genes were also found in other vertebrates, but that does not negate the possibility that humans acquired these genes from bacteria.[48])

Darwinists have begun to embrace the idea that evolution is driven much more by the wholesale acquisition of genes than by genetic copying errors. "We believe random mutation is wildly overemphasized as a source of hereditary variation," write Lynn Margulis and Dorion Sagan. "Mutation accumulation does not lead to new species or even to new organs or new tissues. . . . Rather the important transmitted variation that leads to evolutionary novelty comes from the acquisition of genomes."[49] Margulis and Sagan argue that although point mutations are not generally useful to higher organisms, higher organisms can acquire useful traits by incorporating foreign genes into their own genomes. Foreign genes can be acquired by accumulating viruses, plasmids, or other short pieces of DNA, or by acquiring entire foreign genomes, sometimes as the end result of a symbiotic relationship with another species. The plant or animal can then pass these traits on to its offspring through its own, now modified, genome.[50]

Horizontal gene transfer has not escaped the notice of creationist researchers. Todd Wood, of Bryan College in Dayton, Tennessee, has recognized the potential of horizontal gene transfer to explain rapid post-Flood speciation. Wood argues that rapid post-Flood diversification within the created kinds occurred as a result of designed mobile DNA sequences he calls *altruistic genetic elements* (AGEs). Wood hypothesizes that AGEs are capable of jumping to different locations in their chromosomes, moving from the genome of one species to that of another, and activating or inactivating genetic potential already present in an organism's genome, perhaps in the "junk DNA"—i.e., the 97 percent of the genome whose function is so complex and non-obvious that scientists have yet to decode it. Wood's model could help explain many phenomena, including rapid diversification and speciation, pseudogenes, relic viral sequences, cancer-causing retroviruses, genetic throwbacks, and convergent evolution.[51]

Both creationists and Darwinists now acknowledge that horizontal gene transfer has played an important role in speciation. I predict that most of the evolutionary change acknowledged by creationists will eventually come to be understood as the result of horizontal gene transfer and other as yet undiscovered genetic mechanisms—not genetic copying errors.[52] Horizontal gene transfer will help creationists explain the extremely rapid post-Flood diversification and speciation that creationist models assume, but it will not help Darwinists support the larger claims of Darwinism. To make good on its larger claims, Darwinism must explain the original creation of new genetic information, and while horizontal gene transfer can help explain the diffusion of genetic information among species, it cannot explain the creation of new genetic information.

Horizontal gene transfer is similar to genetic engineering, but there are crucial differences. Genetic engineering is when people take DNA from one type of organism and insert it into the genome of another type. There are theoretically no boundaries on the extent of organic change that could be accomplished in this way. Horizontal gene transfer also involves moving DNA from one species to another, but it happens naturally, not by human intervention, and there appear to be limits upon the amount of genetic change it can accomplish.

But the main concept of both horizontal gene transfer and genetic engineering is moving DNA among and between different species. Before nucleic acid was discovered and long before DNA was known to be the carrier of genetic information, someone to whom these concepts were revealed might use the same word to describe both types of transfers of genetic information. Quite possibly, Ellen White used the same term, "amalgamation," to describe both the pre-Flood genetic engineering and the post-Flood genetic diversification.

We are now in a position to make sense of Ellen White's comment that, "Since the flood, there has been amalgamation of man and beast, as may be seen in the almost endless varieties of species of animals, and in certain races of men." Before the Flood, the antediluvians artificially combined characteristics of widely dissimilar species. That type of amalgamation was sinful and resulted in species that were combinations of entirely different classes of animals. It substantially deranged and confused the created order and defaced the image of God. That type of amalgamation was first among the reasons why God decided to wipe the slate clean with a worldwide deluge and start over with only the originally created kinds.

By contrast, the amalgamation that occurred after the Flood was a natural process in which mobile genetic elements moved among living organisms, accelerating speciation and diversification, perhaps turning on or off existing genetic potential, helping species to adapt to new ecological niches and altered climates, creating "almost endless varieties of species of animals," and promoting the racial diversification of humanity. This process is still going on today, but it occurred at a much faster rate in the centuries immediately following the Flood.

Post-Flood amalgamation helped plants and animals exploit new environmental niches and adapt to the radically changed environmental conditions of the post-Flood world. The adaptive nature of post-Flood amalgamation can be seen in the racial diversification of humanity. Races that live in cold northern climates tended to be lighter-skinned, whereas races living near the equator developed dark skin as a protection against the sun. Post-Flood amalgamation was and is a natural process that was provided for by a benevolent Creator. It is not a sinful process, except to the extent that nature, like man, is fallen and bears the curse of the fall.

Thus Ellen White seems to have written of three types of organic "amalgamation." The first was Satan's work of introducing tares. After Adam sinned and Satan became the god of this world,[53] Satan used "ingenious methods of amalgamation" to modify God's creation, making predatory and poisonous plants and animals, making parasites, biting and stinging insects, pests of all descriptions, disease-causing microbes, noxious weeds and tares, etc. These were not part of the original creation that God declared "very good."[54] The second type of amalgamation was the genetic engineering practiced by the antediluvians, which produced the "confused species, which God did not create" which "were destroyed in the flood." The third type of amalgamation was the natural process of horizontal gene transfer that led to "almost endless varieties of species" and to the racial diversification of man.

Ellen White apparently used the term amalgamation to describe all genetic recombinations, whether manipulated or natural. The exact mechanics of molecular genetics would not begin to be understood for a century after Ellen White's time, and they are a long way from being fully understood today. But the genetic details were not important for Ellen White's purposes, only the broad contours.

The amalgamation theory of the dinosaurs is, of course, very speculative, but all theories about origins are speculative. All that is known about dinosaurs is what the fossils look like, where they are found in the ground, and what the rocks around them look like. Those things are facts. Everything else is speculation. The age of the rocks, the manner in which the rocks were formed, the time when the dinosaurs lived, and the manner in which the dinosaurs lived and died are all matters of speculation. The sciences that deal with origins are like a huge inverted pyramid of speculation resting upon a tiny patch of fact.

Paleontologists speculate along lines dictated by Lyellism and Darwinism. Their speculations are built upon assumptions so old and so taken for granted that most scientists do not even realize that they are assumptions. If scientists can make guesses about the dinosaurs based upon the assumed truth of Lyellism and Darwinism, we should be allowed to speculate based upon the assumed truthfulness of Scripture and the inspired writings of Ellen G. White.

---

[1] Mendel, Punnett, Bateson, Correns, Tschermak-Seysenegg, de Vries, Flemming, Boveri, Sutton, Stevens, Wilson, Morgan, Sturtevant, Bridges, Shull, Garrod, Miescher, Levene, Beadle, Tatum, Avery, Hershey, Muller, Franklin, McClintock, Watson, and Crick, to name some famous geneticists.

[2] Horner, John R., and James Gorman, *Digging Dinosaurs* (New York: Workman Pub., 1988), p. 14, 167-186.

[3] White, *Patriarchs and Prophets,* pp. 82-83 (italics mine).

[4] White, *Patriarchs and Prophets,* pp. 50-51 (italics mine).

[5] White, *Spiritual Gifts,* vol. IV, p. 154 (italics mine).

[6] White, *Spiritual Gifts,* pp. 155, 156.

[7] Noorbergen, Rene, *Secrets of the Lost Races* (New York: The Bobbs-Merrill Company, Inc., 1977), pp. 48-50; Childress, David Hatcher, *Technology of the Gods: The Incredible Sciences of the Ancients* (Kempton, IL: Adventures Unlimited Press, 2000), pp. 118, 119; "Riddle of 'Baghdad's Batteries'" BBC News, February 27, 2003 http://news.bbc.co.uk/1/hi/sci/tech/2804257.stm.

[8] Temple, Robert, *The Crystal Sun* (London: Random House/Arrow Books, Ltd., 1999), pp. 4, 5.

[9] Temple, at pp. 10-49 and plates 34 to 37.

[10] Temple, at pp. 120-123 and plate 38.

[11] *Popol Vuh, The Sacred Book of the Ancient Quiche Maya,* English version by Delia Goetz and Sylvanus G. Morley from the Translation of Adrian Recinos (Norman, OK: University of Oklahoma Press, 1950), part III, ch. 2 (p. 168).

[12] Because DNA and its function were unknown when Ellen White wrote these statements, she could not possibly have understood the mechanics of "amalgamation." But her inability to exactly understand the amalgamation vision does not detract from her prophetic authority. The apostle Paul says, "For we know in part and we prophesy in part, but when perfection comes, the imperfect disappears. . . . Now we see but a poor reflection as in a mirror; then we shall see face to face. (1 Cor. 13:9-10, 12). Reportedly, her son Willie White once said, "oftentimes when we go to mother and ask her to explain the things she has written, she will say, "I cannot explain it; you should understand it better than I. If you do not understand it, pray to the Lord and He will help you." W.C. White, "The Integrity of the Testimonies to the Church,"

remarks at College View, Nebraska, November 25, 1905, as recalled by F. C. Gilbert.

13 In support of post-Flood genetic engineering, there is evidence that the post-Flood ancients engaged in nuclear warfare. Noorbergen, at pp. 137-150. See also Childress, at pp. 211-257. Moreover, some have suggested that certain fossil hominoid genera, such as *Australopithecus*, which appear to combine ape-like and human-like features, might have resulted from amalgamation. (See April 15, 1991 letter from Clyde L. Webster, Jr. to Rex DeGolyer.)

14 See, e.g., *The Portable Abraham Lincoln*, edited by Andrew Delbanco (New York: Viking Press, 1992) pp. 51, 114, 234-239, 249, 261, 266-267.

15 "Address on Colonization to a Committee of Colored Men, Washington, D.C." *The Portable Abraham Lincoln*, p. 235.

16 Gould, Stephen Jay, *Ontogeny and Phylogeny* (Cambridge, MA: The Belknap Press of Harvard University Press, 1977), p. 127.

17 Ernst Mayr admits that Darwinists consider some geographic races to be incipient species. Ernst Mayr, *The Growth of Biological Thought* (Cambridge, MA: Harvard University Press, 1982), p. 548.

18 Darwin, Charles, *The Descent of Man* (London: John Murray, 1901), pp. 241, 242.

19 Osborn, Henry F., "The Evolution of Human Races," *Natural History*, January/February 1926, reprinted April 1980, p. 129.

20 Gould, *Ontogeny and Phylogeny*, supra, at pp. 77, 78 (". . . Haeckel's greatest influence was, ultimately, in another tragic direction—National Socialism [Nazism]. His evolutionary racism; his call to the German people for racial purity and unflinching devotion to a 'just' state; his belief that harsh, inexorable laws of evolution ruled human civilization and nature alike, conferring upon favored races the right to dominate others; . . ."); Bergman, Jerry, "Darwinism and the Nazi Race Holocaust," *Creation Ex Nihilo Technical Journal* 13(2):101-111 (1999), on the web at: http://www.answersingenesis.org/docs/4162.asp; Sarfati, Jonathan, "The Holocaust and Evolution," *Creation Ex Nihilo*, 22(1):4, 1999-2000, on the web at: http://www.answersingenesis.org/docs/4184cen_d1999.asp. See also Weikart, Richard, *From Darwin to Hitler: Evolutionary Ethics, Eugenics, and Racism in Germany* (New York: Palgrave Macmillan, 2004). (Weikart demonstrates that many leading Darwinian biologists and social thinkers in Germany believed that Darwinism overturned traditional Judeo-Christian ethics, especially the view that human life is sacred. Many of these thinkers supported moral relativism, yet simultaneously exalted evolutionary "fitness" [especially intelligence and health] to the highest

arbiter of morality. Darwinism played a key role in the rise not only of eugenics [a movement to control human reproduction in order to improve the human species] but also euthanasia, infanticide, abortion, and racial extermination. This was especially important in Germany, since Hitler built his view of ethics on Darwinian principles. Hitler was not amoral but upheld a consistent ethic based upon Darwinian foundations.)

[21] Ellen G. White, *Selected Messages*, 3 vols. (Washington, D.C.: Review and Herald Publishing Association, 1958, 1980), vol. 2, p. 488.

[22] Manuscript 107, 1908, p. 7 (*Manuscript Releases*, vol. 4, p. 33, par. 2.

[23] White, *Manuscript Releases* vol. 4, p. 8, par. 1.

[24] White, *Testimonies for the Church*, vol. 9, p. 204, par. 2. Several similar passages can be found: "Much might have been accomplished by the people of America if adequate efforts in behalf of the freedmen had been put forth by the Government and by the Christian churches immediately after the emancipation of the slaves. Money should have been used freely to care for and educate them at the time they were so greatly in need of help. But the Government, after a little effort, left the Negro to struggle, unaided, with his burden of difficulties. Some of the strong Christian churches began a good work, but sadly failed to reach more than a comparatively few; and the Seventh-day Adventist Church has failed to act its part." *Testimonies for the Church*, vol. 9, p. 205, par. 1. "Sin rests upon us as a church because we have not made greater effort for the salvation of souls among the colored people." *The Southern Work* (Washington, D.C.: Review and Herald Publishing Association, 1966), p. 15.

[25] *Testimonies for the Church*, vol. 9, p. 202, par. 2. Her burden for educating Blacks led to the founding of Oakwood College, on a site she approved. She had much to say on the subject: "One of the difficulties attending the work is that many of the white people living where the colored people are numerous are not willing that special efforts should be put forth to uplift them. When they see schools established for them, when they see them being taught to be self-supporting, to follow trades, to provide themselves with comfortable homes instead of continuing to live in hovels, they see the possibility that selfish plans will be interfered with—that they will no longer be able to hire the Negro for a mere pittance; and their enmity is aroused. They feel that they are injured and abused. Some act as if slavery had never been abolished. This spirit is growing stronger as the Spirit of God is being withdrawn from the world, and in many places it is impossible now to do that work which could have been done for the colored people in past years." *Testimonies for the Church*, vol. 9, p. 204, par. 3. See also Benjamin Baker, "The Woman who Wouldn't be Quiet," *Adventist Review*, February 21, 2002. http://www.adventistre-view.org/2002-1508/story2.html.

26 Unfortunately, Uriah Smith once "defended" the amalgamation statements by arguing that "such cases as the wild Bushmen of Africa, some tribes of the Hottentots, and perhaps the Digger Indians of our own country, etc." might have resulted from human-animal crosses. See Gordon Shigley, "Amalgamation of Man and Beast: What did Ellen White Mean?" *Spectrum* (the journal of the Association of Adventist Forums) 12(4):10-19 (1982).

27 See, e.g., George R. Knight, *Ellen White's World* (Hagerstown, MD: Review and Herald, 2005), pp. 118-119.

28 White, Arthur L., *Ellen G. White: The Later Elmshaven Years* (Hagerstown, MD: Review and Herald, 1982), p. 326. Willie White also indicated that his mother had changed wording to avoid needless offense: "Regarding changes in forms of expression, mother has often said: 'essential truths must be plainly told; but so far as possible they should be told in language that will win, rather than offend.'" W.C. White letter, July 25, 1911, as quoted in *Selected Messages, Book Three,* Appendix A, p. 444.

29 Another reason may be that *Patriarchs and Prophets* was intended for a wider audience than *Spiritual Gifts* and the *Spirit of Prophecy,* and with the meaning of "amalgamation" unresolved (and at that time un-resolvable) Ellen White may have thought it best to omit them. We find an example of this type of concern in the preparation of *The Great Controversy,* as reported by White's secretary Marian Davis, to Willie White:

"In *Great Controversy,* volume iv, published in 1885, in the chapter "Snares of Satan," there are three pages or more of matter that was not used in the later editions, which were prepared to be sold to the multitudes by our canvassers. It is most excellent and interesting reading for Sabbathkeepers, as it points out the work that Satan will do in persuading popular ministers and church members to elevate the Sunday Sabbath, and to persecute Sabbathkeepers. [currently found in *testimonies to ministers,* pp. 472-475.]

It was not left out because it was less true in 1888 than in 1885, but because mother thought it was not wisdom to say these things to the multitudes to whom the book would be sold in future years. . . . With reference to this, and to other passages in her writings that have been omitted in later editions, she has often said: "These statements are true, and they are useful to our people; but to the general public, for whom this book is now being prepared, they are out of place. Christ said, even to his disciples, 'I have many things to say unto you, but ye cannot bear them now.' and Christ taught his disciples to be 'wise as serpents, and harmless as doves.' therefore, as it is probable that more souls will be won to Christ by the book without this passage than with it, let it be omitted." W.C. White letter, July 25, 1911, as quoted in *Selected Messages, Book Three,* Appendix A, p. 443-444.

[30] Ellen White, *Selected Messages*, vol. 2, p. 288.

[31] See, e.g., Wheeler, Gerald W., *The Two-Taled Dinosaur* (Nashville, TN: Southern Publishing, 1975), pp. 26, 27.

[32] See, e.g., Frank Lewis Marsh, *Life, Man, and Time* (Escondido, CA: Outdoor Pictures, 1967), pp. 162-180; Harold Clark, *New Creationism* (Nashville, TN: Southern Publishing Assoc., 1980), pp. 21-39; Harold Coffin, *Origin by Design* (Hagerstown, MD: Review and Herald, 1983), pp. 19, 20, 384-395; Ariel Roth, *Origins* (Hagerstown, MD: Review and Herald Pub., 1998), pp. 86-88; Leonard Brand, *Faith, Reason, and Earth History* (Berrien Springs, MI: Andrews University Press, 1997), pp. 111-126.

[33] Wise, Kurt P., *Faith, Form, and Time* (Nashville, TN: Broadman & Holman, 2002), pp. 216-220 ("In perhaps as few as three centuries, scores of new species arose within most mammal baramins, and thousands of species arose within many of the insect and plant baramins."); Wise "*Australopithecus ramidus* and the fossil record," *Creation Ex Nihilo Technical Journal*, 8:160-165 (1994); Wise, "North American Paleontology Convention '96," *Creation Ex Nihilo Technical Journal*, 10:315-321 (1996); Brand, Leonard, *Faith, Reason, & Earth History* (Berrien Springs, MI: Andrews University Press, 1997), pp. 198, 199 ("According to the theory presented here, much of our current taxonomic diversity has been the result of limited evolutionary change after a worldwide catastrophe. The original groups of plants and animals have diversified into multitudes of species as they adapted to fill specific niches in the changed conditions after the catastrophe.")

Rapid post-Flood speciation has long been a belief of creationists. In his first book, *Back to Creationism*, published in 1929, Harold Clark had advocated rapid post-Flood speciation. See *Back to Creationism*, pp. 99-110, in *Creationism in Twentieth Century America, Vol. 8: The Early Writings of Harold W. Clark and Frank Lewis Marsh*, ed. Ronald L. Numbers (New York & London: Garland Publishers, 1995). In a letter to Clark praising him for his book, George McCready Price wrote, "If the Seventh-day Adventist people will all get behind these two ideas, Flood geology and plenty of species-making since the Flood, . . . I believe it would not be long before the scientific world would sit up and take notice." Numbers, at p. 124.

[34] Wood, Todd Charles, "The AGEing process: rapid Post-Flood Intrabaraminic Diversification caused by Altruistic Genetic Elements (AGEs)," *Origins*, No. 54 (2003), p. 6, citing, A. J. Jones, "How many Animals in the Ark?" *Creation Research Society Quarterly*, 19:13-18 (1973).

[35] Milton, Richard, *Shattering the Myths of Darwinism* (Rochester, VT: Park Street Press, 1997), pp. 147-150.

[36] Wise, Kurt P. (1990) "Baraminology: a young-earth creation biosystematic method," in *Proceedings of the Second International Conference on Creationism, Volume II, Technical Symposium,* R. E. Walsh, ed., (Creation Science Fellowship, P.O. Box 99303, Pittsburgh, PA 15233-4303), pp. 345–360; Wayne Frair, "Baraminology: Classification of Created Organisms," *Creation Research Society Quarterly,* 37:2, p. 82-91 (September, 2000).

[37] See, e.g., Wood, Todd Charles, and David P. Cavanaugh, "A Baraminological analysis of subtribe Flaveriinae (Asteraceae: Helenieae) and the Origin of Biological Complexity," *Origins,* no. 52, pp. 21, 22 (2001) ("Superficially, the conclusion that C4 photosynthesis has arisen since creation seems to deny the basic creationist belief that no increase in biological complexity has occurred, that all evolution has been degenerative [citations omitted]").

[38] See, e.g., Price, George McCready, *Evolutionary Geology and the New Catastrophism* (Pacific Press, 1926), pp. 258-274 ("All the fossils . . . give us proofs of an almost eternal spring's having prevailed in the arctic regions, and semitropical conditions in north temperate latitudes . . . give us proofs of a singular uniformity of climate over the globe, which we can hardly conceive possible let alone account for." Price quotes Henry Howarth as stating, "we shall find that the differentiation of the earth's climate into tropical and arctic zones is comparatively modern, and that in past ages not only were the climates more uniform, but more evenly distributed over the whole world," and James Geikie to the effect that "within so recent a period as the Miocene, a climate existed within the arctic regions so mild and genial as to nourish there beeches, oaks, planes, poplars, walnuts, limes, magnolias, hazel, holly, blackthorn, logwood, hawthorn, ivy, vines and many evergreens, besides numerous conifers, among which was the Sequoia . . . This ancient vegetation has been traced up to within eleven degrees of the pole."); Interestingly, many dinosaur fossils have been found in northern Alaska, within the arctic circle, including duckbills, ceratopsians, tyrannosaurids, and small theropods. Obviously, these could not have survived in the current climate of northern Alaska, but the plant fossils buried with them indicate that a much warmer climate prevailed when they were alive. See, e.g.,"The Dinosaurs of Arctic Alaska," *Scientific American* (December, 2004).

[39] See, e.g., "Dressing for Winter" http://www.enature.com/articles/detail.asp?storyID=552. (The ptarmigan (a bird), weasels, Arctic Foxes, and Snowshoe Hares change with the seasons, turning various shades of brown when the snow melts and white in areas of snow cover. By contrast, the Polar Bear, living in year-round snow and ice, remains white in all seasons.

[40] Travis, John, "Pass the Genes, Please," *Science News,* July 22, 2000. http://www.sciencenews.org/articles/20000722/bob10.asp.

[41] Travis, quoting Gary J. Olsen, of the University of Illinois at Urbana-Champaign

[42] Travis, quoting Howard Ochman.

[43] Travis, supra.

[44] Delwiche, Charles F., "Griffins and Chimeras: Evolution and Horizontal Gene Transfer," *BioScience*, Jan. 2000, *a review of, Horizontal Gene Transfer*, Michael Syvanen and Clarence I. Kado, eds. (London: Chapman & Hall, 1998).

[45] Delwiche, supra.

[46] Syvanen, Michael., "Horizontal Gene Transfer: Evidences and Possible Consequences," *Annual Review of Genetics*, 28:237-261 (1994). See also Ulfar Bergthorsson, et al., "Widespread horizontal transfer of mitochondrial genes in flowering plants," *Nature*, 424:197-201, July 10, 2003; C.L. Bishop, "Plant to Plant Horizontal Gene Transfer: Frequent horizontal gene transfer has occurred between distantly related angiosperms," *The Scientist Daily News*, July 10, 2003; F. de la Cruz, J. Davies, "Horizontal gene transfer and the origin of species: lessons from bacteria," *Trends in Microbiology*, 8:128-133, March 2000.

[47] Margulis, Lynn and Dorion Sagan, *Acquiring Genomes: A Theory of the Origins of Species* (New York: Basic Books, 2002) p. 76 ("When the complete sequencing of the human genome was announced at the beginning of this millennium, many were quite surprised to learn that some 250 of the more than 30,000 human genes of our bodies have come directly from bacteria. These genes, long sequences of DNA that code for proteins, are as recognizably of bacterial origin as a feather is recognizably from a bird rather than say, a shark's mouth.")

[48] Stanhope, Michael J., Andrei Lupas, Michael J. Italia, Kristin K. Koretke, Craig Volker & James R. Brown, "Phylogenetic analyses do not support horizontal gene transfers from bacteria to vertebrates," *Nature*, 411:940-944 (June 21, 2001); Steve Bunk, "Lateral Thinking," *The Scientist*, July 9, 2001; S.L. Salzberg, et al., "Microbial genes in the human genome: Lateral transfer or gene loss?" *Science*, 292:1903-6, May 18, 2001.

[49] Margulis, Lynn and Dorion Sagan, *Acquiring Genomes: A Theory of the Origins of Species* (New York: Basic Books, 2002) pp. 11, 12. (See, at p. 15: "Random DNA base changes of course play a role in the evolutionary process. They are like printers' errors that crop up in the copying of books. They rarely clarify or enhance meaning. Such small random changes are nearly always inconsequential or detrimental to the work as a whole. We do not deny the importance of mutations. Rather we insist that random mutation, a small part of the evolutionary saga, has been dogmatically overemphasized." And at p. 29: "Mutations, in summary, tend to induce sickness, death, or deficiencies.

No evidence in the vast literature of heredity change shows unambiguous evidence that random mutation itself, even with geographical isolation of populations, leads to speciation.")

[50] Margulis, et al., at pp. 71-77, 101, 145.

[51] Wood, Todd Charles, "The AGEing process: Rapid post-Flood intrabaraminic diversification caused by altruistic genetic elements (AGEs)" *Origins*, no. 54, pp. 5-34 (2002)

[52] Another concept that may ultimately prove to explain aspects of species change and diversification is "epigenetics." See, e.g., "Epigenetics means what we eat, how we live and love, alters how our genes behave." *Science Daily* (October 27, 2005).

[53] "Not only man but the earth had by sin come under the power of the wicked one, and was to be restored by the plan of redemption. At his creation Adam was placed in dominion over the earth. But by yielding to temptation, he was brought under the power of Satan. 'Of whom a man is overcome, of the same is he brought in bondage.' 2 Peter 2:19. When man became Satan's captive, the dominion which he held, passed to his conqueror. Thus Satan became 'the god of this world.' 2 Corinthians 4:4. He had usurped that dominion over the earth which had been originally given to Adam. But Christ, by His sacrifice paying the penalty of sin, would not only redeem man, but recover the dominion which he had forfeited. All that was lost by the first Adam will be restored by the second. Says the prophet, 'O tower of the flock, the stronghold of the daughter of Zion, unto thee shall it come, even the first dominion.' Micah 4:8. And the apostle Paul points forward to the 'redemption of the purchased possession.' Ephesians 1:14. God created the earth to be the abode of holy, happy beings. The Lord 'formed the earth and made it; He hath established it, He created it not in vain, He formed it to be inhabited.' Isaiah 45:18. That purpose will be fulfilled, when, renewed by the power of God, and freed from sin and sorrow, it shall become the eternal abode of the redeemed. 'The righteous shall inherit the land, and dwell therein forever.' 'And there shall be no more curse: but the throne of God and of the Lamb shall be in it; and His servants shall serve Him.' Psalm 37:29; Revelation 22:3." White, *Patriarchs and Prophets*, p. 67.

[54] 2 Cor. 4:4; Job 1:6-12; 2:1-7; John 12:31; 14:30; 16:11.

CHAPTER 23

# The Asteroid Theory

The island is a tropical paradise covered with lush vegetation. The sun has just set, and the sky toward the west is a beautiful canvass of reds, pinks, and purples. Quietly at first, meteors begin to streak across the evening sky, catching the attention of an iguanodon and several lemurs. Soon a meteor shower lights up the sky. A flock of birds is disturbed and takes flight; a lemur sniffs the air, sensing something wrong.

Suddenly, an enormous chunk of rock plunges toward the earth, glowing white-hot, trailing fire and a cloud of smoke. There is a blinding flash of white light, then a mushroom cloud, as though there had been a nuclear explosion. On impact, the asteroid disintegrates into millions of smaller pieces, which shoot out in all directions, glowing and trailing smoke, landing like a scattered artillery barrage. A circle of devastation around the impact site expands in waves, as when a stone is tossed into a calm pond.

Soon the first shock waves hit the animals, knocking the lemurs off their feet. From the site of the impact, a tidal wave and a wall of fire are racing toward the island. The iguanodon runs for his life. He runs across the island and plunges into the sea on the side opposite the cataclysm. The iguanodon and a few of his lemur friends survive the initial impact, but the world that emerges is not the tropical paradise that existed before. The post-impact world is a desert, almost devoid of plant life and drinkable water.

If you have children, you may have recognized the foregoing as one of the early scenes in the Disney animated feature, "*Dinosaur.*"[1] It reflects the currently fashionable belief that a giant asteroid struck the earth 65 million years ago, wiping out the dinosaurs.

The demise of the dinosaurs is a mystery for which many solutions have been proposed. Paleontologist Robert T. Bakker kept a file of published explanations for the dinosaur extinction. The dinosaurs disappeared—according to these varied theories—because:

- The weather got too hot
- The weather got too cold
- The weather got too dry
- The weather got too wet
- The weather became too hot in the summer and too cold in the winter
- The land became too hilly
- New kinds of plants evolved which poisoned all the dinosaurs
- New kinds of insects evolved which spread deadly diseases
- New kinds of mammals evolved which competed for food
- New kinds of mammals ate the dinosaurs' eggs
- A giant meteor smashed into the earth
- A supernova exploded near the earth
- Cosmic rays bombarded the earth
- Massive volcanoes exploded all around the earth.[2]

The asteroid theory is one of the more recent explanations. It grew out of research conducted by Walter Alvarez, a geologist with the University of California. Alvarez and his colleagues were looking for geological evidence of shifts in the Earth's polarity. While working near Gubbio, Italy, they noticed a layer of reddish-gray clay a couple of centimeters thick right at the boundary between the Cretaceous and the

Tertiary. Alvarez told his father, Luis Alvarez, a Nobel prizewinning physicist also at Cal-Berkeley, about the strange layer of clay.

The elder Alvarez was an adventurous scientist who had once tried to x-ray the pyramids of Egypt to reveal hidden chambers. He tackled the problem of the boundary clay with relish. While testing the clay to see if it could be radiometrically dated, Alvarez discovered it had thirty times the iridium content of the rocks above and below it. Because iridium is vastly more abundant in meteorites than in rocks formed on the earth, the Alvarezes began to theorize that the extra iridium had resulted from the explosion of a supernova.[3]

They obtained samples of the boundary clay from several other locations, as far away as New Zealand and Denmark, and found that these also contained high levels of iridium. They abandoned the supernova theory and, in 1980, published their conclusion that a giant asteroid struck the earth at the end of the Cretaceous period. The resulting cloud of dust blocked out the sun, they theorized, killing the plants. The lack of plant food killed the herbivorous dinosaurs, and the lack of prey animals killed the carnivorous dinosaurs.[4]

The news media responded very favorably to the asteroid theory, implying that the great dinosaur mystery had at last been solved. Even the normally reserved scientific community was unusually open to and interested in the theory. Almost immediately, scientists from a variety of disciplines began working out the details of the asteroid model. A large crater at Chicxulub, in the Yucatan peninsula of Mexico, was claimed as the remnant of the fatal impact. The asteroid theory has now become the most widely accepted theory to explain the "mass extinction" at the Cretaceous-Tertiary boundary, although it is probably not accepted by even a majority of scientists.

The first thing to note about the asteroid theory is that it is not a uniformitarian theory. It relies upon a postulated global catastrophe the like of which has never been observed. Charles Lyell established uniformitarian geology by insisting that only presently observable causes, acting at presently observable rates, could permissibly be employed to explain the geological facts. To invoke catastrophes was unscientific, Lyell insisted. He berated the geologists of the seventeenth and eighteenth centuries for thinking in terms of catastrophe:

> They employed themselves in conjecturing what might have been the course of Nature at a remote period, rather than in the investigation of what was the course of Nature in their own times. . . . It appeared to them far more philosophical to speculate on the possibilities of the past, than patiently to explore the realities of the present; . . . the claims of each new hypothesis to credibility appeared enhanced by the great contrast, in kind or intensity, of the causes referred to and those now in operation. Never was there a dogma more calculated to foster indolence, and to blunt the keen edge of curiosity, than this assumption of the discordance between the ancient and existing causes of change.[5]

Thus did Lyell dismiss the efforts, however scholarly, of all geologists who were open to the idea that catastrophes had helped to form the earth's crust.

Lyell specifically rejected global catastrophism:

> For this reason all theories are rejected which involve the assumption of sudden and violent catastrophes and revolutions of the whole earth, and its inhabitants—theories which are restrained by no reference to

> existing analogies, and in which a desire is manifested to cut, rather than patiently to untie, the Gordian knot.[6]

With this type of rhetoric, Lyell convinced the geologists of his day to abandon all inquiries into global catastrophic events.

Many contemporary scientists, by contrast, do not hesitate to embrace the asteroid theory, even though it posits a global catastrophe and a case where the present is decidedly *not* the key to the past. In fact, contemporary scientists have been so enthusiastic about catastrophic explanations for mass extinctions in the fossil record that some of them, as we shall see, confabulated a star in an effort to explain a perceived pattern of periodic catastrophic mass extinctions.

Catastrophism has made a rapid comeback in modern geology. Geologists still adhere to the vast ages of "geologic time" that were arrived at through Lyellian assumptions, but they are increasingly acknowledging that the rocks contain evidence of tremendous catastrophes. The coinage *neocatastrophism* is used to describe the newer non-uniformitarian school of geologic interpretation.[7]

In the early days of geology, catastrophes were accepted as valid interpretive concepts. But for more than a century after Charles Lyell, catastrophe was eradicated from all geologic interpretations. Now geologists again accept catastrophism.[8] It seems that fashion in geologic interpretation is like fashion in men's ties. Don't be too quick to throw out the old ones.

If theories of global catastrophe are now considered good science—and the acceptance of the asteroid theory shows that they are—on what basis can Flood geology be dismissed as unscientific? When Lyellism dominated geological thinking, geologists had a principled reason for rejecting Flood geology, because they rejected *any* global catastrophe. But now, the only reason not to consider the

Genesis Flood as an explanation for the sedimentary strata is science's philosophical commitment to naturalism.

Although the renaissance of catastrophism is a positive development, the asteroid theory has serious problems. According to the theory, dust from the impact blocked out the sun long enough to kill the plants, which left the dinosaurs without a source of food. But unlike dinosaurs, most plants came through the Cretaceous-Tertiary boundary just fine. "The land plants of the Late Cretaceous did suffer some extinctions," writes Bakker, "but nothing compared to the wholesale devastation experienced by the plant-eating dinosaurs."[9]

North American plants appear to have been hardest hit, especially evergreens, but South American plants were largely unaffected, and there are many dinosaur fossils from South America. So why did the dinosaurs die if the plants did not? According to enthusiasts of the asteroid theory, the plants were snuffed out temporarily but survived in dormant states—as seeds, spores, underground tubers, and bulbs—and grew and bloomed again after the dinosaurs died.[10] Perhaps. But perhaps nothing much happened to the plants, which if true would be fatal to the asteroid theory.

Second, assuming there was an asteroid impact, it was weirdly selective about the species it drove to extinction. There does not seem to be any pattern. Some cold-blooded and some warm-blooded animals came through the Cretaceous-Tertiary boundary, and some of each did not. Some large and some small species survived, but neither large nor small dinosaurs made it across the Cretaceous-Tertiary boundary. Dinosaurs did not make it, but other reptiles, such as crocodiles, turtles, and champosaurs, made it just fine, as did amphibians such as frogs and salamanders.[11]

Many marine vertebrates, including ichthyosaurs, plesiosaurs and mosasaurs, are not found above the Cretaceous-Tertiary boundary.[12] The asteroid theory assumes that they, too, became extinct as a result of the impact. Yet *freshwater* animals, whose ecosystem the putative impact should have damaged much more severely, appear to have been nearly untouched.[13]

Likewise, pterosaurs, the flying reptiles of the Mesozoic, are not found above the Cretaceous-Tertiary (K-T)[14] boundary, but birds are.[15] The survival of birds is particularly telling against the asteroid impact theory. Birds are small and warm-blooded, with high metabolic rates and small energy stores. Sudden storms and severe winters cause high mortality among bird populations, which is why they fly south for the winter. It is difficult to see how they could have survived an extended period of cold, dark, and scarcity of food.[16]

Third, the asteroid impact theory does not fit the fossil record. The theory predicts that we should see most of the dinosaur families continuing up through the fossil record and then stopping, very suddenly, at the K-T boundary. But this is not the observed pattern. Fewer and fewer dinosaur fossils are found as one moves up through the Cretaceous strata. In fact, there are no dinosaur fossils in the upper Cretaceous of South America, southern Europe, and perhaps northern Europe.[17]

In strata notionally formed eleven million years before the end of the Cretaceous, there are fossils of thirty-six dinosaur genera, whereas in strata of the last one million years of the Cretaceous, there are only fifteen to seventeen dinosaur genera.[28] Thus the number of dinosaur genera had already been cut in half long before the end of the Cretaceous. Moreover, the ecological mix of dinosaur species is not nearly as well balanced in the upper Cretaceous as in the lower.[19] Many paleontologists infer from these facts that the dinosaurs had begun to die out millions of years before the end of the Cretaceous, and that the asteroid impact, if there was one, was at best the final blow, the coup de grace that finished a long decline into extinction.

Moreover, if the asteroid impact theory were correct, we

would expect to find dinosaur fossils right on the K-T boundary, and slightly above it, because the iridium dust would be settling out of the atmosphere and into the sediment right as the last dinosaurs were being buried. But most of the dinosaur fossils lie at least three meters (ten feet) below the K-T boundary.[20] Ten feet represents hundreds of thousands of years of "geological time." It would thus appear that almost all of the dinosaurs died out thousands of years before the asteroid hit the earth.

Fourth, the K-T "mass extinction" is not the only one revealed by a Lyellian interpretations of the fossil record.[21] Paleontologists find evidence of eight mass extinctions, including at the top of the Permian, at the top of the Triassic, at the top of the Jurassic, and at the end of the Ice Age, when most of the giant land mammals became extinct.[22] The K-T extinction does not appear to have been the most severe of these. Half of all fossil *families* (families, not species), including over 90 percent of all marine animals, found in the Permian strata do not appear in the Triassic.[23] Each time one of these "mass extinctions" hit land animals, sea creatures went extinct as well. Cuvier noticed this phenomenon, and it is known as "Cuvier's law of land-sea simultaneity."

It would seem that no single event, such an asteroid impact, could be the culprit behind all eight "mass extinctions" separated, in Lyellian thinking, by many millions of years. Or could it? In 1983, scientists studying extinctions of marine animals argued that such extinctions occurred at regular intervals of 26.2 million years. This led some to wonder if an astronomical cycle was correlated to the cycle of extinctions. It was suggested that our sun has a companion star whose orbit brings it close to our solar system every 26.2 million years, dislodging interstellar debris from a hypothetical debris field called the "Oort cloud," causing comets to bombard the earth, causing mass extinctions.[24] This hypothetical death star, which its inventers dubbed "Nemesis," after the Greek goddess of retribution, generated great excitement when it was first proposed.[25]

The Nemesis hypothesis has since been discarded.[26] First, the 26.2 million-year cycle of extinctions turned out to be, as its discoverer candidly admitted, "very much in the eye of the beholder."[27] Several studies that reexamined the data on extinctions found no such cycle. Second, the proposed orbit of Nemesis would take it so far away from our sun that our sun's gravity would not hold it. It would be so weakly bound to the sun that small perturbations from passing objects could cause it to drift away.[28] Third, Nemesis remains hypothetical. In 1984, scientists believed they had a 50 percent chance of discovering Nemesis within three years if it existed.[29] But it has evaded detection for over two decades, strongly suggesting that it never existed.

The Nemesis affair stands as a remarkable example of bold but over-enthusiastic science: a star that apparently does not exist (Nemesis) was invented to explain, poorly, a perceived pattern in the fossil data that probably does not exist (the 26.2 million-year extinction cycle). The asteroid impact theory, and its far out spin-off, the Nemesis hypothesis, do not explain the observed data.

But if there was no asteroid impact, how do we explain the presence of iridium at the K-T boundary? Iridium, it turns out, does not come *only* from outer space. The earth's mantle and core contain iridium, and it is emitted by volcanic activity. The iridium-rich layer of sediment at the K-T boundary could be the result of volcanic activity.[30] The Deccan Traps of western India is an enormous pile of lava that erupted from a crustal fissure, apparently created by rapid plate movement during the Flood.[31] The Deccan Traps lava covers 500,000 square kilometers—an area as large as the states of Oregon and Washington combined—and is a mile and a half thick in places. About ninety percent (90%) of this lava correlates to the K-T boundary.[32] When it erupted, the Deccan Traps would have sent massive amounts of dust and aerosols, including iridium, into the upper atmosphere.[33] Upon settling out, this dust would have created a worldwide layer of iridium-rich sediment. Clearly, there is no need to invoke an extraterrestrial impact to explain the iridium at the K-T boundary.

If an asteroid did not kill the dinosaurs, what happened to them? Robert Bakker compares the death of the dinosaurs to a mass murder. When the police try to solve a murder, they examine the crime scene very carefully. Criminalists photograph the body, look for hair and fibers, and dust for fingerprints. They look for blood and tissue that will yield DNA, which can be used to uniquely identify a person. Let us examine the dinosaur "crime scene," and see what it can tell us about how the dinosaurs died.

[1] Walt Disney Pictures, 2000. Screenplay by John Harrison and Robert Nelson Jacobs, based on an original screenplay by Walon Green.

[2] Bakker, Robert T., *The Dinosaur Heresies* (New York: William Morrow, 1986 [softcover by Zebra Books, Kensington Pub. Corp., 1996]), p. 425.

[3] Wilford, John Noble, *The Riddle of the Dinosaur* (New York: Alfred A. Knopf, 1985 [softcover by Vintage Press, 1987]), p. 263-277.

[4] Wilford, supra. See also Alvarez, L. W., Alvarez, W., Asaro, F., and Michel, H. V., "Extraterrestrial cause for the Cretaceous-Tertiary extinction," *Science*, 208:1095-1108 (June 6, 1980); Bakker, at 434.

[5] Lyell, Charles, "Uniformity of Change," *Scientific Papers*. Vol. XXXVIII, Part 8, The Harvard Classics, (New York: P.F. Collier & Son, 1909–14); Bartleby.com, 2001, found at: http://www.bartleby.com/38/8/2.html.

[6] Lyell, supra.

[7] Throughout his recent book, Leonard Brand refers to modern geology as "neocatastrophism." See Brand, Leonard, *Faith, Reason, and Earth History* (Berrien Springs, MI: Andrews University Press, 1997).

[8] Roth, Ariel, *Origins* (Hagerstown, MD: Review and Herald, 1998), pp. 195-203; Wilford, at 278, 279.

[9] Bakker, at 431.

[10] Bakker, at 434.

[11] See Dobb, Edwin, "What Wiped out the Dinosaurs? (It may be time to stop blaming that asteroid)" *Discover*, June, 2002. ("Present-day alligators and crocodiles are very sensitive to cold temperatures," says paleoecologist Nan Crystal Arens, "and acidification disrupts reproduction among frogs and salamanders.")

[12] Bakker, at 428-430.

[13] Bakker, at 431.

[14] The Cretaceous is named for *creta*, the Latin word for "chalk." Cretaceous rocks were first studied at the famously chalky white cliffs of Dover, along the English Channel; for many years, English geologists referred to the Cretaceous as simply "the chalk." In "K-T," the initial "K" stands for *keta*, the Greek for "chalk."

[15] Wilford, at 279-280. See also Dobb, supra ("most major groups of modern birds originated and began to diversify before the K-T boundary. Yet they, too, survived the extinction event."); Alan Cooper, David Penny, "Mass Survival of Birds across the Cretaceous-Tertiary boundary: Molecular Evidence," *Science*, vol. 275 (Feb. 21, 1997) ("At least 22 avian lineages of modern birds cross the Cretaceous-Tertiary boundary").

[16] Cowen, Richard, "The K-T Extinction," at http://www.ucmp.berkeley.edu/education/events/cowen3b.html.

[17] Wilford, at 280.

[18] Wilford, at 280, 281.

[19] Bakker, at 436-438.

[20] Wilford, at 283, 286. But see Lockley and Meyer, *Dinosaur Tracks and other Fossil Footprints of Europe* (New York: Columbia University Press, 2000), p. 239, who report finding dinosaur tracks just thirty-seven centimeters (fourteen inches) below the K-T boundary in Colorado. There is also a report of a Triceratops skeleton 1.4 meters below the K-T boundary (still thousands of years in geological time). Johnson, K., D. Nichols, C. Labandeira, and D. Pearson, "Devastation of terrestrial ecosystems at the K-T boundary in North America: The first calibrated record of plant and animal response to the Chicxulub impact," in *Catastrophic Events and Mass Extinctions: Impacts and Beyond* (Catastrophic Events Conference, Vienna, Austria, 2000). *Geological Society of America Special Paper.*

[21] I am assuming these "mass extinctions," for purposes of argument. The biblical Creationist view is that most of the strata were deposited relatively quickly during the Genesis flood and that the animals found at all these levels lived contemporaneously and died during the Flood year. The successive "mass extinctions" envisioned by Lyellian geology were part of a single catastrophe, the Genesis Flood. (The extinction of the giant mammals at the end of the Ice Age is an exception; it was a true extinction, probably caused by post-flood environmental changes.)

[22] Bakker, at 428.

[23] Wilford, at 253, 254. See also Stephen Jay Gould, *Wonderful Life* (New York: W.W. Norton, 1989), p. 229.

[24] Wilford, at 298-316.

[25] Bakker, at 431, 432. See, e.g., Raup, David M. *The Nemesis Affair: A Story of the Death of Dinosaurs and the Ways of Science* (New York: W.W. Norton & Co., 1986).

26 Britt, Robert Roy, "Nemesis: Does the Sun have a 'Companion'?" Space.com, April 3, 2001, found at http://www.space.com/scienceastronomy/solarsystem/nemesis_010320-1.html.

27 Britt, supra ("Meanwhile, many scientists see little or no credibility to the studies alleging periodicity in mass extinctions, and hence no need for a Nemesis theory. . . . 'Some of the suggested periodicities are more likely to represent statistical artifacts than robust observations,' Glikson said." J. John Sepkoski, one of the proponents of the theory, stated, "Of the 15 or so re-analyses of our data published since the original paper, about half support periodicity and half reject it. It is still very much in the eye of the beholder."). See also Wilford, at 311, 312, quoting Stephen M. Stanley.

28 Wilford, at 310, quoting Gene Shoemaker, an astrogeologist. See also Britt, supra ("The orbit assumed for Nemesis is an unusual one, Muller admits. No star has ever been found to orbit so far from a companion. 'And that really bothers people,' he said. 'It makes them think that this is a really far-out idea, literally.'").

29 Wilford, at 311.

30 Bakker, at 438, 439.

31 No one knows what causes continental flood basalts. A popular theory posits "mantle plumes" in which teardrop-shaped plumes of hotter material rise from the bottom of the mantle and erupt through fissures at the surface. Others relate the flood basalts to continental rifting, and claim that non-plume, plate tectonics models are capable of explaining them, which is the view here adopted.

32 McLean, Dewey, "The Asteroid Impact vs. Volcano-Greenhouse Dinosaur extinction debate," found at: http://filebox.vt.edu/artsci/geology/mclean/Dinosaur_Volcano_Extinction/index.html., citing McLean, D. M., "A terminal Mesozoic 'greenhouse': lessons from the past," *Science*, 201:401-406 (1978); McLean, D. M., "Terminal Cretaceous extinctions and volcanism," American Association for the Advancement of Science, 147th National Meeting, Abstracts of papers, (1981) p. 128; McLean, D. M., "Deccan volcanism and the Cretaceous-Tertiary transition scenario: a unifying causal mechanism," abstract in Russell, D. A., and Rice, G., eds., *K-Tec II Cretaceous-Tertiary Extinctions and Possible Terrestrial and Extraterrestrial Causes*, Syllogeus Series 39, National Museums of Canada (Proceedings, May 1981 workshop), p. 143, 144. See also Wilford, at 288-290; Richard Cowen, "The K-T Extinction."

33 McLean, supra, citing, Basu, A. R., Renne, P. R., DasGupta, D. K., Teichmann, F., and Poreda, R. J., "Early and late alkali igneous pulses and a high 3He plume

origin for the Deccan flood basalts," *Science*, 261:902-906 (1993). See also Wilford, at 288; Duncan, R. A., and D. G. Pyle, "Rapid eruption of the Deccan flood basalts at the Cretaceous/Tertiary boundary," *Nature*, 334: 841-843 (1988); Toutain, J., and Meyer, G., "Iridium-bearing sublimates at the hot-spot volcano (Piton de la Fournaise, Indian Ocean)," Geophysical Research Letters, v. 16, p. 1391-1394 (1989).

CHAPTER 24

# What Killed the Dinosaurs?

In 1978, Jack Horner was working as a *preparator.* A preparator extricates fossils from the rock in which they are usually embedded, by chipping away at the rock or dissolving it with chemicals, preparing the fossils for display and study. Horner wanted to be a paleontologist, however, not just a preparator slaving away anonymously in a museum basement. But most universities and museums want their paleontologists to have earned a Ph.D, and, although Horner had an excellent education in paleontology, he had neglected to earn any academic degrees.[1] The job did have one important perk: preparators got to spend their summers searching for fossils, and Horner had a passion for hunting dinosaur fossils.

Horner specialized in duckbill dinosaurs. His research indicated that many duckbill dinosaurs, including baby duckbills, had been found in marine sediments such as shale. Baby dinosaurs were rare and valuable, so Horner and his friend Bob Makela, a high school teacher, set out to find some. Horner's library research indicated that Earl Douglass, a famous dinosaur hunter of the early 1900s, had found baby dinosaurs in the Bear Paw shale, near Billings, Montana.

But when they arrived at that location, three days of continuous rain had turned the shale into thick, viscous mud. So instead of squishing around in the mud, they went to help a friend look for mammal fossils.

That friend told them that the owner of a rock shop in Bynum, Montana, wanted a dinosaur bone identified. On their way to yet a third dig, Horner and Makela stopped by the rock shop. They arrived in Bynum on a Sunday, but the store was open; it closed on Saturdays and opened on Sundays because its owners, John and Marion Brandvold, were Seventh-day Adventists. That Sunday, Marion Brandvold was minding the store. She showed them the fossil she wanted identified, which turned out to be part of the backbone of a duckbill. Horner and Makela were in no hurry, so they wandered around the store examining all the fossils and correctly identifying those that had been mislabeled.

Marion Brandvold, who was delighted to benefit from this free expertise, remembered she had some small bones back at her house, and asked the men if they would like to take a look at them. They agreed, and Mrs. Brandvold retrieved a couple of specimens. Horner recognized one of them right away. It was a fragment of duckbill dinosaur femur. Only instead of being four feet long, it was about three inches long. "What I had in my hand," wrote Horner, "was a bone from a baby dinosaur, a duckbill—exactly what I wanted, in a place I never expected to find it."[2] Back at the house, the Brandvolds had a card table covered with these baby duckbill bones, which Marion agreed to surrender to Horner.

More importantly, she told Horner where they had found the bones: the Two Medicine Formation near Choteau, Montana. The planned Bear Paw shale dig was abandoned:

> And we were perfectly happy to give up our plan to go back to the Bear Paw shale. In paleontology, as in

> all sciences, as in all of life, you don't argue with luck. Marion Brandvold had discovered a lovely little window on the late Cretaceous. What we did was to open that window and climb through it.[3]

Climb through it they did. Horner dug for seven summers near Choteau, discovered several duckbill "nests," with clutches of eggs and juvenile dinosaur fossils in close proximity. Based upon his belief that the mother dinosaur cared for the babies for an extended period after they hatched, Horner named a new duckbill species *Maiasaura*, or "good mother lizard." Horner wrote a book, became famous, and became a technical advisor for the "Jurassic Park" movies, and the University of Montana granted him an honorary doctorate. As of this writing, he is curator of paleontology at the Museum of the Rockies, at Montana State University in Bozeman, Montana.

## A. THE CASE OF THE SUICIDAL DINOSAUR HERDS

During Horner's second summer at the Choteau site, the Brandvolds started their own dig. They were looking for an articulated dinosaur skeleton to mount as a tourist attraction for their rock shop, which they subsequently moved from Bynum to Choteau. Horner and his crew were working in the same area, and removed the topsoil from the spot where the Brandvolds wanted to dig.

After a season of digging and carefully mapping and photographing the site, it was clear that the Brandvolds would not find an articulated dinosaur skeleton. The bones were all from duckbills, but they were mostly long bones. The type and condition of the bones from the Brandvold site was the same as that of the bones Horner's crew had uncovered in three other sites in the vicinity.

In 1984, a geologist specializing in stratigraphy noticed that there was a layer of volcanic ash about eighteen inches above the dinosaur bones at one of the sites. He and Horner traced this layer of ash across the Brandvold site and three other dig sites, and the ash layer was always in about the same relation to the dinosaur bones. The four sites all tapped into the same stratum of disarticulated duckbill bones:

> We had one huge bed of maiasaur bones—and nothing but maiasaur bones—stretching a mile and a quarter east to west and a quarter-mile north to south. Judging from the concentration of bones in various pits, there were up to 30 million fossil fragments in the area. At a conservative estimate, we had discovered the tomb of 10,000 dinosaurs.[4]

Horner wondered what could have created such a deposit. No predators had chewed the bones, and they were pointing in the same direction, indicating that they had been transported by water and had oriented in the current. Horner eventually concluded that a volcanic eruption had emitted deadly gases, killing the dinosaurs, which then decomposed and de-fleshed. The bones were later swept away and re-deposited by floodwaters. "This was no ordinary spring flood from one of the streams in the area," wrote Horner, "but a catastrophic inundation."[5] The volcanic ash perhaps rose to the top of the slurry that transported and deposited the bones, or perhaps settled out after they were deposited, creating a layer of volcanic sediment overlying the dinosaur stratum.

At that time, Horner's site was the only single-species "bone bed" with such a vast number of bones.[6] That is no longer true. Dinosaur Provincial Park, in Alberta, Canada, contains several single-species bone beds. The *Centrosaurus* (a horned dinosaur similar to *Triceratops*) bone bed contains between twenty and sixty bones per square meter and represents a herd of young, juvenile, and middle-aged animals, all of which died at the same time.[7] When the *Centrosaurus* bed was first discovered, Phil Currie and Peter Dodson thought it had been deposited when a flash river flood killed a single herd of dinosaurs. Dodson theorized that the leader of the herd had mistakenly thought the river could safely be crossed, leading the herd out into the swollen stream, with disastrous results.[8]

But Currie knew that there were about fifty other bone beds in or near the park. Some of these contained a mixture of species, but there were several other single-species beds, including beds of *Anchiceratops, Chasmosaurus, Monoclonius, Pachyrhinosaurus* and *Styracosaurus* (all ceratopsians).[9] What sort of mass hysteria or suicidal ideation would cause the leaders of all of these different herds to march into catastrophic floodwaters? It soon transpired, moreover, that the *Centrosaurus* bone bed was much larger than originally thought. Its vast extent finally forced Dodson to abandon his river flood theory:

> I worked [the *Centrosaurus* bone bed] with crews from the Royal Tyrrell Museum in the early 1980s, but as work proceeded thereafter, the bone bed continued to increase in size. When I visited with Phil [Currie] in the summer of 1994, what he showed me was staggering. It now appears the bone bed stretches for nearly 10 km, almost horizon to horizon within the confines of Deadlodge Canyon, which is chiseled out of soft Judith River sediments on either side of the Red Deer River. In 1984, Phil and I described our interpretation that this deposit probably documented the drowning of a

> herd of animals in a river in flood. The deposit averages twenty bones per square meter excavated, although in some quadrants the density reaches sixty bones per square meter. We estimated in 1984 that the deposit contained the remains of at least three hundred individuals of *Centrosaurus* from juvenile, subadult, and adult size classes. Today, it is clear the catastrophe, whatever it was, brought down a herd of ten thousand or more individuals. There is no volcanic ash associated with the deposit to suggest that that sort of cataclysm played a role. Not to belabor a cliché, this is a *mystery*![10]

It is even more mysterious when one acknowledges the true number of animals involved. Remember that Horner's bone bed was only a mile and a quarter long, and he *conservatively* estimated that it contained the remains of 10,000 dinosaurs. The Alberta site is six miles long, with a bone density of up to sixty bones per square meter. Assuming the bone bed is one kilometer wide and ten kilometers long, then it is ten million square meters. If the average density is twenty bones per square meter, the bed contains some 200 million bones. Assuming that each Centrosaurus had about 350 bones, the bone bed represents the remains of well over half a million animals.

Clearly, the size of this herd was not normal, and the catastrophe that annihilated it was not normal. The *Centrosaurus* population of an entire region was herded together and finally destroyed by rising floodwaters. The bodies were not immediately buried but decayed to the point where the bones became disarticulated. Water then deposited the bones in a vast bone bed. The same process created the other single-species bone beds in Dinosaur Provincial Park, as well as Horner's *Maiasaurus* bed.

## B. THE CASE OF THE EGGS THAT DIDN'T HATCH

In recent years, many dinosaur eggs and "nests" have been found. Some argue that this proves that the Mesozoic strata represent normal ecosystems, rather than sediments deposited during the Genesis Flood. If one's model of the Flood has floodwaters rising slowly and steadily, as in a bathtub, and never receding having once deluged a given area, it is difficult to explain the presence of dinosaur eggs and "nests." But even dinosaur tracks, of which there are uncounted billions, falsify the "bathtub" model. As we discussed in Chapter Eight, however, both scriptural and geologic evidence indicate that floodwaters advanced and retreated repeatedly over the course of the Flood year.

During the periods between inundations, when previously laid sediments were exposed to the air, female dinosaurs laid eggs. The dinosaurs could hardly have avoided laying eggs during the year-long catastrophe that was the Genesis Flood.[11] Mothers of all species know that babies come when they want to, not when it is convenient.

Many fossilized dinosaur eggs have abnormalities that indicate that the mother was under stress when she laid the eggs. A common abnormality found among dinosaur eggs is a double shell. Based on analogies to modern birds, scientists believe the double shells were caused by the retention of the egg in the mother's oviduct because the animal was under stress. Abnormal dinosaur eggs have been found in France, India, Argentina, China, and in the Cleveland-Lloyd Dinosaur Quarry in Utah.[12] Could the female dinosaurs have been under stress because they were in the midst of world-ending cataclysm when they laid the eggs?

Jack Horner and his crew found some eggs and nests that we will discuss in detail, to illustrate how paleontologists interpret what they find. In 1979, the crew found several eggs on a small hill near one of the sites at which they had been digging. The eggs were lying in plain sight. As they did more work on this hill—which came to be known as "Egg Mountain"—they found two different kinds of eggs, one with a bumpy surface, one with a smooth surface. The smoother eggs were laid in spiral clutches; the bumpy-surfaced eggs seemed to have been laid in paired lines.

By the end of 1980, they had found six clutches of smooth eggs and six individual bumpy eggs, and they had found each kind of egg on each of three different stratigraphical horizons.[13] On each horizon, they also found the skeletal remains of young dinosaurs. In 1983, Horner discovered another area with spiral egg clutches, which they dubbed "Egg Island." Two "nests" were found there, one with nine eggs and one with nineteen. Each of the nineteen eggs, which were x-rayed and CT-scanned, contained embryonic dinosaurs. Most of the bones had become disarticulated and had fallen to the bottom of the eggs, but one egg contained a fully articulated embryonic skeleton.

The animal that laid the smooth eggs in spiral clutches was believed to be an Ornithischian dinosaur similar to *Hypsilophodont.* Horner and colleague David Weishampel named it *Orodromeus* ("Egg Mountain Runner") *makelai,* in honor of Bob Makela, whose skillful wielding of a jackhammer broke up the extremely hard rock of Egg Mountain, exposing the eggs in the lower strata.[14]

Horner began to imagine that Egg Mountain had once been a low-lying island in a shallow, milky green, alkaline lake. Based upon root casts in the rocks, Horner imagined that the edges of the island were thick with brushy plant life, rendering the interior invisible to predators. The lake was large, about ten square miles, but so alkaline that there were few animals in it—mostly clams, snails, and colonies of algae. During nesting season, the Orodromeus makelai went to the "island" and laid eggs in spiral clutches. Several teeth from the carnivorous Saurischian

*Troodon* indicated that, despite the plant screen, predators discovered the nesting site and dined on some of the herbivorous dinosaurs. Because there were three horizons containing clutches, Horner imagined that this scene was repeated at nesting season each year.

Horner's is an interesting interpretation of the site, but it has its weaknesses. First, it does not account for the bumpy eggs. What if the animal that left the teeth—which, when Horner's book was written, was thought to be a carnivorous *Troodon*—also left the bumpy eggs? Second, in the normal course of nature, eggs hatch. Why did so many of these eggs fail to hatch? Third, why are dead juvenile dinosaurs found in the nesting ground? Fourth, if this was an island in a shallow lake, how did the three levels of nests come to be buried so deeply, with almost two feet of sediment between each level?

On this point, Horner's explanation seems particularly unsatisfactory. He reasoned that when the lake was dry, streams and creeks ran through it, and when these flooded in the spring, they deposited the mud that buried the eggs.[15] It is hard to imagine "spring floods" occurring *when the lake was dry*, or streams and creeks running through a dry lakebed depositing sediment on what is normally an *island*, i.e., elevated ground.

How might a creationist interpret the site? The island was not a normal island but temporary high ground, formed by the clumping of uprooted vegetation. The high ground is surrounded by water, but the water is not a normal lake and does not have a normal ecosystem. It is an area into which floodwaters, full of eroded silica and calcium carbonate, have temporarily drained. Female dinosaurs, heeding nature's call, wade through the water and converge on the high ground to lay eggs. Two different types of dinosaurs lay eggs there—one is an herbivore, the other a carnivore. The catastrophe has created strange bedfellows who would not otherwise have laid their eggs in the same spot.

But not without casualties. The carnivorous *Troodons* are not interested in tangling with the larger adult *Orodromeus makelai*, but some smaller juvenile *Orodromeus*—who have tagged along with their mothers—are easy prey. The *Troodons* drop some teeth in the process of killing these. Rising floodwaters then force both types of dinosaurs off the site, and the resulting sediment dump buries the dead juveniles as well the eggs. All of the *Troodon* eggs and many of the *Orodromeus* eggs are buried in two feet of sediment and suffocate long before they can hatch. At another site nearby, a whole clutch of nineteen eggs is prevented from hatching because it is buried too deeply in Flood-laid sediment. This scenario repeats itself two more times before the whole area becomes too deeply inundated for the "island" to be above water.[16]

Obviously, both of these interpretations are exercises in imagination. Neither interpretation can ever be verified. We should also note that a reconstruction of the contents of one of the eggs (completed eight years after Horner's 1988 book) has shown that the eggs previously attributed to *Orodromeus* actually belong to *Troodon*.[17]

The crucial point is that rocks and fossils do not interpret themselves. How one interprets them depends upon the beliefs, assumptions, and expectations that inform one's imagination. A mainstream paleontologist like Horner assumes that the sedimentary strata were formed slowly. This means that most sedimentary horizons should represent a normal ecology. Horner thus expects to find clues about how the dinosaurs lived their lives.

But when biblical creationists interpret Horner's finds, we bring a different set of assumptions to the data. We do not believe that each sedimentary horizon represents a normal ecology. We believe that God intervened in earth's history to destroy the earth in a violent, worldwide deluge. When we look at dinosaur fossils, we find clues not about how they lived but about how they died.

Some sites are so strongly suggestive of catastrophe that no one can deny their catastrophic origin. The single-species bone beds of Montana and Alberta are cases where *only* catastrophic destruction can explain the fossil evidence. A site in northwestern Patagonia, Argentina, provides another example. An expedition sponsored by the American Museum of Natural History and led by paleontologists Luis Chiappe and Lowell Dingus discovered the site in 1997. The expedition had been searching for fossil birds, but on the second day, they stumbled across thousands of dinosaur eggs, in what appeared to have been a sauropod nesting ground. "In many places the fragments of eggs were so abundant that we couldn't walk without stepping on pieces of fossil eggshell."[18]

The site contains so many embryonic remains that "it appears catastrophe struck the nesting ground, keeping many eggs from hatching. Floods penetrated the porous shells and drowned the embryos."[19] "Obviously, all the embryos that we found were not part of a natural rate of embryonic mortality but must have been the result of a catastrophe," Chiappe observed. "[W]e can say with confidence that the catastrophe was a flood."[20] Chiappe and Dingus believe the flood in question was a local catastrophe—a river flood. They believe that the dinosaurs deliberately chose a floodplain to use as their nesting ground. Perhaps these animals needed larger brains after all. Or maybe Chiappe and Dingus need to enlarge their conception of the catastrophe that buried these eggs.

Like most dinosaur egg sites, the one in Patagonia is in upper Cretaceous strata. Ariel Roth notes that most dinosaur eggs have been found in the upper Cretaceous; they are not distributed equally through the Mesozoic. If the eggs had been buried by random catastrophic events spread out over 150 million years, we would expect to find the eggs more or less evenly spread throughout the strata in which dinosaur bones are found. Instead, we find most of the eggs in the upper

Cretaceous. "Could the dinosaurs have laid their eggs during a quieter period of the Genesis Flood," writes Roth, "even allowing time in some places for some development?"[21]

## C. TRACKING THE DINOSAURS

The dinosaurs also left tracks—a lot of tracks. Scientists estimate that there are billions of dinosaur tracks at one site in Utah known as the Moab mega-tracksite.[22] Everyone has seen the playful pictures of a trench coat-clad detective holding a magnifying glass and examining a footprint. Let's put the dinosaur tracks under a magnifying glass and see what they can tell us about how the dinosaurs met their fate.

First of all, how were the tracks formed, and how is it that they are still around for us to gaze upon? The tracks had to be 1) formed and 2) preserved. A track is formed when an animal steps in soft, moist earth.[23] If it steps on rocks or very hard ground, no track is formed; if it steps on dry sand, as in a desert or beach dune, an indistinct impression is formed. To form a track, you need wet dirt or mud.

Making a track is the easy part. Preserving the track is the hard part. Under normal conditions, the track will not last. It will weather away; rain will wash it away, other animals will trample it, plants will grow over it, etc. In order to be preserved, the track must be exposed to the air for a short while to allow it to firm up, then be covered by more sediment, preferably a contrasting type of sediment, which will allow the layers to separate when later exposed.[24] The overlying layer of sediment protects the track, allowing it to harden into rock. Much later, the overlying layer erodes away, exposing the track-bearing stratum.

Under what conditions might these things occur? Interestingly, the experts now believe that mega-tracksites such as Moab, Utah, were formed as a result of a rising sea level, as newer sediment was dumped on top of an older surface, preserving the tracks.[25] The only difference between conventional ichnologists and creationists is the time element. Creationists believe that the "sea level" was rising very rapidly, indeed, over the course of a few days or weeks.

Some believe that the Moab mega-tracksite was formed in a low-lying, coastal plain environment over a long period of time, under more or less normal conditions.[26] A moment's reflection will suffice to show the falsity of that view. Tide and time will wear a coastline into either a beach or a barrier of hard rock. Neither is a suitable place to preserve tracks. Tracks cannot be preserved on a beach, because the would-be track maker is either on a dune, where it will leave only an indistinct impression, or on the wet sand, where its traces will be washed away by the tide within a few hours.

It is possible to imagine near-shore environments—for example, tidal flats, lakeshores, or riverbanks, at which the water is lower than normal or has recently been higher than normal—that create mud. But tracks made in those environments will soon weather away. Under no *normal* circumstances could tracks made at the seashore be permanently covered with enough sediment to protect them from high tides, winter storms, and many thousands of years of erosion.

Only a massive sediment dump, occurring soon after the dinosaur tracks were made, could have preserved them for posterity. And they could be preserved only if that protective sediment layer was not eroded away before yet more sediment was deposited on top of that. Such massive, repeated, and rapidly accumulating deposits do not happen under normal conditions. Although most paleontologists and ichnologists seem oblivious to this fact, the dinosaur tracks tell us that they were formed during a flood—and not just any flood.[27] Such a flood is the only way the tracks could have been both formed and permanently preserved.[28]

## D. THE CASE OF THE MISSING BODIES

There is a very interesting fact about dinosaur tracks: we find the tracks lower in the sedimentary pile than we find the bones of the dinosaurs that made them. The lowest dinosaur *bones* are found in the upper Triassic, but the lowest dinosaur *tracks* are found in the lower and middle Triassic strata.[29] The tracks predate the bones by millions of years of "geologic time." This is a riddle to paleontologists, and several have responded by trying to cast doubt on these tracks, either as to their nature as tracks or as to their having been made by dinosaurs.[30]

But ichnologists are confident that these tracks really are dinosaur tracks. Tracks from the middle Triassic of France, for example, are extremely dinosaurian in appearance, and so well preserved that there can be no doubt that they are footprints. A three-toed, bipedal animal made the tracks—an animal very different from the quadrupedal, five-toed reptiles of the Permian and Triassic. The three-toed tracks closely resemble those Jurassic theropod tracks to which Edward Hitchcock assigned the name *Grallator.* Georges Demathieu, a French researcher who studied the tracks, concluded that these "dinosauroid" tracks are in fact dinosaur tracks.[31] The tracks reveal a foot structure and gait very like that of dinosaurs.[32]

Moreover, the tracks of identifiable species precede the bones of those species into the fossil record by huge chunks of "geologic time." Very good *Stegosaurus* track ways are found in the Dordogne region of France, in lower Jurassic strata conventionally dated to about 204 to 208 million years ago. The actual *bones* of a *Stegosaurus* are not found until the middle Jurassic, usually dated to between 165 and 175 million years ago.[33] This means that *Stegosaurus* was walking around leaving tracks for thirty million years before any of its bones managed to get buried and fossilized.

These tracks wreak havoc on evolutionary "family trees," as *Stegosaurus* was leaving tracks long before his own ancestor, the hypothetical common ancestor of the *Stegosauridae* and the *Ankylosauria*, is thought to have existed, and, in fact, even before that ancestor's hypothetical ancestor is thought to have existed.[34] Perhaps the mysterious triangular plates on *Stegosaurus*' back were used for time travel, so that it could leave tracks thirty million years before its own time! More likely, the thirty million years are not real. They are a product of the Lyellian imagination. In reality, very little time passed between the making of the tracks and the burial of the track maker.

The mystery of the missing dinosaur bodies is not unique to Europe. Since the days of Cope and Marsh, the Jurassic Morrison Formation of North America has been one of the dinosaur hunters' most fertile fields. The Morrison Formation is in the upper Jurassic, which geologists believe to have been deposited between 157 and 144 million years ago. The lower and middle Jurassic (believed to have been deposited between 208 and 157 million years ago), by contrast, contain no dinosaur bones. But these strata have yielded many millions of dinosaur tracks. *Grallator*-like theropod footprints, and sauropod tracks, have been found in abundance in the lower Jurassic Wingate Formation.[35]

There are also dinosaur tracks in the Kayenta and Navajo Formations of the lower Jurassic but, again, no bones to match the tracks.[36] The middle Jurassic Entrada formations are riddled with dinosaur footprints, especially theropod footprints of various sizes.[37] Yet there is not one dinosaur skeleton in the middle Jurassic of the entire western United States, and no skeleton of any description except for one small crocodilian.[38]

Consider the implications of these facts. We are asked to believe that dinosaurs were walking around leaving tracks for fifty million

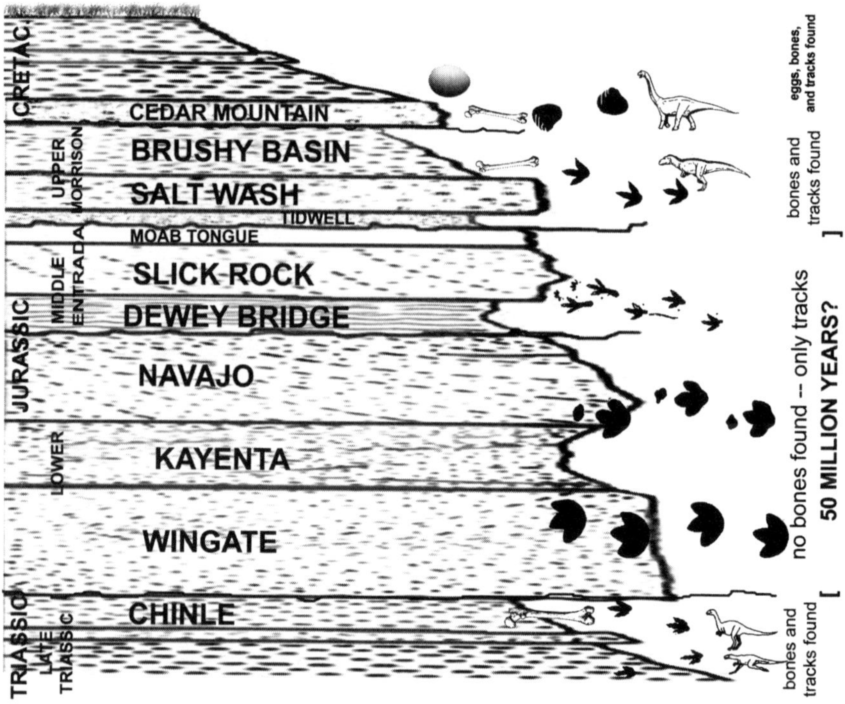


**Figure 24.1: Absence of dinosaur bones in the middle and lower Jurassic strata. These thick beds of sedimentary rock represent fifty million years of "geologic time." But how likely is it that not one dinosaur would be fossilized during this purported fifty-million-year period, even though dinosaurs left millions of tracks?**

years—"a substantial chunk of Jurassic time," as Lockley and Hunt modestly describe it[39]—but during all those millions of years not one unlucky dinosaur was buried in sediments and preserved as a fossil. For the Lyellian, this is truly a mystery. Not so for the creationist.

The story these tracks would tell can only be understood by discarding the notion that the strata were deposited over vast eons of time. The strata were deposited rapidly during the Genesis Flood. While being herded toward higher ground, the dinosaurs left tracks on recently laid, muddy deposits.[40] After the dinosaurs had evacuated the area, floodwaters returned, dumping sediment on top of the tracks and thus preserving them. This happened repeatedly as the Jurassic rocks were being deposited, and explains why we find dinosaur tracks, but no bones, in large vertical sections of strata. It also explains why the tracks enter the fossil record long before the bones. Eventually, the dinosaurs themselves were swept away by the returning waters, which deposited their bodies in still higher sedimentary strata—namely, the upper Jurassic Morrison Formation.

Another piece of evidence confirms our solution to the case of the missing bodies. Track sites tend to be associated with the tops of formations.[41] The top of a formation represents a hiatus or break in sediment accumulation; geologists refer to these breaks as "unconformities." The Moab mega-track site is right at the boundary between the Entrada Formation and the Summerville formation. The tracks are on top of the Entrada and just below Summerville.[42] (The Summerville is not depicted in Figure 24.1, but it would be above Entrada but below Morrison.[43])

Paleontologists are puzzled that there should be evidence of so much dinosaur activity during a hiatus in the accumulation of sediment. "This is a paradox, in a way, almost a contradiction," write Lockley and Hunt, "a record of dinosaurs where the rock is missing."[44] But it is only a paradox if one assumes that formations such as the Entrada accumulated slowly, at the rate of one fifth of a millimeter—the thickness of a human hair—per year.[45] If the Entrada Formation accumulated anywhere near that slowly, it makes no sense that the dinosaur tracks are all at the top, rather than distributed more or less evenly throughout.

But if the entire formation was deposited rapidly, virtually at once, it makes perfect sense that the tracks are found at the top. The dinosaurs could not have left tracks anywhere else in the formation, because the formation was deposited while the area was underwater. After the waters retreated, "during a hiatus or break in sediment accumulation," the dinosaurs left tracks in the mud at the top of the formation. At this time, the formation was either exposed to the air or covered only in shallow water, through which the dinosaurs could wade and leave tracks. Hence, we have a cluster of dinosaur tracks right at the top of the Entrada Formation. When the flooding resumed, the dinosaurs were caught up and their bodies deposited in a higher formation, the Morrison.

## E. THE CASE OF THE SEAGOING DINOSAURS

All dinosaurs are terrestrial, land-dwelling creatures. They are not marine creatures. Yet some dinosaurs have mysteriously turned up in "marine sediments." Hadrosaurs (duckbill dinosaurs) are a case in point. Many Hadrosaurs have been found in sediments supposedly deposited at the bottom of the sea.[46] How do paleontologists explain this? Those who believe Hadrosaurs were strictly land-dwelling creatures assume that they were washed out to sea from inland habitats. Other paleontologists suggest that Hadrosaurs were semi-aquatic, despite the fact that they appear to have been no more adept at swimming than deer or horses.[47] These believe that duckbills may have foraged in near-shore marine environments.[48]

The really odd thing, however, is that nearly all specimens from marine habitats are found with their skeletons articulated. That means that nothing messed with the body after the animal died. In at least one remarkable instance, even the skin of a *Hadrosaurus* was found preserved in the rock.

Hadrosaurs were not the only dinosaurs buried at sea. The same is true of the armored *Nodosaurus,* which had thick bony plates along its back. At Como Bluff, Marsh's collectors found a nodosaur lying on its back embedded in sea-floor sediment. That the nodosaur flipped over in the water and settled on its back is not surprising; the thick armor plates on its back must have made it top heavy. But that is not what needs explaining. The mystery is what a land-dwelling nodosaur was doing out at sea in the first place.

"The problem of ocean going nodosaurs is especially perplexing because the Como carcass, upside down at the bottom of the Benton Sea, is not an isolated instance," writes Robert Bakker. "Nodosaur carcasses lying on their backs cropped up in marine beds in Kansas in 1909 and several times since in similar sedimentary circumstances."[49] Bakker suggests that perhaps the dinosaurs died on shore, dried up in a drought, were then carried out to sea by a river flood, and buried before anything could scatter their remains.

Now, let me see if I have this straight. A nodosaur dies, but for some reason its body is not disposed of in the normal course of nature, i.e., no scavenger comes along and eats it. It dries up and mummifies in desert-like drought conditions. In spite of the drought, "flood swollen river waters" wash it out to sea. It is now out in the middle of the sea, with lots of toothy plesiosaurs, pliosaurs, ichthyosaurs, mosasaurs, and sharks, yet nothing eats it. It settles to the bottom, underbelly exposed, armor-plated back resting on the bottom, yet nothing comes along and eats it or *even disarticulates its remains.* This happens repeatedly; several nodosaurs have been found in this position. Does anyone find this explanation credible?

The explanation for the Hadrosaurs is even farther fetched. Hadrosaurs had no body armor; they were just sitting ducks (sitting duckbills, anyway). They are supposed to have wandered, or been washed, out to sea, drowned, and settled to the bottom without being molested by any of the fierce predators mentioned above. On the bottom of the sea, they did not decompose and were not scattered by the scavengers and bottom feeders whose ecological niche is to dispose of dead bodies, but instead remained articulated, with even the skin of one specimen being preserved.

These explanations, although offered in good faith, are wildly improbable. Nature does not work that way. A catastrophe prevented the predators, scavengers, and bottom feeders from playing their normal roles. The hadrosaurs and nodosaurs were buried in Flood-laid sediment, very rapidly, before nature could run its course. That the rocks where they are found are thought to be of marine origin reflects catastrophic mixing of marine and land fauna. It suggests a tidal wave of seawater sweeping away land animals and depositing them, along with marine animals, in a mixed sedimentary layer.

This interpretation is supported by the fact that half of the duckbills found in marine sediment were juveniles. (Recall that when Jack Horner went looking for juvenile dinosaurs, he headed for the marine sediments of the Bear Paw shale.[50]) Tidal waves and surges were more likely to sweep away juvenile dinosaurs than to carry off the much larger, heavier adults. Joe Taylor writes of a fascinating trackway in which a large dinosaur, possibly a hadrosaur, slipped backward in the mud and fell on its left hip, leaving a clear skin impression on the mud.[51] In these conditions, the larger hadrosaurs would have been able to regain their footing, whereas the juveniles would have been swept away into the ever-expanding "sea."

## F. THE CASE OF THE FASTING DINOSAURS

Let us take out our magnifying glasses again and examine a portion of the Morrison Formation near Delta, Colorado, known as Dry Mesa Quarry. Dry Mesa is the most prolific of all the Morrison quarries, containing seventeen of the twenty-one dinosaur genera found in the Morrison, including theropods, sauropods, and ornithischians.[52] The bones are found in a "lens" of sediment ranging from about two feet to four and a half feet thick. At average rates of sedimentation, four and a half feet represents thousands of years of geologic time.

*Yet no plant fossils of any kind have been found at Dry Mesa Quarry.* "Because no micro or macro plant fossils have been found in the quarry locale nothing can be stated about the flora of the Morrison Formation in the quarry region."[53]

Dry Mesa Quarry is not an isolated example. The entire Morrison Formation, covering one million square kilometers with an average thickness of 100 meters (330 feet), contains very little in the way of edible plants.[54] One researcher comments that, "although the Morrison plain was an area of reasonably rapid accumulation of sediment, identifiable plant fossils are practically nonexistent."[55] Another states that the Morrison Formation of Montana "is practically barren of plant fossils throughout most of its sequence."[56]

A paper by four leading paleontologists and dinosaur specialists notes that the "absence of evidence for abundant plant life in the form of coal beds and organic-rich clays in much of the Morrison formation is puzzling."[57] These experts also expressed frustration at not finding "palynomorphs," fossilized pollen and spores. Even these microscopic indicators of the presence of plants are relatively rare in the Morrison.[58] Everyone seems to agree that the climate of this purported ecosystem was dry or semi-arid.

Again, let me see if I have this straight. First, geologists tell us it took between four and seven million years to deposit the Morrison Formation. Second, the Morrison is famous for its sauropod remains, and a sauropod like *Brontosaurus* would consume more than a ton of green plants every day. (Estimates of the daily food intake of a large sauropod have ranged as high as three and a half tons.[59]) Third, the fossil evidence suggests that the climate was dry and there was not an abundance of greenery. Do you see the problem? *What were these dinosaurs eating?* Were they eating out of higher strata that would not be formed for millions of years—or lower strata that had already turned to stone millions of years before?

If the strata reflect normal ecologies over the course of thousands and millions of years, there should be evidence of abundant plant life as well as dinosaur bones. The Lyellian has an "anomalous and baffling" problem whenever he finds a substantial thickness of sedimentary rock that contains animal fossils but not plant fossils.[60]

The creationist has no such problem. We assume these formations were formed catastrophically and do not reflect a complete ecosystem. Plants could easily have been sorted by the floodwaters and buried at locations and levels different from those in which the dinosaurs were buried. Thus, we do not necessarily find plant fossils in the same strata and formations in which we find dinosaur bones.[61] The solution to the case of the fasting dinosaurs is clear: The dinosaurs were destroyed and buried by catastrophic flooding.

## G. WHISTLING PAST THE GRAVEYARD

Dinosaur bones are typically found in great clusters called "dinosaur graveyards." There are many such sites in North America; they were the gold mines of the early dinosaur prospectors. Paleontologists believe that most of these graveyards were formed when dinosaurs were overtaken and deposited by floodwaters. They do not, of course, believe that a universal Flood created these deposits, just a series of local floods or river floods.

Ghost Ranch, near Abiquiu, New Mexico, is a dinosaur graveyard first discovered in 1881 by David Baldwin, a fossil hunter who had worked for Marsh but was then working for Cope. Baldwin found the bones of an animal that Cope named *Coelophysis*.[62] (**Plate 31**) (Cope, being a "splitter," named several different species, but more recent research holds that these specimens were within the normal range of variation for one species.[63]) Using Baldwin's notes, Edwin Colbert rediscovered the site in 1947. Colbert's and subsequent expeditions dug up the remains of many fine specimens of *Coelophysis*; paleontologists estimate that there may be an additional 9,000 still in the ground.

The *Coelophysis* deposit is a remarkable concentration, suggesting that "the dinosaurs must have been together in some sort of herd when they were overtaken by disaster." The animals had been transported and deposited by strong currents. There was no evidence of predation, and the bones were generally unbroken, except that the tails were often broken off. The heads were often bent over the backs. "It seems likely," writes Spalding, "that the living *Coelophysis* were probably trapped unexpectedly by floodwaters or quicksand."[64] That 10,000 animals would simultaneously mire themselves in quicksand seems unlikely. The animals were obviously destroyed and buried in a flood.

Paleontologists believe that a river flood formed the fifteen-acre fossil graveyard at Hell Creek, Montana, where Barnum Brown found the first *Tyrannosaurus rex* and where Horner has recently found several more good specimens. At Hell Creek, paleontologists have found a fossil impression of dinosaur skin, which could only have been formed by very rapid, anaerobic burial. Paleontologists believe the dinosaurs at Dry Mesa Quarry were buried by water-deposited sediment in a river channel.[65] In Fruita Quarry, near Fruita, Colorado, thirty different fossil species have been quarried from sandstone obviously deposited by flowing water.

Dinosaur National Monument, in Vernal, Utah, is a vast dinosaur graveyard. Earl Douglass discovered the site in 1909. Over the course of fifteen years, Douglass collected about a dozen well-articulated sauropod skeletons from the site. In 1958, the site was opened to the public. In one of the quarries, the fossils were left embedded in the face of a hill and the entire site enclosed by a building.

The animals appear to have been thrown together with great

violence. Many of them were dismembered, with only portions of the vertebral columns still intact. Mixed in with the dinosaurs are large numbers of freshwater clams and snails, indication that the assemblage was deposited by water. Paleontologists have theorized that the animals drowned trying to ford a stream or were washed away during floods.[66] But the sediment is not typical river sediment, either in its mineral content or in the way it was deposited. The graveyard is the result of a complex catastrophic water-burial event.[67]

**Figure 24.2: The Quarry at Dinosaur National Monument**

Some of the graveyards seem to show living dinosaurs becoming trapped in deep mud. Samuel Williston, who collected for Marsh at Como Bluff, wrote that most of the quarries "are found containing remains of numerous individuals mingled together in the most inextricable confusion, and in every conceivable position, with connected limb bones standing nearly upright, connected vertebrae describing vertical curves, etc."[68]

The fact that many dinosaur graveyards contain articulated

skeletons indicates that the animals were buried quickly. If an animal dies and is not buried, scavengers and bacteria move in to dispose of the body. The body is de-fleshed, and the bones quickly become disarticulated and scattered. Eventually, even the bones disintegrate.

Malcolm Coe studied this phenomenon in Kenya in the early 1970s, when a prolonged drought killed thousands of elephants. The elephants' bones flaked and cracked within five weeks of death. Within a year, the bones were scattered far and wide. Only if bones were quickly and permanently buried in sediments carried by a flooding river did they stand much chance of being preserved.[69] The degree of skeletal disarticulation correlates to the time elapsed between mortality and burial. The better the condition of the preserved skeleton, the faster it was buried. Only raging floodwaters are able to deliver sufficient quantities of sediment quickly enough to preserve in articulation the bones of large animals.[70]

"Most fossil bones owe their preservation to quick burial by sediment right after the death of the owner," writes Bob Bakker.[71] Sue, the eight-million-dollar *T. rex*, illustrates this principle. "She died in or near a river and was quickly buried," says paleontologist Bill Simpson, "which is wonderful for us because that means the skeleton didn't rot away."[72] The same was true of a *T. rex* excavated by John Horner's Museum of the Rockies in 1990: "The large size and articulated state of the skeleton . . . suggest minimal transport within the channel and rapid burial. Lack of bone surface modification by scavenging, transport, or weathering, combined with the high degree of articulation, indicate rapid in-channel burial of an essentially intact carcass."[73]

One can imagine how much sediment was needed to bury a forty-five-foot "king of the tyrant lizards," and it would have taken much more to bury a huge sauropod, such as a *Brachiosaurus*, *Camarasaurus*, or *Brontosaurus*. "Big bones, such as those of dinosaurs," writes Bakker, "required big floods of mud to cover them, . . ."[74] *Big floods of mud.* That raises an interesting issue. The Morrison Formation in the area of Dry Mesa Quarry is about 125 meters (412 feet) thick, and geologists believe it took anywhere from four to seven million years to be deposited.[75] Dividing the thickness of the formation by the number of years it supposedly took to form—let's be conservative and take the lower figure of four million (4,000,000) years—the average annual sedimentation was about .031 millimeters. That is about 1/30th of a millimeter per year, or .0012 inches, about 1/900th of an inch, per year.

Obviously, 1/900th of an inch of sediment will not cover a dinosaur carcass. Several feet of mud would have been required to bury a large sauropod. The fact that many dinosaur skeletons are preserved in articulation shows that the sediment was not deposited slowly and gradually but suddenly and in great torrents—"big floods of mud." What basis is there, then, for insisting that it took millions of years to form the sedimentary deposits?

Paleontologists acknowledge, of course, that it would have taken huge quantities of sediment to bury the dinosaurs, but they argue that the episodes of catastrophic deposition were separated by thousands of years of non-deposition and erosion. Bakker, whom we interrupted in mid-sentence, writes, "Big bones, such as those of dinosaurs required big floods of mud to cover them, and these events didn't happen except at long intervals, perhaps hundreds or thousands of years apart."[76] There is no way to disprove that belief. But that is all it is: a belief, an interpolation imposed upon the observable evidence.

The burial of the dinosaurs in their numerous and vast graveyards suggests a catastrophic flood depositing unimaginably vast quantities of sediment. Biblical creationists accept this fossil record. Catastrophists are, as Stephen Jay Gould wrote, "the true champions of a geological record read as directly seen."[77] Lyellians, by contrast, impose their theory on the fossil record, in this case by imagining vast periods of non-deposition separating the great cataracts of sedimentation necessary to bury the dinosaurs.

## H. THE CASE OF THE DROWNING DINOSAURS

Some dinosaurs are found in a pose indicating that they were drowning when they were buried. Of the duckbill mummies found in Wyoming's Lance Creek beds, Bakker writes: "The Wyoming mummies . . . were nearly entire bodies, preserved lying on their backs, their chests expanded *as if in a last gasp for breath,* both fore and hind feet extended outward in an agonized pose."[78]

Many dinosaurs have been found with their necks arched back over their bodies, just as if, in their final moments, they were straining to keep their heads above water. Paleontologists have encountered this pose so frequently that they have developed an explanation for it. They believe that *rigor mortis* tightened the tendons on the backs of the dinosaur's necks, pulling the neck and head back over the body.[79] But is this really a good explanation? If the animals were buried in several feet of mud, as their articulated skeletons indicate, the weight of the sediment would have prevented any post-mortem movement of the body.

We now have more than enough evidence to indict the dinosaurs' killer: catastrophic flooding. The mass destruction of incredibly vast herds of dinosaurs, the dinosaur graveyards, the numerous articulated skeletons, the preservation and stratigraphic arrangement of dinosaur tracks, and even the eggs that didn't hatch, all testify that catastrophic flooding killed the dinosaurs. The dinosaur fossils are dramatic evidence of a former world destroyed by water. The Bible is reliable, trustworthy, and true when it states that, "the world of that time was deluged and destroyed." (2 Peter 2:6)

---

[1] Horner, John R., and James Gorman, *Digging Dinosaurs* (New York: Workman Publishing, 1988), p. 14 (in the introduction by James Gorman).

[2] Horner and Gorman, at 34.

[3] Horner and Gorman, at 49.

[4] Horner and Gorman, at 128. See also Weishampel, David B. and John R. Horner, "Hadrosauridae," in *The Dinosauria*, Weishampel, Dodson, Osmolka, Eds. (Berkeley, CA: University of California Press, 1990), p. 561.

[5] Horner and Gorman, at 131.

[6] Horner and Gorman, at 128.

[7] Spalding, David A.E., *Dinosaur Hunters* (Rocklin, CA: Prima Publishing, 1993), pp. 193, 194.

[8] Spalding, at 194, 195.

[9] Spalding, at 192, 193.

[10] Dodson, Peter, *The Horned Dinosaurs* (Princeton: Princeton University Press, 1996), p. 261 (italics in the original).

[11] Some dinosaurs are estimated to have lain as many as 100 eggs per year. Roth, Ariel A., *Origins: Linking Science and Scripture* (Hagerstown, MD: Review and Herald, 1998), p. 242, citing Paul, G.S., "Dinosaur reproduction in the fast lane: implications for size, success, and extinction," in *Dinosaur Eggs and Babies*, Carpenter, Hirsch, and Horner, eds. (Cambridge University Press, 1994), pp. 244-255.

[12] Roth, at 243, citing Hirsch, K.F., "Upper Jurassic eggshells from the western interior of North America," in Carpenter, Hirsch, and Horner, supra; Hirsch, K.F., Stadtman, K.L., Miller, W.E., and Madsen,J.H, "Upper Jurassic dinosaur egg from Utah," *Science*, 243:1711-1713 (1989); Erben, H.K., Hoefs, J., Wedepohl, K.H. "Paleobiological and isotopic studies of eggshells from a declining dinosaur species," *Paleobiology*, 5(4):380-414 (1979); Zhao, Z-K, "Dinosaur eggs in China: on the structure and evolution of eggshells," in Carpenter, Hirsch, and Horner, supra.

[13] Horner and Gorman, at 142-145.

[14] Horner and Gorman, at 156, citing J.R. Horner and D. B. Weishampel, "A Comparative Embryological Study of Two Ornithischian Dinosaurs," *Nature*, 332:256-257 (1988). Bob Makela passed away at a dig site in 1987.

15 Horner and Gorman, at 146, 161.

16 Leonard Brand interprets the Horner "nests" along these lines. See, Leonard Brand, *Faith, Reason, & Earth History* (Berrien Springs, MI: Andrews University Press, 1997), p. 293.

17 Varricchio, Jackson, Borkowski and Horner, "Nest and egg clutches of the dinosaur *Troodon formosus* and the evolution of avian reproductive traits," *Nature* 385:247-250 (Jan. 16, 1997).

18 Chiappe, Luis M., Lowell Dingus, and Nicholas Frankfurt, *Walking on Eggs: The Astonishing Discovery of Thousands of Dinosaur Eggs in the Badlands of Patagonia* (New York: Scribner, 2001), p. 23.

19 Associated Press, "First embryos of big dinosaurs found in Argentina," November 17, 1998, quoting Luis Chiappe.

20 "First dinosaur embryos found with fossilized skin: A firsthand report of an expedition to Patagonia" Exhibition at the American Museum of Natural History; on the web at: http://www.amnh.org/exhibitions/expeditions/dinosaur/patagonia. See also Chiappe, et al., *Walking on Eggs*, supra, at pp. 117-123.

21 Roth, at 243, citing, Carpenter, K. Alf, K., "Global distribution of dinosaur eggs, nests, and babies," in Carpenter, Hirsch, and Horner, supra.

22 Lockley, Martin, and Adrian P. Hunt, *Dinosaur Tracks and other Fossil Footprints of the Western U.S.* (New York: Columbia University Press, 1995), p. 155.

23 See, e.g., Glen J. Kuban, "An overview of Dinosaur Tracking," *Mid-America Paleontology Society Digest*, April 1994. On the web at http://paleo.cc/paluxy/ovrdino.htm.

24 Kuban, supra. There would be no need for a covering layer of sediment if the mud in which the track was left had the quality of quick-drying cement, and the resulting stone was exceptionally hard.

25 Lockley, Martin G., "The Paleoecological and Paleoenvironmental Utility of Dinosaur Tracks," in *The Complete Dinosaur*, James O. Farlow and M.K. Brett-Surman, Eds. (Bloomington, IN: Indiana University Press, 1997), at p. 569 ("However, further work indicates that megatracksites form during rising sea level, which causes sediments along the coastline to aggrade or accumulate. . . . Sequence stratigraphic studies show that rising sea level causes aggradation and leads to enhanced preservation of sediments and the tracks and other fossils they contain." "The bottom diagram shows how such aggradation of sedimentary rock layers during sea level rise will result in the accumula-

tion of series of rock layers containing footprints, each layer younger than the one below it, and each formed later in the course of flooding of the continental margin.")

[26] Lockley and Hunt, at 155, 183.

[27] See, e.g., Brand, Leonard, and James Florence, "Stratigraphic distribution of vertebrate fossil footprints compared with body fossils," *Origins* 9(2):67-74 (1982) ("rapid sedimentary processes during the global flood were usually conducive to the burial and preservation of footprints before they were destroyed by weathering processes. Thus during the flood there would have seen a great diversity of reptiles and amphibians making tracks under conditions uniquely suitable for preserving those tracks.")

[28] Kuban, supra, writes that in addition to the "classic" method of track formation and preservation, "recent research suggests another mechanism of print formation, which involves a dinosaur walking on a very soft surface. In such a case, the animal's feet may push into firmer layers below the surface. The soupy surface material may then rush back over the upper depressions, simultaneously covering the prints made in the lower layers. The subsurface prints are known as underprints, undertracks, or ghost tracks. Because they are buried as soon as they are made, any erosion or other destructive forces occurring at the surface would pose no threat to them, increasing their chances of being preserved."

Interestingly, Kuban uses this method to explain some of the Paluxy River tracks that some creationists have claimed are giant man tracks. But this method of track formation is, if anything, even more consistent with flood conditions than the "classic" method. It obviates the need for the track to be exposed to air before being covered by more waterborne sediment. It is strongly suggestive of dinosaurs wading through shallow water, leaving tracks on top of the previously laid sediment. Leonard Brand has done field research and conducted experiments suggesting that most of the Coconino sandstone amphibian tracks were made in shallow water conditions. See Brand, Leonard, "Footprints in the Grand Canyon," *Origins*, 5(2):64-82 (1978)

[29] Lockley, Martin, and Christian Meyer, *Dinosaur Tracks and other Fossil Footprints of Europe* (New York: Columbia University Press, 2000), p. 67.

[30] Lockley and Meyer, at 67-69.

[31] Lockley and Meyer, at 71.

[32] Lockley, Martin, and Adrian P. Hunt, *Dinosaur Tracks and other Fossil Footprints of the Western U.S.* (New York: Columbia University Press, 1995), pp. 75, 76.

[33] Lockley and Meyer, at 122-125. Stegosaurus is not the only example. Ornithomimid

tracks have been identified in lower Jurassic sediments, but the bones of an ornithomimid have not been found below the lower Cretaceous. So in that case, the tracks appear some fifty million years before the bones. See Lockley, "The Paleoecological and Paleoenvironmental Utility of Dinosaur Tracks," in *The Complete Dinosaur*, supra, p. 557.

34 See Paul Sereno's dinosaur family tree (**Plate 15**).

35 Lockley and Hunt, at 112-116.

36 Lockley and Hunt, at 112-116; 134, 135; 149-153.

37 Lockley and Hunt, at 150.

38 Lockley and Hunt, at 150. The absence of bodies in the middle Jurassic is not a local phenomenon, but a universal one. "[T]he Aalenian (mid-Jurassic) has shown no known tetrapod fossils anywhere in the world . . ." Kathleen Hunt, "Transitional Vertebrate Fossils FAQ," at Talk.Origins website: http://www.talkorigins.org/faqs/faq-transitional/part1a.html. See also, "New rare dinosaur tracksite found in northern Wyoming," Indiana University office of Communications and Marketing, November 16, 2000, quoting Indiana University geologist Erik Kvale.

39 Lockley and Hunt, at 113 (in their caption to the diagram that is our Figure 24.1).

40 Commenting on the hundreds of animal trackways found in the lower half of the Coconino Sandstone of northern Arizona, Ariel Roth notes that almost all of the trackways in the Coconino, as well as trackways in the de Chelly sandstone to the east, indicate that the animals were traveling uphill when they made the tracks. "Is it possible," asks Roth, "that animals seeking to escape the waters of the deluge left all those uphill tracks?" Roth, at 221, citing C.W. Gilmore, "Fossil footprints from the Grand Canyon: Second contribution," Smithsonian Miscellaneous Collections 80(3):1-78 (1927); Lockley, Hunt and Lucas, "Abundant ichnofaunas from the Permian De Chelly Sandstone, northeast Arizona: implications for dunefield paleoecology," Geological Society of America Abstracts With Programs, 26(7):A374 (1994); P.P. Vaughn, "Vertebrates form the Cutler Group of Monument Valley and vicinity," in *Guidebook of Monument Valley and Vicinity, Arizona and Utah*, H.L James, ed., (New Mexico Geological Society, 1973), pp. 99-105. See also Brand, Leonard, "Footprints in the Grand Canyon," *Origins*, 5(2):64-82 (1978); Roth, Ariel, "Incomplete Ecosystems," *Origins*, 21(1):51-56 (1994).

41 Lockley and Hunt, at 152-155.

[42] Lockley and Hunt, at 153-155.

[43] See, e.g., Britt, Brooks B., "Theropods of the Dry Mesa Quarry," *Brigham Young University Geology Studies*, vol. 37 (1991) pp. 1-72. ("The Morrison Formation conformably overlies the uppermost, beach-type, laminar sand unit of the Summerville Formation. The 33 meters [107 feet] of Summerville Formation lies above the cliff forming, large scale cross-bedded Entrada Sandstone and below the Morrison Formation.").

[44] Lockley and Hunt, at 154.

[45] Milton, Richard, *Shattering the Myths of Darwinism* (Rochester, VT: Park Street Press, 1997), pp. 23, 68, 79.

[46] Weishampel, David B. and John R. Horner, "Hadrosauridae," in *The Dinosauria*, supra, p. 561, citing Horner, J.R. "Upper Cretaceous dinosaurs from the Bearpaw Shale (marine) of south-central Montana with a checklist of Upper Cretaceous dinosaur remains from marine sediments of North America," *Journal of Paleontology*, 53:566-578 (1979). See also Horner and Gorman, at 23-25.

[47] Brett-Surman, M.K., "Ornithopods," in *The Complete Dinosaur*, supra, at p. 342-344.

[48] Weishampel, Horner, supra, at pp. 560, 561.

[49] Bakker, Robert T., *The Dinosaur Heresies* (New York: Kensington Pub., 1986), p. 40.

[50] Horner and Gorman, at 23-25, 32.

[51] Taylor, Joe, *Fossil Facts and Fantasies* (Crosbyton, TX: Mt. Blanco Publishing Co., 1999) p. 45.

[52] Britt, supra.

[53] Britt, supra.

[54] Roth, supra, at 219; Britt, supra, citing, Dodson, P., Behrensmeyer, A.K., Bakker, R.K., and McIntosh, J.S., "Taphonomy and paleoecology of the dinosaur beds of the Jurassic Morrison Formation," *Paleobiology*, 6[2]:208-232 (1980); Stokes, W.L., "Morrison Formation and related deposits in and adjacent to the Colorado Plateau," *Bulletin of the Geological Society of America*, 55:951-992 (1944).

[55] Roth, at 220, quoting Theodore White, "The Dinosaur Quarry," in *Guidebook to the geology and mineral resources of the Uinta Basin*, E.F. Sabatka, ed. (Salt Lake City: Intermountain Association of Geologists, 1964), pp. 21-28.

56 Roth, at 220, quoting R.W. Brown, "Fossil plants and Jurassic-Cretaceous boundary in Montana and Alberta," *American Association of Petroleum Geologists Bulletin*, 30:238-248 (1946).

57 Roth, at 220, quoting Dodson, Behrensmeyer, Bakker and McIntosh, supra;

58 Britt, citing Tschudy, B.D., Van Loenen, S. and Doher G., "Illustrations of plant microfossils from the Morrison formation; I, Plant microfossils from the Brushy Basin Member," *United States Geological Survey Open-File Report*, 81-0035 p. 46 (1981) and Dodson, et al., (1980).

59 Roth, at 220, citing Theodore White, "The Dinosaur Quarry."

60 The Morrison is not the only case of an incomplete ecosystem. Ariel Roth notes that the *Protoceratops* fossils found in the Cretaceous of Mongolia are also found in strata that do not yield many plant fossils. Researchers call the absence of well-developed plant colonization "anomalous and baffling." Roth, citing Fastovsky, D.E., Badamgarav D., Ishimoto H., Watabe M., Weishampel, D.B., "The paleoenvironments of Tugrikin-Shireh (Gobi Desert, Mongolia) and aspects of the taphonomy and paleoecology of *Protoceratops* (Dinosauria: Ornithischia)," *Palaios* 12:59-70 (1997).

61 Roth, at 221, 222. See also Roth, "Incomplete Ecosystems," *Origins*, 21(1):51-56 (1994). Found at: http://www.grisda.org/origins/21051.htm.

62 Spalding, at 171, 172.

63 Spalding, at 179.

64 Spalding, at 179.

65 Britt, supra.

66 Whitcomb, John C., and Henry Morris, *The Genesis Flood* (Phillipsburg, NJ: P & R Publishing, 1961), p. 280, citing J. M. Good, T.E. White, and G. F. Stucker: "The Dinosaur Quarry," U.S. Government Printing Office, 1958, p. 26.

67 Hoesch, William A, and Steven A. Austin, "Dinosaur National Monument: Jurassic Park or Jurassic jumble? ICR Impact no. 370, April 2004.

68 Wilford, John Noble, *The Riddle of the Dinosaur* (New York: Vintage Books, 1985) p. 120.

69 Helder, Margaret, "Dinosaur Disasters" Creation Science Association of Alberta website, citing *Discover* January 1981, p. 85.

[70] Helder, supra.

[71] Bakker, at 44.

[72] "It's not easy rebuilding Sue: Museum readies to assemble its T-Rex" CNN Interactive, October 23, 1997, Lisa Price, contributing correspondent. http://www.cnn.com/TECH/9710/23/operation.sue/

[73] Mary H. Schweitzer, James G. Schmitt, John R. Horner, "Paleoenvironmental and taphonomic parameters of an exceptionally preserved *Tyrannosaurus rex.*" GSA Annual Meeting, Nov. 5-9, 2001, Paper 160-0.

[74] Bakker, at 45.

[75] Britt, supra. ("The total thickness of the Morrison Formation in the area is approximately 125 meters [412 feet], but the exact thickness is not known because the formation is difficult to distinguish from the overlying Burro Canyon Formation.")

[76] Bakker, at 45.

[77] Gould, Stephen Jay, *Time's Arrow, Time's Cycle* (Cambridge, MA: Harvard University Press, 1987) pp. 133, 134.

[78] Bakker, at 148 (Italics added).

[79] Bakker, at 427; Spalding, at 179. Early in the movie "Jurassic Park," Dr. Alan Grant (a fictional John Horner, played by actor Sam Neill) and his crew are digging dinosaurs in the badlands of Montana. They are using some sort of ground-penetrating radar or seismic reflection technology to generate a computer image of an articulated Velociraptor skeleton that is still buried underground. The Velociraptor is in the familiar pose with its neck bent backward, and Dr. Ellie Satler, the paleobotanist (played by actress Laura Dern), comments, "post-mortem contraction of the posterior neck ligaments."

CHAPTER 25

# Were All Dinosaurs Destroyed in the Flood?

There is no physical evidence that dinosaurs survived the Flood. Their bones have never been found in strata deposited after the Flood. Many other species of animals survived the Flood but subsequently became extinct. Had dinosaurs been taken aboard the ark and, after the Flood, reproduced at roughly the same rate as the other species, we would expect to find their remains in post-Flood strata such as the Pleistocene rocks, which were deposited during the Ice Age. That no dinosaur bones or tracks have ever been found in post-Flood strata is compelling evidence that the dinosaurs did not survive the Flood.

If dinosaurs were the result of antediluvian amalgamation, God would not have allowed them to survive the Flood. One of the purposes of the Flood was to wipe out the amalgamated creatures, giving the world a fresh start:

> Every species of animal which God had created were preserved in the ark. The confused species which God did not create, which were the result of amalgamation, were destroyed by the Flood.[1]

The dinosaurs were not the only confused species produced by antediluvian genetic engineering. Think of the countless species of mammal-like reptiles, sea-going reptiles, and flying reptiles. Imagine a world in which all of these were living at the same time, and you can understand why Ellen White used the phrase "confusion everywhere." To take these amalgamated creatures aboard the ark, and thereby preserve them, would have defeated one of the purposes of the Flood.

It is possible that God created some of the dinosaurs during creation week. These, too, were destroyed in the Flood because of their size. Had the largest dinosaurs survived the Flood, man could not have controlled them. Antediluvian men were much taller and stronger than modern men. Ellen White states that because of the post-Flood decrease in man's size and strength, men simply could not have controlled the dinosaurs:

> Soon after the flood the race began to rapidly decrease in size, and in length of years. *There was a class of very large animals which perished at the flood.* God knew that the strength of man would decrease, and these mammoth animals could not be controlled by feeble man.[2]

Would anyone really want to come across a live *Tyrannosaurus rex* out in the wild? Even the presumably gentle, herbivorous *Brontosaurus* would be way too much to handle—he was eighty-five feet long and weighed over thirty tons!

But God's decision to destroy the dinosaurs in the Flood—and not to save any representatives on the ark—cannot be explained solely on the basis that the dinosaurs were too large for modern man to handle. Most of the dinosaurs were eight feet long, or smaller. Modern men are able to cope with wild animals of this size. It would seem that most of the dinosaurs, along with many other Paleozoic and Mesozoic vertebrates, were destroyed because they were amalgamated creatures.

I believe that all of the dinosaurs were destroyed in the Flood, and that none were saved on the ark. This belief is based upon the stratigraphical location of the dinosaur bones and tracks. We find no trace of dinosaurs above the Cretaceous, which was surely deposited during the Flood. The prophetic ministry of Ellen White explains why these creatures were not brought aboard the ark: either because they were confused species that God did not create, or because they were too large and God knew that drastically smaller and weaker post-Flood men would not be able to control them.

In our conviction that dinosaurs were not on the ark, we part company with most biblical creationists. Many creationists believe that dinosaurs were brought aboard the ark, either as juvenile specimens or as eggs, but became extinct after the Flood as a result of the more severe post-Flood climate. This note from Whitcomb and Morris' "The Genesis Flood" is typical of biblical creationist thought on this issue:

> If representative dinosaurs were taken on the Ark (presumably young ones), then it is likely that their final extinction is accounted for by the sharp changes in climate after the flood. On the other hand, some may have persisted for a long time, possibly accounting for the universal occurrence of 'dragons' in ancient mythologies.[3]

We also part company with our own Seventh-day Adventist scholar, Frank Lewis Marsh, who asserted that some dinosaurs, such as *Compsognathus*, were brought aboard the ark, whereas others, such as the carnivorous *Tyrannosaurus* or the ungainly *Brontosaurus*, were not.[4] But although small, Compsognathus was an amalgamated creature, combining elements of birds and reptiles, and was thus slated for destruction in the Flood.[5]

It is worth exploring why most creationists believe that dinosaurs survived the Flood, and why a few cling to the increasingly forlorn hope that someday, somewhere, someone will find a living dinosaur, or at least the fossil remains of a dinosaur in post-Flood strata.[6]

## A. WERE ALL ANIMALS THAT LIVED BEFORE THE FLOOD CARRIED ON THE ARK?

The main reason many biblical creationists maintain that dinosaurs were brought aboard the ark is that Genesis 6:19, 20 indicates that at least one pair of each kind of animal was taken aboard:

> 19. And of every living thing of all flesh, you shall bring two of every sort into the ark, to keep them alive with you: they shall be male and female.
>
> 20. Of the birds according to their kinds, and of the animals according to their kinds, of every creeping thing of the ground according to its kind, two of every sort shall come in to you, to keep them alive (RSV).

In verse 19, the use of the term *all flesh* seems to be categorical, and allow no exceptions. Verse 20, however, is a more specific description. It specifies birds, "according to their kinds," animals, "according to their kinds," and every creeping thing, "according to its kind." An important rule of construction is that if there is an apparent conflict between a specific definition and a more general definition, the specific definition is preferred to the general. Verse twenty's specific description of the animals as "according to their kinds" takes precedence over the more general "all flesh" in verse nineteen.

And the phrase *according to its kind* is important. It is the same Hebrew terminology used in Genesis 1:11, 12, 21, 24, and 25, in describing the original creation:

> 21. So God created the great creatures of the sea and every living and moving thing with which the water teems, *according to their kinds,* and every winged bird *according to its kind.* And God saw that it was good. . .
>
> 24. And God said, "Let the land produce living creatures *according to their kinds*: livestock, creatures that move along the ground, and wild animals, each *according to its kind.*" And it was so.
>
> 25. God made the wild animals *according to their kinds,* the livestock *according to their kinds,* and all the creatures that move along the ground *according to their kinds.* And God saw that it was good (NIV, emphasis added.)

By using the same language to describe the animals brought aboard the ark as was used to describe the animals originally created, Genesis 6:20 implies that *only the created "kinds" were brought aboard the ark.*[7]

Genesis 6:20 indicates that only the originally created "kinds" were to be taken aboard the ark, and Ellen White testifies that "every species of animal which God had created were preserved in the ark." Thus, there is no conflict between the Bible and the prophetic ministry of Ellen White. There is perfect harmony.

As already noted, moreover, were we to believe that literally every animal that lived before the Flood was preserved on the ark, we are at a loss to explain the total absence of several classes of animal remains from post-Flood strata. The truth that amalgamated creatures were left off the ark and destroyed in the Flood solves a serious problem for creationists.

## B. WHAT WERE THE "BEHEMOTH" AND "LEVIATHAN" OF JOB?

A theory popular in creationist circles today is that the creature or creatures described in Job chapters 40 and 41, and called "behemoth" and "leviathan," were dinosaurs. Because Job is a post-Flood book (as are all of the books of the Bible), dinosaurs must, it is argued, have survived into the post-Flood world. The description of the "behemoth" in Job 40:15-24 is as follows:

> Look at the behemoth, which I made along with you and which feeds on grass like an ox. What strength he has in his loins, what power in the muscles of his belly! His tail sways like a cedar; the sinews of his thighs are close-knit. His bones are tubes of bronze, his limbs like rods of iron. He ranks first among the works of God, yet his Maker can approach him with his sword. The hills bring him their produce, and all the wild animals play nearby. Under the lotus plants he lies, hidden among the reeds in the marsh. The lotuses conceal him in their shadow; the poplars by the stream surround him. When the river rages, he is not alarmed; he is secure, though the Jordan should surge against his mouth. Can anyone capture him by the eyes, or trap him and pierce his nose? (NIV.)

The word *behemoth* means "beast par excellence." Most commentators believe that this term is intended to describe a large land animal, either an elephant or a hippopotamus. Some commentators believe that the word behemoth is derived from Egyptian and means "water-ox," or "Nilehorse." Advocates of the theory that "behemoth" is a dinosaur point to the description of the tail as being "like a cedar," which is not consistent with a hippopotamus but is evocative of the thick, bony and powerful tails of the dinosaurs.

On the other hand, the animal in question lies "under the lotus plant, hidden among the reeds in the marsh. The lotuses conceal him in their shadow; the poplars by the stream surround him." That he is "hidden by the reeds in the marsh" seems to indicate that the animal in question was not as large as a sauropod dinosaur, which towered above the trees. The phrase, "can anyone capture him by the eyes?" can also be translated, "can anyone capture him by a water hole?" Although it was once believed that sauropod dinosaurs had to stand in water to help support their staggering weight—and they were depicted as swamp dwellers by famous dinosaur illustrators Charles R. Knight (1874-1953) and Rudolph F. Zallinger (1919-1985)—paleontologists no longer believe that the sauropods were swamp dwellers.[8] Thus, although a hippo lacks a tail that "sways like a cedar," the rest of the passage does not describe a dinosaur.

While the meaning of "behemoth" involves guesswork, the majority of commentators and translators believe that the "leviathan" of Job 41 was a crocodile. The creature described has much in common with the crocodile:

> Can you make a pet of him like a bird or put him on a leash for your girls? . . . If you lay a hand on him, you will remember the struggle and never do it again! . . . Any hope of subduing him is false; the mere sight of him is overpowering. . . . Who dares

> open the doors of his mouth, ringed about with his fearsome teeth? His back has rows of shields tightly sealed together; each is so close to the next that no air can pass between. They are joined fast to one another; they cling together and cannot be parted. . . . Strength resides in his neck . . . the folds of his flesh are tightly joined, they are firm and immovable. When he rises up, the mighty are terrified; they retreat before his thrashing. . . . His undersides are jagged potsherds, leaving a trail in the mud like a threshing sledge. He makes the depths churn like a boiling cauldron . . . behind him, he leaves a glistening wake.

Some of the convergences between the "leviathan" and the crocodile are the teeth, the toughness of the back of the animal, and the fact that it leaves a trail in the mud, yet also moves through the water, leaving a wake.

The author of Job seems to indulge in some poetic exaggeration in his description of leviathan:

> . . . his snorting throws out flashes of light; his eyes are like the rays of dawn. Firebrands stream from his mouth; sparks of fire shoot out. Smoke pours from his nostrils as from a boiling pot over a fire of reeds. His breath sets coals ablaze, and flames dart from his mouth. Job 41:18-21 (NIV)

That passage describes what we would call a dragon. As far as I am aware, paleontologists do not believe that dinosaurs could belch fire and smoke.

The "leviathan" of Job 41 cannot be a dinosaur because the term "leviathan" always describes a sea creature. The term is sometimes used in a figurative or symbolic sense, but the setting is always the ocean.[9] For example, leviathan is referred to in Isaiah 27:1 as "the monster of the sea." Because the dinosaurs were land-dwelling animals, the term "leviathan" clearly does not describe a dinosaur.

It might still be argued that the passage refers to one of the marine reptiles of the Mesozoic, such as a plesiosaur, implying that such creatures survived the Flood. But the animal described in Job 41 is formidable both on the land and in the water. The crocodile fits the bill; the Mesozoic marine reptiles do not. Ichthyosaurs and mosasaurs would have been immobile and defenseless when beached. Plesiosaurs and pliosaurs had flippers, not legs, and while some have argued that they could, like sea turtles, have crawled on shore to lay eggs, they would have been slow, awkward, and ungainly out of the water.

A saltwater crocodile could fill the role of "sea monster" implied by the term "leviathan." Saltwater crocs have grown to twenty-five feet in the wild—which certainly qualifies as a "sea monster"—and one grew to over eighteen feet in captivity—easily large enough to be lethal to man. While the passage is somewhat mysterious, most of the evidence points to a crocodilian. There is no compelling reason to believe the passage refers to a dinosaur, or any other extinct Mesozoic vertebrate.

The translators of the New English Bible have taken an interesting approach to this passage. They treat Job 40:15 through 41:34 as one continuous description of a single animal—that animal being a crocodile. The chapter divisions, added more than a thousand years after the book was first committed to writing, affect our reading of the text. Because there are two different chapters, we subconsciously expect that that they describe two different animals. Take away the chapter divisions and there is no reason to think that a new discussion of a different creature has begun. Job 40:15-24 perfectly describes the habitat and characteristics of a crocodile—provided "feeds on oxen like grass" is substituted for "feeds on grass like an ox," as the translators of the New English Bible have done—and finally makes sense of the statement that the tail "sways like a cedar." The crocodile is both behemoth and leviathan, both a "beast par excellence" and a "sea monster."

## C. DO POST-FLOOD ARTIFACTS PROVE DINOSAURS SURVIVED THE FLOOD?

Artifacts occasionally surface that seem to indicate that post-Flood man coexisted with dinosaurs. There is a carving on a wall in the Havasupai Canyon in northern Arizona that could be interpreted as a theropod.[10] There is an etching at Natural Bridges National Monument near Blanding, Utah, that looks like a sauropod.[11] There are also a couple of large collections of artifacts that warrant further discussion because they include depictions of humans coexisting with dinosaurs.

In 1945, Waldemar Julsrud, a local hardware merchant, discovered a curious pottery figurine near Acambaro, Guanajuato, Mexico. He agreed to pay a man one peso for each such figurine the man could find and deliver in one piece to Julsrud. Julsrud eventually amassed a huge collection of about 33,500 statues and figurines; it occupied every room of his twelve-room house. A number of these figurines bear an uncanny resemblance to various types of dinosaurs. Samples from the figurines were dated by radiocarbon, yielding dates between 1,100 B.C. and 4,500 B.C. The figurines have also been dated by thermoluminescence, a technique used to date pottery, yielding dates around 2,500 B.C.[12]

A similar discovery was made in Peru, in the province of Ica. The Nazca people of Peru are famous for having produced various shapes and figures, discernable only from the air, on the flat desert floor. The Nazca often buried their dead with carved stones. In

1969, some of these stones came to the attention of Dr. Javier Cabrera Darquea, an instructor of medicine at the University of Lima, Peru. Dr. Cabrera collected over 11,000 of these engraved stones and displayed his collection at his house on the Plaza de Armas in the city of Ica.[13]

**Figure 25.1: A Stegosaurus-like figurine from Julsrud's collection**

Some of these carvings accurately depict dinosaurs. For example, there is a depiction of two theropod dinosaurs fighting with a man; another stone shows a man riding a *Triceratops*. Two of the stones are displayed by the Creation Evidences Museum in Glen Rose, Texas. Skeptics charge that the anomalous stones are recent forgeries. Some of the stones clearly are fakes; a local man admitted to having carved some of them. But people familiar with the stones claim that the recent forgeries can be distinguished from the genuine article by a patina of age that covers the carved grooves of the genuine stones. There is no way to date them, radiometrically or otherwise, except by cultural affiliation.

Both the Julsrud figurines and the Ica stones depict sauropod dinosaurs with protrusions along their backs, visually similar to the triangular plates along the backs of *Stegosaurus*. Until recently, scientists did not believe that sauropods had triangular frills along their spines. But in 1992, Stephen Czerkas made a discovery that may explain the curious sauropod depictions. He found the fossilized impressions of skin from a *Diplodocus* at Howe Ranch Quarry. The animal had dermal (skin) spines projecting upward from the dorsal mid-line of the end of the tail.

Czerkas believes the spines may have continued up the tail

and along the back and neck of *Diplodocus.* The spines were never shown in artists' depictions of sauropods before 1992 because they were not known to exist.[14] If sauropods commonly had dermal frills along their spines, it would indicate that the Julsrud figurines and the Ica stones are not forgeries. It would take an uncannily prescient forger to anticipate, in 1945 or 1969, an anatomical feature that was not even suspected until 1992.

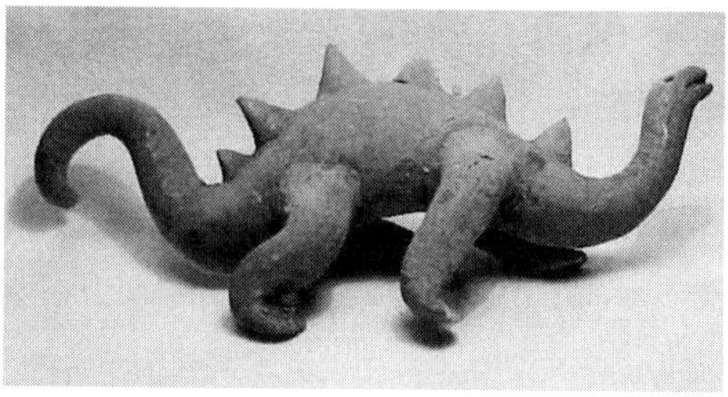

**Figure 25.2: A sauropod figurine from the Julsrud collection—but what are those triangular things on the spine?**

If these artifacts are genuine, at least one of two things must be true. Either 1) post-Flood humans were able to remember dinosaurs as part of their history, or 2) dinosaurs lived to fairly recent times. Mainstream scientists, holding that men and dinosaurs never lived together and were separated by sixty-five million years, must utterly debunk all of these artifacts. We do not find it necessary to do that. If and to whatever extent they are genuine, they do not necessarily prove that dinosaurs survived the Flood. They could merely reflect a vibrant cultural memory of a time when men and dinosaurs shared the earth.

Remember that the early post-Flood peoples had extremely high intelligence and very strong memories. Ice Age cave art indicates that they also had excellent artistic abilities. Early post-Flood explorers of the Western Hemisphere could have left the carvings at Havasupai and Natural Bridges.[15] The thermoluminescence dating of the Julsrud figurines at 2,500 B.C. is consistent with the figurines having been made by early post-Flood settlers of the Western Hemisphere, who would still have had a cultural memory of the history of pre-Flood dinosaurs.

The Peruvian burial stones, which seem to indicate a much

later culture, are more problematic but fall short of proving the survival of dinosaurs into post-Flood times. If anthropology and archeology are ever seriously pursued from a biblical creationist perspective, it will be established that some of the first post-Babel pioneers to the Western Hemisphere landed on Peru's Pacific shore and from there radiated into South and Central America.[16] Some of the most remarkable archeological sites in the world (the Nazca lines, Tiahuanaco, Sacsayhuaman, Ollantaytambo, Machu Picchu) are in this general vicinity. We noted how some of the most startlingly accurate flood legends are American—the most astonishing example being the Toltec flood legend. It is possible that the knowledge that dinosaurs and men co-existed could have persisted for a long time in the cultures of this region.

**Figure 25.3: Ica stone depicting a human fighting two theropods**

---

[1] White, Ellen, *Spiritual Gifts*, vol. III p. 75 (1864). (I Spirit of Prophecy, p. 78).

[2] White, Ellen, *Spiritual Gifts*, vol. IV-A, p. 121 (Italics mine).

[3] Whitcomb, John C., and Henry Morris, *The Genesis Flood* (Philipsburg, N.J.: Presbyterian and Reformed, 1961) p. 280, note 2.

[4] Marsh, Frank Lewis, *Life, Man, and Time* (Escondido, CA: Outdoor Pictures, 1967), p. 106 ("There can be no question that dinosaurs were represented [on the ark], possibly by forms like *Compsognathus,* which was no larger than a rooster. But we would expect the terrible flesh-eating *Tyrannosaurus* to be left outside, along with his vegetarian but tremendously ungainly fellow reptiles, *Diplodocus* and *Brontosaurus,* whose very bulk, thirty-five and fifty tons respectively, would make them a hazard around the houses of postdiluvian man.") See also James L. Hayword, "Noah's Ark or Jurassic Park?" *Spectrum,* vol. 23, no. 2 (August, 1993).

[5] *Compsognathus* has recently suffered some bad press. At the beginning of the second *Jurassic Park* movie, *The Lost World,* a young girl is mauled by a pack of ravenous compys. Later in the movie, a man who is separated from his party is killed and eaten by a flock of the vicious little creatures. Most who have seen *The Lost World* are not sorry that *Compsognathus* is extinct.

[6] The coelacanth—a lobe-finned bony fish of the order Crossopterygii—is often mentioned in this connection. Coelacanths were thought to have become extinct about sixty million years ago, but in 1938 a living coelacanth was netted in the Indian Ocean near the southern coast of Africa. Based upon this find, some creationists have argued that dinosaurs or other extinct Mesozoic creatures might also be found alive. But the Coelacanth dwells in the depths of the oceans, which explains why it was not found sooner (although it was recently sighted at only 320 feet [104 meters] below the surface). ("Fossil fish in dramatic sighting" BBC News, December 1, 2000). It is very unlikely that land-dwelling dinosaurs could have survived to the present time, and yet have escaped detection.

[7] Ken Ham, founder and head of the tremendously dynamic and helpful Answers in Genesis Ministries (http://www.answersingenesis.org/), agrees that it was only the originally created "kinds" that were taken aboard the ark: "The Bible makes it clear in Genesis 6, however, that representatives of all the 'kinds' of land animals were on Noah's Ark. It's interesting that the list of animals given in Genesis 6 is the *same* list of animals given in Genesis 1 concerning the kinds of animals that God created." Ken Ham, "Dinosaurs on the Ark?" *Answers Update,* vol. 9, issue 9 (italics in the original). Ham assumes, however, that dinosaurs must have been included in the groups of animals God created, so he sees no reason that they would not have been on the ark.

[8] Bakker, Robert T., *The Dinosaur Heresies* (New York: Kensington Pub., 1986), p. 124 ("brontosaur orthodoxy had it completely incorrect. Brontosaurs didn't require deep swamps to buoy their bulk; they didn't even like to be near swamps"). For a typical Charles R. Knight illustration of a brontosaurus in a swamp, see the cover of the paperback edition of Gould's *Bully for Brontosaurus* (New York: W.W. Norton & Co., 1991).

[9] The Old Testament writers use the term *leviathan* in both a literal and a figurative sense. In its literal sense, the term describes a sea creature. For example, Psalms 104:25–26 states, "There is the sea, vast and spacious, teeming with creatures beyond number—living things both large and small. There the ships go to and fro, and the leviathan, which you formed to frolic there." The figurative sense is drawn from the fact that ancient middle-eastern peoples believed that deep waters held monsters, waiting to gobble up any hapless sailor who found himself in the deep. These mythical monsters were called "Leviathan" or "Rahab." The symbolic sense of the term leviathan draws on these myths and refers to something huge, monstrous, chaotic and/or frightening, such as the forces of chaos (Psalms 74:12-14; Job 3:3-10; Job 7:12) or wicked nations that God will judge at the end of time (Isaiah 27:1). See, e.g., Younker, Randall, *God's Creation, Exploring the Genesis Story* (Nampa, ID: Pacific Press, 1999), pp. 7, 8.

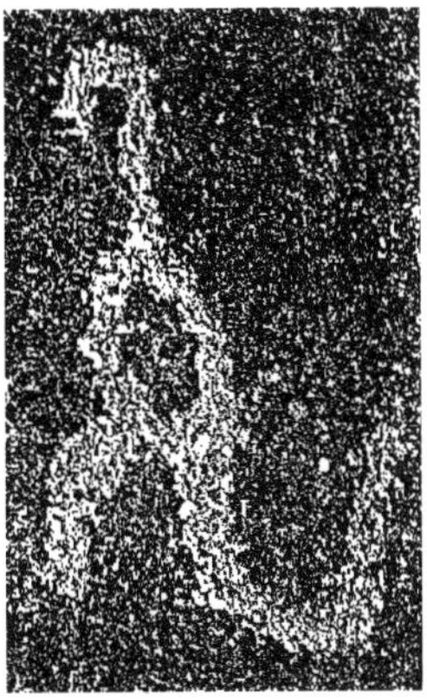

**Figure 25.4: figure of theropod etched on rock at Havasupai Canyon**

[10] According to sources, in October and November 1924, a scientific expedition led by Samuel Hubbard, curator of archaeology at the Oakland Museum, and Charles W. Gilmore, curator of vertebrate paleontology at the United States National Museum, and funded by Los Angeles oil magnate E. L. Doheny, went to Havasupai Canyon to search for evidence of prehistoric man. Hubbard and Doheny had visited the area before, Doheny as a prospector and Hubbard as a scientist. The red sandstone contains iron, which forms a thin black scale on the surface of the stone, called "desert varnish." One can draw a picture by cutting through the varnish with any sharp object. The only way such a pictograph can disappear is to weather off. The accompanying halftone was taken from a photograph made by Robert L. Carson of San Gabriel, California (See below). The dimensions of the figure are as follows: Total height 11.2 inches; greatest width, 7 inches; length of leg, 3.8 inches; length of body, 3.9 inches; width of body, 3 inches; length of neck to top of curve, 3.5 inches; length of tail (approximately) 9.1 inches; length of neck (approx-

imately) 5.' inches. The under cut which outlined this figure was so deep that an excellent cast was made of it by J. F. Roop and Fred V. Shaw. On the same wall were a number of other figures of goat-like creatures, serpents, and unknown forms. The most remarkable of these was a row of symbols, deeply incised, which resembled the Greek sign of Mars showing shield and spear. Harold W. Clark discusses this find in his 1929 book *Back to Creationism*, at pp. 26-28.

[11] Taylor, Joe, *Fossil Facts and Fantasies* (Crosbyton, TX: Mt. Blanco Publishing Co., 1999), p. 75. See also http://www.bible.ca/tracks/native-american-dino-art.htm.

**Figure 25.5: Etching of man and dinosaur at Natural Bridges National Monument**

[12] Hapgood, Charles, *Mystery in Acambaro, An Account of the Ceramic Collection of the late Waldemar Julsrud, in Acambaro, Guanajuato, Mexico,* republished, with additional material, by David Hatcher Childress, (Kempton, IL: Adventures Unlimited Press, 2000). See also *The Dinosaur Figurines of Acambaro, Mexico,* at http://www.bible.ca/tracks/tracks-acambaro.htm.

[13] Benitz, J.J., *Existio Otra Humanidad,* (Barcelona: Plaza & Janes, 1975, 1994) Cabrera, Javier, *El Mensaje de las Piedras Grabadas de Ica.* (Lima: INTI-Sol Editores, 1976, 1980); see also "Dinosaur art from ancient tombs in Peru" at http://www.bible.ca/tracks/peru-tomb-art.htm.

[14] See Stephan A. Czerkas, "Discovery of dermal spines reveals new look for sauropod dinosaurs," *Geology,* 20(12):1068-1070 (1992).

[15] This view is supported by the fact that the Supai Indians attribute the carvings at Havasupai Canyon to "Couda-pahvich," which, freely translated, means "big Indian long time dead." This may reflect a dim tribal memory that early post-Flood man was of taller stature than modern man.

[16] See, e.g., Jacobs, James Q., "Early Monumental Architecture on the Peruvian Coast," (June 14, 2001) (graduate paper at the University of Arizona). Found on the web at: http://www.jqjacobs.net/andes/coast.html. ("Along the central and north coast of Peru the earliest monumental architecture and the earliest complex societies in the Americas are evidenced.").

CHAPTER 26

# A Choice of Paradigms

> *Without faith, it is impossible to please God,*
> *because anyone who comes to Him*
> *must believe that He exists and that*
> *He rewards those who earnestly seek Him.*
> Hebrews 11:6.

God will not force belief. The fossil record is susceptible to more than one interpretation; it does not compel belief in the biblical account of Earth's history. If one chooses to believe in a naturalistic theory of origins, the earth will yield evidence to support that belief. But if one is willing to accept the truth of inspired history, the rocks reveal evidence of God's judgment of the antediluvian world.

Other Christian apologists, going back 100 years to George McCready Price, have reached this same conclusion:

> It requires no faith to believe in a mathematical demonstration or to believe in the existence of Tibet or Kabul, though few of us have been to either place.

> In such instances, the mind is coerced; it cannot refuse assent. Evidently God does not want to deal with us in this way regarding the great moral truths and the truths of religion. All these are of such a nature that we can refuse to believe them if we wish to do so or if we decide to do so. At the same time, ample evidence is granted to all those who are willing to examine the evidence with an absolutely candid mind.[1]

Price's statement echoes Ellen White:

> While God has given ample evidence for faith, He will never remove all excuse for unbelief. All who look for hooks to hang their doubts upon will find them. And those who refuse to accept and obey God's word until every objection has been removed, and there is no longer an opportunity for doubt, will never come to the light.[2]

Some things have changed since Price's day, however. Price imagined himself to be reforming the science of geology to make it more genuinely Baconian, or inductive, and less driven by theory. Of Lyellian geology, he wrote:

> It requires no scientific instruments of precision to recognize in all this a mere travesty on the methods of Bacon and Newton, and it requires but little training in the latter to suggest the better way. Let us throw to the winds all speculations as to how the world was made and hold our restive imaginations within the

> bounds of legitimate science by the strong rein of demonstrable fact, content if we can really explain a few things that have happened since the beginning.[3]

Today, this sounds both naïve and dishonest.

It sounds naïve because we know that science does not—and never did—proceed by gathering facts and waiting until all the facts had been carefully sifted before forming theories. Theory comes first. A pre-selected hypothesis, theory, model, or paradigm determines what data are gathered, what observations are made, what experiments are conducted, and how the results of those experiments are interpreted. Theory completely determines the day-to-day work of what Kuhn called "normal science."[4] "We often hear that [natural laws] are found by examining measurements undertaken for their own sake and without theoretical commitment," wrote Kuhn, "but history offers no support for so excessively Baconian a method."[5]

Price also sounds dishonest, because we creationists do not allow our supernaturalistic model of origins to be falsified. Price was not about to "throw to the wind" all pre-conceived ideas as to how the world was made. To the contrary, he was committed to vindicating the biblical model of Earth's history and toppling the uniformitarian model. He devoted his entire life to pointing out weaknesses in Lyellian geology and arguing that the facts might better fit within a catastrophic model. Thus, his pleas for a more inductive geology ring false.

Kurt Wise, the biblical creationist with a Harvard Ph.D., is more candid about where a creationist must stand:

> Although there are scientific reasons for accepting a young earth, I am a young-age creationist because that is my understanding of the Scripture. As I shared

> with my professors years ago when I was in college, if all the evidence in the universe turns against creationism, I would be the first to admit it, but I would still be a creationist because that is what the Word of God seems to indicate. Here I must stand.[6]

For creationists, the biblical paradigm of Earth's history is not negotiable. If we are honest, we will admit that, regardless of the evidence, we are not going to change our minds.

*But in this, we are no different from the Darwinists!* To them, the naturalistic model of Earth's history is not negotiable—not falsifiable under any circumstances. It is an axiom, a first principle:

> We take the side of science in spite of the patent absurdity of some of its constructs, . . . in spite of the tolerance of the scientific community for unsubstantiated just-so stories, because we have a prior commitment, a commitment to materialism. It is not that the methods and institutions of science somehow compel us to accept a material explanation of the phenomenal world, but, on the contrary, that we are forced by our a priori adherence to material causes to create an apparatus of investigation and a set of concepts that produce material explanations, no matter how counter-intuitive, no matter how mystifying to the uninitiated. Moreover, that materialism is an absolute, for we cannot allow a Divine Foot in the door.[7]

Clearly, the naturalism of the Darwinist is every bit as absolute as the supernaturalism of the creationist.

It is not possible to approach the facts of nature in a neutral and unbiased manner, and allow the facts to inform us whether we were created or evolved. There are too many facts, far too much data, to ever sort out in a hundred lifetimes. One chooses one's model—creation or evolution—before approaching the data. That choice is a religious choice. It is made with the heart, not the intellect.

"If someone seriously did doubt the design of a modern airliner, that person could be convinced by taking him into an aircraft factory and introducing him to the teams of design engineers. In the same way, man's prejudice against design in creation can only really be answered by a radical change of heart and by personally meeting the Author of all."[8] Doubts about creation can only be removed by getting to know, love, and trust the Creator God.

---

[1] Price, George McCready, *Evolutionary Geology and the New Catastrophism* (Nampa, ID: Pacific Press Pub. Assoc., 1926), pp. 342, 343.

[2] White, Ellen G., *The Great Controversy*, p. 527.

[3] Price, at p. 323.

[4] Kuhn, Thomas S., *The Structure of Scientific Revolutions*, 2nd Ed. (Chicago: The University of Chicago Press, 1962), pp. 23-51.

[5] Kuhn, at p. 28.

[6] Wise, Kurt, *In Six Days, Why 50 Scientists Choose to Believe in Creation*, Ed. John F. Ashton, (Sydney: New Holland Publishers, 1999 [published in the U.S. by Master Books, Green Forest, AK]), p. 332. See also Richard Dawkins, "Sadly, an Honest Creationist," *Free Inquiry*, 21(4) (fall, 2001), online at: http://www.secularhumanism.org/library/fi/dawkins_21_4.html.

[7] Lewontin, Richard, "Billions and Billions of Demons," *New York Review of Books*, vol. 44, No. 1, January 9, 1997, a review of Carl Sagan's *The Demon-Haunted World* (New York: Random House, 1996).

[8] Andrew McIntosh, in, *In Six Days, Why 50 Scientists Choose to Believe in Creation*, supra, pp. 154, 155.

# Appendix—The Dating Game[i]

## 1. INORGANIC RADIOMETRIC DATING

Radiometric dating, sometimes called isotopic dating, is an attempt to determine when igneous rocks were formed based upon the rocks' content of certain chemical elements. The technique is based on the principle that an unstable, radioactive parent element "decays" into a stable daughter element at a known average rate. For example, potassium-40 changes into the gas argon-40 at the rate of one half, every 1,280 million years.[2] This means that after 1,280 m.y., half of the potassium-40 has changed into argon-40, and after 2,560 m.y., half of the remainder, or three-quarters of the original, has changed into argon, and so on.[2]

Proponents of radiometric dating argue that by measuring

[i] With apologies to Marvin Lubenow, who attached an appendix entitled "The Dating Game" to his 1992 book *Bones of Contention.* Lubenow can't really complain, however, having borrowed the title *Bones of Contention* from a 1987 book on the same subject by English author Roger Lewin.

the ratio of parent element to daughter element and then applying the rate of decay, one can calculate how long the parent has been decaying into the daughter. For example, if one can determine the relative amounts of potassium-40 and Argon-40 and then apply the 1,280 m.y. half-life figure, one should be able to calculate how long the radioactive potassium has been decaying into argon, which, in turn, is supposed to indicate how much time has passed since the rock was formed. It's a beautiful theory but, like so many beautiful theories, it does not work in reality.

On those occasions when we can test radiometrically derived dates against known history, the radiometric results have often proven wildly unreliable. Lava known to have originated from a volcanic eruption in 1901 yielded a radiometric date of 1.1 billion years old. Other lava originating in known, historically dated eruptions of Mt. Kilauea in Hawaii has produced radiometric ages as great as forty-three million years. Volcanic rock from Alaska deposited in 1912 gave a radiometric date of four million years old.[3] Recently deposited seafloor sediments in the Ross Sea near Antarctica gave a 250 million-year rubidium strontium age.[4]

Isotopic dating techniques sometimes yield dates much older than the conventionally accepted age of the Earth (4.5 billion years). Igneous rock from the Tertiary (no more than sixty-six million years old, according to the accepted chronology) yielded a date of ten billion years old.[5] Precambrian rocks in California yielded a rubidium strontium date of thirty-four billion years old—more than seven times the accepted age of the Earth and older than some estimates of the age of the universe.[6] The potassium argon method has yielded dates in the seven- to fifteen-billion-year range.[7] Other radiometric results include 8.75 billion years.[8] Whatever these numbers mean, they do not reflect the time when the rocks were formed and cannot be used to "date" the rocks.

Radiometric dating is based upon several assumptions, all of which must be true for the method to work. First, the average rate of decay must be known and must have remained constant since the rock was formed. Second, the proportion of daughter product in the rocks at the time the rocks were formed must be known. If there was unaccounted-for daughter product in the rock to begin with, the rocks will have a built-in appearance of age.

Third, the rock must be a "closed system." An hourglass provides a useful example of a closed system. No sand gets into or out of an hourglass; the sand merely falls from the top to the bottom. The same must be true of the parent and daughter elements in tested rocks, if they are to keep time like an hourglass. None of the parent or daughter elements must have been migrating into or out of the rock sample tested. The isotopic system must have been as sealed off as an hourglass.

It is easy to appreciate why, if the method is to work, these assumptions must reflect reality. If sand has been escaping from the bottom of an hourglass, an observer noticing the small amount of sand in the bottom will conclude that the sand has not been running as long as it really has been. By contrast, if sand has been added to the bottom of an hourglass from some source other than the top, an observer would be fooled into believing that the sand had been running much longer than it actually had. If the isotopic system is an open system, radiometric dating will not work.

Astonishingly, dating scientists freely admit that the rocks are *open systems.* In fact, the openness of the rocks is one of the most frequently employed explanations for the many "wrong" dates yielded by the isotopes. It is claimed that the rocks lose or retain daughter product for any number of reasons, including recrystallization, leaching, heating, permeability, depth of burial, grain morphology, weathering, secondary alteration, and metamorphism.[9] We are indebted to John

Woodmorappe for reviewing the technical literature and collecting several admissions that the rocks are not closed systems:

> . . . the assumption that during the whole life of the rock volume being analyzed, neither the radioactive element nor its decay products have moved into or out of the volume is practically *unlikely to be realized in nature at all or, if it is, it occurs only in exceptional cases.* Therefore, differences (often essential) between data determining absolute age [radiometric dating results] and geologic data [index fossils] are *a usual phenomenon that surprises nobody*; it has always been solved in favor of the latter. In the absence of geological data, determination of absolute age [radiometric dating] often leads us into great errors.[10]

> Thirdly, to obtain the age of formation of a rock or mineral, the material must have remained a closed chemical system since its formation, with neither gain nor loss of radioactive parent or daughter atoms or, in the case of complex decay chains, of intermediate members. *Unfortunately, geological materials and environments do not often meet this requirement.*[11]

> Most, *if not all,* isotopic systems of Paleozoic age have been open, to one degree or another, for some period of time since their formation.[12]

> Several factors were studied which could correlate with open-system behavior: grain morphology, extent of weathering, percent expandable layers, recrystallization

> caused by former deep burial, and permeability of the host rock . . ."[13]
>
> Highly altered rocks were thought to give discrepant dates, but "fresh" rock was no guarantee that daughter product had not been lost or excessively retained.[14]
>
> Secondly, the potassium argon method is complicated by the possibility *if not the likelihood* of argon loss.[15]
>
> The Rb-Sr [rubidium strontium] whole-rock method was widely used as a dating tool for igneous crystallization during the 1960's and 1970's, but lost credibility during the 1980's *as evidence of whole-rock open-system behavior mounted.*[16]
>
> The evidence for *open-system* Rb-Sr [rubidium strontium] systematics in numerous environments has now discredited the Rb-Sr isochron method as a dating tool for igneous crystallization.[17]
>
> Unfortunately, the U-Pb [uranium lead] and Th-Pb [thorium lead] systems *rarely stay closed* in silicate rocks . . .[18]

Clearly, the rocks are not closed systems.

Enthusiasts of radiometric dating argue that the technique works most of the time, and that bad dates are the exception rather than the rule. But an examination of the literature shows that bad dates are not rare. In 1978, Woodmorappe compiled a list of 350 grossly discordant dates, that is, dates that differed from the accepted chronology by more than 20 percent.[19] Coffin found twenty-two examples of radiometric ages from Tertiary strata that ranged between seventy million years and 3.34 billion years.[20] For example, shale from the Miocene (conventionally thought to be around twenty-five million years old) gave an average potassium argon age of 254 million years old, off by an order of magnitude.[21]

Obviously, many radiometric dates do not match current geo-chronological notions. In fact, "geologists currently interpret most mineral ages as cooling ages . . ."[22] In other words, most dates yielded by the isotopes are too young to fit the currently accepted chronology; hence, dating scientists believe that the radiometric "clock" must have started running not when the rock was formed but when it cooled.

The most popular isotopic system used in radiometric dating is potassium argon, but there is a major problem with this method. Argon is the 12th-most-abundant chemical element on Earth, and more than 99 percent of it is argon-40. Even assuming the Earth is five billion years old, only about one percent (1%) of Earth's argon-40 could possibly be the product of radioactive decay.[23] Obviously, there is a lot of non-radiogenic argon-40 in the environment. And argon-40 is argon-40; there is no way to distinguish the radiogenic from non-radiogenic. There is no way to tell whether the argon-40 in a rock sample was present when the rock was formed, or whether it was the product of radioactive decay subsequent to the formation of the rock.

Enthusiasts of radiometric dating assert that the argon in igneous rocks is purely radiogenic. They argue that any gaseous argon present in molten magma would diffuse out before the magma hardened into crystalline rock. But there is abundant evidence that this assumption does not reflect reality. For example, solidified dacite from Mount St. Helens' lava dome (which was created by eruptions between 1980 and 1986) gave a potassium argon age of 350,000 years old.[24] Other potassium argon tests of historically dated lava flows have yielded similar results: Hualalai basalt (Hawaii, A.D. 1800-1801) dated to 1.6 m.y.a.; Mt. Etna basalt (Sicily, 122 B.C.) dated to 250,000 y.a.; Mt. Etna basalt (Sicily, A.D. 1792) dated to 350,000 y.a.; Mt. Lassen plagioclase (California, 1915) dated to 110,000 y.a.; Sunset Crater basalt (Arizona, A.D. 1064) dated to 270,000 y.a.[25] These results demonstrate that not all non-radiogenic argon-40 is driven off prior to crystallization.

Moreover, even assuming that all argon is driven off at the formation of the rock, there is nothing to prevent non-radiogenic argon from entering the rock at a later time. "Argon diffuses from mineral to mineral with great ease. It leaks out of rocks very readily and can move from down deep in the earth, where the pressure is large, and accumulate in an abnormally large amount in the surface where rock samples for dating are found."[26]

A revealing fact about argon-40 is that there is a gradient of it in the Earth's crust, with more in the lower strata and less in the upper strata, regardless of the presence or absence of potassium. Just add potassium, and you have an instant time scale that will appear to show that the lower levels are older and the higher levels younger.[27] To whatever extent a complete tally of potassium argon dates would not reveal a totally random scatter, this is probably the explanation.

Bad dates are very common, but there is no way to know how common, because most bad dates are never published in scientific journals. They are just tossed into the laboratory wastebasket without any explanation:[28]

> In general, dates in the 'correct ball park' are assumed to be correct and are published, but those in disagreement with other data are *seldom published* nor are discrepancies fully explained.[29]

> Sm/Nd is partitioned between these two minerals [garnet and clinopyroxene] in a way that this isotopic system should give ideal possibilities for the age measurements. Only in a few cases geologically meaningful ages were obtained. *In the majority of cases the ages are clearly off and the data disappear in a lab-datafile.*[30]

> Often it is not easy to reconcile different isotopic dates, and many age determinations which do not agree with currently accepted time scales *are simply rejected as wrong* without our fully understanding why.[31]

> There is a minimal spread of 40 [million years before present] for the base of the Permian period and 20 MY for the top of the Permian, for example, and *these figures are only obtained by omitting, with no objective reason, the much broader deviations.*[32]

"If all the rejected dates were retrieved from the waste basket and added to the published dates," argues Richard Milton, "the combined results would show that the dates produced are the scatter that one would expect by chance alone."[33]

Since many radiometric dates are wrong—possibly most of them, who knows?—and it is acknowledged that most rocks are open systems and therefore unsuitable for radiometric dating, it is vital that there be some way to tell, *prior to initial radiometric testing*, whether a given rock sample has become "open." Otherwise, dates can be accepted or rejected on a *post hoc* basis depending upon whether they fit with preexisting expectations. Those dates that seemed right based upon preexisting expectations would be accepted, and those that seemed wrong would be rejected. In other words, unless there is some way to tell, prior to initial testing, whether a rock sample is suitable for testing, the entire "science" of radiometric dating could be based on circular reasoning.

There are numerous admissions in the literature, however, that it is *not* possible to determine, prior to preliminary dating, whether a given rock sample has become an open system:

> There may be no mineralogical or textural evidence to warn the geochronologist that an igneous rock or any of its minerals he is analyzing for an age determination has been altered.[34]

> It is quite common for a metamorphic event detected by Rb-Sr dating to leave no obvious imprint on the hand specimen or in thin section.[35]

> An anomalously low Rb-Sr age . . . was found for the biotite. Although no effects of metamorphism or alteration are visible in the syenite, some such process undoubtedly disturbed this . . . radiometric system.[36]

> . . . we suggest the younger dates in the earlier study . . . reflect partial loss of Argon-40 from some specimens. Since all specimens in the earlier K-Ar study passed detailed petrographic examination, it is apparent that rigorous thin-section examination cannot unequivocally eliminate whole-rock basalts that have suffered partial post-crystallization loss of Argon-40.[37]

> The anomalously older date is petrographically, geochemically, and structurally the same as any other sample of this segment of SCPS. The Early Jurassic date is not meaningful and may indicate excess Ar in the hornblende.[38]

Even under a microscope, open system behavior can be undetectable:

> TEM investigations reveal that high-K biotites, though microscopically appearing homogenous, may contain considerable amounts of submicroscopical alteration products.[39]

Simply put, there is no reliable way to tell if a rock sample has become an open system. If the date derived is within the expected range of results, the rock is assumed to have been a closed system. If the date is "wrong," it is assumed that the rock was not a closed system, and one of the standard *post hoc* rationalizations—such as "cooling age," reheating, leaching, inherited argon, argon loss (or other daughter product loss), metamorphism, etc.—is employed to explained the "bad" results.

Clearly, radiometric dating is hopelessly mired in circular reasoning. *Radiometric results are never allowed to falsify prior expectations.* Scientists realize, of course, that this is circular reasoning, but they are not bothered by it because they assume that their expectations are supported by a great mass of evidence.[40]

But is there really a mass of evidence, or just a long history of selective publication? The chronology upon which today's dating scientists base their expectations has been determined by dates that other dating scientists published in the past. Those previously published dates were, in turn, heavily influenced by even earlier published dates, and so on, back to the first published radiometric results. But what about the very first radiometrically derived dates? How could preexisting expectations have skewed those results?

It turns out that "absolute" dates were assigned to the geologic column long before radiometric dating was invented. We have already noted Lyell's estimated age of the K-T boundary of eighty million years ago, which is not far off from the current consensus of about sixty-five million years ago.[41] More remarkably, a geologist named Thomas Mellard Reade (1832-1909) estimated in 1878 that the Cambrian-Precambrian boundary was about 600 million years old.[42] This is within the range of currently accepted *radiometric* dates for the Cambrian-Precambrian boundary—530 million years to 610 million years.[43]

In fact, there never was a time when those attempting radiometric dating did not have an accepted "ballpark" to guide their efforts. Right from the beginning, radiometric daters could rely upon the consensus that geologists had arrived at by observing current rates of sedimentation and applying those rates to the geologic column. Ever since then, dating scientists have had previously published radiometric dates to delineate the acceptable range of results. A consensus, herd, or "groupthink," mentality determined which radiometric results were acceptable and which were "bad." John Woodmorappe refers to this as the "tracking effect."

Those dates that fit within the accepted range were published; those that did not were often made to "disappear in a lab data-file," i.e., tossed in the wastebasket. To the extent that the "bad" results are dealt with in the literature, they are explained away with explanations that sound reasonable but that are unverifiable and un-testable. Thus, it will appear to anyone reviewing the scientific literature that the accepted chronology is supported by an overwhelming mass of data and that scientists understand and can explain the few "bad" results. But this appearance is an artifact of several generations of circular reasoning, selective publication, and *post hoc* rationalizations. In reality, radiometric dating has *never* been an *independent* check on the chronological scheme derived from uniformitarian assumptions.

This is not to imply that the dating scientists were or are dishonest. There is no reason to believe that they have ever been acting other than in good faith. Richard Milton explains the phenomenon as follows:

> Take for example a rock sample from the late Cretaceous, a period which is universally believed to date from some 65 million years ago. Any dating scientist who obtained a date from the sample of, say 10 million years or 150 million years, would not publish such a result because he or she will, quite sincerely, assume it was in error. On the other hand, any dating scientist who did obtain a date for 65 million years would hasten to publish it as widely as possible. . . . If all the rejected dates were retrieved from the waste basket and added to the published dates, the combined results would show that the dates produced are the scatter that one would expect by chance alone.[44]

There is a fascinating story that illustrates many of the problems with radiometric dating. It concerns attempts to date a layer of rock in Kenya. Known as the "KBS tuff controversy," this episode involved several of the leading authorities on radiometric dating.[45] The scientists involved became antagonistic toward each other, and almost everyone in anthropology during the 1970s took one side or the other. It is, as one Darwinist commentator wrote, "a story that demonstrates how very unscientific the process of scientific inquiry sometimes can be."[46]

Anthropologists have long combed through the Lake Turkana (formerly known as Lake Rudolf) area in Kenya, searching for the putative missing links between apes and humans. In 1969, Richard Leakey, son of Louis and Mary Leakey and a leading anthropologist in his own right, had set up camp on the shore of the lake. Kay Behrensmeyer, then a graduate student in paleontology at Harvard and a member of Leakey's team, found several primitive stone tools embedded in a layer of "tuff," which is rock formed from volcanic ash. This particular rock layer became known as the KBS (Kay Behrensmeyer Site) tuff.

Leakey supplied samples from the KBS tuff to Jack Miller, a geophysicist at Cambridge University, to determine the feasibility of radiometrically dating it. Jack Miller and his partner Frank Fitch preliminarily dated the rocks at 212 to 230 million years old.[47] In conventional geochronology, those dates correspond to the Triassic (the lower dinosaur strata), but Leakey was working near the boundary of the Tertiary and the Quaternary, which is thought to be only around two million years old. "From these results," wrote Fitch and Miller, "it was clear that an extraneous argon age discrepancy was present."[48]

Fitch and Miller believed that the tuff, instead of being deposited when ash settled out of the air, had been formed when rivers or streams had washed the ash down from highlands into a floodplain. They hypothesized that during this process the ash had picked up minerals from older deposits. But they did not identify any exposed Triassic rock that could have contributed material to the KBS tuff, which suggests that this was a *post hoc* rationalization concocted after the initial results missed the anticipated "ballpark" by two orders of magnitude.

Fitch and Miller sent back for more samples from the KBS tuff. They tested these and reported a preliminary age of 2.4 million years. They told Leakey that there were two techniques available to give a more exact radiometric age: 1) the traditional potassium argon technique and 2) a newer, more sophisticated technique called argon-40 argon-39. This latter technique involves converting the radiogenic potassium to the gas argon-39, then simultaneously measuring both types of argon in a machine called a mass spectrometer. This method was about twice as expensive as the older method, but it could work with smaller samples and would yield a much more exact measurement. The newer technique, they told Leakey, "would result in the tuff being *incontrovertibly* dated and with greater accuracy than any other site in Africa or elsewhere."[49] Leakey chose the technique that promised the "incontrovertible" date. Fitch and Miller ran the tests and reported that the tuff was 2.61 million years old, plus or minus 0.26 million years.

That result was within the "ballpark," and thus not very controversial.[50] A few years later, however, it was to become extremely controversial. In 1972, one of Leakey's African associates, Bernard Ngeneo, found a skull *below* the KBS tuff. This was skull KNM-ER 1470. As we discussed in Chapter 17, Darwinists can point to relatively few fossils as possible transitional forms between apes and humans. As a result, individual fossils can loom large in attempts to construct an evolutionary lineage or "family tree." Individual hominid fossils can, and often do, change the whole evolutionary picture.[51]

Such was the case with skull 1470. Although it had a very small cranial capacity, in general morphology (at least as reconstructed by Leakey)it resembled a modern human skull much more closely than did the *Australopithecines* ("southern apes"). If the KBS tuff was 2.6 million years old, then skull 1470, which was found below the tuff, must be even older than that—probably almost three million years old. But the Australopithecines had been dated at about 1.9 million years old—one million years *younger* than the much more human looking skull 1470.

The relative ages gave the impression that, over the course of a million years, a more human looking primate had evolved into a more ape-like primate. *Evolution was going in the wrong direction!* Dated as it was, skull 1470 exploded the then-current theory about how and when humans had evolved. "Either we toss out this skull," said Leakey, "or we toss out our theories of early man."[52] Anthropologists were not about to toss out their theories of early man. Something had to give.

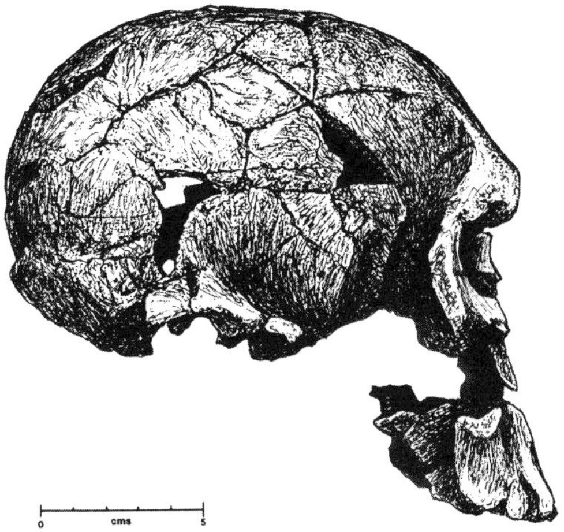


**Figure A1: KNM-ER 1470. KNM-ER stands for Kenya National Museum—East Rudolph**

It did. Garniss Curtis, of the University of California at Berkeley, was called upon to re-date the (already *incontrovertibly* dated) KBS tuff. He dated samples at 1.6 million years old and 1.8 million years old, about a million years younger than Fitch and Miller's date. Curtis' result was much more in line with the Darwinian assumption that fossil hominid crania should be moving from a more ape-like to a more human-like condition, not the other way around. Had Curtis been diplomatic toward Fitch and Miller, the controversy might have ended there, but, adding insult to injury, Curtis criticized Fitch and Miller's samples, their methodology, and their laboratory techniques. The gauntlet had been thrown down.

Taking up the gauntlet, Fitch and Miller redid their tests, using a new and supposedly more accurate rate of decay, and found that the KBS tuff was 2.42 m.y. old. Most tellingly, Fitch published the fact that Curtis had gotten a "scatter" of results ranging from 1.5

to 6.9 million years. How had Curtis settled on 1.6 million years when he could have selected 6.9 million years? Obviously, 1.6 million coincided with previous expectations, and 6.9 million did not.

Fitch reported that his scatter of results had only ranged from 0.5 to 2.64 million years, implying that his technique was better because his scatter wasn't as large. In understated but deliciously snide language, Fitch wrote of Curtis:

> . . . K-Ar apparent ages in the range 1.6-1.8 Myr obtained from the KBS Tuff by other workers are regarded as discrepant, and may have been obtained from samples affected by argon loss.[53]

Fitch's suggestion that Curtis' samples were affected by argon loss illustrates the ease with which the open system excuse is invoked after the fact to explain away unwanted results.

In 1980, Ian McDougall of Australian National University weighed in on the KBS tuff controversy, reporting a date of 1.88 million years. Again, much more interestingly, McDougall revealed that Fitch and Miller had achieved a scatter of 0.52 to 2.64 million years on one set of samples, but on another set, they had gotten a scatter of *8.43 to 17.5 million years!* Thus, the total scatter was from half a million years to 17.5 million years, *a range of 17 million years* (not including the preliminary finding of 230 million years) *on a layer of rock supposedly around two million years old.* Said Garniss Curtis, of Fitch and Miller's work: "Their choice of 2.6+ for the age of the KBS Tuff seemed to be by their reaching into a hat filled with all the numbers they had obtained and coming out with 2.6 m.y."[54]

The enormous range of results recorded by the competing groups of dating scientists illustrates Richard Milton's point that if all results were published, not just those in the expected ballpark, the combined results would show the scatter that could be expected from random chance.

Meanwhile, Ian McDougall's group had some scatter of its own to deal with, as one of their samples gave an age of 4.11 million years and another sample gave an age of 7.46 million years. These dates were of course rejected:

> We attribute these poorly reproducible ages to the presence of variable but small amounts of old detrital K-feldspar in the aliquants used in the argon extractions. *Careful petrographic examination of the mineral concentrate, however, did not lead to positive identification of detrital K-feldspar.* Nevertheless, there is no doubt that old detrital material was being brought into the East Turkana Basin during deposition of the sediments.[55]

In other words, McDougall believed that *his* bad dates were a result of older material being included in the tuff when it was deposited, or re-deposited (which, recall, was the same story Fitch and Miller came up with to explain their 230-million-year date).

McDougall's group could find no trace of this putative older mineral material, even after very carefully examining the rock, but they know it was there because . . . well, because the samples gave bad dates. This illustrates that there is no way to tell, by examining a sample, whether it is suitable for radiometric dating. If the result is in the expected ballpark, everything is assumed to be okay. If not, then the sample must have been unsuitable, even though a careful examination revealed no problems.

The controversy was ultimately settled at around 1.8 million years, not on the basis of radiometric dating—any date between 500,000 and 17.5 million years could have been selected with equal justification based purely on the isotopes—but on the basis of pig fossils.[56] *Here, as in every other case where the radiometric dates have conflicted with fossil evidence, fossil evidence prevailed, and conflicting radiometric dates were ignored or explained away.*[57] But the faunal remains were not clear, either. Correlation of elephant fossils seemed to support Fitch and Miller's date; some researchers claimed that the pig fossils supported a younger date, while others claimed that the pig fossils supported Fitch and Miller's date.[58]

But neither the faunal remains nor the isotopic results were really driving the KBS Tuff debate. Rather, it was the mess that skull 1470 made of the human evolution story that created the pressure to re-date the Tuff. Before the problems created by skull 1470 became too urgent, everything seemed to support Fitch and Miller's date. Faunal correlation, paleomagnetism, and fission-track dating all impressively and concertedly supported their date.[59] After it became clear what mischief the initial dating of the KBS tuff would do to the human evolution story, everything changed. It turned out that everything was negotiable, *except* the need to come up with a plausible presentation of the putative apes-to-humans transition.

Because this controversy was fought in print, the public learned how many different dates were tossed into the wastebasket. Several factors came together, including: 1) the discovery of a fossil that looked like it might fit into the evolutionary story of human origins; 2) the relevant layer of rock had already been dated before the fossil was found; 3) the dating of this fossil was crucial because, in terms of evolutionary theory, the fossil seemed to be out of sequence; 4) three different groups of dating scientists worked on the problem; and 5) these groups were not diplomatic in commenting on each other's work.

Were it not for this combination of circumstances, we would

never have known how many, and how widely varying, were the discarded dates, not to mention the facile *post hoc* explanations for them. Regarding this episode, Roger Lewin wrote, "[T]here is a degree of uncertainly in science that is not often made public, because it is contrary to the mythology of what science is supposed to be like."[60] The KBS tuff controversy shows that the reality of science resembles politics—maybe even theology!—more closely than it resembles its own carefully nurtured public image of detachment and objectivity.[61]

By now it should be obvious that radiometric dating is not what we have been led to believe. It is nothing like an exact science. Even after taking pains to test only samples that will yield a preconceived result, there is typically a large scatter of results for the dating scientist to choose from.[62] Based upon prior expectations, which, in turn, are based upon geology, stratigraphy, index fossils, professional consensus, and what seems reasonable—none of which have anything to do with the isotopes *per se*—one of the dates is selected as "accurate." After the winning candidate is chosen, the other dates are ignored or explained away by appealing to one (or more) of a staggering array of excuses. The *post hoc* rationalization chosen depends only on the dating scientist's imagination. The entire process is an exercise in prejudice confirmation. As a method of getting at truth, it is utterly worthless.

Lyellians have a couple of replies to the creationist critique of radiometric dating. First, they state that, regardless of the scatter of results that they often get, all of the results are in the millions to billions range and thus inconsistent with young-earth creationism. But the million-year ages obtained from historically dated lava flows suggest that the numbers are simply not meaningful. The most basic assumptions underlying the radiometric technique are wrong.

Second, Lyellians argue that creationists have developed no general theory about the radiometric results and have provided no alternative to radiometric dating. But it is the Lyellians who are claiming that the isotopes can be used to date rocks. The onus is on them to prove that radiometric dating works, and *there has never been any attempt to prove that radiometric dating works.* Creationists, by contrast, are under no obligation to develop a general theory or explanation of radiometric dating. It is the Lyellians who are trying to sell us this airplane. They have to demonstrate that it flies. By pointing out that it has crashed, creationists assume no obligation to determine why it crashed or to design one that flies.

## 2. ORGANIC RADIOMETRIC DATING (RADIOCARBON/CARBON-14 DATING)

There is a type of isotopic dating that deals not with rocks but with organic material—things that were alive. This method is called radiocarbon or carbon-14 dating. There is a certain proportion of radioactive carbon-14 in the carbon dioxide of the Earth's atmosphere. Plants absorb this carbon-14, and when animals and humans eat plants, we absorb it into our bodies in the same proportion. Carbon-14 degrades into nitrogen-14, but because living things are constantly reabsorbing carbon-14, the proportion of carbon-14 in a living organism stays the same until it dies. This is where carbon-14 dating has a decisive advantage over inorganic radiometric techniques: We know when the clock starts running. It starts when the organism dies. The half-life of carbon-14 is approximately 5,730 years. It is claimed that radiocarbon dating works accurately for about 50,000 years after the death of the organic material. After that, there is not enough radiocarbon remaining to accurately measure.[63]

When scientists claim that organic material can be accurately dated for at least 50,000 years, they make a uniformitarian assumption with which creationists disagree. They assume that the proportion of carbon-14 in the atmosphere has been uniform during this entire time. Creationists believe that before the Flood and in the early post-Flood centuries, the Earth's atmosphere did not contain as much radioactive carbon-14. Organic material originating from before the Flood and in the centuries following the Flood absorbed less carbon-14 than plants and animals do today.[64] Organic material from those times will give a carbon-14 date that is much too old: Dating scientists assume that the missing carbon-14 has degraded into nitrogen when, under creationist assumptions, it was never there to begin with.[65] Most creationists do agree, however, that carbon-14 dating can often be roughly accurate on organic materials not more than about two or three thousand years old.

In fact, many biblical creationists are downright enthusiastic about carbon dating, because, it turns out, carbon dating is of enormous value in debunking other types of radiometric dating. Any significant quantity of radiocarbon detected in organic material indicates that the material is thousands of years old, not millions.[66] Radiocarbon has frequently been detected in organic samples that are supposed to be much too old to have any. Near Auckland, New Zealand, for example, stands Mt. Rongitoto, an inactive volcano. Back when it was active, lava from an eruption buried a nearby forest. Scientists dated the lava using the potassium argon method and concluded that the lava was extruded 485,000 years ago. But the carbon-14 content of trees buried in the lava indicated that the trees died fewer than 300 years ago.[67]

Another example concerns fossilized wood taken from the Marlstone Rock Bed in southern England, a formation assigned by the index fossil method to the lower Jurassic, about 189 million years old, according to conventional chronology. Samples of the fossilized wood were sent for radiocarbon analysis to reputable, internationally recognized laboratories in Cambridge, Massachusetts, and Sydney, Australia. The laboratories' staffs were not told where the samples came from or their assumed age. The labs calculated carbon-14 ages for the samples ranging from 20,700 ± 1,200 to 28,820 ± 350 years before the present.[68] Most geologists would not have bothered to run a radiocarbon test on the fossilized wood, because they would have assumed that all the carbon-14 had decayed into nitrogen some 188 million years ago.

Yet another example concerns the Hawkesbury Sandstone of Sidney, Australia, which has been assigned to the middle Triassic, notionally around 225 to 230 million years old. Because of its hardness and durability, the Hawkesbury Sandstone provides a solid foundation for downtown Sydney's skyscrapers, and has often been quarried for building material. In 1997, a finger-sized piece of fossil wood was discovered in a Hawkesbury Sandstone slab just cut from the quarry face. Though petrified, the original character of the wood was still evident. A piece of the fossil wood was sent for radiocarbon analysis to Geochron Laboratories in Cambridge, Massachusetts. Again, to ensure there would be no resultant bias, the laboratory workers were not told the notional age of the rock from which the wood was taken. The laboratory reported that detectable radiocarbon had been found in the fossil wood, yielding an age of 33,720 ± 430 years before the present. Clearly, the wood and the rock formation in which it was found are thousands of years old, not hundreds of millions.[69]

These are not isolated examples. Radiocarbon has been detected repeatedly in samples from Tertiary (thought to be at least 1.8 million years old) or lower strata. Paul Giem catalogs seventy-seven instances where radiocarbon was detected in these types of samples. These were all relatively recent tests (tests from 1984 or later) conducted with sophisticated equipment.[70] According to the conventional Lyellian chronology, all of this carbon should long since have been "dead." That there is still radioactive carbon-14 in these samples indicates that the rock is not as old as the Lyellian chronology would claim. As Giem summarizes the results of his research, "if there is residual [radiocarbon] activity in material considered to be 350 million years old, or 2 million years old, or even 300,000 years old, the material in question simply is not that old. . . . it is probably not even 100,000 years old."[71]

Another conflict between carbon-14 and inorganic dating methods came to light during attempts to date "human tracks" found in Mexico. In 2003, Silvia Gonzalez discovered over 250 footprints at the bottom of an abandoned quarry near Puebla, Mexico. The tracks were made in a layer of volcanic ash that was soft at the time but today is as hard as concrete. Gonzalez' team extensively examined the tracks and determined that many of them were human tracks. "We are sure that they are human footprints because we have done detailed mapping of the prints together with laser scanning. This technique gives a 3D image of a human footprint. So we don't have any doubts," Gonzalez said.[72]

Gonzalez' team naturally wanted to date the track-bearing layer. They found shells in a layer of sediment just above the layer of ash; these shells were carbon-dated to 38,000 years ago. A second dating technique, optically stimulated luminescence (OSL), was used to date sand grains baked into the volcanic ash layer itself, and this method confirmed a date of approximately 40,000 years before the present. Careful dating using two different dating techniques had led to convergent results.

The date of 40,000 years before the present was, however, a controversial challenge to the prevailing orthodoxy on human migration into the Americas. The "Clovis model" holds that humans did not reach the Americas until about 12,000 years ago, when they crossed the Bering Strait (which, when sea levels were lower during the Ice Age, is thought to have been the Bering Land Bridge) from Siberia. Defenders of the orthodox Clovis faith soon came forward to challenge Gonzalez' findings. Paul Renne, a scientist from Cal-Berkeley, re-dated the ash layer using the inorganic argon-40/argon-39 technique. He arrived a date of 1.3 million years before the present, more than 32 times older than the date arrived at through carbon dating, and his date was reportedly confirmed by magnetic reversals.[73]

Which date, if either, is correct? It is worth noting that Renne was on the record as doubting that the tracks were of human origin long before he conducted the radiometric dating. He was not a neutral observer.[74]

Let us close our discussion with one more case study in which radiocarbon results confounded Lyellian expectations. In 1994, Tolo Lake in Idaho was drained so that dirt blown into the lake from nearby agricultural fields could be shoveled up and hauled away. Bulldozers struck fossil bone and mammoth tusk at several spots along the shore, and a scientific expedition was organized to study the site. The scientists recovered a well-preserved, nearly complete skeleton of a large, adult Columbian mammoth (*Mammuthus columbi*).[75]

Mammoth remains are found in Pleistocene sediments, and conventional chronology holds that the Pleistocene ended about 10,000 years ago. The scientists were therefore expecting carbon dating to confirm that these mammoths died at least 10,000 year before the present, perhaps much longer ago than that. They were disappointed. The radiocarbon results showed that the animals had lived around 2,300 to 3,000 B.C., about when many creationists believe that the Ice Age actually occurred. "Radiocarbon dates obtained on sediments and mammoth bone so far have been anomalously young—around 4,300 to 5,000 years [before the present]. Contamination is suspected as the cause of recent dates."[76]

As with inorganic dating methods, whenever carbon dating reaches a "wrong" result, the scientists start making excuses. With radiocarbon dates that are "too young," by far the most popular plea is that the sample was contaminated, despite the fact that labs take elaborate precautions to prevent contamination.

The scientists hoped that stratigraphy or paleomagnetism would allow them to disregard the radiocarbon result: "The sedimentary history and geology of Tolo Lake should provide a relative dating sequence to place the mammoth fossils in time. Results from coring done in the fall of 1994 identified some features useful to interpreting the depositional history and a possible paleomagnetic reversal."[77] They were also searching for index fossils that would allow them to establish a conventional date and disregard the radiocarbon results:

> The spectacular preservation of the megafauna gave rise to optimism that other species critical to dating the mammoths and characterizing the paleoecology of Tolo Lake would be recovered. But in spite of continuous screen washing of tons of sediment throughout the summer, almost no vertebrate or invertebrate microfossils were recovered.[78]

Clearly, if these scientists could find any index fossils, paleomagnetic reversals, or stratigraphic evidence to hang their hats on, they were going to disregard the radiocarbon results indicating that these animals lived around 2,300 B.C.

In the historical sciences, the Lyellian long-ages chronology is the ultimate fact. Within narrow limits, that chronology can be refined, but the basic scheme is not considered to be falsifiable.[79] Radiometric results never trump index fossils. Never. Index fossils always trump radiometric results. Always. Radiometric results that do not fit the accepted chronology always have been, and always will be, ignored or explained away. And that is all you need to know about "radiometric dating."

---

[1] Roth, Ariel A. *Origins: Linking Science and Scripture* (Hagerstown, MD: Review and Herald, 1998) p. 251.

[2] Actually, most of the K-40, about 89 percent of it, degrades into Ca-40; only 11 percent degrades into Ar-40.

[3] Brown, R. H. "Radioisotope age," Modified from *Geoscience Reports*, Spring 1996, no. 20, pp. 1-3. On the web at: http://www.grisda.org/georpts/gr20_01.htm.

[4] Harold Coffin, *Origin by Design* (Hagerstown, MD: Review and Herald, 1983), p. 306, citing Shaffer and Faure, "Regional variation of Sr/Sr ratios and mineral compositions of sediment from the floor of the Ross Sea, Antarctica," *Geological Society of America Bulletin*, 87:1491-1500 (1976).

[5] Woodmorappe, John, "Radiometric Geochronology Re-appraised," in *Studies in Flood Geology* (El Cajon, CA: Institute for Creation Research, 1999), p. 167, citing Mark, Lee-Hu, Bowman, and McKee, "Recently depleted radiogenic mantle source of ocean ridge-like Thoileitte, Northern Great Basin," *Geological Society of America, Abstracts with Programs*, 6:456 (1974). (Woodmorappe is a *nom de plume*. The individual worked as a geologist at a secular university and feared a backlash if his creationist activities became known.)

[6] Woodmorappe, "Radiometric Geochronology Re-appraised," supra, at 167, citing Faure, G. and J.L. Powell, *Strontium Isotope Geology* (New York: Springer-Verlag Publishers, 1972) p. 92.

[7] Woodmorappe, "Radiometric Geochronology Re-appraised," supra, at p. 167, citing Gerling, E.K., Morozova, I.M., and V.D. Sprintsson, "on the nature of the excess argon in some minerals," *Geochemistry International* 9(6):1090 (1968).

[8] Woodmorappe, John, *The Mythology of Modern Dating Methods* (El Cajon, CA: Institute for Creation Research, 1999), p. 25. While on the subject of the age of the Earth, we note that the conventionally accepted age of 4.5 billion years was arrived at by the uranium lead method. When radioactive uranium decays, it turns not only into lead but also into helium gas. But there is several thousand times less helium in the Earth's atmosphere than we would expect if radioactive uranium really had decayed for 4.5 billion years. What happened to all the helium? Defenders of the conventional chronology believe 99.96 percent of it escaped the Earth's atmosphere and blew off into space. But the only evidence of helium loss at such a rapid rate is that the amount of atmospheric helium that "should" exist, if the assumptions behind uranium lead dating are correct, does not exist. (Milton, Richard, *Shattering the Myths of Darwinism*

(Rochester, VT: Park Street Press, 1997), pp. 40-46, citing Melvin A. Cook, *Prehistory and Earth Models* (London: Max Parrish, 1966); Cook, "Do radiological clocks need repair?" *Creation Research Society Quarterly* 5:70 (Oct. 1968); Cook, "Where is the Earth's Radiogenic Helium?" *Nature*, 179:213 (Jan. 1957). See also Larry Vardiman, "Up, Up, and Away! The Helium Escape Problem," Institute for Creation Research, Impact no. 143, May, 1985; Vardiman, *The Age of the Earth's Atmosphere: A Study of the Helium Flux through the Atmosphere* (San Diego, CA: Institute for Creation Research, 1990), p. 28.) Creationist scientists conducting original research on this issue have found that most of the radiogenic helium is still in the rocks. In fact, the amount of such helium still trapped in the structure of minerals leads to the conclusion that radiogenic helium has been escaping from the rocks for only 4,000 to 14,000 years. Humphries, D. Russell, "Nuclear Decay: Evidence for a Young World," Institute for Creation Research, Impact no. 352, October 2002. http://www.icr.org/article/302/.

[9] Woodmorappe, *Studies in Flood Geology*, pp. 158-160.

[10] Skobelin, Sharapov and Bugayov, "Deliberations of state and ways of Perestroika in geology," in Barto-Kyriadkidis, *Critical Aspects of Plate Tectonics Theory* (Athens: Theophrastas Publications, 1990), p. 25 (italics added), as quoted in Woodmorappe, *The Mythology of Modern Dating Methods*, p. 27.

[11] Durrance, *Radioactivity in Geology* (Chister, UK: Ellis Horwood, Ltd., 1986), p. 287, as cited by Woodmorappe, *The Mythology of Modern Dating Methods*, supra, at p. 28 (emphasis added).

[12] Kunk and Sutter, "Ar/Ar age spectrum dating of biotite from Middle Ordovician bentonites—eastern North America," in Bruton, ed., *Aspects of the Ordovician System* (University of Oslo Press, 1984), p. 13, as quoted in Woodmorappe, *The Mythology of Modern Dating Methods*, p. 27 (emphasis added).

[13] Morton and Long, "Rb-Sr Ages of Paleozoic Glauconites from the Llano Region of Central Texas," *Geology Society of America Abstracts with Programs* 10(1):22-23 (1978), as quoted in Woodmorappe, *Studies in Flood Geology*, p. 159.

[14] Woodmorappe, *Studies in Flood Geology*, p. 160, citing Durant, Dobson, Kokelaar, MacIntyre, and Rea, "Preliminary report on the nature and age of the Blackstones Bank Igneous Centre, western Scotland, *Geochemistry International*, 132:325 (1976).

[15] Hallam, et al., "Jurassic to Paleocene: Part I, in Snelling, N. J., ed. *The Chronology of the Geological Record*, Geological Society of London, Memoir 10, (1985), as quoted in Woodmorappe, *The Mythology of Modern Dating Methods*, p. 27 (emphasis added).

[16] Dickin, A. P., *Radiogenic Isotope Geology* (New York: Cambridge University Press, 1997), p. 51, as quoted in Woodmorappe, *The Mythology of Modern Dating Methods,* p. 35 (emphasis added).

[17] Dickin, supra, at p. 53, as quoted in Woodmorappe, *The Mythology of Modern Dating Methods,* p. 27 (emphasis added).

[18] Dickin, supra, at p. 105, as quoted in Woodmorappe, *The Mythology of Modern Dating Methods,* p. 33 (emphasis added).

[19] Woodmorappe, *Studies in Flood Geology,* pp. 148-158.

[20] Coffin, supra, at p. 306.

[21] Coffin, supra, at 304, citing Perry, "Diagenesis and K-Ar dating of shales and clay minerals," *GSA Bulletin,* 85:827-830 (1974).

[22] Zeitler, P.K., et al., "U-Th-He dating of apatite: A potential thermochronometer," *Geochimica et Cosmochimica Acta* 51:2865-2868 (1987), as cited by Woodmorappe, *The Mythology of Modern Dating Methods,* supra, at 28.

[23] Milton, Richard, *Shattering the Myths of Darwinism* (Rochester, VT: Park Street Press, 1997), p. 47.

[24] Austin, Steven A., "Excess argon within mineral concentrates from the new dacite lava dome at Mount St. Helens volcano," *Technical Journal* 10(3):335–343 (December 1996) on the web at Answers in Genesis: http://www.answersingenesis.org/tj/v10/i3/argon.asp.

[25] Austin, supra.

[26] Plaisted, David, "The radiometric dating game," http://www.cs.unc.edu/~plaisted/ce/dating.html#Why%20K-Ar%20dating%20is%20inaccurate.

[27] Giem, Paul, in In Six Days: *Why 50 Scientists Believe in Creation* (Sidney: New Holland Publishers, 1999 (published in the U.S. by Master Books, Green Forest, AK), p. 48. See also Paul Giem, *Scientific Theology* (La Sierra University Press, 1997).

[28] Woodmorappe, "Radiometric Chronology Re-appraised," supra, at pp. 159, 160; Woodmorappe, *The Mythology of Modern Dating Methods,* supra, at pp. 13, 14, 16.

[29] R. L. Mauger, "K-Ar Ages of Biotites from the Tuffs in Eocene Rocks of the Green River, Washakie, and Uinta Basins, Utah, Wyoming and Colorado," *Contributions to Geology* (University of Wyoming), 15(1):37 (1977), as quoted in Woodmorappe, "Radiometric Geochronology Re-appraised," supra, at p. 159, and by Roth, at p. 252 (emphasis added).

[30] Jagoutz, E. "Isotopic systematics of metamorphic rocks" in Lamphere, M.A., G. B. Dalrymple, and B.D. Turrin, *Abstracts of the Eighth international conference on Geochronology, Cosmochronology, and Isotope Geology.* U.S. Geologic Survey Circular 1107 (1994), p. 156, as quoted in Woodmorappe, *The Mythology of Modern Dating Methods,* at p. 16.

[31] Paul, *The Natural History of Fossils* (New York: Holmes and Meier Pub. Co., 1980), p. 184, as quoted in Woodmorappe, John, *The Mythology of Modern Dating Methods,* (El Cajon, CA: Institute for Creation Research, 1999), p. 13 (emphasis added).

[32] J.B. Waterhouse, "Chronologic, ecologic, and evolutionary significance of the Phylum Brachiopoda" in E.G. Kauffman, and J.E. Hazel, eds. *Concepts and Methods of Bio-stratigraphy,* (PA: Dowden, Hutchinson, and Ross, 1979), as quoted in Woodmorappe, John, *The Mythology of Modern Dating Methods,* at p. 13 (first italics in original, second italics added).

[33] Milton, at p. 51.

[34] Faure and Powell, *Strontium Isotope Geology* (Berlin, Heidelberg, and New York: Springer Verlag Pub. Co., 1972), p. 92, as quoted in Woodmorappe, *Studies in Flood Geology,* p. 160.

[35] Hamilton, *Applied Geochronology* (London and New York: Academic Press, 1965), p. 105, as quoted in Woodmorappe, *Studies in Flood Geology,* p. 160.

[36] Zartman, Brock, Heyl, and Thomas, "K-Ar and Rb-Sr ages of some alkalic intrusive rocks from central and eastern United States," *American Journal of Science,* 265:863 (1967), as quoted in Woodmorappe, *Studies in Flood Geology,* p. 160.

[37] Baksi, Hoffman, and McWilliams, "Testing the accuracy of the geomagmatic polarity time-scale (GPTS) at 2-5 Ma, utilizing Argon-40/Argon-39 incremental heating data on whole-rock basalts," *Earth and Planetary Science Letters,* 118:135-144 (1993), as quoted in Woodmorappe, *The Mythology of Modern Dating Methods,* pp. 46, 47.

[38] Hunt and Roddick, "A compilation of K-Ar ages, Report 20, in Radiogenic Age and isotopic Studies: Report 4" *Geological Survey of Canada,* Paper 92-2, p. 207-261, as quoted in Woodmorappe, *The Mythology of Modern Dating Methods,* p. 46.

[39] Hess, Lippolt, and Wirth, "Interpretation of Argon-40/Argon-39 spectra of biotites," *Chemical Geology* 66:137-149 (1987), as quoted in Woodmorappe, *The Mythology of Modern Dating Methods,* p. 47.

[40] See, e.g., Andrew MacRae, "Radiometric Dating and the Geological Time Scale: Circular Reasoning or Reliable Tools?" Talkorigins.org. On the web at: http://www.talkorigins.org/faqs/dating.html#Circularity. McRae admits that

radiometric results are judged as reliable or discrepant based upon how well they fit preexisting expectations: "If an inconsistent data point is found, geologists ask the question: 'Is this date wrong, or is it saying the current geological time scale is wrong?' In general, the former is more likely, because there is such a vast amount of data behind the current understanding of the time scale, and because every rock is not expected to preserve an isotopic system for millions of years." McRae also admits that, "there is almost zero chance that the broad understanding of geological history (e.g., that the Earth is billions of years old) will change."

[41] Milton, Richard, supra, p. 50. ("Ever since Charles Lyell estimated that the end of the Cretaceous was 80 million years ago, the accepted value has been in this ballpark. Any dating scientist who suggested looking outside the ballpark, at 20 million years or 10 million or 5 million, would be looked on as a crackpot by his colleagues. More significantly, perhaps, he would not be able to get any funding for his research.")

[42] Reade, T. Mellard, "Limestone as an Index of Geological Time," *Proceedings of the Royal Society* (1878) 28:281.

[43] Woodmorappe, *The Mythology of Modern Dating Methods*, p. 13, citing W. Compston and I.S. Williams "Ion Probe ages for the British Ordovician and Silurian stratotypes," (pp. 59-67), in B.D. Webby and J.R. Laurie, eds., *Global Perspectives on Ordovician Geology*, Balkem, Roterdam, Netherlands, (1992); Jie-Dong, et al, "Sm-Nd isotopic age of the Precambrian-Cambrian boundary in China," *Geological* Magazine 33(1):53-61 (1996); Schuchert, C., "Geochronology," *Bulletin of the National Research Council*, 80:10-64 (1931).

[44] Milton, Richard, supra, at p. 51.

[45] See, generally, Lubenow, Marvin L., *Bones of Contention* (Grand Rapids, MI: Baker Book House, 1992), pp. 247-266, appendix entitled, "The Dating Game." See also Roger Lewin, *Bones of Contention* (New York: Simon and Schuster, 1987), pp. 189-252; Richard Milton, supra, at pp. 53-56; Michael A. Cremo, Richard L. Thompson, *Forbidden Archeology: The Hidden History of the Human Race* (Los Angeles: Bhaktivedanta Book Publishing, Inc., 1993), pp. 693-702. F.J. Fitch and J.A. Miller, "Radioisotopic Age Determinations of Lake Rudolf Artifact Site," *Nature*, 226:226-228 (April 18, 1970).

[46] Lewin, Roger, *Bones of Contention* (New York: Simon and Schuster, 1987), p. 190.

[47] Lewin, at pp. 192, 194, 195; Lubenow, at p. 249.

[48] F.J. Fitch and J.A. Miller, "Radioisotopic Age Determinations of Lake Rudolf Artifact Site," *Nature*, 226:226-228 (April 18, 1970).

[49] Lewin, at p. 192 (emphasis added).

[50] There was one note of doubt, however. During the late Pliocene/early Pleistocene, pig molars were increasing in length and height. Basil Cooke, an expert on fossil pig teeth, noted that the pig teeth found in the area of the KBS tuff seemed to indicate that the tuff should be dated younger than 2.6 million years. Lewin, at pp. 195-197.

[51] In *Ever Since Darwin*, Gould wrote that each year when the topic of human evolution "comes up in my courses, I simply open my old folder and dump the contents into the nearest circular file. And here we go again." In his next book, *The Panda's Thumb*, he writes, "I'm mighty glad I wrote [those words] because I now want to invoke that passage to recant an argument made later in the same article." See Gould, Ever Since Darwin (New York: W.W. Norton & Co., 1977), p. 56; *The Panda's Thumb* (New York: W.W. Norton & Co., 1980), p. 125. Another example concerns Neanderthal Man. In *The Flamingo's Smile* (New York: W.W. Norton, 1985), p. 41, Gould wrote that Neanderthal was "a race of our own species, not an ancestor or any form of 'missing link'." But in *Leonardo's Mountain of Clams and the Diet of Worms* (New York: Harmony Books, 1998), p. 210, Gould wrote that the Earth of 40,000 years ago contained three coexisting human species *Homo neanderthalensis*, *Homo erectus*, and *Homo sapiens*. Evidently he had come to believe, in the meantime, that Neanderthal Man was a separate species. Clearly, human evolution is not a subject where there is a lot of stability or continuity of concepts.

[52] Lubenow, at p. 253, quoting Richard Leakey, "Skull 1470" *National Geographic*, June 1973: 819.

[53] Fitch, Hooker, and Miller, "Argon 40/Argon 39 dating of the KBS Tuff in Koobi Fora Formation, East Rudolf, Kenya," *Nature* 263:741 (October 28, 1976, italics added) as quoted in Lubenow, p. 256. Later, Fitch also suggested "overprinting" as an explanation for the younger dates. Lewin, at p. 203.

[54] Lewin, at p. 233, quoting Garniss Curtiss.

[55] McDougall, Maier, Sutherland-Hawkes, and Gleadow, "K-Ar age estimate for the KBS tuff, East Turkana, Kenya," *Nature* 284:230-31 (March 20, 1980) as quoted in Lubenow, p. 263 (emphasis added).

[56] Lubenow, at pp. 255-266 ("The pigs won. . . . The pigs won over the elephants. The pigs won over K-Ar dating. The pigs won over Ar-40 Ar-39 dating. The pigs won over fission-track dating. They won over paleomagnetism. The pigs took it all."); Lewin, pp. 195-252. Regarding scatter, Lewin writes that "at a conference in Nairobi held in September 1973 [Fitch and Miller] presented 41 separate age determinations on the KBS tuff, which varied between 223 million and 0.91 million years." Lewin, at pp. 194, 195

[57] Differences between radiometric data and fossil data "are *a usual phenomenon that surprises nobody*" and are always resolved in favor of the fossil data. Skobelin, Sharapov, and Bugayov, "Deliberations of state and ways of Perestroika in geology," in Barto-Kyriadkidis, *Critical Aspects of Plate Tectonics Theory* (Athens: Theophrastas Publications, 1990), p. 25 (italics added), as quoted in Woodmorappe, *The Mythology of Modern Dating Methods,* p. 27.

[58] Lewin, at p. 235. ("When Harris and White first examined the pig fossils, in 1974, their initial impression was that a 2.61 million-year date was correct for the KBS Tuff."); Lubenow, at p. 262.

[59] Lubenow, at p. 265. The fission-track dating of the KBS tuff at first confirmed Fitch and Miller, then later, remarkably, confirmed the younger date. Every time an atom of radioactive U-238 atom explodes, its nucleus is torn in half and the two halves are propelled through the crystal in opposite directions at great speed, boring a tiny tunnel through the crystal lattice. The idea behind fission track dating is to count the number of these fission tracks in the zircon crystal. When researchers Tony Hurford and Andy Gleadow did this on KBS tuff crystals, they independently reached a result of 2.62 m.y. plus or minus .40, a stunning confirmation of Fitch and Millers number of 2.61 m.y. plus or minus .26. (Lewin, at p. 231.) Later, after the tide of opinion had turned against Fitch and Miller, Hurford and Gleadow re-counted the fission tracks and came up with 1.8 m.y., a stunning confirmation of Garniss Curtis' results. After they changed their tune, they claimed that Fitch had subtly pressured them to reach a number that confirmed his radiometric results, but it is hard to avoid the conclusion that, later on, the subtle pressure to support Curtis and McDougall became just as great. (Lewin, at 243-247.) Hurford stated, "[Y]ou can bias your results 10 percent either way, easily. You go crystal by crystal, and you begin to see where the rolling average is going. If you need the count to be higher with the crystal you're working on, so that it will fit in, you might include something that is a doubtful track. If you want the count to be lower, you don't include it." (Lewin, at p. 246.) But, of course, Hurford and Gleadow changed their results not by ten percent, but by *more than 30%*.

[60] Lewin, p. 235.

[61] As a sadder and wiser Richard Leakey pointed out, "One realizes that even in the most pure of sciences, which geophysics should be, there is a potential to identify careers, status, and results—and there's a strong political element, too." Lewin, at p. 251.

[62] The problem of "scatter" has not been ameliorated by more sophisticated equipment. "Improved laboratory techniques and improved constants have not reduced the

scatter in recent years." Waterhouse, J. B. "Chronostratigraphy for the World Permian," *Contributions to the Geologic Time Scale* (American Association of Petroleum Geologists Studies in Geology no. 6, 1978), p. 316, as quoted by Woodmorappe, *Studies in Flood Geology*, p. 147.

63 The precision of the measuring instruments has advanced to the point that one atom of carbon-14 might be present and detectable for up to 300,000 years.

64 Carbon-14 is created by cosmic rays bombarding the upper atmosphere. It seems likely that before the Flood, the Earth's atmosphere was partially protected from bombardment by cosmic rays and thus from radioactive Carbon-14. Roth lists the following explanations for the lower concentration of C-14: 1) a larger carbon reservoir diluting C-14 before the flood; 2) a stronger magnetic field before the flood, deflecting the cosmic rays that produce C-14; 3) a rate of mixing of C-14 into the oceans after the flood that would affect both atmospheric and oceanic concentrations of C-14; and 4) a change in the intensity of the source of cosmic rays that create the C-14. Roth, at p. 250. We might add that the earth's atmosphere may have been protected from bombardment by cosmic rays by the water vapor canopy that some believe existed before the Flood. (See Chapter 5, footnote 16.)

65 See, e.g., R. H. Brown, "The Interpretation of C-14 Dates," *Origins* 6(1):30-44 (1979). On the web at: http://www.grisda.org/origins/06030.htm.

66 Giem, Paul, "Carbon-14 Content of Fossil Carbon," *Origins* 51:6-30 (2001)

67 Brown, R. H. "Radioisotope age," Modified from *Geoscience Reports*, Spring 1996, no. 20, pp. 1-3. See also Coffin, p. 304, citing McDougall, Polach, and Stipp, "Excess radiogenic argon in young subaerial basalts from the Auckland volcanic field, New Zealand," *Geochimica et Cosmochimica Acta* 33:1485-1520 (1969).

68 Snelling, Andrew, "Geological conflict: Young radiocarbon date for ancient fossil wood challenges fossil dating," condensed from an article first published in *Creation Ex Nihilo* 22(2):44–47, March–May 2000. Found at AIG website: http://www.answersingenesis.org/home/area/magazines/docs/v22n2_geology.asp. With regard to claims that the fossil wood was contaminated with extraneous carbon, Snelling writes the following:

a. Pieces of the same sample were sent to the two laboratories and they both independently obtained similar results. Furthermore, three separate samples were sent to the same laboratory in two batches and again similar results were obtained. This rules out contamination.

b. The radiocarbon "dates" depend on the amounts of radiocarbon, originally in the living plants, now left in the fossil wood samples. In these samples, the radiocarbon left was between about 2.5 percent and 7.5 percent of the amount in living plants today. Any unavoidable contamination (e.g., dust, fungal spores) would be minuscule and would amount to at most 0.2 percent, which would have a negligible effect on these radiocarbon "dates".

c. The results are consistent with the analyzed carbon in the fossil wood representing organic carbon from the wood of land plants.

d. Such a claim would, by implication, cast a slur on the Ph.D. scientific staff of two radiocarbon laboratories, who, as qualified routine practitioners, understand the potential for contamination and how to avoid it in sample processing.

[69] Snelling, Andrew, "Dating Dilemma: Fossil wood in 'ancient' sandstone," condensation of an article first published in *Creation* 21(3):39–41, June–August 1999. Found at AIG website: http://www.answersingenesis.org/home/area/magazines/docs/v21n3_date-dilemma.asp Again, on the subject of contamination by modern C-14, Snelling writes:

> Anticipating objections that the minute quantity of detected radiocarbon in this fossil wood might still be due to contamination, the question of contamination by recent microbial and fungal activity, long after the wood was buried, was raised with the staff at this, and another, radiocarbon laboratory. Both labs unhesitatingly replied that there would be no such contamination problem. Modern fungi or bacteria derive their carbon from the organic material they live on and don't get it from the atmosphere, so they have the same 'age' as their host. Furthermore, the lab procedure followed (as already outlined) would remove the cellular tissues and any waste products from either fungi or bacteria.

[70] Giem, Paul, "Carbon-14 Content of Fossil Carbon," *Origins*, no. 51, pp. 6-30 (2001). At GRI website at: http://www.grisda.org/origins/51006.htm.

[71] Giem, "Carbon-14 Content of Fossil Carbon," supra.

[72] Rosella Lorenzi, "Ancient Footprints May Re-write American History," *Discovery News*, July 6, 2005.

[73] See, e.g., "Alleged 40,000 year-old human footprints in Mexico much, much older than thought" Science Daily, December 1, 2005.

[74] Lorenzi, *Discovery News*, supra.

[75] "Excavation in Idaho Lake Bed Yields Mammoth, Bison Remains," *Mammoth Trumpet*, 11(2) (1996). (*Mammoth Trumpet* is a newsletter on paleo-American research published by Robson Bonnichsen, an archeologist at Oregon State University). Field excavation of the Tolo Lake site was directed by Susanne Miller, Lockheed/Idaho National Engineering Laboratory, Idaho Falls, in coordination with principal investigators Robert Yohe, State Archaeologist, Idaho State Historical Society, Boise; and William A. Akersten, Curator of Paleontology, at the Idaho Museum of Natural History in Pocatello. The fossils were to be housed at the Idaho Museum of Natural History on the Idaho State University campus in Pocatello.

[76] Ibid.

[77] Ibid.

[78] Ibid.

[79] See, e.g., Andrew MacRae, supra, "there is almost zero chance that the broad understanding of geological history (e.g., that the Earth is billions of years old) will change."

# Selected Bibliography


Ager, Derek V., *The Nature of the Stratigraphical Record* (New York: John Wiley & Sons, 1973, 1981).

Aldridge, Susan, *The Thread of Life, the Story of Genes and Genetic Engineering* (Cambridge and New York: Cambridge University Press, 1996).

Ashton, John F., (Editor), *In Six Days, Why 50 Scientists Choose to Believe in Creation* (Sydney: New Holland Publishers, 1999 [U.S. publisher: Master Books, Green Forest, AK]).

Bakker, Robert T., *The Dinosaur Heresies* (New York: Kensington Publishing, 1986).

Behe, Michael, *Darwin's Black Box: The Biochemical Challenge to Evolution* (New York: Free Press, 1996).

Blinderman, Charles, *The Piltdown Inquest* (Buffalo, NY: Prometheus Books, 1986).

Brand, Leonard, *Faith, Reason, and Earth History* (Berrien Springs, MI: Andrews University Press, 1997).

Camp, Ashby L., "Reappraising the Crown Jewel," *Creation Matters* (September/October 1998).

Chadwick, Arthur V., "Megatrends in North American Paleocurrents," 2000 (Internet).

________"Preliminary Creation/Flood Model," (Internet).

Charig, Alan, *A New Look at the Dinosaurs* (New York: Facts on File, Inc., 1983).

Clark, Harold W., *Fossils, Flood, and Fire* (Escondido, CA: Outdoor Pictures, 1968).

Clark, Robert T. and James D. Bales, *Why Scientists Accept Evolution* (Grand Rapids, MI: Baker Book House, 1966).

Coffin, Harold, *Origin by Design* (Washington, D.C.: Review and Herald, 1983).

Colbert, Edwin H., *The Great Dinosaur Hunters and their Discoveries* (New York: Dover Publications, 1968).

Cooper, Bill, *After the Flood: The Early Post-Flood History of Europe Traced Back to Noah* (Chichester, UK: New Wine Press, 1995).

Cremo, Michael A., and Richard L Thompson, *Forbidden Archeology: The Hidden History of the Human Race* (Los Angeles: Bhaktivedanta Book Pub., Inc., 1993).

Currie, Philip J., and Eva B. Koppelhus, *101 Questions About Dinosaurs* (Mineola, NY: Dover, 1996).

Currie, Philip J., and Kevin Padian, editors, *Encyclopedia of Dinosaurs* (New York: Academic Press, 1997).

Darwin, Charles, *Origin of Species* (1859, Reprint: New York: Gramercy Books, 1979).

Denton, Michael, *Evolution: A Theory in Crisis* (Bethesda, MD: Adler & Adler, 1985).

Dobzhansky, Theodosius, *Genetics and the Origin of Species*, 3rd ed. (New York: Columbia University Press, 1951).

DeSalle, Rob, and David Lindley, *The Science of Jurassic Park and the Lost World or, How to Build a Dinosaur* (New York: Basic Books, 1997).

Dodson, Peter, *The Horned Dinosaurs* (Princeton NJ: Princeton University Press, 1996).

Donnelly, Ignatius, *Atlantis: The Antediluvian World*, 1882, Modern Revised Edition, edited by Egerton Sykes (New York: Gramercy Pub. Co., 1949).

Erickson, Jon, *Ice Ages: Past and Future* (Blue Ridge Summit, PA: TAB Books, 1990).

Farlow, James O., and M.K. Brett-Surman (editors), *The Complete Dinosaur* (Bloomington, IN: Indiana University Press, 1997).

Fox, Michael W., *Beyond Evolution: The Genetically Altered Future of Plants, Animals, the Earth, and Humans* (New York: The Lyons Press, 1999).

"Harvest of Fear: Exploring the Growing Fight over Genetically Modified Foods," *Frontline/NOVA*, 2001 (Television documentary).

Gaddis, Vincent H., *American Indian Myths & Mysteries* (Radnor, PA: Chilton Book Co., 1977).

Giem, Paul, "Carbon-14 Content of Fossil Carbon," *Origins* 51:6-30 (2001).

Gould, Stephen Jay, *Bully for Brontosaurus* (New York: W.W. Norton & Co., 1991).

________ *Eight Little Piggies* (New York: W.W. Norton, 1993).

________ *Ever Since Darwin* (New York: W.W. Norton & Co., 1977).

________ *Hen's Teeth and Horse's Toes* (New York: W.W. Norton & Co., 1983).

________ *I Have Landed* (New York: Harmony Books, 2002).

________ *Leonardo's Mountain of Clams and the Diet of Worms* (New York: Harmony Books, 1998).

________ *Ontogeny and Phylogeny* (Cambridge, MA: The Belknap Press of Harvard University Press, 1977).

________ *Time's Arrow, Time's Cycle* (Cambridge, MA: Harvard University Press, 1987).

________ *Wonderful Life* (New York: W.W. Norton & Co., 1989).

Hapgood, Charles, *Mystery in Acambaro* (Kempton, IL: Adventures Unlimited Press, 2000).

Hasel, Gerhard F., "The Meaning of the Chronogenealogies of Genesis 5 and 11," *Origins* 7(2):53-70 (1980).

Hawkins, Gerald S., with John B. White, *Stonehenge Decoded* (Garden City, NY: Doubleday, 1965).

Horner, John R., and James Gorman, *Digging Dinosaurs* (New York: Workman Pub., 1988).

Horner, John R., and Don Lessem, *The Complete T. rex* (New York: Simon & Schuster, 1993).

Johnson, Phillip, *Darwin on Trial* (Washington, D.C.: Regnery Gateway, 1991).

Kaku, Michio, *Visions: How Science Will Revolutionize the 21st Century* (New York: Doubleday Anchor Books, 1997).

Keesey, T. Mike, *The Dinosauricon* (Internet)

Kuhn, Thomas S., *The Structure of Scientific Revolutions*, 2d Edition (Chicago: University of Chicago Press, 1962, 1970).

Lancaster-Brown, Peter, *Megaliths, Myths, and Men* (New York: Taplinger Pub. Co., 1976).

Levine, Joseph, and David Suzuki, *The Secret of Life: Redesigning the Living World* (Boston, MA: WGBH, 1993).

Lewin, Roger, *Bones of Contention* (New York: Simon & Schuster, 1987).

Lockley, Martin, and Adrian P. Hunt, *Dinosaur Tracks and other Fossil Footprints of the Western United States* (New York: Columbia University Press, 1995).

Lockley, Martin, and Christian Meyer, *Dinosaur Tracks and Other Fossil Footprints of Europe* (New York: Columbia University Press, 2000).

Lubenow, Marvin L., *Bones of Contention* (Grand Rapids, MI: Baker Book House, 1992).

Margulis, Lynn, and Dorion Sagan, *Acquiring Genomes* (New York: Basic Books, 2002).

Marsh, Frank Lewis, *Life, Man, and Time* (Escondido, CA: Outdoor Pictures, 1967).

Mayr, Ernst, *The Growth of Biological Thought* (Cambridge MA: Harvard University Press, 1982).

McGowan, Christopher, *Dinosaurs, Spitfires, and Sea Dragons* (Cambridge, MA: Harvard University Press, 1991).

________*The Dragon Seekers* (Cambridge, MA: Perseus Publishing, 2001).

________*The Successful Dragons* (Toronto: Samuel Stevens, 1983).

McLoughlin, John C., *Synapsida: A New Look into the Origin of Mammals* (New York: Viking Press, 1980).

Milton, Richard, *Shattering the Myths of Darwinism* (Rochester, VT: Park Street Press, 1997).

Mitchell, Colin, *The Case for Creationism* (Alma Park, Grantham, Lincolnshire, UK: Autumn House, Ltd., 1994).

________*Creationism Revisited* (Alma Park, UK: Autumn House, Ltd., 1999).

Nelson, Byron C., *The Deluge Story in Stone* (Minneapolis, MN: Bethany Fellowship, 1931, 1968).

# SELECTED BILBIOGRAPHY

Noorbergen, Rene, *Secrets of the Lost Races* (Indianapolis, IN: The Bobbs-Merrill Company, 1977).

Norman, David B., *Dinosaur!* (New York: Prentice Hall, 1991).

________*Prehistoric Life: The Rise of the Vertebrates* (New York: Macmillan, 1994).

Numbers, Ronald L., *The Creationists* (New York: Alfred Knopf, 1992 [paperback: University of California Press, 1993]).

Oard, Michael J., *An Ice Age Caused by the Genesis Flood* (El Cajon, CA: Institute for Creation Research, 1990).

O'Kelley, Michael J., *Newgrange: Archeology, Art, and Legend* (London: Thames and Hudson, 1982).

Ovid, *The Metamorphoses,* translated by Horace Gregory (New York: The Viking Press, 1958).

Preston, Douglas J., *Dinosaurs in the Attic* (New York: St. Martin's Press, 1986).

Price, George McCready, *Evolutionary Geology and the New Catastrophism* (Mountain View, CA: Pacific Press, 1926).

________*Genesis Vindicated* (Washington, D.C.: Review and Herald, 1941).

Rappaport, Rhoda, *When Geologists were Historians,* 1665-1750 (Ithaca, NY: Cornell University Press, 1997).

Ritland, Richard M., *A Search for Meaning in Nature* (Mountain View, CA: Pacific Press, 1970).

Roth, Ariel, *Origins* (Hagerstown, MD: Review and Herald, 1998).

Rudwick, Martin J.S., *The Meaning of Fossils* (New York: Neale Watson Academic Publications, 1972, 1976).

Ruse, Michael, *Can a Darwinian Be a Christian?* (Cambridge and New York: Cambridge University Press, 2001).

Setterfield, Barry, *Creation and Catastrophe Chronology,* 1999 (Internet).

Shigley, Gordon, "Amalgamation of Man and Beast: What did Ellen White Mean?" *Spectrum* (Journal of the Association of Adventist Forums) 12(4):10-19 (1982).

Spalding, David A.E., *Dinosaur Hunters* (Rocklin, CA: Prima Publishing, 1993).

Spetner, Lee, *Not by Chance* (Brooklyn, NY: The Judaica Press, Inc., 1997).

Stanley, Steven M., *Earth and Life through Time,* 2d ed. (New York: W. H. Freeman and Co., 1986, 1989).

Stringer, Christopher, and Clive Gamble, *In Search of the Neanderthals* (New York: Thames and Hudson, 1993).

Temple, Robert, *The Crystal Sun* (London: Random House/Arrow Books, Ltd., 1999).

Vardiman, Larry, "Numerical simulation of precipitation induced by hot mid-ocean ridges," Fourth International Conference on Creationism, Pittsburgh, PA, August 3-8, 1998.

Wallace, David Rains, *The Bonehunters' Revenge* (New York: Houghton Mifflin, 1999).

Weishampel, Peter Dodson, and Halszka Osmolska, editors, *The Dinosauria* (Berkeley: University of California Press, 1990).

Wells, Jonathan, *Icons of Evolution* (Washington, D.C.: Regnery, 2000).

Wheeler, Gerald W., *The Two-Taled Dinosaur* (Nashville, TN: Southern Publishing, 1975).

Whitcomb, John C., and Henry Morris, *The Genesis Flood* (Phillipsburg, NJ: P & R Publishing, 1961).

White, Ellen, *Patriarchs and Prophets* (Washington, D.C.: Review and Herald).

________*Spiritual Gifts* (Washington, D.C.: Review and Herald).

Wieland, Carl, "Sensational Dinosaur Blood Report," from *Creation Ex Nihilo* 19(4):42, 43, September-November 1997.

Wilford, John Noble, *The Riddle of the Dinosaur* (New York: Alfred A. Knopf, 1985).

Wise, Kurt P., *Faith, Form and Time* (Nashville: Broadman and Holman, 2002).

Woodmorappe, John, "Mammal-like reptiles: major trait reversals and discontinuities," *Creation ex Nihilo Technical Journal* 15(1):44-52 (2001).

________*The Mythology of Modern Dating Methods* (El Cajon, CA: Institute for Creation Research, 1999).

________*Studies in Flood Geology* (El Cajon, CA: Institute for Creation Research, 1999).

Younker, Randall W., *God's Creation* (Nampa, ID: Pacific Press, 1999).

# Photography & Illustration Credits

## PLATES:

**Plate 1:** Portrait of Georges Cuvier, by Mathieu-Ignace van Bree (1773-1839), used by permission of Muséum National d'Histoire Naturelle, Bibliothèque Centrale, Paris.

**Plate 2:** *Mosasaurus missouriensis* skull, © Mike Everhart, Oceans of Kansas Paleontology, http://www.oceansofkansas.com., used by permission.

**Plate 3:** Mosasaurus, © De Agostini/ Natural History Museum Picture Library (NHMPL), used by permission

**Plate 4:** Pteranodon, artwork © 1996 Joe Tucciarone, used by permission

**Plate 5:** Mary Anning, by B. J. Donne (1847), used by permission of Geological Society/NHMPL.

**Plate 6:** Dr. Gideon Algernon Mantell, used by permission of Geological Society/NHMPL.

**Plate 7:** Sir Richard Owen. Photograph by Maull & Polyblank, c. 1855, NHMPL, used by permission.

**Plate 8:** Camarasaurus, © De Agostini/ NHMPL, used by permission.

**Plate 9:** Stegosaurus, artwork © Joe Tucciarone, used by permission.

**Plate 10:** Allosaurus, artwork © Joe Tucciarone, used by permission.

**Plate 11:** Brontosaurus, artwork © Joe Tucciarone, used by permission.

**Plate 12:** *Tyrannosaurus rex*, © The Natural History Museum, London/ Kokoro Dreams, used by permission.

**Plate 13:** *Tyrannosaurus rex*, "Sue," Photograph by Fritz Geller-Grimm, used pursuant to Creative Commons Attribution ShareAlike 2.5 License.

**Plate 14:** Grand Canyon, photograph by the author.

**Plate 15:** Dinosaur Family Tree, illustration by Christine Cummings-Weis, after Dr. Paul Sereno, University of Chicago, used by permission.

**Plate 16:** DNA in test tube, from Levine, Joseph, and David Suzuki, *The Secret of Life: Redesigning the Living World* (Boston, MA: WGBH, 1993), used by permission.

**Plate 17:** Structure of DNA, courtesy the National Library of Medicine.

**Plate 18:** Triceratops, artwork © 1996 Joe Tucciarone, used by permission.

**Plate 19:** Parasaurolophus, artwork © 1995 Joe Tucciarone, used by permission.

**Plate 20:** Seismosaurus, artwork © 1996 Joe Tucciarone, used by permission.

**Plate 21:** Deinonychus, artwork © 1996 Joe Tucciarone, used by permission.

**Plate 22:** Ichthyosaur from Holzmanden, Germany, © NHMPL, used by permission.

**Plate 23:** Berlin Archaeopteryx, NHMPL, used by permission.

**Plate 24:** Dimetrodon, © Michael Long/NHMPL, used by permission.

**Plate 25:** Estemmenosuchus, © John Sibbick/NHMPL, used by permission.

**Plate 26:** *Inostrancevia alexandri* assaulting *Scutosaurus karpinski*, © Kelly Taylor, used by permission.

**Plate 27:** Lystrosaurus, © De Agostini/NHMPL, used by permission.

**Plate 28:** Child of Gibraltar, reconstruction by Elisabeth Daynes, photo © Philippe Plailly/Eurelios, used by permission.

**Plate 29:** Stonehenge construction phases, from artwork by W Brouard, English Heritage Photo Library, used by permission.

**Plate 30:** Central vault at Maeshowe, photograph © Charles Tait, used by permission.

**Plate 31:** Coelophysis, artwork © 1998 Joe Tucciarone, used by permission.

## FIGURES:

**Fig. 1.1:** Ichthyosaur, © Dover Publications, used by permission.

**Fig. 1.2:** Plesiosaur, © Dover Publications, used by permission.

**Fig. 1.3:** Iguanodon tooth, 1825 lithograph.

**Fig. 1.4:** Iguanodon, © Dover Publications, used by permission.

**Fig. 1.5:** William Buckland, NHMPL, used by permission.

**Fig. 1.6:** Megalosaurus jaw, first published by William Buckland, "Notice on the Megalosaurus or great fossil lizard of Stonesfield," *Transactions of the Geological Society of London*, (1824).

**Fig. 1.7:** Megalosaurus, by Neave Parker, © NHMPL, used by permission.

**Fig. 2.1:** Hadrosaurus, © Victoria Edwards/NHMPL, used by permission.

**Fig. 2.2:** Othniel Charles Marsh, circa late 1800s.

**Fig. 2.3:** E.D. Cope, circa 1895.

**Fig. 2.4:** Triceratops, © Dover Publications, used by permisson.

**Fig. 4.1:** Major divisions of geological strata, permission of U.S. Geological Survey.

**Fig 4.2:** Plot's fossil, from *The Natural History of Oxfordshire* by Robert Plot, Oxford, 1676.

**Fig 4.3:** Charles Lyell, Courtesy of the National Library of Medicine.

**Fig. 8.1:** Major divisions of Strata with Lyellian Chronology, illustration by Christine Cummings-Weis.

**Fig. 8.2:** Avant and Apres, published by Antonio Snider-Pellegrini, 1858.

**Fig. 9.1:** Louis Agassiz, circa 1840s.

**Fig. 9.2:** Ice Age in North America, from *Origin by Design* by Harold Coffin (Review and Herald, 1983). Used by permission of Review and Herald, and Harold Coffin.

**Fig. 10.1:** Courtyard Edfu Temple, photo by Benjamin E. Wood, © 2003, www.healinglight.org, used by permission.

**Fig. 10.2:** Hieroglyphics on the Edfu Temple, photo by Benjamin E. Wood, © 2003, www.healinglight.org, used by permission.

**Fig. 10.3:** Pictographic writing, Courtesy of www.creationism.org.

**Fig. 11.1:** Subdivision of Triassic Strata, illustration by Christine Cummings-Weis.

**Fig. 11.2:** Cladogram of the Archosaurs, Parrish, J. Michael, "Evolution of the Archosaurs," in *The Complete Dinosaur*, James O. Farlow and M.K. Brett-Surman, Eds. (Bloomington, IN: Indiana University Press, 1997), p. 198, permission of Indiana University.

**Fig. 12.1:** A simplified diagram of the replication of the DNA double helix, from Aldridge, Susan, *The Thread of Life: The Story of Genes and Genetic Engineering* (Cambridge and New York: Cambridge University Press, 1996), p. 26, permission of Cambridge University Press.

**Fig. 13.1:** Darwin's tree of life, from original 1859 edition of *Origin of Species*.

**Fig. 14.1:** Triceratops dental batteries, circa 1907.

**Fig. 14.2:** Dinosaur classification based upon hip structure, illustration by Sleepy Read.

**Fig. 14.3:** *Tyrannosaurus rex*, "Sue," at The Field Museum, courtesy of Melissa Squires.

**Fig. 14.4:** Duckbills drinking at a stream, artist unknown.

**Fig. 14.5:** Cross section of parasaurolophus, © De Agostini/NHMPL, used by permission.

**Fig. 14.6:** Zonal bone growth, reprinted from *Modern Geology*, 18, R. E. H. Reid, "Apparent zonation and slowed late growth in a small Cretaceous theropod," pp 391-406, 1993, with permission from Elsevier.

**Fig. 14.7:** Fibro-lamellar bone growth, reprinted from *Nature*, 292, 5818, R. E. H. Reid, "Lamellar-zonal bone with zones and annuli in the pelvis of a sauropod dinosaur," pp 49-51, 1981, with permission from Elsevier.

**Fig. 17.1:** Boule's Neanderthal, "An Ancestor: the Man of Twenty Thousand Years Ago" by Frantisek Kupka, based on the work of Marcellin Boule, published in *L'Illustration* in 1909.

**Fig. 17.2:** *Hesperopithicus*, from *Illustrated London News*, 1922.

**Fig. 19.1:** Sunrise over the heelstone, photograph by Les Still, used by permission.

**Fig 19.2:** Sarsen Circle from the Heel stone, illustration by Christine Cummings-Weis.

**Fig. 19.3:** Alignments of station stones, illustration by Christine Cummings-Weis, after Hawkins.

**Fig. 19.4:** Megalithic alignments, illustration by Christine Cummings-Weis, after Hawkins.

**Fig. 19.5:** Alignments of Stonehenge I, illustration by Christine Cummings-Weis, after Hawkins.

**Fig. 19.6:** Sarsen lintel, © Mjausson, used by permission.

**Fig. 19.7:** Cross-section of passage grave at Newgrange, from *Newgrange: Archeology, Art, and Legend*, Michael J. O'Kelly and Claire O'Kelly, (London: Thames and Hudson, 1982), p. 88, used by permission of Eve O'Kelly.

**Fig. 22.1:** Baghdad Battery, illustration by Christine Cummings-Weis.

**Fig. 24.1:** Mid and lower Jurassic tracks and bodies, illustration by Christine Cummings-Weis.

**Fig. 24.2:** Dinosaur National Monument, © John Crossley, www.americansouthwest.net, used by permission.

**Fig. 25.1:** Julsrud stegosaur figurine, photo by Don R. Patton, used by permission.

**Fig. 25.2:** Julsrud sauropod figurine, photo by Don R. Patton, used by permission.

**Fig. 25.3:** Ica Stone depicting man fighting dinosaur, photo by Don R. Patton, used by permission.

**Fig. 25.4:** Footnote 12, Etched sauropod at Havasupai, photo by Don R. Patton, used by permission.

**Fig. 25.5:** Footnote 13, Man and sauropod etched on rock, photo by Don R. Patton, used by permission.

**Fig. A1:** Profile of skull 1470, Reprinted by permission of Waveland Press, Inc. from Clark Spencer Larsen, Robert M. Matter, and Daniel L. Gebo, *Human Origins: The Fossil Record, Third Edition.* (Long Grove, IL: Waveland Press, Inc., 1998). All rights reserved.

# Index

## A

Abel, 405
Abishai, 397
Accelerator mass spectrometry, 391
Accommodation theories, 92–100, 136–38
Active gap theory, see Gap theory
Acts, Book of, 136
*Acts of the Apostles* (White), 516
Adam, 89–91, 117–20, 394, 398, 400, 512, 516, 530
Adam and Eve story, 119–20, 400
Adams, Henry, 187–88
Agassiz, Louis, 183–86, 184f, 188
Ager, Derek, 71–72
Agricultural uses of genetic engineering, 452–56
Ahiman, 396
Albertosaurus, 295
Allosaurus, 29, 293, Plate 10
  "Big Al," 494
Alpha-1-antitrypsin (AAT), 459–60
Altruistic genetic elements (AGEs), 528
Alvarez, Luis, 543
Alvarez, Walter, 542–43
Amalgamation:
  post-Flood, 517, 518, 522–31
  pre-Flood, see Amalgamation theory
  Satan's, 522–23, 530–31
Amalgamation theory, 483–503
  debate on meaning of amalgamation, 483–90
  defense of, 509–31
    diversification and species change, 523–31
    scientific knowledge of antediluvians, 509–16
    White (Ellen) and different amalgamation contexts, 516–23, 529–31
  dinosaur extinction and, 589–94
  dinosaurs as product of genetic engineering, 492–502
  explanatory power of, 496–502
  human/animal combinations, 502–3

American Museum of Natural History, 35–39, 297, 358, 359, 565
*American Naturalist,* 21
Amino acids, 249
Amphibians, 338
Amphicyonids, 501
Anakites, 395–97, 401
Anatomy, principles of, 2–3
Anchiceratops, 559
*Ancient Hunters* (Sollas), 359
Ancient man:
- cave paintings of, 387–92
- Darwinian view of, 353–77
  - misinterpretations and hoaxes in, 353–64, 371–74
  - nature of paleoanthropology, 374–76
  - theoretical models and facts, 364–71
- early post-Flood, 415–38
  - British megaliths and, 415–34, 438
  - Egyptian civilization as legacy and, 434–38
- pre-Flood, see Antediluvians

Andrews, Roy Chapman, 38–39, 91
Angels, 402
Animals:
- amalgamation theory and, see Amalgamation theory
- destruction by antediluvians, 494
- transgenic, 456–62
- See also Dinosaurs; names of specific animals

Ankylosaurus, 493–94, 569
Anning, Joseph, 4
Anning, Mary, 3–6, Plate 5
Anning, Richard, 4
Anomoepus, 286, 287
Antediluvians, 387–410
- created perfect, 393–94
- defined, 394n
- destruction of animals by, 494
- eyesight of, 515–16
- genetic engineering by, see Amalgamation theory
- intelligence and technology of, 405–9, 509–16
- life spans of, 394, 510
- physical stature of, 395–404
- unity of language of, 510

Antibiotic resistance, 253–54
*Antiquity of Man, The* (Keith), 359
*Antiquity of Man in East Anglia, The* (Moir), 360
Apache flood legend, 220
Apatosaurus, 29
Apennine Mountains, Italy, 53
Aqua Bounty Technologies, 457–58
AquaAdvantage salmon, 458
*Archaeology of Sussex, The* (Curwen), 360
Archaeopteryx, 315–16, 328, 329, 333, 335, 497, Plate 23
Archaeoraptor liaonengensis, 335–36
Archosaurs, 229, 230f, 231
Arctic flounder, antifreeze gene of, 456, 457
Arduino, Giovanni, 60
Argon-40, 611, 612, 615–17, 619–20, 624, 625, 631
Argon-40 argon-39 method, 624, 633–34
Aristotle, 51, 523

Ark:
presence of dinosaurs issue, 591–601
White (Ellen) on animals being supernaturally led to, 501
See also Genesis Flood
Ark of the Testimony (Ark of the Covenant), 87–88
Armored dinosaurs, 493–94
Arntzen, Charles, 454–55
Art, Ice Age, 388–90, 600
Artifacts, post-Flood, 598–601
Ashurbanipal, 207, 208
Asteroid theory, 541–50
Astronomical theory of Ice Age, 188
Atheism, 121–23
Atkinson, Richard, 430
Atlantis, 436
Atlantosaurus, 25, 26
Atonement, 88–89, 117
Aubrey, John, 418
Aubrey Holes, of Stonehenge, 416, 424–25, 427, 430
Augustine, 523
Austin, Steven A., 195
Australopithecines, 624
Australopithecus, 370
Avery, Oswald, 248
Avian characteristics:
of Archaeopteryx, 315–16
of dinosaurs, 285–304, 497
of pterosaurs, 313–15
Avimimus, 334
Azendohsaurus, 231

# B

Babel, Tower of, 213, 219, 437
Babylonian flood legends, 204–8
Bacillus thuringiensis, 452–53
Bacteria, horizontal gene transfer and, 526–27
Bakker, Robert T., 287, 288, 292, 297, 302, 315, 542, 546, 550, 573, 579–81
Baldwin, David, 576
Ballard, Bob, 97
Bananas, transgenic, 454–55
Baraminology, 524–25
Baramins, 488, 490, 524–25
"Barney" (television character), 42
Barrows, 433–34
Bathtub model (of Genesis Flood), 561
Batteries, Baghdad, 514, 514f
Baumgardner, John, 164
Beaks:
of Archaeopteryx, 315
of dinosaurs, 289, 291–92, 295–96
Bear dogs, 501
Bear Paw Shale, 555–56, 574
Beasley, Bruce, 388
Behe, Michael J., 104, 259
Behemoth, 594–95, 598
Behrensmeyer, Kay, 623
Beipiaosaurus, 337
Belgian Royal Museum of Natural History, 10
Benton, Michael, 233
Berg, Paul, 447, 448, 450
Bergounioux, F. M., 373–74
Bernissart, Belgium, 9–10, 41

Berosus, 95, 204–6, 208–9
Bestiality, 486
*Beyond 2050: Angels in America* (Kaku), 464
Bible:
accommodation theories and, 92–100, 136–38
ancient man and, see Antediluvians
chronologies in, 168–73
on creation of animals, 488
Darwinism and undermining of, 115–27
geologic column and, 149–73
See also Genesis Flood
King James translation of, 169
Lyellism and undermining of, 86–91
See also specific books of the Bible
Bijou Creek flood, 151–52
Biome succession theory, 500
Biotechnology, see Genetic engineering
Bird, Roland T., 299
Birds:
claim of evolution of dinosaurs into, 301, 322, 327–37
See also Avian characteristics
"Bison alticornis", 30
Blacks, 518–21
Blinderman, Charles, 364
Bone Cabin Quarry, 36, 37
Bones:
abnormalities in herbivore dinosaurs, 494–95
of Archaeopteryx, 315, 328
of coelurosaurs, 301
fossilization of, 46–47
of Ichthyosaurus, 312
periosteal, growth patterns of, 302–4, 302f
of pterosaurs, 313–14
of sauropods, 298–99
of Triceratops, 290–92
of Tyrannosaurus rex, 293–94
unfossilized, 149–51, 466–67
See also Dinosaur fossils; Fossils
Booth, Ernest S., 146
Boule, Marcellin, 354–56, 367
Boyer, Herbert, 450
Brachiosaurus, 579
Brachylophosaurus, 297
Brand, Leonard, 146
Brandvold, John, 556–58
Brandvold, Marion, 556–58
Breastfeeding mechanism, 342–43
Breuil, Henri, 391–92
Briggs, Derek, 275
British Association for the Advancement of Science, 5, 14
British megaliths, 415–34, 438
British Museum, 15, 360–61, 389
Broadberry, Richard, 211–12
Bromoxynil, transgenic crops with resistance to, 454
Brontosaurus, 27, 29–30, 36–37, 41, 233, 575, 579, 590, 591, Plate 11
Bronze Age barrows, 433–34
Brown, Barnum, 36–38, 150, 299, 577
Brown, Robert H., 146
Brunel, Éliette, 387–88
Bryan, William Jennings, 357, 358, 368
Buckland, Mary, 185
Buckland, William, 10–12, 10f, 14, 66, 67, 92, 96, 185

Burgess Shale, Canada, 154, 275–76
Burkitt, M. C., 360
Burnet, Thomas, 56–57
Buttners, D. S., 58–59

# C

Cabrera Darquea, Javier, 599
Cain, 400, 403–5, 484–89
Caleb, 396
Calgene, Inc., 456
Callanish stones, Isle of Lewis, 426
Camarasaurus, 26, 30, 233, 579, Plate 8
Cambrian explosion, 156, 273–79
Cambrian-Precambrian boundary, 621
Cambrian strata, 48f, 49, 155f, 155–56, 161
Campbell, Keith, 470
*Can a Darwinian Be a Christian?* (Ruse), 119–20
Canaanites, 395, 396, 401
Cañon City, Colorado, 26
Carbon 14 dating, 630–35
Carboniferous Redwall Formation, 161
Carboniferous strata, 155f, 156, 159–60, 316, 317
Carlin, William Edward, 26–29
Carnegie Museum, Pittsburgh, 37
Carnian strata, 230–31, 230f, 233
Carp, transgenic, 457
Cary, Elizabeth Cabot, 186
Catastrophic plate tectonics, 164–65, 190–92
Catastrophism, 66, 67, 70, 72, 75
  asteroid theory, 541–50
  Lyell, (Sir Charles) on, 543–45
  vs. Lyellism (uniformitarianism), 66, 67, 70, 72, 75, 543–44
  renaissance of, 545, 546
  See also Genesis Flood
Catastrophists, 66, 67, 70, 71, 580
Catcott, Alexander, 60
Caudipteryx, 337
Cave, A. J. E., 354–55
Cave paintings, 387–92
Cenozoic strata, 48f, 49, 102
  correlation with stages of Genesis Flood, 166–68
  in Lyellian chronology, 155f
  mammal fossils in, 500
Central Asiatic Expedition, 38–39
Centrosaurus, 559–60
Ceratopsians, 30, 292, 559
  See also Horned dinosaurs
Cetiosaurus, 13, 298
Chadwick, Arthur V., 146, 160, 166
Chalmers, Thomas, 93
Chambers, Robert, 100
Charig, Alan, 320–21
Chasmosaurus, 290, 559
Chauvet, Jean-Marie, 387–88
Chauvet-Pont-d'Arc cave, 388, 391–93
Cheese, vegetarian, 455–56
Chiappe, Luis, 335, 565
Chicxulub crater, Yucatan, 543
Chimeras, 457, 463, 465, 503
Chinese:
  flood legends, 210–12
  "old earth" traditions, 95
Choctaw Indian flood legend, 217–20
Christian Church (Disciples of Christ), 124

Christianity:
- accommodation theories, 92–100
- enlightenment rejection of, 62–64
- implications of Darwinism for, 115–27
- implications of Lyellism for, 86–91

Chromosomes, 248
1 Chronicles, Book of, 398
2 Chronicles, Book of, 169
Cladeidon, 13
Clark, Harold W., 146, 167, 483–86, 488–90
Clarke, Arthur C., 465
Clift, Caroline, 14
Clift, William, 13, 15
Cloning, 470, 471
Clovis model, 633
Cluver, Michael A., 340
Coal geology, 156–59
Coal measures, 156–58
Cockburn, Patrick, 59
Codex Alexandrinus, 172
Codex Vaticanus, 172
Codon, 249
Cody, "Buffalo Bill," 23
Coe, Malcolm, 579
Coelophysis, 576–77, Plate 31
Coelurosaurs, 300–301, 327–28
Coffin, Harold G., 146, 169, 616
Cohen, Stanley, 450
Colbert, Edwin, 576–77
Commentary on Genesis (Luther), 55–56
Como Bluff, Wyoming, 26–30, 36, 572–73
*Complaint and Vindication of the Fishes* (Scheuchzer), 58
Compsognathus, 300, 329, 335, 591–92
"Conflict of the Ages" series (White), 516
Continental drift, 154, 162–64
Conway-Morris, Simon, 275
Conybeare, William, 4, 8, 11–12, 312
Cook, Harold, 356
Cooling, volcanic activity and, 191–94
Cope, Edward Drinker, 19–22, 23f, 24–31, 35, 298, 576
Copying error theory of evolution, 239–60, 525
- credibility of, 259–60
- Darwin's theory, 239–46
- modern theory, 246–56
- problem of irreducible complexity, 257–59
- See, also, Darwinism, evolution

1 Corinthians, Book of, 90, 136, 171
2 Corinthians, Book of, 215
Corn, transgenic, 453, 454
Cotton, transgenic, 453
Cows, transgenic, 459, 460
Crazy Horse, 24
*Création et ses Mystéres Dévoilés, La* ("Creation and Its Mysteries Revealed) (Snider-Pellegrini), 161–62
Creation Evidences Museum, 599
Cretaceous strata, 48f, 49, 314
- asteroid theory and, 547
- claim of dinosaur-to-bird transition and, 334, 335
- correlated with stages of Genesis Flood, 165–67
- cyclical coal deposits in, 157
- dinosaur eggs in, 565–66
- in Lyellian chronology, 155f

"scientific" view of dinosaur extinction and, 76
Cretaceous-Tertiary boundary, 167, 546–50, 621
Creutzfeldt-Jakob disease, 451
Crichton, Michael, 465, 466, 469, 471
Cro Magnon cave artists, 388–89
Crocodile, behemoth/leviathan as, 595–98
Croll, James, 188
Crops, transgenic, 452–56
Crow Indians, 24
Crustal divergence, 154, 161–64, 163f
Cuozzo, Jack, 373–74
Currie, Phil, 335, 559–60
Curtis, Garniss, 625–26
Curwen, E. Cecil, 360
Custer, George Armstrong, 23–24
Cuvier, Georges, 1–3, 7–8, 12, 66–67, 70, 73, 311–12, 548, Plate 1
Cuvier's law of land-sea simultaneity, 548
Cyclothems, 156–58
Cynodonts, 319, 339–40
Czerkas, Stephen, 335–36, 599–600

# D

Daily Telegraph, 207
d'Alva Ixtilxochitl, Ferdinand, 220–22
Darwin, Charles, 77, 86n, 117
- on Cambrian explosion, 273–74
- on Christianity, 122
- on evolution of eyes, 257
- on Lyell (Sir Charles), 101–4
- on peacock feathers, 333
- race issue and, 519
- theory of evolution of, 239–46
- on transitional forms, 267–68, 273
- Tree of Life of, 276–77, 276f
- See also Darwinism

Darwinism, 85, 86n
- Agassiz (Louis) and, 183
- as axiom, 608
- Cambrian explosion and, 273–79
- claim of dinosaur-to-bird transition, 301, 322, 327–37
- claim of reptile-to-mammal transition, 337–43
- copying error theory of evolution, see Copying error theory of evolution
- defined, 115
- dinosaur origins and, 229–34
- fossil record and, 267–80
- implications for Christianity, 115–27
- Owen (Richard) and, 14, 15
- race issue and, 519–20
- relationship between Lyellism and, 100–107
- speciation and, 523–24
- view of ancient man, 353–77
  - misinterpretations and hoaxes in, 353–64, 371–74
  - nature of paleoanthropology, 374–76
  - theoretical models and facts, 364–71
- White (Ellen) on, 144–45
- See, also, evolution, Copying error theory of evolution

Dating, see Radiometric dating
David, 397
Dawkins, Richard, 252

Dawson, Charles, 358, 361
Day/age theory, 94–96, 99, 136–38
Day of Atonement, 88
DDT, 254
De La Beche, Henry, 312
De Luc, Jean-Andre, 60
de Vries, Hugo, 246
Death as result of the Fall:
  local flood theory and, 97
  Lyellism and, 88, 90, 101
  tranquil theory and, 96
Deccan traps, India, 549–50
Deinonychus, 301, 328, Plate 21
Delano, Columbus, 23
Deluge, the, see Genesis Flood
*Deluge Story in Stone, The* (Nelson), 62
Dembski, William, 104
Dendrochronology, 172
Dental batteries:
  of duckbill dinosaur, 296
  of Triceratops, 289, 290f
Derham, William, 58
DeSalle, Rob, 469–70
*Desire of Ages* (White), 516
Deucalian, 216
Deuteronomy, Book of, 87, 88, 171, 395, 396
Devonian strata, 48f, 49, 155f, 156, 157, 161
Dicynodonts, 319
Diluvium theory, 92, 99
Dimetrodon, 317, 339, Plate 24
Dingus, Lowell, 565
*Dinosaur* (film), 42, 541–42
Dinosaur fossils:
  asteroid theory and, 547–48
  claim of dinosaur-to-bird evolution and, 301, 322, 327–37
  discovery of, see Fossilists
  distribution of, 40
  eggs, 39, 561–66
  in geologic column, see Genesis Flood; Geologic column; “Scientific” view of geologic column
  intermediate forms, lack of, 229–34, 267–74, 277, 278
  recent finds, 40
  species and genera, difficulties in determining number of, 40–42
  See also Dinosaurs
*Dinosaur Heresies, The* (Bakker), 302
Dinosaur National Monument, 161, 577–78, 578f
Dinosaur Provincial Park, Alberta, Canada, 558–60
Dinosaurs:
  alternative theory of, see Amalgamation theory
  classification of, 288, 290–91, 290f
  coining of term, 14–15, 287
  Darwinian view of origins of, 229–34
  discovery of unfossilized bones, 149–51
  extinction of:
    asteroid theory, 541–50
    by Genesis Flood, see Genesis Flood, evidence of dinosaur extinction by
    list of published explanations of, 542
    “scientific” view of, 76

"family tree," 232, Plate 15
fossils of, see Dinosaur fossils
as mixture of modern vertebrate classes, 285–304, 320–22, 496–97
physiology of, 285–304
as product of genetic engineering, 492–502
See also Amalgamation theory
resurrection of, 465–72
thermal characteristics of, 301–4, 497
tracks of, 158–59, 285–88, 292–93, 299, 566–72
White, Ellen on, 144, 590, 591
Diodorus Siculus, 421–22
Diplodocus, 27, 30, 599–600
*Discourse on the revolutions of the surface of the globe* (Cuvier), 66
*Discovery of Genesis, The* (Kang and Nelson), 211
*Discovery of Primitive Man in China* (Elliot Smith), 360
Diseases, inheritable genetic mutations and, 252–53
DNA (deoxyribonucleic acid):
copying error theory of evolution and, 248–56, 250f, Plate 16, Plate 17
horizontal gene transfer and, 526–29
"junk," 459–60, 528
recombinant, see Genetic engineering
DNA ligase, 449
DNA polymerase, 250
Dobzhansky, Theodosius, 247–48, 487
Docodonta, 341–42
Dodson, Peter, 289, 559–60
Dollo, Louis, 9–10
"Dolly" (cloned sheep), 470, 471
Dominance, Law of, 244
Dominant paradigm, 365–71
Donnelly, Ignatius, 436
Double helix, 249
Douglass, Earl, 30, 577
Doyle, Sir Arthur Conan, 362
Dromaeosaurus, 300–301
Dry Mesa Quarry, 574–75, 577, 580
Dubois, Eugene, 375–76
Duckbill dinosaurs, see Hadrosaurs
Duckbill platypus, 320

# E

Earthquake zones, 163
Easter Island, 436
Ecclesiastes, Book of, 415, 483
Echidna, 320
Eclipse prediction, 416–31
Ecological zonation theory, 500
Edaphosaurus, 317
Edfu, Temple of, 209f
hieroglyphics at, 209–10, 210f
Ediacaran fauna, 48f, 154–55
"Egg Island," 562
"Egg Mountain," 562
Eggs, dinosaur, 39, 561–66
Egyptian:
civilization, 434–38
flood legends, 209–10
Einstein, Albert, 272–73
El Niño, 191
Elasmosaurus, 22
Elhanan, 397
Elliot Smith, Sir Grafton, 356f, 359–60, 362
Emites, 396

Encyclopedia Britannica, 40, 389
Enlightenment, the, 62–64
Enoch, 394
*Enquiry Concerning Human Understanding* (Hume), 63–64
*Enquiry Into the Truth and Certainty of the Mosaic Deluge, An* (Cockburn), 59
Entrada Formation, 569, 571, 572
Enzymes, restriction, 449
*Eoanthropus dawsoni* ("Piltdown Man"), 358–64
Eocene epoch, 49
Eoraptor, 231, 233
Epic of Gilgamesh, 207–8
Epihippus, 268
Episcopal Church, 124
Equus, 268
Erratics, 184
Erythropoietin, transgenic, 451
Escherichia coli, 447–48, 450, 526–27
*Essay Towards a Natural History of the Earth* (Woodward), 58
Estemmenosuchus, 318, Plate 25
Etna, Mt., 617
*Études sur les Glaciers* (Studies on Glaciers) (Agassiz), 184
Euparkeria, 230
European corn borer, 452
Evaporation, rate of, 190–91
Eve, 119–20, 398, 400, 512
Evolution, theory of:
    Cuvier, Georges and, 3
    Darwinism, see Darwinism
    micro-evolution, 525
    See, also, copying error theory of evolution
*Evolution, Creation, and Science* (Marsh), 487
*Evolution of Man* (Elliot Smith), 360
Exodus, Book of, 86–88
Explanatory elegance, 496
Extinctions, 1–2, 6
    dinosaur, see Dinosaurs, extinction of
    plant, asteroid theory and, 546
Eyesight, of antediluvians, 515–16
Ezekiel, Book of, 88, 93

# F

Factor VIII, transgenic, 451
Faith, 605–9
Fall of mankind:
    Darwinism and, 117–20
    death as result of, see Death as result of the Fall
    dinosaur violence and, 495, 496
    Lyellism and, 88–90, 96, 97, 101
Fallen angels, 402
Feathers, 315, 329–33, 331f, 336–37
"Feathers for T. rex?" (*National Geographic*), 336
Fibro-lamellar bone growth, 302f, 303–4
Field Museum, Chicago, 40
Fish, transgenic, 457–58, 461–62
Fisher, Ronald, 247
Fitch, Frank, 623–27
"Flaming Cliffs" site, Mongolia, 39
Floods:
    Bijou Creek, 151–52
    Genesis, see Genesis Flood
    Lucifer, 93
Fluorine test, 360–61

Foods, transgenic, 452–56
Forestier, Amédée, 356f
Fossil Corals (Buttners), 59
Fossiliferous strata:
  major divisions of, 48–50, 48f
  See also Geologic column
Fossilists:
  American Museum of Natural History, 35–39, 297, 358, 359, 565
  Andrews (Roy Chapman), 38–39, 91
  Anning (Mary), 3–6, Plate 5
  Brown (Barnum), 36–38, 150, 299, 577
  Buckland (William), 10–12, 10f, 14
  Cope (Edward Drinker)/Marsh (Othniel Charles) feud, 19–31, 298
  Currie (Phil), 335, 559–60
  Cuvier (Georges), 1–3, 7–8, 12, 66–67, 70, 73, 311–12, 548, Plate 1
  Dodson (Peter), 289, 559–60
  Horner (John R. "Dinosaur Jack"), 149, 150, 302, 510, 555–58, 561–64, 574, 577, 579
  Mantell (Gideon Algernon), 6–9, 12, 14, Plate 6
  Osborn (Henry Fairfield), 35–38, 356–58, 367–68
  Owen (Richard), 13–15, 287–88, 298, 315, Plate 7
  Parkinson (James), 3
  Sternberg family, 296–97
Fossilization, 46–47, 498
Fossils:
  of ancient man, see Ancient man, Darwinian view of
  claim of reptile-to-mammal transition and, 337–43
  dating of, see Radiometric dating
  dinosaur, see Dinosaur fossils
  Ediacaran fauna, 48f, 154–55
  formation of, 46–47
  in geologic column, see Genesis Flood; Geologic column; "Scientific" view of geologic column
  index or guide, 47, 49, 635
  intermediate forms, lack of, 229–34, 267–74, 277, 278, 366
  mammal, 499–502
  of mammal-like reptiles, 316–19
  of Mesozoic reptiles other than dinosaurs, 311–16
  as mixture of modern vertebrate classes, 285–304, 320–22, 496–97
  scarcity of, 374–75
  use of term, 46
Foulke, William Parker, 19
*Founders of Geology, The* (Geike), 70
Fourth Commandment, 87, 88, 138
Frame shift mutations, 251
Fruit fly (Drosophila melanogaster) experiments, 245, 247, 256, 463–64
Fruita Quarry, Colorado, 577
*Fundamentals of Christian Education* (White), 487–88
Funerary architecture, 431–34

## G

Gallup Poll, 123
Gap theory, 92–94, 99
Geike, Archibald, 70
Gene pool, 245–46

Gene splicing, see Genetic engineering
Gene transfer, horizontal, 526–29
Genentech, 450
Genera, difficulties in determining number of, 41–42
Genes:
  homeobox, 463–64
  homeotic, 322
  horizontal transfer of, 526–29
  master, 332
  See also Genetic engineering; Genetics
*Genes and Genesis* (Clark), 485
Genesis, Book of:
  accommodation theories and, 92–100
  on animals brought into the ark, 592–93
  on creation of man, 393
  decline in intellect, physical strength, and length of life in, 437
  dinosaur violence as verification of, 495–96
  the Flood, see Genesis Flood
  as foundational to Christianity, 118–19
  gap theory and, 93
  genealogies of, 168–70, 172, 173
  life spans of antediluvians in, 394
  Lyellism vs., 86, 89, 90
  physical stature of antediluvians in, 395, 396, 399–404
  tower of Babel story, 437
  White (Ellen) on creation week, 136–40
*Genesis and the Mystery Confucius Couldn't Solve* (Nelson and Broadberry), 212
Genesis Flood, 75, 76
  aftermath of, see Ice Age
  back and forth motion of, 152–53
  biblical chronologies and, 168–73
  comeback of catastrophism in modern geology and, 545–46
  configuration of fossils and, 278–80
  continental plate movement during, 154, 162–64
  correlation of conventional geological period names with stages of, 154–68
  "crossing over" phenomenon during, 154
  Cuvier (Georges) and, 66
  day/age theory and, 95, 96
  early geologists and, 53–63
  evidence of dinosaur extinction by:
    amalgamation theory, 589–94
    behemoth/leviathan and, 594–98
    burials in marine sediments, 572–74
    dinosaur graveyards, 576–81
    drowning poses, 581
    lack of plant fossils, 574–76
    lack of post-Flood remains, 497–98, 589, 591
    post-Flood artifacts and, 598–601
    suicidal herds, 557–60
    tracks, 566–67
    unhatched eggs, 561–66
    violence, 495–96
  gap theory and, 94
  in history, 203–23
  local flood theory and, 97–100

Lyellism and denial of, 89, 91
marine mammals, survival of during, 499–500
model of, 151–68
tranquil theory and, 92, 96–97
White (Ellen):
on amalgamation and, 484
on fossils as relics of, 143–44
*Genesis Flood, The* (Whitcomb and Morris), 591
Genetic engineering, 447–72, 531
agricultural uses of, 452–56
as big business, 450
controversy over, 447–48, 450
as "cut and paste" operation, 449
dinosaur resurrection and, 465–72
dinosaurs as product of, see Amalgamation theory
future of, 462–65
vs. horizontal gene transfer, 529
medical uses of, 450–52
transgenic animals, 456–62
Genetic homeostasis, 245
Genetics, 243–56, 509
horizontal gene transfer, 526–29
See also Genetic engineering
*Genetics and the Origin of Species* (Dobzhansky), 487
Genomics, 463, 471
Geochron Laboratories, 632
Geologic column:
Biblical view of, see Genesis Flood
major divisions of, 47–50, 48f
"scientific" view of, see "Scientific" view of geologic column
value of, 49–50
Geological Society of London, 5, 7
Geoscience Research Institute, 146
Germinal mutations, 251–52
Ghost Ranch, New Mexico, 576–77
Giants, 395
Giem, Paul, 632–33
Glaciation, 183–95
See also Ice Age
GloFish, 461–62
Glufosinate, transgenic crops with resistance to, 454
Glyphosate, transgenic crops with resistance to, 454
Golden rice, 454
Goliath, 397
Goliath the Gittite, 397
Gonzalez, Silvia, 633
Gorgosaurus libratus, 295
Gould, Stephen Jay, 85
on amphibian/reptile differences, 338
on Cambrian explosion, 274–75, 277
on catastrophists, 580
on cave paintings, 390–91
on Haeckel (Ernst), 372
on lack of intermediate fossil forms, 269
on Lyell (Sir Charles), 68–71, 73–74
on Nebraska man, 368
on Piltdown hoax, 361
on racism and Darwinism, 519
on science and religion, 120–21
Grand Canyon, 160–61, Plate 14
Granger, Walter, 39
Grant, Ulysses S., 23
Graveyards, dinosaur, 576–81

Gray, Willard F. M., 514
*Great Controversy, The* (White), 516, 521
Greek flood legends, 212–16
Gregory, William King, 358
Griffins, 503
Griffith, Fred, 248, 525–26
Guide fossils, see Index fossils

# H

Haddonfield, New Jersey, 19, 22
Hadrosaurs (duckbill dinosaurs), 295–98, 296f, 494–95, 555–58, 572–74, 581
Hadrosaurus, 20f, 572
  foulkii, 19
Haeckel, Ernst, 371–72, 520
Haldane, J. B. S., 247
Harris, John, 54
Harvard University, 457
Hasel, Gerhard, 98, 169, 173
Hatcher, John Bell, 30
Havasupai Canyon, Arizona, 598, 600
Hawaiian flood legend, 216
Hawkesbury Sandstone, Australia, 632
Hawkins, Gerald S., 418–24, 428
Hebrews, Book of, 88, 89, 117, 172, 402, 605
Heel stone, at Stonehenge, 417, 417f, 418, 418f, 430
Heilmann, Gerhard, 327–28, 330
Hell Creek Formation, 37, 150, 577
Hemoglobin gene, 460–61
Hepatitis B vaccine, transgenic, 451, 455
Herbicides, transgenic technology and, 454
Herrerasaurus, 231, 233
"Hesperopithecus, the First Anthropoid Primate Found in America" (Osborn), 357
"Hesperopithecus Apparently not an Ape nor a Man" (Gregory), 358
Hesperopithecus haroldcookii, 356–58, 356f
Hess, Harry, 162
Heterodox nature of dinosaurs, 285–304, 320–22, 496–97
Hieronymus the Egyptian, 209
Hillaire, Christian, 387–88
Hindu "old earth" traditions, 95
Hipbone structure, Saurischians v. Ornithischians, 290–91, 290f
Hitchcock, Charles, 287, 288, 320
Hitchcock, Edward, 90, 285–87, 496, 568
Holocene epoch, 49
Homeobox genes, 463–64
Homeotic genes, 322
Hooton, Ernest, 359
Hooves, 288–89
Horizontal gene transfer, 526–29
Horn, Siegfried, 170, 171
Horned dinosaurs, 288–93
Horner, John R. "Dinosaur Jack," 149, 150, 302, 510, 555–58, 561–64, 574, 577, 579
Horns, 288–89
Howe Ranch Quarry, Wyoming, 37–38, 599
Howells, W. W., 369–70
Hoyle, Sir Fred, 427–29, 431
Hualalai basalt, 617
Human genome, DNA map of, 465
Human Genome Project, 527
Human growth hormone (hGH), transgenic, 451

Humans:
- amalgamation with animals, 502–3
- horizontal gene transfer and, 527
- See also Ancient man; Antediluvians; Races

Hume, David, 63–64
Humphreys, D. Russell, 164
Hunt, Adrian, 292, 570, 571
Hunter, John, 8, 13
Hunterian Museum of the Royal College of Surgeons, 8, 13–14
Hutchinson, John, 59
Hutton, James, 64–67, 70–71
Huxley, Julian, 247
Huxley, Thomas Henry, 316, 329
Hylaeosaurus, 12
Hyracotherium, 268

## I

Ica stones, 598–601, 601f
Ice Age:
- art, 387–92, 600
- astronomical theory of, 188
- Carboniferous-Permian, 157
- creationist model of, 189–95
- glacial theory and, 183–86
- little, 189
- Milankovitch theory of, 188–89
- in North America, 186–87, 186f

Ichnology, 285
*Ichnology of New England* (Hitchcock), 287
Ichthyosaurus, 4, 4f, 311–13, 321–22, 497, 499, 546, 597, Plate 22
*Icons of Evolution* (Wells), 105
Ideographs, 211
Iguanodon, 8–11, 9f, 14, 41
Imidazolinone, transgenic crops with resistance to, 454
Independent Assortment, Law of, 244
Index fossils, 47, 49, 635
"Indian Ring, the," 23–24
Inheritance, Mendel's laws of, 244, 248
Inorganic radiometric dating, 611–30
Insect species, number of, 524
Insecticides, transgenic technology and, 452–53
Insulin, transgenic, 450–51
Intellectual ability:
- of antediluvians, 405–9, 510–13
- decline of, 429–31, 434, 435–38, 516

Intelligent Design movement, 104–5
Interleukin-2, transgenic, 451
Intermediate fossil forms, lack of, 229–34, 267–74, 277, 278, 366
Interval theory, see Gap theory
Introns, 459–60
Iridium, 543, 549
Irreducible complexity, 257–59, 331
Isaiah, Book of, 90, 93, 597
Ishbi-Benob, 397
Isotopic dating, see Radiometric dating
Israelites, 395–98, 401

## J

Jamnia, council of rabbis at, 170, 171
Jared, 394
"Java Man," 375–76
Jefferson, Thomas, 119
Jesup, Morris K., 35
Jesus Christ:
- on angels, 402
- as Creator, 117

on historical reality of the Flood, 89
as Redeemer of mankind, 117, 118, 120
on writings of Moses, 119
Job, Book of, 357, 402, 512, 594–98
Joel, Book of, 136
John, Book of, 88, 117, 119
Johnson, Phillip, 104–6, 121
Jonathan, 397
Josephus, Flavius, 171, 173, 208–9
Joshua, 396
Joshua, Book of, 396
Jove, 212–16
Julsrud, Waldemar, 598
figurines of, 598–600, 599f, 600f
"Junk DNA," 459–60, 528
*Jurassic Park* (Crichton), 458, 465, 466
*Jurassic Park* (film), 37, 42, 91, 302, 328, 465–72, 557
Jurassic strata, 48f, 49
claim of dinosaur-to-bird transition and, 334, 335
correlated with stages of Genesis Flood, 166
dinosaur tracks in, 568, 569
in Lyellian chronology, 155f
mammal-like reptiles in, 316, 337
mammals in, 341–42
middle and lower, absence of dinosaur bones in, 570f
"scientific" view of dinosaur extinction and, 76

# K

Kaku, Michio, 464, 465, 503
Kanapoi, Kenya, 369–70
Kane (Hawaiian deity), 216
Kang, C. H., 211
Kayenta Formation, 569, 570f
KBS tuff controversy, 622–29
Keith, Sir Arthur, 359, 360, 362
Kelley, Jonathan, 85
Kelley, Patricia, 85, 120–21
Kenan, 394
Kentrosaurus, 233–34
Kern, Milton E., 483–84
Kettlewell, Bernard, 372–73
Khujut Rabu, Iraq, 514
Kilauea, Mt., 612
Kimbrell, Andrew, 461
Knight, Charles R., 595
KNM-ER 1470 skull, 624–25, 625f, 628
Konig, Wilhelm, 514
KP-271 fossil, 369–70
Kuhn, Thomas S., 365, 369, 607
Kuyûnjik, mound of, 206–7

# L

Lagosuchus, 230–31
Lake Tahoe, 219–20
Lake Turkana, Kenya, 623
Lakes, Arthur, 25
Lamarck, Jean Baptiste, 100
Lance Creek beds, Wyoming, 581
Lassen, Mt., 617
Lateral gene transfer, 526–29
Lawrence, Jeffrey, 526
Layard, Austen Henry, 514
Leakey, Richard, 374, 623–25
Lehmann, Johann Gottlob, 61
Leidy, Joseph, 19, 21, 22

Leigh, Charles, 58
Leonardo da Vinci, 51, 65
Leroi-Gourhan, Andre, 391–92
Leviathan, 594–98
Leviticus, Book of, 88
Lewin, Roger, 367, 375, 611n, 629
Lewis Overthrust, 164–65
Liaoning Province, China, 336
Life spans, antediluvian, 394, 510
Lincoln, Abraham, 518–19
Lindley, David, 469–70
Linne, Carl (Carolus Linnaeus), 96
Lithosphere, 163
Little Ice Age, 189
Local flood theory, 97–99
Locke, John, 57, 62–63
Lockley, Martin, 292, 570, 571
Lockyer, Sir Norman, 417–18
Long barrows, 433
*Longisquama insignus*, 328–29, 333
Lubenow, Marvin, 611n
Lucas, O. W., 25–26
Lucifer, 93, 94
Lucifer flood, 93, 94
Luciferase, 461
Luke, Book of, 89, 119, 145, 173, 402
Lusi naturae ("games of nature"), 51
Luther, Martin, 55–56
Lyell, Sir Charles, 7, 67–76, 69f, 86n, 96
  on catastrophism, 543–45
  Darwin, Charles on, 101–4
  glacial theory and, 185–86
  support of Darwinism by, 103–4
Lyellism (uniformitarianism), 64–76, 85–107
  Christian reaction to, 92–100
  chronology of major divisions of strata, 155f, 635
  coal measures and, 156–57
  confusion of methodological principles with substantive claims, 73–75
  continental drift and, 164
  creationist criticism of radiometric dating and, 629–30
  Darwinism and, 77
  Hutton (James) and, 64–67
  Ice Age and, 187–88, 192–93
  implications for Christianity, 86–91
  Lyell (Sir Charles) and, 67–76, 69f
  paleocurrents and, 160
  Price (George McCready) on, 606–7
  relationship between Darwinism and, 100–107
Lystrosaurus, 319, Plate 27

# M

McDougall, Ian, 626, 627
McElhany, James Lamar, 483
McGonigle, Joseph, 458
Machu Picchu, Peru, 436, 601
Macrodontophion, 12
Maes Howe tomb, Orkney Islands, Scotland, 431, 432, Plate 30
Mahalalel, 394
Maiasaura, 557, 558, 560
Makela, Bob, 555–56, 562
Malaria, 255

Mammal-like reptiles:
- claim of reptile-to-mammal transition and, 337–43
- physiology of, 316–19

Mammalian characteristics:
- of dinosaurs, 287–89, 292, 293, 299–300, 302–4
- of Ichthyosaurus, 312–13
- of pterosaurs, 314
- of synapsids, 316–19

Mammoth fossils, 2, 498, 634–35

Mammuthus columbi, 634–35

Man, ancient, see Ancient man

Manases, 209

Mantell, Gideon Algernon, 6–9, 12, 14, Plate 6

Mantell, Mary Ann, 6, 7

Marasuchus, 230–31

Margulis, Lynn, 527–28

Mark, Book of, 402

Marlstone Rock Bed, England, 631–32

Marsh, Frank Lewis, 146, 483–90, 524, 591

Marsh, Othniel Charles, 20–31, 22f, 35, 288–89, 298, 299, 572

Martin, Larry, 329

Mask of the Jaguar (North), 363–64

Masoretic text, 170–73

Massetognathus, 319

Master genes, 332

Mastodon fossils, 2

Matteole flood legend, 217

Matthew, Book of, 89, 169, 402, 522

Mayr, Ernst, 40, 245, 247, 487

McLoughlin, John C., 339

Medical uses of genetic engineering, 450–52

Medici, Ferdinando de, 52

Medicine Bow incline, Wyoming, 36

Megaliths, British, 415–38

Megalosaurus, 11–12, 14, 303

Melville, Herman, 119

Mendel, Johann Gregor, 244, 248

Mesohippus, 268

Mesopotamia, 99

Mesozoic strata, 48f, 49, 102
- correlated with stages of Genesis Flood, 161–66
- dinosaur fossils in, see Dinosaur fossils
- dinosaur tracks in, 158–59
- in Lyellian chronology, 155f
- mammal fossils in, 500
- reptile fossils other than dinosaurs in, 311–16
- "scientific" view of dinosaur extinction and, 76

*Metamorphoses, The* (Ovid), 212–16

Methodist Church, 124–26

Methodological principles, 73–75

Methuselah, 394

Meyer, Hermann von, 13

Miao, 210–11

Mice, transgenic, 457, 461

Micro-evolution, 525

Microbes, horizontal gene transfer and, 526–27

Microinjection, 456–57

Microraptor, 337
- gui, 337

Mid-Atlantic Ridge, 162

Mid Clyth stone rows, Scotland, 427

Miescher, Friedrich, 248

Milankovitch, Milutin, 188–89

Miller, Jack, 623–27
Milton, Richard, 268–69, 618, 622, 626–27
Miocene strata, 49, 157
Miohippus, 268
Miracles, Biblical, 125
Mississippian strata, 48f, 49
Mitochondria, 468
Moab mega-tracksite, 566–67, 571
Moabites, 396, 401
Moir, J. Reid, 360
Monoclonius, 559
Monsanto Company, 453–55
Montezuma, 220
Moody, Pliny, 285
Moon positions, megaliths and, 420–29, 421f–423f
Moraines, 184, 185
Morgan, Thomas Hunt, 247
Morland, Mary, 11
Morris, Henry, 591
Morrison Formation, 25, 569, 571, 574–75, 580
Morton, John, 58
Mosaics, 457
Mosasaurus, 2, 4, 11, 499, 546, 597, Plate 2, Plate 3
    copeanus, 21
Moses, 87, 98, 119, 141, 208, 209, 398
*Moses' Principia* (Hutchinson), 59
Most Holy Place, 88
Mudge, Benjamin, 25
Muller, Hermann J., 252
Multituberculata, 341
Museum of the Rockies, 579
Mutations, 246–56, 258, 525, 527–28
Mythology, human/animal combinations in, 503

# N

Nachidsheuan, 208
National Academy of Sciences, 448
*National Geographic*, 336
Native American flood legends, 204, 216–23
Natural Bridges National Monument, 598, 600
*Natural History of Oxfordshire, The* (Plot, 51
*Natural History of the Mineral Kingdom* (Williams), 61
Natural selection, 239–47, 525
Natural theology, 2
Naturalism, 100, 105, 106, 138–41, 605, 608–9
    See also Darwinism; Lyellism (uniformitarianism)
*Nature*, (scientific journal) 21, 420, 424, 426, 427
Navajo Formation, 569, 570f
Nazca lines, 436, 601
Nazca people, 598–99
Nazi genocide, 520
Neander, Joachem, 353
Neanderthals, 353–56, 354f, 367, 373–74, Plate 28
"Nebraska Man," 356–58, 368
Nehemiah, Book of, 396
Neill, Sam, 302
Nelson, Byron C., 62
Nelson, Ethel R., 211–12
Nemesis hypothesis, 549
Neocatastrophism, 545
Neogene strata, 155f
Nephilim, 395, 398, 401
New English Bible, 597

*New Theory of the Earth, A* (Whiston), 59
*New York Herald*, 21, 22, 31
*New York Times*, The, 357–59
Newgrange tomb, Ireland, 431–32, 432f
Newhall, R. S., 421–22
Newham, C. A. "Peter," 420–21, 428–29
Ngeneo, Bernard, 624
Nichol, Francis D., 489
Nicholas of Damascus, 209
Nineveh, 206–7
Noah, 89, 98–99, 400, 404, 437, 495, 503, 523, 524
Noah's Flood, see Genesis Flood
Nodosaurus, 572–74
Noorbergen, Rene, 169, 438
Norman, David, 314
North, Jessica, 363–64
North Slope, Alaska, 46
"Notice on the Iguanodon, a newly-discovered Fossil Reptile, from the Sandstone of Tilgate Forest, in Sussex" (Mantell), 9
"Notice on the Megalosaurus or Great Fossil Lizard of Stonesfield" (Buckland), 12
Novartis AG, 454
Nuclear transfer, 460
Nucoton, 453
Numbers, Book of, 395, 401, 402

# O

Oakley, Kenneth, 362–63
Oard, Michael J., 189, 191, 194
Occipital condyle, 294
Oceans evaporation, 190–91
Ochman, Howard, 526
Og, King, 396
O'Kelley, Claire, 432–33
O'Kelley, Michael, 432–33
Oligocene strata, 49
Ollantaytambo, Peru, 436, 601
Olsen, Gary, 526
Olson, Storrs, 336
*On the Origin of Species by Means of Natural Selection* (Darwin), 77, 101–2, 239, 276, 277, 523
*On the Relation Between the Holy Scriptures and Some Parts of Geological Science* (Smith), 97
Oort cloud, 548
Optical technology, of ancient man, 514–15
Optically stimulated luminescence (OSL), 633
Ordovician strata, 48f, 49, 155f, 156, 161
Organic radiometric dating, 630–35
*Organic Remains of a Former World* (Parkinson), 3
*Origin of Birds, The* (Heilmann), 327–28, 330
Ornithischians, 232–34, 291–92, 295, 322
Ornitholestes, 301
Ornithomimus, 300, 334
Ornithopods, 234, 287, 295, 322
*Orodromeus makelai*, 562–63
Orohippus, 268
Ortelius, Abraham, 161
Osborn, Henry Fairfield, 35–38, 356–58, 367–68, 520
Ostrom, John, 328, 329
Ovid, 212–16
Oviraptor, 300, 334
Owen, Richard, 13–15, 287–88, 298, 315, Plate 7
Oxytocin, 342

# P


Pachyrhinosaurus, 559

Papagos flood legend, 220

Palaeosarus, 13

Paleoanthropology:

- nature of, 374–76
- See also Ancient man, Darwinian view of

Paleocene strata, 49

Paleocurrents, 160, 167

Paleogene strata, 155f

Paleozoic strata, 48f, 49, 102

- correlated with stages of Genesis Flood, 155–61
- in Lyellian chronology, 155f

Palynomorphs, 575

Pantotheria, 341–42

Paradigm, dominant, 365–71

Parasaurolophus, 297–98, 298f, Plate 19

Pareiasaurs, 318–19

Parkinson, James, 3

*Patriarchs and Prophets* (White), 516, 521

Patterson, Brian, 369–70

Paul, the Apostle, 171

Pax-6 (gene), 463

Peabody, George, 20

Peabody Museum of Natural History, Yale University, 20, 25, 29

*Pedigree of the Human Race, The* (Wilder), 360

Pelycosaurs, 317, 338

Pennsylvanian strata, 48f, 49

Peppered moths, 372–73

Periosteal bone, growth patterns of, 302–3, 302f

Permian strata, 48f, 49, 155f, 316, 317, 337

Permineralization, 46

Peter, the Apostle, 89, 100

2 Peter, Book of, 89, 100, 215, 581

Petrifaction, 46

Pharmaceutical Proteins Limited (PPL), 459–60

Pharming, 459–60

Pharming Group, 460

Philistines, 396–97

*Philosophical Transactions* (Royal Society of London), 8–9

Physical stature of antediluvians, 395–404

Physiology:

- Archaeopteryx, 315–16
- dinosaur, 285–304
- ichthyosaur, 311–13
- mammal-like reptile, 316–19
- pterosaur, 313–15

Pictographic writing, 211–12, 212f

Pilbeam, David, 374

*Piltdown Inquest, The* (Blinderman), 364

"Piltdown Man," 358–64, 368–69, 371, 374, 376

Pisanosaurus, 231

Plant extinctions, asteroid theory and, 546

Plant fossils, 574–76

Plasmids, 449, 526, 528

Plastic virtue concept, 50–51

Plate tectonics, 57, 366

- catastrophic, 164–65, 190–92
- explanatory power of, 163–64
- during Genesis Flood, 154, 164–66

Plateosaurus, 13

Plato, 51, 95, 436

Playfair, John, 66

Pleisiosaurus, 4, 4f, 11, 499, 546, 597
Pleistocene strata, 49, 63, 192, 194
Pleurocoels, 294
Pliocene strata, 49
Pliosaurs, 597
Plot, Robert, 51, 52f
Pneumatic foramina, 314
Pollack, Robert, 447
Polyphyletic origin of dinosaurs, 232
Popol Vuh, 515–16
Potassium-argon dating method, 611, 612, 615–17, 619–20, 624, 625, 631
Potrykus, Ingo, 454
"Praise to the Lord, the Almighty, the King of Creation" (hymn), 353
Pre-Edenic ruin theory, see Gap theory
Precambrian strata, 48f, 102, 154–55, 274, 275
Precipitation, in Ice Age, 189–91
Predation:
  local flood theory and, 97
  Lyellism and, 90–91
Predentary bone, 291
*Prehistoric Sussex* (Curwen), 360
*Prehistory* (Burkitt), 360
Preparators, 555
Presbyterian Church, 124
Price, George McCready, 146, 165, 605–7
Price, Utah, 158
*Principles of Geology* (Lyell), 68, 73, 101–2
Probelesodon, 319
*Prodromus* ("Preliminary discourse to a dissertation on a solid body naturally contained within a solid,") (Steno), 54
Progressive creationism, 106
*Prophets and Kings* (White), 516
Prosauropods, 233
Protarchaeopteryx, 337
Protestant denominations, 124–26
Protoceratops, 39, 91
Proverbs, Book of, 358
Provine, Will, 121
Psalms, Book of, 88, 139, 396
Pteranodon, 313
Pterosaurs (Pterodactyls), 2, 313–15, 497, 547, Plate 4
Ptyonius marshii, 21
Pyramid texts, 210
Pyramids, 416, 436
Pyrrha, 216

# Q

*Quarterly Journal of the Geological Society*, 5
Quaternary strata, 48f, 49, 60–61, 76
Quiche Maya, 515

# R

Races:
  amalgamation and, 518–21
  diversification of, 523
Radiocarbon dating, 630–35
Radiometric dating, 77, 611–35
  inorganic, 611–30
  organic, 630–35
Rainbow trout, transgenic, 457
Rapha, sons of, 397–98
Rappaport, Rhoda, 62
Rb-Sr (rubidium-strontium) dating method, 612, 615, 619

Reade, Thomas Mellard, 621
Recent epoch, 49
Recombinant DNA, 448
Recombinant DNA Advisory Committee, 448
Red Cloud, 23
Redemption, 117–18, 120
Reed, William Harlow, 26–30
Reid, Robin, 303
Renne, Paul, 633–34
Rephaites, 396
Reptiles:
    -to-mammals transition, claim of, 337–43
    fossils other than dinosaurs in Mesozoic strata, 311–16
    mammal-like, see Mammal-like reptiles
    marine, death of, in Genesis Flood, 499–500
Reptilian characteristics:
    of Archaeopteryx, 315–16
    of dinosaurs, 287, 288, 290, 292–95, 299–301, 497
    of Ichthyosaurus, 287, 288, 290, 292–95, 299–301, 312
    of therapsids, 319
Restriction enzymes, 449
Resurrection, 125
*Review and Herald* (now *Adventist Review*), 489
Rhamphorhynchus, 313
Rice, transgenic, 454
Ricqles, Armand de, 303, 510
Ridley, Mark, 271
Riggs, Elmer, 29
Riley, Henry, 12–13
Robinson, D. E., 486
Roemer, Ferdinand, 22
Roman flood legends, 212–16
Romans, Book of, 88, 89, 399
Rongitoto, Mt., 631
Roslin Institute, 459, 470
Roth, Ariel A., 146, 565–66
Round barrows, 433
Roundup, 454
Royal College of Surgeons, 8, 14
Ruin and reconstruction theory, see Gap theory
Ruin and restoration theory, see Gap theory
Ruse, Michael, 119–20
Ruskin, John, 100

# S

Sabbath, 87, 89, 91, 135–37
Saber-tooth cat fossils, 498
*Sacred Theory of the Earth, A* (Burnet), 56
Sacsayhuaman, Peru, 436, 601
Sagan, Dorothy, 528
St. Helens, Mt., 616–17
Salmon, transgenic, 457, 458
Samaritan Pentateuch, 170–73
1 Samuel, Book of, 397
2 Samuel, Book of, 88, 397
Saph, 397
Sarsens, at Stonehenge, 416, 417, 428, 429f
Satan, 522–23, 530–31
Saturation mutagenesis, 256
Saurischians, 232, 233, 291, 294, 300

Sauropods, 26, 41–42, 595, Plate 8, Plate 11, Plate 20
artifact depictions of, 599–600, 600f
evolutionary ancestry of, 233, Plate 15
graveyard of, 577–78, 578f
physiology of, 298–300
tracks of, 569
Scheuchzer, Jean-Jacques, 58
Schweitzer, Mary, 149–51, 467
*Science*, 463
*Science of Jurassic Park and the Lost World, The* (DeSalle and Lindley), 469–70
Scientific knowledge:
of antediluvians, 405–9, 509–16
years to reach current level of, 509–10
"Scientific" view of geologic column, 45–77
age of the earth and, 76–77
Darwinism and, see Darwinism
early conceptions, 50–51
enlightenment and, 62–64
Genesis Flood and, 53–63, 66, 75, 76
Lyellism (uniformitarianism), see Lyellism (uniformitarianism)
Steno (Nicolaus Stenonis) and, 52–54
White, Ellen on, 138–44
Sclerotic ossicles, 312
Scofield, Cyrus Ingersoll, 93
Scottish Enlightenment, 63
Scutosaurus, 318–19, Plate 26
Seafloor spreading, 162–63, 165
*Search for Man's Ancestors* (Elliot Smith), 360
*Secrets of the Lost Races* (Noorbergen), 169
Sedimentary rock:
formation of, 45–46
fossils and, see Dinosaur fossils; Fossils
in geologic column, see Geologic column
paleocurrents and, 160
Seeley, Harry Govier, 290–91
Segisaurus, 328
Segregation, Law of, 244
Seismosaurus, 298, Plate 20
*Selected Messages* (White), 516–17
Selective breeding, 239–46, 489
Septuagint (LXX), 170–73
Sereno, Paul, 232
Seth, 394, 403, 404, 484–89
Sharov, Alexander, 314
Sheep:
cloning of, 470
transgenic, 459–60
Shekinah Glory, 88
Sheridan, Philip, 22
Sherman, William Tecumseh, 22–23
Sheshai, 396
Sibbecai, 397, 398
Siccar Point, Scotland, 65
Sickle cell trait, 254–56
*Signs and Witnesses of the Flood* (Buttners), 58–59
Silliman, Benjamin, 66
Silurian strata, 48f, 49, 155f, 156, 161
Simpson, George Gaylord, 247, 269
Sin, 89, 117, 401, 485, 486, 495
Sinclair Oil Company, 29, 38
Sinornithosaurus, 337

Sinosauropteryx prima, 336
Sippai, 398
Sitting Bull, 24
*Sketches From the Life of Paul* (White), 516
Skin:
of duckbill dinosaurs, 296–97
of Triceratops, 290
of Tyrannosaurus rex, 295
Smith, Adam, 64
Smith, George, 207
Smith, John Pye, 97
Smith, Sam, 24
Smith, William, 47
Smith Woodward, Arthur, 358, 359, 361
Snider-Pellegrini, Antonio, 161–62, 163f
Snowfall, in Ice Age, 189–91
Soft body parts, fossilization of, 47
Sollas, William J., 359, 362
Somatic mutations, 251
"Sons of God," 399–404
Souciet, Etienne-Augustin, 59
Soybeans, transgenic, 453, 454
Species:
belief in fixity of, 523
changes in (speciation), 523–31
difficulties in determining number of, 40–42
evolution of, see Darwinism
Sphenacodonts, 339
Spielberg, Steven, 465, 469
Spirit Lodge, 220
*Spirit of Prophecy* (White), 517
*Spiritual Gifts, Important Facts of Faith in Connection with the History of Holy Men of Old* (White), 484, 516, 517
Stahl, Barbara, 331
Staurikosaurus, 231
Stebbins, G. L., 247
Stegosaurus, 27, 233–34, 295, 568–69, Plate 9
Steno (Nicolaus Stenonis), 52–54
Stensen, Nils, see Steno (Nicolaus Stenonis)
*Steps to Christ* (White), 135
Sternberg, Charles H., 296
Sternberg, George, 296
Sternberg, Levi, 296
Stomach stones (gastroliths), 300
Stonehenge, 416–31, 417f, 418f, 421f–423f, 429f, 436–38, 513, Plate 29
"Stonehenge—An Eclipse Predictor" (Hoyle), 427
Stonesfield quarry, Great Britain, 11–12
Stratigraphy, 47
Stratomorphic intermediates, 229–34, 268
See also Intermediate forms
lack of
Straus, William L., 354–55
Striations, 184
*Structure of Scientific Revolutions, The* (Kuhn), 365
Stukeley, William, 417, 418
Stutchbury, Samuel, 8, 12–13
Styracosaurus, 559
Suarophthirus, 314
Subduction zones, 162–64, 192
Substantive claims (of Lyell), 73–75
Substitution mutations, 251, 254
Sulfonylurea, transgenic crops with resistance to, 454

Sun positions, Stonehenge and, 419–27, 421f, 422, 422f, 423f
Sunset Crater basalt, 617
Survival of the fittest, 240
Sweet potatoes, transgenic, 455
Symmetrodonta, 341
Synapsids, 316–19

# T

Tablets of the Testimony, 87–88
Talmai, 396
Taylor, Bert L., 27–28
Taylor, Joe, 574
Technology:
  of the ancients, 513–15
  of antediluvians, 405–9, 509–16
  years to reach current level of, 509
Teeth:
  of Archaeopteryx, 315
  of coelurosaurs, 301
  of duckbill dinosaurs, 296
  of Ichthyosaurus, 312
  of pterosaurs, 313
  of sauropods, 300
  of therapsids, 318, 319
  of Triceratops, 289, 290f
  of Tyrannosaurus rex, 295
Teilhard de Chardin, Marie-Joseph Pierre, 358, 361
Telomeres, 467
Temple, Robert, 514, 515
Temple Wood stone circle, Scotland, 427
Ten Commandments, 87–89, 91
Teotihuacan, Mexico, 436
Tertiary strata, 48f, 49, 60–61, 76, 157
  correlation with stages of Genesis Flood, 166–68
  in Lyellian chronology, 155f
  mammal fossils in, 501–2
Testimony, tablets of, 87–88
Thecodontosaurus, 12–13
Theistic evolution, 106
*Theory of the Earth* (Hutton), 66
Therapsids, 317–19, 338–40
Thermal characteristics of dinosaurs, 301–4, 497
Theropods, 91, 569, 599
Thom, Alexander, 426, 427
Tiahuanaco, Bolivia, 436, 601
1 Timothy, Book of, 141
Titanosaurus, 25
Titanosuchians, 318
Tolo Lake, Idaho, 634–35
Toltec flood legend, 220–22, 601
Tomatoes, transgenic, 456
Tracking effect, (isotopic dating), 621
Tracks:
  bird, 334
  dinosaurs, 158–59, 285–88, 292–93, 299, 566–72
Tranquil theory, 92, 96–97, 99, 185
Transforming principle, 525–26
Transgenic technology, see Genetic engineering
Transitional forms, see Intermediate forms, lack of
Transmutation, 100
  See also Darwinism
*Treatise on the Deluge, A* (Catcott), 60

*Trevoux* (18th Century French Journal), 59
Triassic strata, 48f, 49, 155f
bird tracks in, 334
coal deposits in, 157
correlated with stages of Genesis Flood, 166
dinosaur tracks in, 568, 570f
in Lyellian chronology, 155f
mammal-like reptiles in, 316, 337, 338
mammals in, 341
"scientific" view of dinosaur extinction and, 76
subdivisions of, 230f
Triceratops, 289, 290f, 497, 599
beak of, 289
body of, 292
hip-bone structure of, 290–91
horns of, 288–89
horridus, 30, 31f, 288–93, Plate 18
skin of, 290
sprawling forelimbs of, 292–93
Triconodonta, 341
Trilithons, at Stonehenge, 416, 419
Troodon, 300, 563, 564
Twain, Mark, 37
Two Medicine Formation, Montana, 556–57
Tycho Brahe, 426
*Tyrannosaurus rex*, 37, 90–91, 149–51, 467, 500, 577, 590, 591, Plate 12
physiology of, 293–95
"Sue," 40, 293, 294f, 494, 579, Plate 13

## U

Uniformitarianism, see Lyellism (uniformitarianism)
U.S. Geological Survey, 49
*Up from the Ape* (Hooton), 359
Ussher, James, 169
Utnapishtim, 207–8

## V

Vaccines, transgenic technology and, 451, 454–55
Van Heerden, Jacques, 233
Vardiman, Larry, 191
Velociraptor, 39, 91, 293, 328
Vendian strata, 48f, 155f
Verduin, Leonard, 118
*Vertebrate History: Problems in Evolution* (Stahl), 331
Vertebrate classes, dinosaurs as mixture of, 285–304, 320–22, 496–97
Victoria, Queen of England, 15
Volcanic activity:
during Flood year, 168, 191–92
and Hutton's theory of endlessly cycling earth, 65
Ice Age and, 191–94
plate tectonics theory and, 164
Voltaire, 63
Vorlage, 170

## W

Walcott, Charles Doolittle, 275
Wallace, David Rains, 29
Wambugu, Florence, 455
Watt, James, 64

Wegener, Alfred, 162, 366
Weiner, J. S., 363
Weishampel, David, 562
Wells, Jonathan, 105
Whiston, William, 59
Whitcomb, John C., 591
White, Ellen Gould, 43, 135–46
on animals being supernaturally led to the ark, 501
on animals preserved in the ark, 593
on beginning of Genesis Flood, 190
on creation week, 136–40
on Darwinism, 144–45
death of, 485
on destruction of animals by antediluvians, 494
on dinosaurs, 144, 590, 591
on fossils as relics of Genesis Flood, 143–44
on intelligence and technology of antediluvians, 405–9, 510–13
on lust for gold by antediluvians, 213–14
on Lyellism (uniformitarianism), 136–44
on man as perfect creation, 393–94
on physical stature of antediluvians, 398–99, 404
on post-Flood amalgamation, 517, 518, 523–24, 529, 531
on pre-Flood amalgamation, 484–90, 498, 502, 517, 529–31
race issue and, 518–21
on "sons of God," 403–4
on unbelief, 606
White, W. C., 486
Whittington, Harry, 275
Whooping cough vaccine, transgenic, 451
Wilder, Harris, 360
Williams, John, 61–62
Williston, Sam, 26, 28, 35, 578
Wilmut, Ian, 470, 471
Wingate Formation, 569, 570f
Wise, Kurt, 115–16, 229, 607–8
Wishbones, 293, 328, 329
Wood, Todd, 528
Woodhouse, Mary Ann, see Mantell, Mary Ann
Woodmorappe, John, 613–16, 621
Woodward, John, 57–59, 61
Woodward, Scott, 467
Woolley, C. Leonard, 97

# X

Xenophanes, 64
Xisuthros, 204–6

# Y

Yakima flood legend, 217
Yorktown Technologies, 461
Younker, Randall, 496

# Z

Zallinger, Rudolph F., 595
Zamzummites, 396
Zbarzevski, 12
Zonal bone growth, 302–3, 302f
Zuckerman, Solly, 374